Asynchronous Transfer Mode Networks

Performance Issues

Second Edition

For a complete listing of *The Artech House Telecommunications Library,*
turn to the back of this book.

Asynchronous Transfer Mode Networks

Performance Issues

Second Edition

Raif O. Onvural

Artech House
Boston • London

Library of Congress Cataloging-in-Publication Data
Onvural, Raif O., 1959-
Asynchronous transfer mode networks: performance issues/Raif O. Onvural. – 2nd ed.
p. cm.
Includes bibliographical references and index.
ISBN 0-89006-804-6 (alk. paper)
1. Asynchronous transfer mode. 2. Integrated services digital networks. I. Title.
TK5105.35.054 1995 95-33936
004.6'6–dc20 CIP

British Library Cataloguing in Publication Data
Onvural, Raif O.
Asynchronous Transfer Mode Networks: Performance Issues. – 2Rev. ed.
I. Title
621.382

ISBN 0-89006-804-6

685 Canton Street
Norwood, MA 02062

International Standard Book Number: 0-89006-804-6
Library of Congress Catalog Card Number: 95-33936

10 9 8 7 6 5 4 3 2

For their love . . .
Meliha,
Seval, Aytan, and Arzu.
For the colors of the rainbow in my world . . .
Melih and Doruk.
None of this would have been possible without Nur.

Contents

Foreword

The objectives for broadband integrated services digital networks (B-ISDN) services, and their asynchronous transfer mode (ATM) base, are the most far-reaching ever set for a communications system. In his thought-provoking foreword to the first edition of *Asynchronous Transfer Mode Networks: Performance Issues,* Gerald Marin interpreted the international recommendations for B-ISDN as a metamorphosis of the world's telecommunications networks into fast-packet systems. On the occasion of the second edition, can we point to significant progress toward this ambitious goal? In the interval between the first and second editions, the number, quality, and practical value of ATM products, from a large number of competing sources, have been impressive. Testing reveals that the daunting challenges of controlling cell congestion and cell loss are so far, at least, being met. Scaling ATM from tiny to huge systems is becoming workaday reality, with the ATM Forum now sponsoring an ad hoc group for small-business and residential applications.

New products for local communications focus on ATM as a unifying concept for multimedia communications among desktops and mainframes and among clients and servers. Switched Ethernets and Token Rings cooperate with ATM switching to ensure a smooth migration for both network-centric and host-centric systems. Despite continuing questions about routing and delays, early performance measurements in local systems are encouraging.

Soon to be available for wide-area communications in the United States are ATM public services at 155 Mbps. Perhaps more striking are commitments in China and other nations, often considered out of the communications mainstream, to move quickly into the ATM era. To accomplish orderly migration of customers to ATM, carriers have products available that provide multiservice switching: ATM, frame relay, ISDN, and private branch exchange (PBX) networking. Also, facilitating the introduction of ATM into current operational wide-area networks are private-network products such as front-end processors, routers, and bandwidth managers.

Rapid progress to date notwithstanding, tall hurdles remain. Applications and their network support needs are in an early stage. Medical applications, sponsored by the National Information Infrastructure Testbed (NIIT), will determine what demands are placed on ATM to support the vital but complex creation, manipulation, and transmission of computer tomography and magnetic resonance images.

Exchanges of intricate particle detection images among physicists may eventually rely on ATM as the main mechanism for sharing the world's limited accelerator and collider facilities. Detector images and ensuing interactions of scientists across the globe will entail stringent quality and performance targets for ATM.

The Library of Congress virtual-digital-library project will put ATM to the test as a centerpiece of the national data superhighway. Commonplace, accessible-to-all multimedia applications, like video mail, video-on-demand, and interactive classrooms will generate widely varying traffic with unprecedented volumes of information as ATM becomes a ubiquitous infrastructure. Admixtures of short and long transmissions of voice, video, and data from sources in many regions and countries will characterize the new services' performance and connectivity requirements.

As we try new applications, performance subtleties will prove critical. Interactive video, once past its current primitive state, might combine the burstlike nature of data with the bandwidth appetite of moving images in new ways. Standards for traffic management, admission control, flow control, cell delay, and error control have to evolve apace with progress in video and other—many still unforeseen—applications.

ATM is moving into the heart of communications networks, from local to global, at an extraordinary pace. Continued success depends on adapting to emerging applications with their ever-increasing needs for high performance levels and widespread interoperability.

Len Felton
Vice President, Quality and General Development
Networking Hardware Division
IBM

Preface

It has been two years since I completed writing the first edition of this book. At that time, I wrote that "only a few years ago, ATM might have been considered arcane, of interest only to a few experts. Today, ATM is appearing in daily newspaper headlines." After two years, ATM is now a reality in local-area and several wide-area (commercial and research) test beds and trials. The ATM Forum, an industry consortium producing interoperability specifications, has grown from a few dozen companies in 1992 to more than 700 organizations today.

In that first edition I also quoted M. E. Thyfault et al. (*Information Week,* 19 April 1993, pp. 22–25): "Talk to managers and chances are you will find they know a thing or two about pain relievers. Networking today induces headaches in any number of ways, and that is why a lot of people are campaigning for a new way to spell relief: ATM." Well, ATM has not yet been proven to be the remedy for all types of headaches. The early trials have illustrated various benefits of ATM in providing the basic framework for universal networking. Yet there are several issues that need to be resolved before the promises of ATM can be fulfilled. We still do not know how to take full advantage of some of ATM's capabilities. However, if the last two years are any indication of how quickly we can expect the various challenges imposed by ATM to be resolved, it may not be too long before ATM becomes widely available.

Eleven chapters addressing different aspects of ATM networks are included in this book. Chapters 1 and 2 introduce the basic concepts of ATM as the transfer mode of choice for B-ISDN. Chapter 3 explores the various characteristics of B-ISDN applications, including their information transfer parameters and service requirements. Chapter 4 addresses the congestion control problem in ATM networks and presents the various proposed approaches for the controlled use of network resources. Chapter 5 is a review of switch architectures proposed for ATM and their simplistic models for investigating the behavior of various performance metrics of interest. Chapter 6 presents an overview of the current ATM interfaces, including user-to-network, network-to-network, and data exchange interfaces. Chapter 7 discusses signaling in

ATM networks, both at the user-to-network and network-to-network interfaces (private network-to-network interfaces (P-NNI) and B-ISDN user part signaling). An end-to-end connection setup is used as an example to illustrate how an end-to-end connection may be established and how various network services can be provided in the network in a standards-compliant manner. Chapter 8 describes routing in ATM networks and covers the general routing framework and P-NNI routing. A critical factor in the success of ATM is how well it can support legacy applications. Chapter 9 describes various approaches proposed in standards organizations and the ATM Forum to provide connectionless service in ATM, with a view to enable different applications already developed in current packet-switched networks. Chapter 10 covers the underlying concepts of transport protocols, provides a short review and comparison of various lightweight protocols, and discusses reliability in multicast services and multimedia networking. Chapter 11 presents the current status of ATM network management.

A large portion of the book is suitable for self-study. Mathematical portions, however, require a background in stochastic processes in general and queuing theory in particular. Various chapters have been reorganized and mathematical derivations have been moved to the end of the chapters so that readers not necessarily interested in the details of such models can skip those parts without any interruptions in the general flow. A large number of references are provided at the end of each chapter. Although I made every effort to make these lists as complete as possible, valuable contributions reported in the literature may have been unintentionally left out. I would like to request the readers of this book kindly send to me any reference that would be of interest to the general readership.

Acknowledgments

I am indebted to all who contributed to this book. Various sections have greatly benefited from the material provided by Professors Y. Viniotis, H. Perros, and A. Nilsson and their students of North Carolina State University and Mr. J. Russel of IBM. Dr. C. Bisdikian, Mr. R. Cherukuri, and Mr. B. Ellington of IBM provided invaluable discussions, comments, and corrections. I also received encouragement and support from my management chain, particularly Dr. J. Marin.

My wife Nur and our two sons Melih and Doruk have shown a lot of patience over the past several years during the completion of this project. Their love and understanding are perhaps the two things I will cherish most about my work on this book.

1 Broadband Integrated Services Digital Network

The transfer mode defines how information supplied by network users is eventually mapped onto the physical network. The International Telecommunications Union Telecommunication Standardization Sector (ITU-T) defines the transfer mode as a technique used for the transmission, multiplexing, and switching aspects of communications networks. Current communications networks are classified according to the type of transfer mode used as follows:

- Circuit-switched networks;
- Message-switched networks;
- Packet-switched networks:
  - Datagram packet switching;
  - Virtual-circuit packet switching.

1.1 CIRCUIT-SWITCHED NETWORKS

In circuit switching, used mainly for telephone networks, a circuit is established between the calling and called parties to exchange information for the complete duration of the connection. Each circuit has a fixed bandwidth, such as 64 kbps for voice. Several circuits can be multiplexed onto a link, and switching is performed by translating the incoming circuit position to the outgoing circuit position. Before the data can start flowing, a collection of consecutive circuits is reserved at each link along the end-to-end path from source to destination by signaling. The establishment of a circuit, then, is a mapping from the incoming circuit position to the corresponding outgoing circuit position. Once a circuit is established, its bandwidth cannot be used by any other connection, whether there is any traffic flowing on it or not. For example, once two voice-bandwidth circuits are established between the called and the calling party, one circuit in each direction, the respective bandwidths of each circuit, 64 kbps in this case, are dedicated to the two parties for the duration of the connection. When one of the parties is listening to the other party, the listening party does not generate any voice frames (i.e., user traffic). Although the circuit is not used during such silent periods, it cannot be used to transfer any other traffic.

Circuit switching minimizes the end-to-end delay of connections. There is practically no buffering of voice frames at the intermediate switches. Accordingly, the total end-to-end delay in such networks is almost constant (with a negligible variation). This delay is the sum of the propagation delays of links the connection passes through and the switching times at intermediate nodes.

Circuit switching is not flexible enough to efficiently support applications with significantly varying bit rate requirements. Since each circuit has a fixed bit rate, it might be possible to define a basic circuit and allocate as many circuits to different applications as they require. However, this scheme introduces the problem of synchronizing, coordinating, and managing the possibly large numbers of circuits required by these connections. Furthermore, the selection of the basic circuit bit rate is a rather complicated issue. The bit rate requirements of various applications may vary from a few kilobits to several megabits per second. If the basic rate chosen is too small, say 1 kbps, then a connection requiring a 20-Mbps bit rate would require the control and management of 20,000 circuits. If the basic rate chosen is relatively large, then the dedicated bandwidth of each circuit would not be used effectively by small connections. For example, if the basic rate is chosen to be 256 kbps, a voice connection (64 kbps) would not use 192 kbps of a basic rate circuit.

1.2 MESSAGE SWITCHING

Circuit switching dedicates a fixed amount of network resources to each connection for the duration of the call. Although this simplifies the control and management of the network, doing so can waste network resources. For example, during a telephone conversation, a circuit is used less than 50% (typically around 35%) of the time. For data applications such as e-mail, file transfer, and transaction processing, treating each information unit as a logical entity, referred to as a *message*, proves to be much more efficient than dedicating resources. In this case, each message is transmitted in the network independently of each other. This is done by adding to each message a header that defines the destination node. Then each intermediate node stores the incoming messages, processes them to determine how they can be routed toward their destination, and transmits the messages in the corresponding outgoing link. In this framework, no circuit establishment with a dedicated capacity is necessary. Although network resources are used more efficiently than in circuit switching, end-to-end delays are no longer constant in message switching. This is due to the random delays that messages waiting to be processed and transmitted will incur in switch buffers. Accordingly, message switching is not suitable for real-time or delay-sensitive applications such as voice and interactive video (such as video conferencing).

1.3 PACKET SWITCHING

Packet switching is an attempt to combine the advantages of both circuit switching and message switching. In message switching, the entire message must be received before it is processed and transmission at the outgoing link can start. Applications generate messages of different lengths. A short e-mail message may be only a few hundred bytes, whereas a file transfer application may generate messages of several thousands of bytes.

Packet switching is essentially the same as message switching except that the size of the information unit transmitted in the network is restricted to a maximum value on the order of a few thousand bytes. Accordingly, user messages may be segmented into packets before they are transmitted. Doing so allows the reception and transmission of message packets to overlap, thereby reducing the end-to-end delays of messages. For example, assuming for presentation purposes that the incoming and outgoing transmission link speeds are the same and equal to 1,000 kbps, an 8,000-bit message would take 8 ms to arrive completely and another 8 ms to be transmitted. If the processing time is assumed to be negligible, it would take 16 ms for this message to arrive and leave the switch if there is no other packet waiting to be transmitted at the same outgoing link. If the same message is segmented into 1,000-bit packets, the arrival and transmission times of packets may overlap. That is, the third packet arrives during the time the second packet is being transmitted, the fourth packet arrives while the third packet is being transmitted, an so on. In this case, the total message can arrive and leave the switch in a little bit more than 9 ms, compared with the 16 ms it takes in the messaging mode.

This is a very simplified version of the packet-switching concept. In packet switching, each packet is required to include a header for routing purposes (which is perhaps larger than that of message switching, to include things such as packet sequence numbers). Hence, the main disadvantage of packet switching is that transmitting a message requires more overhead per message, thereby reducing effective resource utilization in the network compared with message switching. Each packet in the network is treated as an independent entity. However, they must be reassembled at the receiver to form the original information unit (message) before being passed to the user. Two approaches are used to handle packet streams in the network.

In the datagram approach, all packets are treated independently and may follow different paths to their destination. The main disadvantage of this approach is that packets may arrive at the destination out of sequence; sequencing packets to form information units is a processing-intensive operation. Alternatively, end-to-end logical connections can be established with no dedicated transmission capacity, before the transmission can start and all packets of a message follow (subject to failures and error conditions) the same path in the network. This guarantees the sequential delivery of packets to the receiver but requires a call setup phase. Various characteristics of different transfer modes are summarized in Table 1.1.

Early work in supporting real-time applications together with data applications in integrated networks concentrated on fast packet switching. As the transmission speeds increase, the main challenges introduced in fast packet switching with variable-length packets are the complexity of the switching fabrics and the buffer management scheme.

Table 1.1
Comparison of Communication Switching Techniques

Circuit Switching	*Message Switching*	*Datagram Packet Switching*	*Virtual-Circuit Packet Switching*
Dedicated transmission path	No dedicated path	No dedicated path	No dedicated path
Continuous transmission of data	Transmission of messages	Transmission of packets	Transmission of packets
Fast enough for interactive	Too slow for interactive	Fast enough for interactive	Fast enough for interactive
Messages are not stored	Messages are filed for later retrieval	Packets may be stored until delivered	Packets are stored until delivered
The path is established for entire conversation	Route established per message	Route established per packet	Route may be established for entire conversation
Call setup plus negligible transmission delay	Message transmission delay per hop	Packet transmission delay per hop	Call setup plus packet transmission delay
Busy signal if called party is busy	No busy signal	Sender may be notified if packet not delivered	Sender notified of connection denial
Overload may block call setup; no delay for established calls	Overload increases message delay	Overload increases packet delay	Overload may block call setup; increases packet delay
User responsible for message loss protection	Network responsible for messages	Network may be responsible for individual packets	Network may be responsible for packet sequencing
Usually no speed or code conversion	Speed and code conversion	Speed and code conversion	Speed and code conversion
Fixed bandwidth for call duration	Dynamic use of bandwidth	Dynamic use of bandwidth	Dynamic use of bandwidth
No overhead bits after call setup	Overhead bits in each message	Overhead bits in each packet	Overhead bits in each packet

Source: W. Stallings, *Data and Computer Communications*, New York: Macmillan, 1991, p. 280.

1.4 INTEGRATED SERVICES DIGITAL NETWORK

ITU-T adopted the first set of *Integrated Services Digital Network* (ISDN) recommendations in 1984. ISDN extends the concept of the telephone network by incorporating additional functions and features of circuit-and packet-switching networks to provide existing and new services in an integrated manner.

ISDN is a digital end-to-end telecommunications network supporting a wide range of voice and nonvoice applications in the same network. The network may be provided publicly, privately, or as a combination of both. The network is characterized by its access requirements and service characteristics and provides digital access to digital transmission services, packet data services, and network-provided data services.

ISDN is based on 64-kbps switching technology and is intended to support current voice facilities, existing data services, and low-speed video. Four physical interfaces are specified for ISDN for copper-wire transmission at speeds of 2.048 Mbps and lower. Two of these are the *basic rate interfaces* that carry in a duplex fashion a combined payload of 144 kbps made up of two 64-kbps voice and/or data channels (both circuit switched) and a 16-kbps data channel (packet switched).

One of the two basic rate interfaces defines a premises distribution interface that permits multiple terminals or phones within a home or office to be attached to the same physical wiring arrangement within a reasonable distance from the point of network attachment. Terminals on such an arrangement cannot communicate among themselves directly. They may, however, communicate with each other indirectly using an active switching element such as a private branch exchange (PBX) or a central office switch.

The other basic rate interface provides support for a point-to-point wiring arrangement. It supports existing central office distribution capability as well as the capability that commonly exists for terminals attached to a PBX.

In addition to these two interfaces, two *primary rate interfaces* are defined with total bit rates of 1.544 Mbps (i.e., T1), used in North America and Japan, and 2.048 Mbps (i.e., E1), used in Europe.

Based on this framework, various types of channels available in ISDN are:

- A bearer (B) channel that provides a 64-kbps transmission rate in a circuit mode;
- A high-speed bearer (H) channel that combines several B channels to provide higher bandwidth in circuit mode:
  - H0: 384 kbps;
  - H11: 1,536 kbps (on T1);
  - H12: 1,920 kbps (on E1);
  - H10 (North America only): 1,472 kbps;

- A D channel used for the dialogue between an ISDN terminal and the network to obtain an ISDN connection. Two bit rates are available on the D channel depending on the type of interface:
    - 16 kbps for a basic rate interface (2B + D);
    - 64 kbps for a primary rate interface (23B + D on T1 and 30B + D on E1).

Furthermore, it is possible to combine H channels and B channels at an ISDN interface.

It was soon realized that higher bit rates are required for applications such as the interconnection of local-area networks (LAN), video, and image, bringing the standardization process to the introduction of *broadband ISDN* (B-ISDN) concepts.

1.5 BROADBAND ISDN

B-ISDN was conceived as an all-purpose digital network. Activities currently under way are leading to the development of a worldwide networking technology based on a common set of user interfaces and universal communications. Once deployed throughout the world, B-ISDNs will facilitate worldwide information exchange between any two subscribers without strict limitations imposed by the communication media.

ITU-T Recommendation I.113 [3] defines the term *broadband* as "a service or system requiring transmission channels capable of supporting rates that are greater than the primary access rate." Currently, B-ISDN interfaces support up to 622 Mbps with the possibility of defining higher rates in the future. The concepts of B-ISDN are summarized in ITU-T Recommendation I.121 as follows [2]:

> B-ISDN supports switched, semi-permanent and permanent, point-to-point and point-to-multipoint connections and provides on demand, reserved and permanent services.
> Connections in B-ISDN support both circuit mode and packet mode services of a mono and multi-media type and of a connectionless or connection oriented nature and in a bi-directional and uni-directional configuration.
> A B-ISDN will contain intelligent capabilities for the purpose of providing service characteristics, supporting powerful operation and maintenance tools, network control, and management.

Accordingly, B-ISDNs will support services with both constant and variable bit rates and connection-oriented and connectionless transfers. At least conceptually, B-ISDNs not only support all types of communications applications that we know of, but also provide the framework to support future applications that we do not fully understand, or even know of, today. Hence, the challenge in putting the pieces of this puzzle together is to create the pieces as they fit together while still seeing the whole picture. The fact that this process is proceeding at full steam proves that there are enough people who believe that B-ISDN may be a reality as envisaged.

ITU-T classified possible broadband applications into four categories [1]:

1. Conversational services;
2. Retrieval services;
3. Messaging services;
4. Distribution services:
   a. Without user-individual presentation control;
   b. With user-individual presentation control.

Conversational, retrieval, and messaging services are interactive services, whereas distribution services are classified into two subcategories: distribution with or without user interaction for presentation control. Table 1.2 illustrates a number of examples in each category as presented by ITU-T.

1.5.1 Conversational Services

The first category of conversational services provides the means for communication with real-time end-to-end information transfer. The flow of information can be bidirectional or unidirectional. Some examples of conversational services presented by ITU-T are given in Table 1.2. Video telephony services provide person-to-person communication for the transfer of sound, moving pictures, scanned images, and data in an integrated manner. For example, today's advanced workstations may integrate a video camera, a telephone handset, and a facsimile machine. Using these types of workstations, individuals working on a joint project can collaborate freely, liberated from the tyranny of geography, if a communications network that can support such services exists among them. It may not be that long before video telephony would allow businesses to have their employees work at home while easily maintaining the involvement necessary in the day-to-day activities of a project team.

Video telephony services in a B-ISDN environment can be easily extended to include multipoint communications, referred to as *broadband video conferencing*. Current video conferencing is provided in a point-to-point manner, while audio conferencing is done in a multipoint fashion. The addition of other types of data, such as still images, facilitates the use of this medium for work group and collaborative development applications. Such applications are key in industries' drive to reduce cycle time and improve the quality of products and services. Eliminating geographical restrictions will permit the rapid formation of expert project teams.

Table 1.2
B-ISDN Conversational Services

Type of Information	*Examples of Broadband Services*
Conversational	
Moving pictures and sound	Broadband video telephony
	Broadband video conference
	Video surveillance services
	Video/audio information transmission services
Sound	Multiple sound program signals
	High-speed digital information transmission
Data	High-volume file transfer services
	High-speed teleaction
Document	High-speed telefax
	High-resolution image communication services
	Document communication services
Messaging	
Moving pictures (video) and sound	Video mail service
Document	Document mail service
Retrieval	
Text, data, graphics, sound, and still images	Broadband videotex
	Video retrieval services
	High-resolution image retrieval services
	Date retrieval services
	Document retrieval services
Distribution without user-individual presentation control	
Moving pictures and sound	Video information distribution services
Text, graphics, still images	Document distribution services
Data	High-speed information distribution services
	Existing-quality TV distribution services
Video	Pay TV
	Extended-quality TV distribution services
	High-definition TV distribution services
Distribution with user-individual presentation control	
Text, graphics, sound and still images	Full-channel broadcast videography

Document transfer includes high-resolution facsimile or mixed documents that combine text, voice annotation, video, and/or facsimile. In the data arena, B-ISDN conversational services include interconnection of LANs and metropolitan-area networks (MAN), as well as multisite interactive computer-aided design and manufacturing and distributive computing between research center supercomputers. Virtual supercomputing increases the amount of computational power available to an end user by dynamically sharing the "spare machine cycles" from network-attached supercomputers. Initially, such computing power would be of interest mainly to scientists and engineers. The bandwidth on demand and multipoint communication capabilities of B-ISDNs are ideally suited to the interconnection of supercomputers, providing dynamic load sharing.

Teleaction refers to computer control of a physical device from a remote location. Such services are typically associated with real-time process control. In these types of applications, data are collected from various physical devices, such as sensors and transducers, and are analyzed in real time according to a preestablished set of rules.

Depending on the outcome of the processing, an appropriate set of signals is sent to remote devices, thereby controlling their operations. Examples of such services include the operation of a hydroelectric generation station, computer-controlled welding machines, robotic assembly plants, and environmental quality monitoring.

1.5.2 Messaging Services

Messaging services offer communication between users via storage units with store-and-forward, mailbox, and message handling functions. Accordingly, these services, unlike conversational services, are not real-time services. The ITU-T classifies messaging services (see Table 1.2) according to data type. Video mail services are expected to replace current e-mail and phone messages by integrating the two, and are also expected to replace mailing services for video/audio cassettes. Similarly, document mail services allow the transmission of mixed documents that may include voice, text, image, and video. Another example of a messaging service is the reporting of local weather statistics in the form of data or graphics. They are sent to a central electronic mailbox for periodic retrieval, processing, and redistribution by regional weather services and news organizations.

1.5.3 Retrieval Services

Retrieval services provide users with the capability to retrieve information stored elsewhere in the network (see Table 1.2). The information stored at an information center starts flowing to a user at the user's request. For example, product documentation of highly complex machines, which includes various types of data (e.g., text, graphics, and sound), can currently be distributed by the use of CD-ROMs (compact disc–read-only memory). In this case, a manufacturer could distribute updated documentation, sales catalogs, literature, advertising, education, and so forth by transmitting data to the end user's site, either for playback in real time or for recording to a high-capacity media. This would replace mailing storage media and raise the meaning of product service to a new level.

Videotex is an interactive information retrieval service. Users can access public information networks, such as The Source, Prodigy, and Dow Jones, by using retrieval services to obtain stored information from a database. Since the information that can be provided is no longer only text, one can easily envision future libraries supporting electronic access to all stored information. Other types of videotex applications include e-mail order catalogs, airline reservation without operator interaction, and encyclopedia entries.

Another example of retrieval services is video-on-demand for both entertainment (as a replacement for cable television broadcasts and video rental) and remote education and training purposes.

1.5.4 Distribution Services

Video and audio transmission services are generally among the main driving forces for the deployment of B-ISDNs. Digital transmission of services such as distribution of TV programs to local stations, live TV interviews, and high-quality audio requires much more bandwidth compared to that available in current networks. B-ISDNs are expected to provide cost-effective means for the deployment of such services by using their expected multicasting capabilities.

Distribution services without user-individual presentation control provide a continuous flow of information distributed from a central location to a number of authorized users. The user has no control over the starting time or the order of presentation (see Table 1.2).

Document distribution services might prove to be useful in the distribution of printed materials (e.g., books, newspapers, and magazines) electronically from a central location to local printing sites. As another example, a stock market quotation database can be broadcast cyclically from a central site to a group of distributed servers, which can then be selectively accessed by local users. For reasons of economy and security, it is possible for today's news and weather forecast wire services to turn away from using satellite communications and use B-ISDNs to transmit their data. Another common example of this service is broadcast television. Current technology uses either radio waves or cable television distribution systems. With the introduction of B-ISDN, this service can also be integrated with other telecommunication services.

Distribution services with user-individual presentation control provide broadcast services in which the information is generated as a sequence of frames with cyclical repetition, so that the user has the ability to select individual information entities and can control the start and order of the information. The information distributed in this fashion includes text, graphics, audio, and still images, as well as mixed documents that integrate more than one of these services. An example of this type of service is the electronic replacement of traditional print media (e.g., newspapers, magazines). User-individual presentation control allows the user to browse the information selectively.

The standardization of B-ISDN has been taking place since the mid-1980s. To meet the requirements of supporting a wide range of applications with significantly different characteristics in the same network, the B-ISDN standardization work includes service descriptions, user and network interface specifications, equipment features, and performance requirements.

1.6 B-ISDN PROTOCOL REFERENCE MODEL

B-ISDN standards are being developed in a number of standards bodies around the world and are being finalized by ITU-T. The initial ITU-T recommendation on B-ISDN was published in 1988 [2]. The 13 ITU-T recommendations outlining the fundamental principles and initial specifications for B-ISDN were approved in 1990. These recommendations include the B-ISDN *protocol reference model* (PRM), illustrated in Figure 1.1. The B-ISDN standards and protocol layers are being developed around the PRM.

The physical layer is the underlying transport of the network. The services offered by this layer include bit timing, jitter and wander specifications, network clock, and maximum bit error rate. The control plane is used for connection management, including the connection setup and release functions. Once a connection is established, the user data are transmitted using one of the protocols in the user plane. Both planes use lower layers to transmit their messages and data. The transfer mode defines how the information supplied by higher

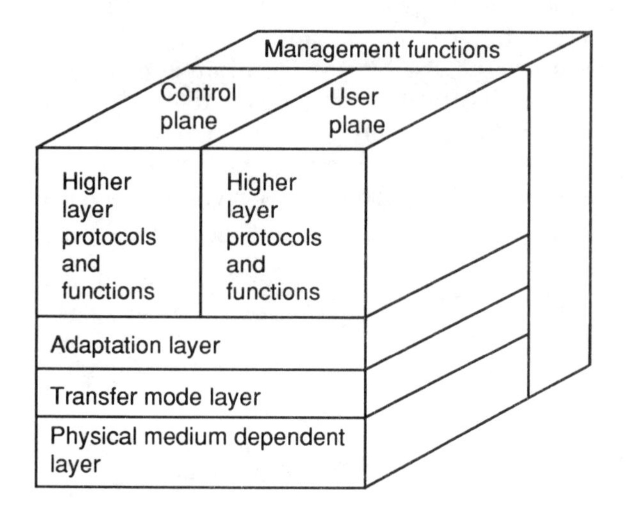

Figure 1.1 B-ISDN protocol reference model.

layers is to be mapped onto the physical layer. The adaptation layer supports the higher layer functions of the user and control planes. Examples of adaptation functions include continuous bit stream–oriented services adaptation functions, connectionless services, and packet mode services. Finally, the ITU-T PRM defines how management functions relate to the user and control planes.

1.6.1 Reference Configurations

Conceptually, B-ISDN consists of a number of customer premises and a network of nodes and links providing connections between these premises. ITU-T Recommendation I.413 [4] defines a number of functional groups: B-TE1, B-TE2, B-TA, B-NT1, B-NT-2, and reference points T_B, S_B, U_B, and R, as illustrated in Figure 1.2.

The B-ISDN reference configuration is deduced from that of ISDN and defines the interfaces between different entities of the network and their functions. B-NT1 performs low-layer functions such as line transmission termination and transmission interface handling associated with the physical and electrical termination of the B-ISDN on the user's premises. B-NT1 may be controlled by the network provider, forming the network boundary. B-NT2 performs higher layer functions, including multiplexing/demultiplexing traffic, bandwidth enforcement, switching internal connections, signaling protocol handling, buffering, and resource allocation.

Terminal equipment refers to user equipment that makes use of B-ISDN. B-TE1 terminates the standard B-ISDN user interface and performs the termination of all end protocols from lower to higher layers. B-TE2 is used for existing nonstandard B-ISDN interfaces. Such equipment requires terminal adapters to plug into a B-ISDN interface. B-TA performs all functions necessary to attach a nonstandard interface to a B-ISDN, including the rate adaptation. The reference point R provides a non-B-ISDN interface between nonstandard user equipment and adapter equipment. The reference point T_B separates the network provider's equipment from the user's equipment. Similarly, the reference point S_B corre-

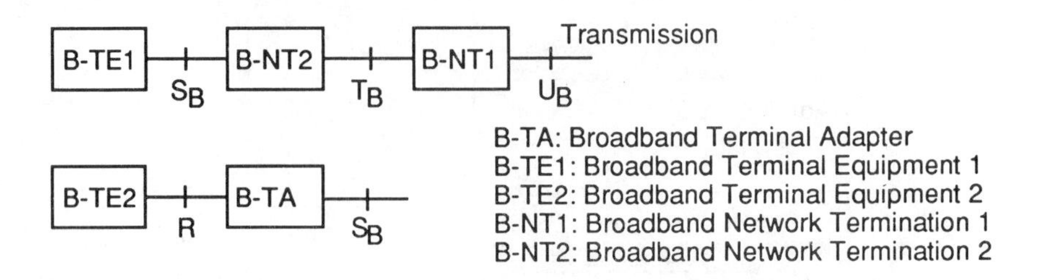

Figure 1.2 B-ISDN reference configuration.

sponds to the interface of individual B-ISDN terminals and separates the user's equipment from network-related communications functions.

Various options defined for the structuring of the physical bit stream at these interfaces include *cell-based* and *synchronous digital hierarchy* (SDH). The T_B interface is physically a point-to-point interface; that is, B-NT2 and B-NT1 are always paired. On the other hand, B-NT2 may support a number of B-TE1s. Moreover, various physical configurations of functional groupings and reference points are realizable, as illustrated in Figure 1.3. In Figure 1.3(a,b), B-NT2 supports multiple physical connections. Existing LANs such as Token Ring and Ethernet may be connected to B-ISDNs via interworking units (IWU), which provide medium-access mechanisms to ensure access to B-ISDNs. For example, the terminal equipment and B-NT2 functions are combined in Figure 1.3(c). An example of this configuration is the case in which a host computer acts as a packet switch in a private packet network. The B-NT1 and B-NT2 terminations are combined in Figure 1.3(d) to reflect the case in which a provider puts together the functions of both into a single device. This is mostly the case in several countries without competitive telecommunications services, where a single provider offers, for example, both LAN and wide-area network (WAN) services. The combined TB and SB interface in Figure 1.3(e) illustrates B-ISDN

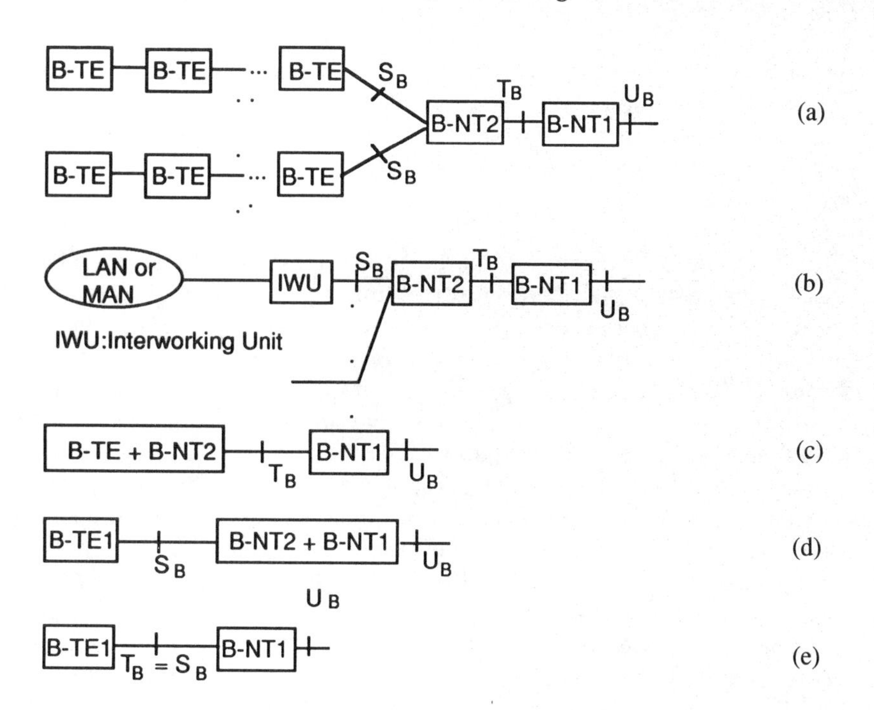

Figure 1.3 Examples of B-ISDN configurations. IWU, interworking unit.

interface compatibility where a B-ISDN user can connect directly to a subscriber loop terminator or to a LAN using the same interface, ensuring portability.

1.6.2 Issues in B-ISDN

Considering the sets of applications given in Table 1.2, B-ISDNs are expected to support interactive and distributive services, bursty and continuous traffic, connection-oriented and connectionless services, and point-to-point and complex communications, all in the same network. The types of services B-ISDNs are envisaged to offer can be characterized by one or more of the following attributes:

- High bandwidth;
- Bandwidth on demand;
- Varying quality of service parameters;
- Guaranteed service levels;
- Point-to-point, point-to-multipoint, and multipoint-to-multipoint connections;
- Constant- or variable-bit-rate services;
- Connection-oriented or connectionless services.

Accordingly, B-ISDN should be capable of assigning usable capacity dynamically on demand. B-ISDN switch fabrics should be capable of switching all types of services. Furthermore, the network should take the bursty nature of variable-bit-rate applications into consideration in allocating the available bandwidth, while guaranteeing the quality of service for all network applications.

The introduction of highly reliable fiber systems into the access network provides the necessary high bandwidth required for B-ISDN. However, there are a number of issues that need to be resolved before B-ISDN networks are widely deployed. As technology advances rapidly to meet the need for high-speed communications, the bottlenecks in communications networks are moving from the transmission media to the communications processors. The throughput and end-to-end delay requirements of applications become limited by the processing power at network nodes, necessitating fast network protocols.

The suitability of current network protocols in B-ISDN has not been fully addressed in the standardization committees. For example, it is not clear whether an existing transport protocol such as transport control protocol (TCP), modified through clever tuning, can meet the requirements of B-ISDN services, or if a new protocol needs to be designed. Adopting a modified version of an existing protocol can significantly reduce the amount of time required for standardization, whereas a new protocol standard may take several years to develop.

Congestion control is another major area that needs to be satisfactorily addressed. B-ISDNs will support a very large number of connections simultaneously in the network. Simple call admission, as in today's telephone systems, and hop-by-hop flow control, used in current packet networks, are no longer effective in B-ISDNs. The problem is much more complicated than can be addressed with the experience gained from traditional packet-switching networks, mainly because of the introduction of high-bandwidth links with relatively large propagation delays into the backbone.

One issue that is resolved is the transfer mode: *asynchronous transfer mode* (ATM) is ITU-T's transfer mode of choice for B-ISDN. In simple terms, ATM is a connection-oriented, packet-switching and multiplexing technique that uses short fixed-size cells to transfer information over a B-ISDN network. The short cell size of ATM at high transmission rates is expected to offer full bandwidth flexibility and provide the basic framework for guaranteeing the quality of service requirements of applications with a wide range of performance metrics, while allowing statistical multiplexing. The term *statistical multiplexing* refers to the fact that several variable-bit-rate connections can share a link with a capacity less than the sum of the their peak (maximum) bit rate requirements, whereas the term *asynchronous* is used to reflect the fact that the cells of an information unit may appear at irregular intervals over the network links. On the other hand, although it can and will support connectionless data transfer, ATM is a connection-oriented technique in that end-to-end paths are required to be established prior to the beginning of the information transfer. The packet-oriented nature of ATM is well suited to applications with bursty traffic characteristics. ATM can carry constant-bit-rate traffic equally well. Hence, ATM is an attempt to utilize the properties of both the packet- and circuit-switch networks in an integrated network.

1.7 ATM IN THE CONTEXT OF OTHER TRANSFER MODES

In general, the transfer mode of B-ISDN is envisaged to have the following properties:

- Should support and integrate all existing network applications as well as those with *yet unknown* characteristics that would emerge in the future;
- Should minimize switching complexity;
- Should minimize the processing load per packet at intermediate switching nodes to be able to support very high transmission speeds;
- Should minimize the buffer requirements at the intermediate nodes to bound the delay and the complexity of buffer management;
- Should provide the basis for guaranteeing application quality of service requirements in the network.

ATM is an attempt to meet all of these objectives. However, it is a compromise in that it is not designed specifically for voice, data, or video. Instead, ATM is intended to allow integration of all these different types of services. ATM has various features that extend the capabilities of current packet-switching networks toward incorporating the most desired features of circuit switching to support real-time traffic most efficiently.

1.8 B-ISDN AND ATM

B-ISDN was originally conceived as a carrier service to provide high-speed communications to end users in an integrated way. Some still view B-ISDN as a carrier interface. Some people do not like the term B-ISDN because it sounds awfully similar to ISDN, which has had limited success in the marketplace. Despite how it all started originally, B-ISDN today should be viewed as a universal networking technology and cannot be attributed to only one part of the industry.

ATM is the technology of choice to achieve universal networking. It is the transport mode of choice for B-ISDN, in which user information is transmitted between communicating entities using fixed-size packets, referred to as *ATM cells.* The universal acceptance of ATM among every segment of the networking industry comes from the fact that it is a compromise. Voice service providers complain because the cell size is too big. Data service providers complain because the cell size is too small. Since everyone in networking has complaints about the choice of ATM, perhaps this means that standards organizations did a very good job: they came up with a transfer mode that is not optimized for only a single application. ATM is a compromise that allows the integration of different services with different characteristics and requirements. ATM does not handle voice as efficiently or as cost-effectively as an isochronous network does. ATM does not handle video as easily as isochronous transfer. ATM does not handle data as efficiently as a packet-switching technology with variable-length packets. But it is the only standard technology that integrates them all.

Is it possible to say that ATM is a carrier technology? A private networking technology? Is B-ISDN designed for public service providers to go into the business of private networking? Does ATM allow private networking vendors to go after the voice business (among others), which is what a public network is all about? The answer to each one of these questions is a clear yes. ATM provides a place for all who want to play the game.

1.9 SUMMARY

B-ISDNs are envisaged to support not only current services, but also new services with varying traffic characteristics and service requirements. Providing services for such a variety of applications, with different characteristics, is a challenge that, if successfully addressed, will change the face of networking. The B-ISDN PRM is proposed to meet this challenge. Standardization is necessary and will provide the basic framework upon which B-ISDNs can be built. However, before B-ISDNs can be deployed, a lot remains to be satisfactorily addressed by network providers beyond what can and will be standardized, such as:

- Traffic management framework to guarantee the service requirements of various applications with different characteristics and requirements;
- Transport protocols that can take advantage of high transmission bandwidth;
- Switch fabrics with multicast capabilities;
- Support of connectionless and connection-oriented services in the same network.

In summary, B-ISDN is envisaged to provide universal networking. ATM is the transfer mode of choice for B-ISDN. The rest of the story is a list of problems. Throughout the rest of this book, we provide a few solid (and many not so solid) solutions that have been proposed to address the challenges of ATM.

References and Bibliography

[1] ITU-T Recommendation I.211, "B-ISDN Service Aspects," 1990.

[2] ITU-T Recommendation I.121, "Broadband Aspects of ISDN," 1990.

[3] ITU-T Recommendation I.113, "Vocabulary of Terms for Broadband Aspects of ISDN," 1991.

[4] ITU-T Recommendation I.413, "Broadband Aspects of ISDN," 1990.

[5] Albanese, A., H. E. Bussey, S. B. Weinstein, and R. S. Wolf, "A Multi-Network Research Testbed for Multimedia Communication Services," *ICC '91,* 1991, pp. 86–91.

[6] Anagnostou, M. E., et al., "Quality of Service Requirements in ATM Based B-ISDNs," *Computer Communications,* Vol. 14 No. 4, 1991, pp. 197–204.

[7] Kano, S., K. Kitami, and M. Kawarasaki, "ISDN Standardization," *Proc. IEEE,* Vol. 79, No. 2, February 1991, pp. 118–124.

[8] Kleinrock, L., "ISDN—The Path to Broadband Networks," *Proc. IEEE,* Vol. 79, No. 2, February 1991, 112–117.

[9] Minzer, S. E., "Broadband ISDN and Asynchronous Transfer Mode (ATM)," *IEEE Comm. Magazine,* September 1989, pp. 17–24.

[10] Nikolaidis, I., and R. O. Onvural, "A Bibliography of Performance Issues in ATM Networks," *ACM Computer Communications Review,* October 1992.

[11] Onvural, R. O., and H. G. Perros, "Performance Issues in High Speed Networks," *Proc. TELECOM '91,* 1991.

[12] Rice, W. O., and D. R. Spears, "What's New in B-ISDN Standards," *High Speed Networks,* H. G. Perros, ed., New York: Plenum, 1992, pp. 125–138.

[13] Sammartino, F., and D. Blackketter, "Desktop Multimedia Communications—Breaking the Chains," *ICC '91,* 1991, pp. 73–77.

2 Asynchronous Transfer Mode

The B-ISDN PRM with ATM is shown in Figure 2.1, which illustrates that the transfer mode of choice for B-ISDN is ATM.

PRM consists of three planes: control, user, and management. The control plane handles all connection-related functions, addressing, and routing. These functions play a particularly important role for connections that are established dynamically (on demand) in the network. The user plane transmits end-to-end user information between two or more communicating entities. The management plane provides for operations and management functions, and also provides the mechanisms to exchange information between the user and the control planes. This plane is further divided into two layers: layer and plane management. Layer management deals with layer-specific management functions that include the detection of failures and protocol abnormalities. Plane management provides management and coordination functions related to the complete system.

All three planes use the physical and the ATM layers. The ATM adaptation layer (AAL) is service-specific and may or may not be used depending on the requirements of the (user, control, and management) applications.

Figure 2.2 summarizes various functions performed at the physical, ATM, and AAL layers. An ATM network provides end-to-end ATM layer connectivity between end stations. The ATM layer mainly provides the switching and multiplexing of traffic. There is no awareness of the specifics of applications at this layer. This simplicity is necessary to keep up with the high-speed network links. Application-specific services are provided at end stations by the adaptation layer. AAL, when used, provides an interface between the user layer and the ATM layer for applications with similar service requirements. AAL is not part of the user plane at the intermediate ATM switches inside the network. Different AALs are defined to support different types of traffic. Finally, the physical layer transports ATM cells between two adjacent ATM layers.

How well the ATM architectural framework addresses the challenges of B-ISDN services and at what expense have been long argued. Despite the arguments made against it, ATM is the transfer mode of choice for B-ISDN, and we will all most likely have to live with it for several years to come.

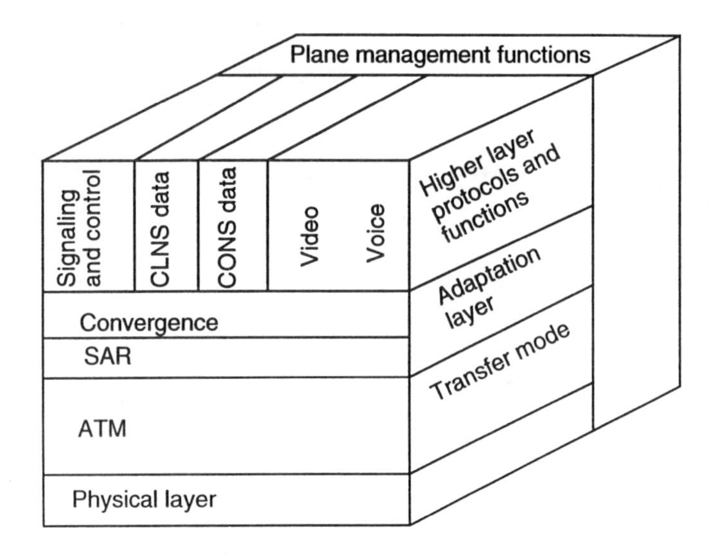

Figure 2.1 ATM protocol reference model.

Higher layers

Layer name		Functions	
Higher layers		Higher layer functions	
AAL	Convergence sublayer	Service specific (SSCS) Common part (CPCS)	Layer Management
	SAR sublayer	Segmentation and reassembly	
ATM		Generic flow control Cell header generation/ extraction Cell VPI/VCI translation Cell multiplexing/ demultiplexing	
Physical	Transmission convergence (TC) sublayer	Cell rate decoupling Cell delineation Transmission frame generation/recovery	
	Physical medium dependent (PMD) sublayer	Bit timing Physical medium	

Figure 2.2 ATM adaptation, ATM, and physical layers.

2.1 ATM CELL HEADER

Before we proceed with the details of various functions performed at each layer of the ATM architectural framework, let us look at the ATM cell itself.

Some of the design objectives used to define ATM include:

- Integrating voice, audio, video, image, and data services;
- Minimizing the switching complexity, the processing needed at intermediate nodes, the buffer management complexity, and the buffering needed at intermediate nodes mainly to keep up with the high-speed transmission links in the network.

These design objectives are met at high transmission speeds by keeping the fundamental unit of ATM transmission—the ATM cell—short and fixed in length. Keeping the cell length short provides a great deal of flexibility in the way the bandwidth can be used. This flexibility in turn provides the basic framework for B-ISDN to support a wide range of services that emerging applications require. Moreover, the use of short, fixed-length cells fosters the efficient use of transmission buffers/bandwidth through statistical multiplexing. With statistical multiplexing, the total bandwidth needed by an aggregate of traffic may be less than the sum of the maximum bandwidth needed by each one of the components of the aggregate (i.e., individual traffic streams), provided that the components generate traffic at rates that fluctuate independently over time.

More technically, the ATM framework is a connection-oriented packet-switching technology that segments application data frames into 48-byte-long cell payloads, adds the associated 5-byte header, transfers these cells through the ATM network, and assembles the cell payloads at their destination to reconstitute the original user data frames.

Characterizing ATM as *asynchronous* indicates that cells may occur at irregular times that are determined by the nature of applications rather than the framing structure of the transmission system. Characterizing ATM as *connection-oriented* arises out of the need to reserve network resources in order to meet the application service requirements. It also indicates that each switch in the network keeps cell-routing tables which tell the switches how to associate incoming cells with the proper outgoing links (i.e., where to switch incoming cells).

The ATM cell header illustrated in Figure 2.3 consists of five fields: *virtual path identifier* (VPI), *virtual channel identifier* (VCI), *payload type* (PT), *reserved field* (Res), *cell loss priority* (CLP), and *header error check* (HEC). At a demarcation point between an ATM end station and the network, the cell header also includes a *generic flow control* (GFC) field. This 4-bit field is part of the VPI inside the network. These fields are defined next, and a more detailed description of each field follows.

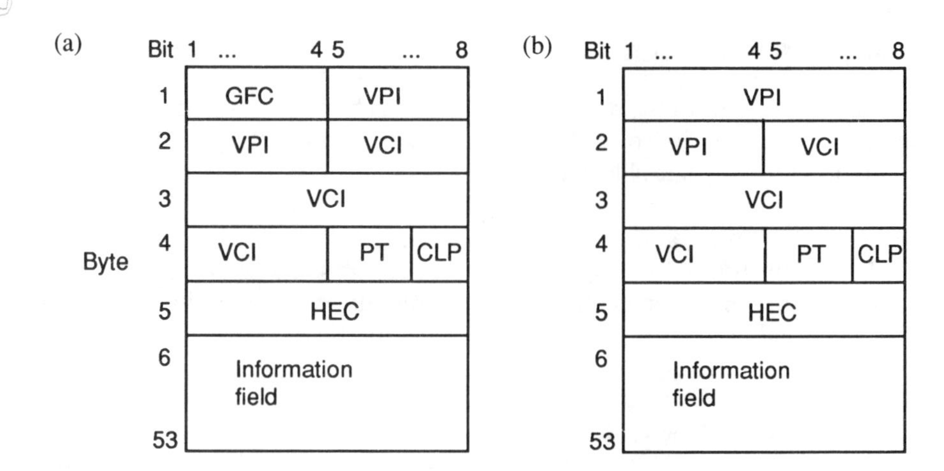

Figure 2.3 ATM cell formats: (a) user-to-network interface; and (b) network-to-network interface.

GFC is a 4-bit field that provides a framework for flow control and fairness to the user-to-network traffic and does not control the traffic in the other direction (i.e., network-to-user traffic flow). The GFC field has no use within the network and is meant to be used by access mechanisms that implement different access levels and priorities.

In ATM, end-to-end *virtual channels* are established between end stations before the traffic can start flowing. Routing of cells in the network is performed at every switch for each arriving cell. The routing information of a cell is included in the two routing fields of the header: VPI and VCI. The two levels of routing hierarchies, virtual paths (VP) and virtual channels (VC), are defined in ITU-T Recommendation I.113 as follows:

- VC: A concept used to describe unidirectional transport of ATM cells associated by a common unique identifier value, referred to as the VCI.
- VP: A concept used to describe the unidirectional transport of cells belonging to VCs that are associated by a common identifier value, referred to as the VPI.

The payload type indicator specifies whether the contents of a payload carries user data or management data. The management data may be used inside ATM networks, whereas the ATM layer is not concerned with the contents of ATM cells that carry user information.

The CLP at the ATM cell header is a 1-bit field used for cell loss priority. Due to the statistical multiplexing of connections, it is unavoidable that cell losses will occur in ATM networks. Cells with CLP bit set (low priority) may be discarded earlier at congested switches than cells with CLP bit not set (high priority).

The HEC field is used mainly for two purposes: for discarding cells with corrupted headers and for cell delineation. The 8-bit field, when used for HEC, provides single-bit error correction and a low probability corrupted cell delivery capabilities. The field is also used to identify the cell delineation (i.e., determining cell boundaries from received bit stream), which is discussed later. The HEC value is equal to the remainder of the division of the product x^8 and the polynomial of order 31. The coefficients of the polynomial are given by the bit values of the 4 bytes of the cell header.

2.2 THE ATM LAYER

The ATM layer transfers cells between peer ATM layer entities. It provides in sequence delivery of cells among ATM layer users by utilizing services provided by the physical layer. At the originating end station, it receives any 48-byte cell payload from an ATM layer user, adds the 4 bytes of the corresponding cell header (excluding the HEC byte), and passes the cell to the physical layer for HEC calculation and transmission. At the destination end stations, the ATM layer receives cells from the physical layer, removes the cell header, and passes the cell payloads to their corresponding ATM layer users. Inside an ATM transport network, there is no ATM layer user for the user traffic (i.e., in the user plane), and cells are passed from the receiving ATM layer entities to the transmitting counterparts at each switching node along their paths between the source and destination end stations. Accordingly, the ATM layer mainly provides the switching function of ATM networks. Cells with errored headers (i.e., bit errors) are discarded at intermediate nodes.

Similarly, cells that arrive at a time when the transmission link buffer is full may be dropped at a switch. However, the ATM layer provides its services unreliably. That is, there is no retransmission of errored and lost cells inside the network and it is up to the end stations to ensure the integrity of the data carried in ATM cell payloads.

The ATM layer specifications elaborate on the following elements:

- Cell structure and encoding;
- Services expected from the physical layer;
- Services provided to ATM layer users;
- ATM layer management;
- Traffic and congestion control.

2.2.1 Cell Structure and Encoding

Cell structure and encoding are illustrated in Figure 2.3. Each of the cell header fields is described in detail below.

2.2.1.1 Generic Flow Control

The GFC mechanism is used to control traffic flow from end stations to the network by limiting their effective ATM layer transport capacity. Two sets of procedures are defined across a user-to-network interface (UNI) (i.e., at the demarcation point between an ATM end station and the network): uncontrolled and controlled, thereby implicitly defining two classes of connections, controlled and uncontrolled. Traffic on uncontrolled connections enters the network without GFC control, while traffic on controlled connections requires GFC control to enter the network. The GFC mechanism does not control the traffic in the network-to-end-station direction (i.e., network-to-user traffic). Furthermore, these mechanisms are not used inside the network or at the demarcation points between networks (i.e., network-to-network interface (NNI)). The GFC control takes place using the first 4 bits of the ATM cell header. This field is used as a part of the VPI field inside the network and the NNIs.

The default mode at an interface provides for a single queue for controlled ATM connections and allows uncontrolled connections. Optionally, an additional controlled queue may be supported at the interface. Hence, in general, the GFC framework allows:

- Uncontrolled connections;
- Connections from controlled queue A;
- Connections from controlled queue B.

The determination of an uncontrolled, queue A, or queue B setting is either made at call setup time or determined by configuration information, and it is identical for all cells for a given value of VPI/VCI. The GFC field of each cell transmitted to the network includes this information, as is illustrated in Figure 2.4.

The traffic flow is controlled by the network (i.e., controlling equipment) using a credit-based schema. The two counters and two flags used for this purpose are defined as follows:

- GO-CNTR specifies the number of credits currently available to the connection group (i.e., queue A or queue B); that is, the number of cells that can be transmitted until the controlling equipment sends another command.

Value	In the direction of end station to network
0000	Terminal is uncontrolled. Cell is an idle cell or on an uncontrolled ATM connection
0001	Terminal is controlled. Cell is an idle cell or on an uncontrolled ATM connection
0101	Terminal is controlled. Cell is on a controlled ATM connection queue A
0011	Terminal is controlled. Cell is on a controlled ATM connection queue B

Figure 2.4 The use of GFC bits in the direction of end station to network.

- GO-VALUE specifies the maximum number of credits available to the connection group (i.e., queue A or queue B).
- TRANSMIT specifies whether cell traffic from the end station to the network is enabled or not. If this flag is set, the end station is free to transmit cells to the network on any uncontrolled connection and from controlled connections if the GO-CNTR is greater than zero. If this flag is null, the end station is not permitted to send any ATM layer cells (i.e., excluding idle cells that are generated at the physical layer) to the network on any connection.
- GFC-ENABLE specifies whether the end station should perform controlled GFC procedures (i.e., the flag is set) or not (i.e., the flag is reset).

The controlling equipment uses four signals to perform its control: SET, NULL, HALT, and NO-HALT. The HALT signal is optional. When used, it causes the traffic flow to the network to stop on all connections, including uncontrolled connections. The SET signal sets the credit counter of the corresponding queue (SET-A for queue A and SET-B for queue B) to a specified integer number (i.e., GO-CNTR = GO-VALUE). The NULL signal does not cause any action while implicitly indicating that the GO-CNTR value should not be reset. The NO-HALT signal negates the HALT signal and thereby allows cell traffic to start flowing.

The use of the field and the combinations of GFC signals in the direction of network to end station are illustrated in Figure 2.5.

Let us consider an ATM end station controlled by the GFC mechanism. As a default, the end station has one "controlled" queue. Optionally, there can also be another controlled queue. The procedure is the same with one or two queues.

The default value for the GFC credit counter (GFC-CNTR) is initialized to 1 (GFC-VALUE = 1). This value can be changed by management procedures. Similarly, GFC-ENABLE is initialized to reset and TRANSMIT = null.

The controlling equipment uses HALT and SET signals to solicit GFC capability for some period (i.e., up to 5 sec). The end station performs uncontrolled GFC procedures until it receives HALT and/or SET signals. Upon receiv-

Value	In the direction of network to end station
0000	NO-HALT, NULL
1000	HALT, NULL-A, NULL-B
0100	NO-HALT, SET-A, SET-B
1100	HALT, SET-A, SET-B
0010	NO-HALT, NULL-A, SET-B
1010	HALT, NULL-A, SET-B
0110	NO-HALT, SET-A, SET-B
1110	HALT, SET-A, SET-B

Figure 2.5 The use of GFC bits in the direction of network to end station.

ing one of these signals, the controlled equipment sets the GFC-ENABLE flag and starts performing GFC procedures.

Upon receiving a HALT signal, the end station nulls the TRANSMIT flag and stops sending any ATM layer cell from any connection. A NO-HALT signal causes the TRANSMIT flag to be set and the cell traffic to flow. The HALT/NO-HALT signals logically limit the effective ATM transport capacity (i.e., decrease the effective transmission capacity). For example, if the HALT signal is in effect 50% of the time, the physical layer transmission capacity as seen by the ATM layer is reduced by half.

Upon receiving the SET signal, the controlled equipment sets its credit counter to GO-VALUE. The NULL signal has no action on GO-CNTR.

Cells from controlled connections can be transmitted only if (1) the TRANSMIT flag is set, (2) there is no cell to transmit from the uncontrolled connections, and (3) the GO-CNTR is greater than zero. Every time a cell from a controlled connection is transmitted, the corresponding GO-CNTR value (i.e., queue A or B) is decreased by 1.

The procedures are the same with one (default) or two (optional) queues. In the latter case, another flag, GROUP-SELECT, is used to define the priority for sending cells from each queue. In particular, given that a cell can be transmitted from a controlled queue (based on the rules above), if the GROUP-SELECT flag is set, then a cell from queue B can be sent only if either GO-CNTR-A = 0 or there is no cell to transmit in queue A. Similarly, if the GROUP-SELECT flag is not set, then a cell from queue A can be sent only if either GO-CNTR-B = 0 or there is no cell to transmit in queue B.

2.2.1.2 Connection Identifiers: VPI and VCI

ATM is connection-oriented and requires end-to-end connections to be established before the traffic can start flowing. ATM connections are either preestablished using management functions or they are set up dynamically on demand using signaling. Preestablished connections are referred to as *permanent virtual connections* (PVC), whereas connections that are managed dynamically using signaling are referred to as *switched virtual connections* (SVC).

The ATM cell header includes a 28-bit routing field consisting of two identifiers: VPI and VCI. The two identifiers, together with the physical link the cells arrive from, uniquely identify connections at each ATM switch. Before proceeding further with the details of the routing hierarchy in ATM, let us consider the 2×2 switch illustrated in Figure 2.6, with traffic given in Table 2.1.

In this example, we have cells arriving from three different connections at physical link 1 and two connections at physical link 2. Note that:

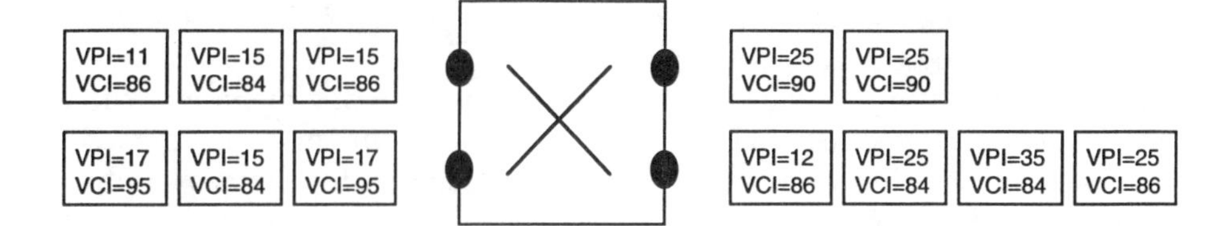

Figure 2.6 Switching ATM cells from incoming to outgoing links.

Table 2.1
Routing Table for the Switch in Figure 2.6

Connection	*Incoming Cells*				*Outgoing Cells*	
	Physical Link	*VPI*	*VCI*	*Physical Link*	*VPI*	*VCI*
1	1	15	86	4	25	86
2	1	15	84	4	25	84
3	1	11	86	4	12	86
4	2	17	95	3	25	90
5	2	15	84	4	35	84

- Cells of different connections are interleaved in the physical link by using different connection identifiers (CI).
- For some cells, both the VPI and the VCI are changed. whereas for some others only the VPI values are changed from the incoming links to the outgoing links.
- Two different connections on the same physical link might have the same VPI value.
- Two cells on different physical links may have the same VPI/VCI value.

In general, VCI values are unique only in a particular VPI value, and VPI values are unique only in a particular physical link. Accordingly, a connection is identified by the CI as a triplet CI = (incoming link identifier, VPI, VCI). Furthermore, the VPI/VCI has local significance only and the VPI/VCI label is translated at every switch the cell traverses. This technique is known as *label swapping*.

A connection within an ATM switch is a simple mapping that translates incoming CIs to the outgoing CIs. The VPI/VCI of a cell used at the outgoing link of a switch is the same as the incoming VPI/VCI at the neighbor switch the cell visits next along its end-to-end path. Any two switches connected by a point-to-point transmission link first negotiate CIs to be used and use the agreed-upon values to set up their routing tables, which define the mapping from the incoming CIs to the outgoing CIs.

An ATM routing framework defines a two-level routing hierarchy: virtual path (VP) and virtual circuit (VC). A VP is a logical *semipermanent* connection established by using network management and/or control functions over physical links in sequence. That is, routing table entries that map an incoming (physical link identifier, VPI) pair to a particular outgoing (physical link identifier, VPI) pair are predefined and are only changed when the network is reprovisioned. Note that CIs for VPs do not include VCIs. This is done intentionally to simplify network control and management functions. In particular, let us consider a VP established over a four-hop path between ATM switches 1 and 4, as illustrated in Figure 2.7.

Since this VP is preestablished, it is a point-to-point logical connection between points A and B. A VP can accommodate up to 65,536 (2^{16}) VC connections, all of which originate at point A and terminate at point B. Accordingly, there is no need to translate their VCIs, since they can be delivered from A to B using only VPI. In this case, only one table entry is required at the four switches for up to 65,536 individual connections. Based on this framework, the VP concept may be used to simplify the routing table management, connection setup process, path selection, and bandwidth reservation. For example, there might now be fewer "logical" links needing to be considered in determining end-to-end paths, compared with selecting paths based on a potentially larger number of physical transmission links.

Let us for a moment assume that a VP is defined between every source and destination in the network. The routing decision is then simply reduced to determining whether a particular VP connecting the two end stations can support the application service requirement. If so, the connection is assigned a VCI to distinguish different connections at the two end stations. If there is no end-to-end VP that can support this service requirement, then the connection request is rejected.

By using end-to-end VPs, the path selection function becomes very simple (and fast), the routing tables are small and network management functions are simplified. However, all these advantages sound too good to be true. Although end-to-end VPs are used in some of the early ATM networks, there is something quite wrong with this approach: utilization of network resources.

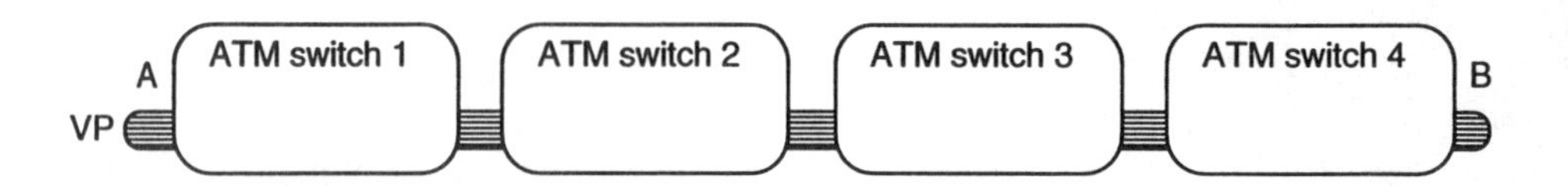

Figure 2.7 An end-to-end VP connection.

The main problem with end-to-end VPs arises out of the fact that most ATM connections require service guarantees, and the network has to allocate resources (i.e., bandwidth and/or buffer) to provide the quality of service requested by network applications. With end-to-end VPs, allocation of a VCI in a VP necessitates confirmation from every switch along the path of the VP that the connection request is accepted, establishing that the service requirement of the connection can be met. This would mean that intermediate nodes would still do processing on a per-connection basis. This is what we wanted to avoid in the first place.

Clearly, there is a need to compromise (as any other solution in ATM). We will present the solution in two parts. First, each VP can be allocated a certain amount of bandwidth at each physical link it traverses (and/or other resources, depending on the internal ATM network implementation) that cannot be used by any other VP (i.e., any other ATM connection) in the network. This would eliminate the need for per-connection resource reservation processing at the intermediate switches. The switch that originates the VP is the owner of the VP connection. It knows the amount of available resources along the VP and makes the connection acceptance decision based on connection traffic and service characteristics. If the required resources are available, the connection is accepted; otherwise it is rejected. Although this approach solves the problem of intervention at the intermediate nodes, it introduces another problem.

As a possibly exaggerated (yet equally realistic!) example, traffic between two end stations may be active for an hour or so a day and no other (or very little) traffic might be generated during the off periods. Provisioning an end-to-end VP based on the busy-hour traffic requirements would cause the VP to be used very little during the off periods. Furthermore, since the resources allocated to a VP cannot be used by the others, this solution would artificially restrict the resource utilization in the network.

High utilization of network resources, on the other hand, would be achieved if VPs are defined over a single physical link (i.e., VPI is just a part of VCI). In this case, the available bandwidth on a VP may be used by any connection between two switches. This eliminates the unavailability of resources on end-to-end VPs to other connections imposed because a VP can be accessed only from its end points (as a connection can be multiplexed on a VP only at the points it originates). With the VP defined between two neighbor switches, however, network management and control functions become processing intensive and relatively more complex.

Hence, with VPs defined between neighbor switches, the resource utilization is maximized at the expense of complex (processing-intensive) management and control functions. With VPs defined end to end, it becomes difficult to achieve high resource utilization, but control and management functions are simplified because there is no need to do any setup in the intermediate switches.

A compromise and possibly fairly good solution is to define VPs between ATM switches that are two or more hops away so that an end-to-end connection may use one or (usually) more VPs to provide connectivity between two ATM end stations. This framework essentially provides a pseudonetwork of switches (such that there is one or more switches between two such switches) that can be used in establishing new connections without incurring processing burdens at the bypassed switches. In this framework, a VP is shared by the traffic generated from a number of source-destination pairs and provides the basic framework for using network resources (relatively) efficiently without introducing large processing burdens at the intermediate nodes for connection setup.

Let us now summarize the connection framework in ATM. ATM requires the establishment of end-to-end connections before the traffic can start flowing. An end-to-end connection is referred to as a *virtual channel connection* (VCC). A VCC is composed of one or more VC links consisting of one or more physical links between the point where the VCI is assigned and the point where it is swapped out (i.e., translated). A VP link is defined between the point where the VPI is assigned and the point where it is removed or translated. A virtual path connection (VPC) is a concatenation of one or more VP links. A VC link consists of one or more VPCs. This hierarchy of "logical" connections in ATM is illustrated in Figure 2.8.

A VP is a collection of VCs between two nodes in an ATM network. A predefined route is associated with each VP in the physical network (i.e., VPs are semipermanent connections) and routing tables for VP switching are preset in ATM networks using network management functions. In order to take full advantage of the VP concept, each VP is allocated a fixed amount of bandwidth. This essentially creates an overlay network with "logical" links on top of a physical network.

In establishing a connection between end stations A and B, the network indicates the two VCIs (a and b) used by the sender and the receiver in both transmitting and receiving the cells. The routing tables are set up at the intermediate nodes that are the end points of VPs (along the path connecting the end stations).

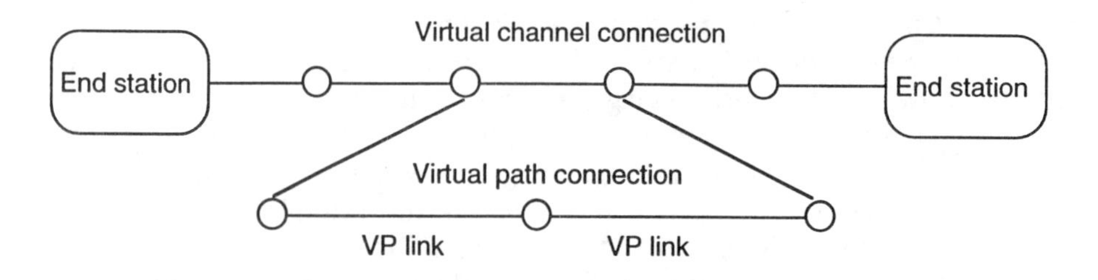

Figure 2.8 Hierarchy of logical connections in ATM networks.

VPIs are used to route packets between two nodes that originate, remove, or terminate the VPs, whereas VCIs (unique only within a VP) are used at the VP end points to distinguish between different connections. Cells can be relayed from one VP to another, or one VC to another, either in the same or a different VP. The VPI/VCI pair used at a switching node has a local meaning only. (Even the VCI does not change within a VPC; the VPI/VCI is translated as the VPI is translated at every switch.) Switching from incoming to outgoing links is done by reading the routing fields of incoming cells, performing a table lookup to determine the outgoing link, placing the new routing identifiers used on the outgoing link, and delivering the cell to the corresponding outgoing link port with the new header. The corresponding table entries must be set up at every switch before traffic can start flowing.

2.2.1.3 Payload Type Indicator

The payload type indicator (PTI) field is 3 bits long. Bit 3, which is the left-most (most significant) bit, is used to specify whether the cell carries user or operations, administration, and maintenance (OAM) data. When this bit is equal to 0 (i.e., user data), bit 2 is used to indicate if the cell has experienced (i.e., passed through) one or more congested switches. Bit 1 with user data is currently used only by Aal 5 to distinguish the last cell of a user frame from the others (see Section 2.3.4). In addition to the specification of two types of OAM flows, PTI coding includes the definition of a resource management (RM) cell. The seven of the eight PTI values defined so far are given in Table 2.2. The one remaining value of PTI coding is reserved for a future function.

The cell payload of a resource management cell is illustrated in Figure 2.9. RM cells may be used to manage link bandwidth at each switch dynamically

Table 2.2
Payload Type Indicators

PTI Coding	*Meaning*
000	User data cell, congestion not experienced, SDU type = 0
001	User data cell, congestion not experienced, SDU type = 1
010	User data cell, congestion experienced, SDU type = 0
011	User data cell, congestion experienced, SDU type = 1
100	Segment OAM flow-related cell
101	End-to-end OAM flow-related cell
110	RM cell
111	Reserved

SDU = Service data unit.

PTI= 110	RM protocol identifier (1 byte)	Function specific field (44 bytes)	Reserved (6 bits)	CRC-10

Figure 2.9 Resource management cell format.

by changing the amount of reserved bandwidth for ATM connection. The RM cell is also used for congestion control of connectionless traffic (see Chapter 6).

An ATM block [6] is a group of cells delineated by two RM cells, one before and another after the last cell of the block. Currently defined ATM block transfer (ABT) service in ITU-T 371 is used to deliver blocks of ATM cells (i.e., complete blocks only with no cell losses) by allocating network resources (i.e., link bandwidth) needed for the transfer of an ATM block dynamically on an ATM block basis. Two ABT capabilities are defined: ABT with delayed transmission (ABT/DT) and ABT with immediate transmission (ABT/IT). In ABT/DT, the user wishing to transmit an ATM block reserves resources by sending the leading RM cell into the network and waits for a response RM cell from the network before transmitting the remainder of the ATM block. In ABT/IT, the ATM block is transmitted immediately after the leading RM cell without waiting for the response from the network.

At connection setup, a point-to-point connection is established between two network users. As part of the connection setup, the network knows the maximum of bandwidth required by the connection. However, no network resources (other than connection labels in the routing tables along the end-to-end path) is allocated to an ABT service connection. During the duration of the call, the maximum bandwidth needed for the transmission of ATM blocks is dynamically negotiated between the network users and the network.

In ABT/DT, a user wishing to transmit (or receive) an ATM block may modify the maximum cell rate by sending an RM cell and by waiting for the response RM cell from the network. The network may send a positive or a negative acknowledgment, depending on the traffic level within the network at the time the request is made. Upon receiving a positive acknowledgment, the user sends its ATM block followed by an RM cell. Intermediate switches along the path of the connection accept the block, switch cells from the incoming link to the outgoing link, and release the reserved resources (i.e., link bandwidth) after receiving the RM cell following the ATM block. If the network user receives a negative acknowledgment, it reschedules its transmission and retries at a later time.

One disadvantage of this service is that the network user is required to wait for at least a round-trip propagation delay before start sending its traffic, restricting the effective throughput that can be achieved by this service. Its main advantage is that ABT/DT guarantees the delivery of the ATM block (subject to bit errors and potential link failures) to its destination, when the block of cells is allowed to enter the network.

In ABT/IT, the network user wishing to transmit an ATM block sends an RM cell immediately followed by the remaining part of the block. The leading RM cell is processed immediately by each network element to check whether the requested resources are available. If resources are available the block passes through the network element, otherwise the whole block is discarded. The last cell of the ATM block is also an RM cell that indicates a new reservation for the next ATM block or the release of resources. A network element that rejects the block due to lack of resource availability may optionally send a negative acknowledgment back to the network user who transmitted the ATM block. However, it is in general the responsibility of the higher-layer protocols used at the end stations to recover from block losses within the network, if the application requires data integrity.

ABT/IT solves the throughput restriction problem since it does not require waiting for any response from the network to start transmitting its ATM blocks. However, the network no longer provides any guarantees to deliver the ATM blocks.

ATM employs an *explicit forward congestion notification* scheme. In this scheme, the queue occupancies for each trunk is monitored. When the queue size reaches a predefined threshold, all user cells passing through that trunk are marked "*congestion experienced*" until the congestion period ends; that is, the queue occupancy falls below another predefined threshold value. Once a cell is marked "congestion experienced" at a node, it cannot be modified back to "congestion-not-experienced" by any downstream node along the route to the destination end station. A cell received at a destination node with this bit set indicates that there is a congested node along the path of this connection. Note that destination nodes have no information as to which nodes in the network are congested.

This scheme is designed to provide a framework for end stations to control the rate at which they submit traffic to the network. The destination station can use this congestion indication to send an end-to-end message to the originating end station and request it to slow down. The semantics and the syntax of this function is not a part of the standards activity. Due to large propagation delays of transmission links compared with the cell transmission times, by the time a marked cell arrives at a receiver, it is possible that the problematic (congested) nodes are no longer congested. In this case, the destination station has no means of knowing that the congestion is cleared and may inform the originating station to slow down. Hence, it is necessary to develop procedures to accurately determine whether there is a momentary or sustained congestion along the path. The accuracy of such procedures may use statistical estimations, which require a large number of cells marked "congestion experienced" before the other end is requested to slow down. The drawback of statistical techniques is that the end stations may react too late to minimize the effect of congestion.

2.2.1.4 Cell Loss Priority

The CLP field of the ATM cell header is a 1-bit field used for cell loss priority—high or low. Due to the statistical multiplexing of connections, it is unavoidable that some cell losses will occur in ATM networks. A cell with CLP bit set (low-priority cell) may be discarded by the network during congestion, whereas cells with CLP bit not set (high-priority cell) have higher priority and will not be discarded to the extent it is possible.

The CLP bit can be used by the network and by applications. When it is used by the network, the process is referred to as *tagging*. During connection establishment time, a traffic contract between the end station that generated the connection request and the network is agreed upon. Based on this contract, the network guarantees the requested service quality to the connection as long as the user traffic stays within the values specified in the contract. To protect itself, the network monitors the cell stream at the edge of the network to detect whether or not the contract has been violated. When a cell is detected to be in violation of the contract, the network can either drop the cell at the interface or decide to accept the cell on a best-effort basis. The network assumes with a calculated risk that this extra cell will not cause service degradation to the other user cells within their contract input parameters. In order to guarantee that contract-violating cells do not cause service degradation to conforming traffic, it is necessary to identify nonconforming cells to the network, so that they can be the first dropped cells. To achieve this, the network sets the CLP bit of nonconforming cells to 1 (turns it on).

For this CLP bit mechanism to work, it is necessary to have in the network a process that will drop tagged cells before they have a chance to cause problems. How the cells are dropped is not defined in any standard or ATM specification; it is left to vendors to implement.

The user can also take advantage of the CLP bit. For example, let us consider a video stream in which it is possible to classify some frames as more important than others. In this case, the application may submit parts of its traffic that are not so important, marked "low priority," whereas important traffic may be sent marked "high priority." This does not mean that the source arbitrarily allows its cells to be discarded by the network. It merely indicates to the network that if a condition arises in which some of the cells of a traffic stream have to be dropped, then the application prefers that cells marked "low priority" are discarded before those marked "high priority."

2.2.1.5 Header Error Check

The HEC field is used mainly for two purposes: (1) the detection of bit errors at the cell header to discard cells with corrupted headers; (2) cell delineation. This 8-bit field provides single-bit error correction and a low probability of a corrupted cell delivery. The HEC field is also used to identify the cell boundaries from the received bit stream. This function is a part of the physical layer and is discussed in Section 2.4.

2.2.2 Services Expected From the Physical Layer

The ATM layer expects the physical layer to transport ATM cells between two communicating ATM entities. Two primitives are defined at the service access point (PHY-SAP) between the ATM layer and the physical layer:

- PHY-UNIT-DATA.request;
- PHY-UNIT-DATA.indication.

The ATM entity passes one cell to the physical layer per *request* primitive and accepts one cell per *indicate* primitive.

2.2.3 Services Provided to the ATM Layer Users

The two primitives used for the exchange of ATM cell payloads and the parameters used in each primitive are defined as follows:

- ATM-DATA.request (ATM-SDU, SDU type, submitted CLP);
- ATM-DATA.indication (ATM-SDU, SDU type, received CLP, congestion experienced).

ATM.DATA.request initiates the transfer of an ATM cell payload (ATM-SDU) and its associated SDU type (i.e., 0 or 1) to its peer entity over an existing connection. ATM-SDU is 48 bytes of data, to be transferred by the ATM layer between peer communicating upper layer entities. The loss priority and the SDU type parameters are used to assign the proper CLP and PTI fields to the corresponding ATM physical data unit (ATM-PDU) generated at this layer. The ATM-PDU is the 53-byte cell, including the cell header values (except the HEC field) and the cell payload. ATM-DATA.indication indicates the arrival of an ATM-SDU over an existing connection, along with a congestion indication (which is set in the network if the received ATM-SDU had passed through one or more network nodes experiencing congestion) and the received SDU type. Table 2.3 summarizes the ATM-SAP parameters.

Table 2.3
ATM-SAP Parameters

Parameter	*Meaning*	*Value*
ATM-SDU	48-byte pattern	Any 48-byte pattern
SDU type	End-to-end cell type indicator	0 or 1
Loss priority	CLP	High or low priority
Congestion experienced	FECN* indication	True or false

*Forward explicit congestion notification.

2.2.4 ATM Layer Management

In general, the OAM framework includes configuration, fault, and performance management and security and billing functions. The current ATM layer OAM specification includes only the performance and fault management functions. In particular, at the ATM layer, network resources are monitored for equipment faults and performance degradation.

If a resource in the network is determined to be operating at a nondesired level, maintenance actions are taken to diagnose the cause and repair it. Operations activities involve continuous coordination of administration and maintenance of network resources based on feedback collected from the network resources and user requirements.

The ATM layer management protocol data unit used to perform OAM functions is fully contained within a single cell payload, as illustrated in Figure 2.10.

The 10-bit cyclic redundancy check code (CRC-10) field is used to detect bit errors in the cell payload. The first two fields identify management and OAM function types. Management functions include fault management, performance management, and activation/deactivation.

Currently, five OAM functions are defined:

- Performance monitoring (PM): Normal functioning of the managed entity is monitored by continuous or periodic checking of functions, and maintenance event information is maintained.

OAM type 4 bits	Function type 4 bits	Function specific field (360 bits)	Reserved (6 bits)	CRC-10

Figure 2.10 ATM layer management protocol data unit.

- Defect and failure detection: Malfunctions are detected by continuous or periodic checking, and maintenance event information or various alarms are produced.
- System protection: The effect of a failure of a managed entity is minimized by blocking or changeover to other entities, and the failed entity is excluded from operation.
- Failure or performance information: Failure information, alarm indications, and response to request are given to corresponding management entities and planes.
- Fault localization: The determination by internal or external test systems of a failed entity of a failure of information is insufficient.

ATM layer OAM functions are performed at five OAM hierarchical levels associated with the ATM and physical layers. Not all of these need to be present. The OAM functions of a missing level are performed at the next higher layer.

A VP connection end point is defined as the switch where a VP originates or terminates. These end points generate and extract VPI values; enforce VP-related traffic parameters; multiplexes and demultiplexes cells preserving VP cell sequence integrity; and generates, inserts, and extracts VP-level OAM cells.

A VP connecting point, on the other hand, translates the VPI value; possibly provides VP-related usage/network parameter control (depending on the location of the VP connecting end point); multiplexes and demultiplexes cells (preserving VP cell sequence integrity); and generates, inserts, and monitors (but does not extract) OAM cells.

A VPC segment (VCC segment) is defined as one VPC (VCC) link or multiple interconnected VPC (VCC) links where all of the links are under the control of a single administration or organization.

The five levels of OAM flows are illustrated in Figure 2.11. The F1, F2, and F3 flows are developed for the physical layer. The mechanisms used to provide OAM functions and to generate physical layer OAM flows depend on the transmission system used (see Section 2.4).

The ATM layer OAM flows are defined at two levels: F4 is the VP level, and F5 is the VC level. General rules that apply to both types of flows include:

- The OAM flows are terminated only at the end points of the connection (i.e., VPC or VCC) or at the connecting points (i.e., end points of a VPC segment or a VCC segment). Each flow is bidirectional.
- The OAM cells for both directions must follow the same physical path for the connecting points in both directions.
- Connecting points may monitor incoming OAM cells or insert new OAM cells, but they cannot terminate the OAM flows.
- The administration that controls the insertion of OAM cells must ensure that OAM flows are properly terminated.

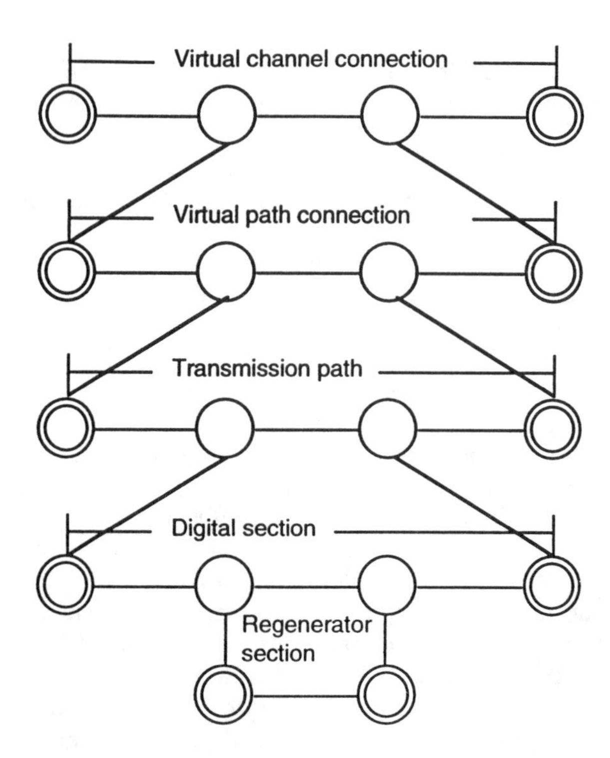

Figure 2.11 OAM hierarchical levels.

F4 flows (i.e., end-to-end and segment) are uniquely identified using VCI values of 3 and 4, respectively. F5 flows have the same VPI/VCI values as the user cells and are identified by the special PTI values (i.e., the end-to-end F5 flow is used for end-to-end VCC operations, and it is identified with PTI = 101, whereas segment F5 flows are identified with PTI = 100). The OAM cell contents used in F4 and F5 flows are given in Table 2.4.

2.2.4.1 OAM Functions for F4 Flow

Various fault and performance management functions described in I.610 for VPCs are:

- VP-level fault management, including VP alarm indication signal (VP-AIS), VP far-end receive failure (VP-FERF) alarms, and VPC continuity check;
- VP-level performance management, including mechanisms to detect errored blocks and loss/misinsertion of cells;
- VP loopback capability.

Table 2.4
OAM Cell Contents

OAM Cell Type	*Value*	*OAM Function Type*	*Value*
Fault management	0001	Alarm indication signal	0000
		Far end receive failure	0001
		OAM cell loopback	0010
		Continuity check	0100
Performance management	0010	Forward monitoring	0000
		Backward monitoring	0001
		Monitoring/reporting	0010
Activation/deactivation	1000	Performance monitoring	0000
		Continuity check	0001

VP-AIS cells are generated as soon as a failure condition is detected and transmitted to all affected active VPCs from the VPC connecting point. These cells are transmitted to downstream switches periodically, one cell per second, as long as the failure condition persists, in order to indicate VPC unavailability. A VPC end point enters the VP-AIS state as soon as a VP-AIS cell indicating a failure along the path is received. Such end points remain at this state until the VP-AIS state is removed, which occurs if either of the following conditions occur:

- Absence of VP-AIS cell for nominally 3 sec;
- Receipt of one valid cell (a user cell or a continuity check cell).

The VPC continuity check cell is sent by a VPC end point when no user cell has been sent for a period of time and there is no VPC failure to indicate that the connection is still active.

VP-FERF is sent to the far end from a VPC end point as soon as it has declared a VP-AIS state or detected a VPC failure. These cells are transmitted periodically, one cell per second, as long as the failure condition persists, in order to indicate VPC unavailability. A VPC end point enters the VP-FERF state as soon as a VP-FERF cell indicating a failure along the path is received. This state is removed when no VP-FERF cell is received during a 3-sec period.

PM of a VPC or VPC segment is performed by inserting monitoring cells at the ends of the VPC or VPC segment. A PM cell is inserted at the first free cell location after *N* user cells. The monitoring cell is used to detect errored blocks, loss/misinsertion of cells, and other related performance metrics of interest.

PM requires activation or deactivation to coordinate the beginning or end of the transmission and to establish agreement on the block size and the direction of transmission to start or stop monitoring. A handshaking procedure is defined for this purpose.

The ATM layer loopback capability allows for operations information to be inserted at one location along a VPC and returned (or looped back) at a different location, without having to take the connection out of service. Loopback cells can be inserted and looped back at connecting and connection end points. The looped-back location is identified by the loopback location identifier included in the cell payload. The loopback indication field is set at the looped-back end point to eliminate infinite looping of loopback cells. Various loopback applications defined in I.610 include the following:

- End-to-end loopback: A VP end-to-end loopback cell is inserted by a VP end point and looped back by the corresponding VP end point.
- Access line loopback: A VP segment loopback cell is inserted by the customer or the network and looped back by the first ATM node in the network or customer equipment, respectively. The segment for this application is defined by a mutual agreement between the two involved parties.
- Interdomain loopback: A VP segment loopback cell is inserted by one network operator and looped back by the first ATM node in an adjacent network operator domain. The segment for this application is defined by a mutual agreement between the two involved parties.
- Network-to-end point loopback: A VP end-to-end loopback cell is inserted by one network operator and looped back by the VP end point in another domain.
- Intradomain loopback: A VP segment loopback is inserted by a VP connection/segment end point or a VP connecting point and looped back by a VP segment or a VP connecting point. For this application, the use of the loopback location identifier is a network operator option.

2.2.4.2 OAM Functions for F5 Flow

The fault and performance management functions for VCCs are defined in the same way as those for F4 flows and are not repeated.

2.2.5 Reserved VPI/VCI Values

Table 2.5 lists the set of cell header values currently reserved for various purposes.

In addition, the VCI values in Table 2.6 are defined for various specific functions such as signaling and LAN emulation. The range of VCI values in a VP service may be determined by the network provider.

Table 2.5
Reserved Cell Header Values at UNI

VPI	*VCI*	*PTI*	*CLP*	*Use*
000000000000	0000000000000000	000	1	Idle cell
pppp00000000	0000000000000000	ppp	p	Physical layer-reserved
000000000000	0000000000000000	001	1	F1-flow
000000000000	0000000000000000	100	1	F3-flow
0000xxxxxxxx	0000000000000001	0a0	c	Metasignaling
0000xxxxxxxx	0000000000000010	0aa	c	Broadcast signaling
0000xxxxxxxx	0000000000000101	0aa	c	pt-pt signaling
0000yyyyyyyy	0000000000000011	0a0	a	Segment OAM F4 flow
0000yyyyyyyy	0000000000000100	0a0	a	End-to-end OAM F4 flow
0000yyyyyyyy	zzzzzzzzzzzzzzzz	100	a	Segment OAM F5 flow
0000yyyyyyyy	zzzzzzzzzzzzzzzz	101	a	End-to-end OAM F5 flow
0000yyyyyyyy	zzzzzzzzzzzzzzzz	110	a	RM
000000000000	0000000000000000	bbb	0	Unassigned cell

a = Bit is available for use by the appropriate ATM layer function.
b = Don't care bit.
c = Set to zero at the originating entity, which can be changed by the network.
xxxxxxxx = Any VPI value. If equal to zero, then used for signaling.
yyyyyyyy = Any VPI value.
zzzzzzzzzzzzzzzz = Any VCI value other than zero.

Table 2.6
The Range of VCI Values

VCI Value	*Reserved For*
0-15	ITU-T-defined functions
16-31	ATM forum-defined functions
32-65535	User traffic

2.3 ATM ADAPTATION LAYER

An ATM network provides an end-to-end ATM layer connectivity between end stations. The ATM layer deals only with the functions of the cell header, regardless of the type of information carried in the payload. This simplicity is necessary to keep up with high-speed transmission links, and it is achieved by leaving out various services required by applications. At the ATM layer, in particular, there is no:

- Information on the frequency of the service clock;
- Detection for misinserted cells;
- Detection for lost cells;
- Means to determine and handle cell delay variation.

The main reason for not providing these functions inside the network (i.e., at the ATM layer) is that not all of these services are required by every application. For example, data traffic does not require any information on the frequency of the service clock, whereas voice may not require any awareness on possible bit errors.

Architecturally, the AAL is used between the ATM layer and the next higher layer in both the user plane, and the control plane is used to enhance the services provided by the ATM layer supporting the functions required by the next higher layer. Accordingly, AAL is service-dependent. It isolates the higher layers from the specific characteristics of the ATM layer by mapping the higher layer protocol data units (PDU) into the ATM cell payload and vice versa.

It is not feasible to address the requirements of all of the applications served by ATM networks either individually or in a single AAL framework. Instead, the functionality required by various applications are grouped into a small number of classes based on the commonality of their service requirements and traffic characteristics. A different AAL is then defined for each class of service.

ITU-T classifies B-ISDN services based on three parameters:

- Timing relationship between source and destination (required or not required);
- Bit rate (constant or variable);
- Connection mode (connection-oriented or connectionless services).

Real-time services such as voice and video require an end-to-end timing relationship among communicating applications. This mainly arises out of the need to receive voice and video frames at the same rate as they are transmitted by the originating end station. Non-real-time services (i.e., data applications), on the other hand, do not have any requirement for an end-to-end timing relationship.

Traffic generated from a source can be submitted to the network either at a constant rate, referred to *as constant bit rate* (CBR) service, or at a varying rate, referred to as *variable bit rate* (VBR) service. The former is the case in current (circuit-switching) telephony networks (i.e., 64-kbps voice), where a channel with a fixed and constant transmission rate is allocated to connections during their duration. The latter is the case in packet networks, where either there is no connection established in the network or, if connection-oriented, connections are not allocated a fixed, dedicated bandwidth. CBR and VBR services are described in Chapter 3.

CBR services in the ATM framework are assumed to require a timing relationship between the communicating entities. VBR services, on the other hand, are further classified into two subcategories depending on whether an end-to-end timing relationship is required or not. The former case is intended for emerging applications such as VBR video and VBR audio. The latter classification is mainly for data services.

Real-time applications and some data services require resources to be reserved in the network to guarantee the services provided to them do not degrade due to traffic generated by other sources. Reserving resources in the network implies connection establishment prior to traffic flow. Most current data applications on LANs are connectionless services that do not require the connection establishment overhead. Accordingly, VBR services that do not require an end-to-end timing relationship are further classified into two categories: connection-oriented and connectionless.

Since B-ISDN services are classified based on three service parameters and each parameter can take two values, this framework categorizes all B-ISDN services in eight different service classes, of which four are defined as follows:

- Class A: This class corresponds to CBR, connection-oriented services with a timing relationship between source and destination. The two typical services of this class are 64-kbps voice and CBR video.
- Class B: This class corresponds to VBR connection-oriented services with a timing relationship between source and destination. VBR-encoded video is a typical example of this service class.
- Class C: This class corresponds to VBR connection-oriented services with no timing relationship between source and destination. A typical service of this class is connection-oriented data transfer.
- Class D: This class corresponds to VBR connectionless services with no timing relationship between source and destination. Connectionless data transfer between two LANs over a WAN is a typical example of this type of service.

In addition to these four classes, two more classes are defined:

- Class X: In this service class, the AAL, traffic type (CBR or VBR), and timing requirements are user defined.
- Class Y: This class is defined to allow the ATM layer transfer characteristics provided by the network to possibly change after the connection establishment. The class Y user specifies the maximum required bandwidth to the network at the connection establishment time. The bandwidth available from the network may become arbitrarily small, with the minimum usable bandwidth being specified on a per-connection basis. This minimum bandwidth may as well be zero.

Class X service is defined as a raw cell service to allow the use of a proprietary AAL with terminal equipment that supports the particular AAL defined by a networking vendor. It can also be used for applications that use the ATM layer directly without using any AAL service. Class Y service is envisaged for use by data applications with no stringent delay constraints. Data communication protocols are designed to adapt to a time-varying available bandwidth in the network. In particular, various transport, network, and link layer protocols vary the rate traffic submitted to the network based on different congestion indicators generated by the network or end user. For example, frame relay uses backward explicit congestion notification to request end stations to regulate their traffic submission rate. Similarly, TCP/IP uses a slow start to regulate its traffic. Class Y service provides the capability to support time-varying traffic behavior in ATM networks.

Corresponding to the four service classes A, B, C, and D, four AAL protocols are currently defined, with others possibly to be defined in the future. These protocols are referred to as AAL 1, AAL 2, AAL 3/4, and AAL 5. The relationship between each AAL type and its service class is summarized in Figure 2.12.

Based on Figure 2.12, it appears that there is no difference between AAL 3/4 and AAL 5. The distinction between the two is clarified later in this section.

The AAL structure, independent of the AAL type, is illustrated in Figure 2.13.

AAL provides the transparent and sequential transfer of data units (AAL service data units (AAL-SDU)) between corresponding upper layer entities with an agreed-on quality of service. Each AAL protocol provides the capabilities to transfer AAL-SDUs from one AAL service access point (SAP) to one or more AAL-SAPs through an ATM network. An AAL user accesses the AAL through a SAP. Each SAP is associated with a set of quality of service parameters such as delay and cell loss. AAL users choose the AAL-SAP that best fits their requirements.

The functions of the AAL are grouped into two sublayers: the segmentation and reassembly (SAR) sublayer and convergence sublayer (CS). The SAR sublayer deals with the segmentation and reassembly of data units and maps them into fixed-length cell payloads. The CS performs a set of AAL service-specific functions. The CS is further divided into a service-specific convergence sublayer (SSCS) and a common part convergence sublayer (CPCS). SSCS may be null for applications that do not require any service-specific function.

Parameter	AAL 1	AAL 2	AAL 3/4	AAL 5
Timing relationship	Required	Required	Not required	Not required
Bit rate	Constant	Variable	Variable	Variable
Connection mode	Connection oriented		Connection oriented or connectionless	

Figure 2.12 Currently defined AALs.

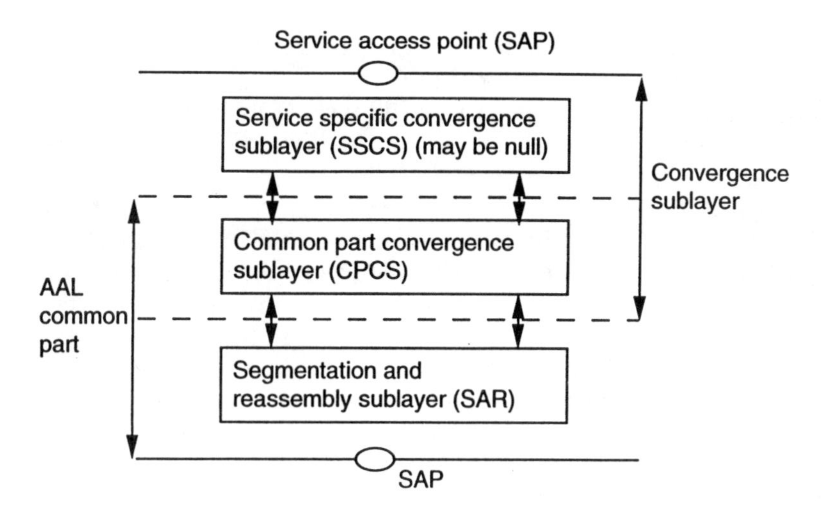

Figure 2.13 AAL structure.

2.3.1 AAL 1

AAL 1 is used for CBR services that require a timing relationship between the end points of connections. Examples of such services include CBR audio, CBR video, and CBR voice.

The services provided to AAL 1 users include the following:

- Transfer of SDUs with a CBR to the destination AAL 1 users with the same bit rate;
- Transfer of timing information between the source and destination;
- Transfer of structure information between the source and destination;
- Indication of lost or errored information.

Data between the AAL user and the AAL 1 are exchanged across a SAP using the following primitives:

- AAL-UNITDATA-REQUEST(data, structure);
- AAL-UNITDATA-INDICATION(data, structure, status);

The REQUEST primitive is used to request the transfer of an AAL-SDU (i.e., contents of the data parameter) from the local AAL entity to its peer AAL entity. Both the length of the AAL-SDU and the time interval between two consecutive primitives are constants (i.e., CBR service). The structure parameter in this primitive is optional. It is used when the user data stream is organized into groups of bits. The length of the structured block is an integer multiple of

8 bits, and it is fixed for each instance of AAL service. The two values of the structure parameter are START and CONTINUATION. START is used when the data are the first part of the structured block. CONTINUATION is used for all the other data of the block. The use of this parameter is discussed next as a part of AAL 1 CS services.

The INDICATION primitive is used to notify the AAL user that an AAL-SDU from its peer AAL entity has arrived. The STATUS parameter identifies whether the AAL-SDU has detected errors (INVALID) or no errors (VALID).

AAL 1 CS may include the following functions:

- Handling of cell delay variation;
- Processing of the sequence count;
- Providing a mechanism to transfer a timing information;
- Providing the transfer of structure information between source and destination;
- Forward error correction (FEC).

In the CBR service, AAL-SDUs are delivered to the AAL user at a constant rate. However, ATM is a packet-switching network and there might be a variation in the interarrival times of ATM cells at the destination end stations. Unless preventive measures are taken, this could potentially mean that there would be a variation in the rate AAL-SDUs are delivered to the AAL user. This clearly violates the fundamental CBR service characteristic. A buffer is used to support the handling of cell delay variation at the AAL. Cell delay variation in ATM networks is discussed in more detail in Chapter 3.

The AAL 1 segmentation and reassembly PDU (SAR-PDU) header is illustrated in Figure 2.14. It is a 1-byte field and is carried in the payload of each ATM cell that carries AAL 1 data. It is composed of a convergence sublayer

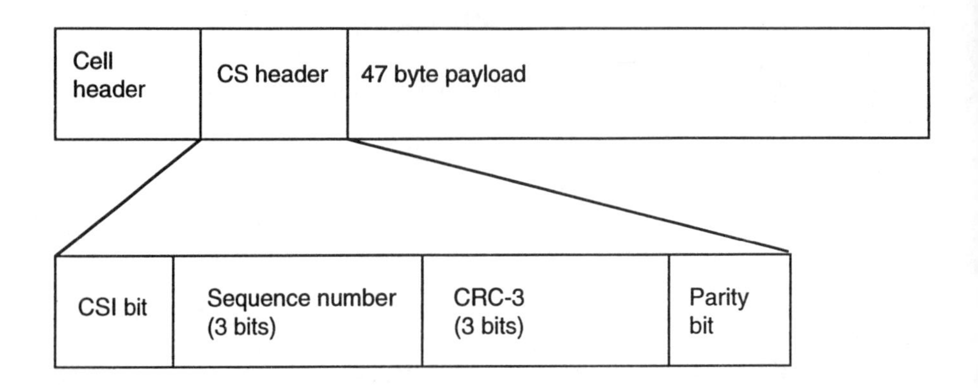

Figure 2.14 AAL type 1 SAR-PDU header.

indicator (CSI) bit, a 3-bit sequence number field, CRC-3, and a parity bit. The CSI bit carries the CS indication. The sequence number field carries the sequence number. The CSI bit and the sequence number fields are protected by a 3-bit CRC code (CRC-3). The resulting 7-bit field is protected by an even parity check bit. The CSI and the sequence number field values are provided by the CS layer.

The SAR sublayer at the transmitting end accepts a 47-byte block of data from the CS and appends a 1-byte SAR-PDU header to each block to form the SAR-PDU. In addition to the CSI and sequence number values it receives from the CS layer, the SAR sublayer calculates the CRC-3 and the parity bit for the 7-bit field. The SAR sublayer at the destination end station receives the 48-byte SAR-PDU. Before passing the 47-byte payload, it performs detection of lost and misinserted cells.

The AAL 1 SAR uses a sequence count to detect lost or misinserted cells. The 3-bit sequence number field provides sequence numbers from 0 to 7 (i.e., modulo 8). For presentation purposes, consider three consecutive CS-PDU arrivals with the sequence number of the first one being equal to 3. If the next two CS-PDUs have the sequence numbers 4 and 5 in that order, then AAL assumes all three are received in the correct order. If these sequence numbers are, say, 5 and 6, then AAL detects that CS-PDU with the sequence number 4 is lost. If these sequence numbers are 2 and 4, then AAL detects that CS-PDU with the sequence number 2 is misinserted. All misinserted cells are discarded. In order to maintain the bit count integrity of the AAL user information, it may be necessary to compensate for the lost cells by inserting dummy CS-PDUs with contents depending on the type of service provided. Note that AAL CS is not able to detect lost cells only when eight consecutive cells are lost in the network, which is rather unlikely to occur.

In order to provide its functions (i.e., delivering AAL-SDUs to an AAL user at a CBR), CS needs a clock. Three methods are defined for the handling of a timing relationship: adaptive clock method, synchronous clock derived directly from the network clock, and synchronous residual time stamp (SRTS) method.

In the adaptive clock method, the receiver writes the received information into a buffer and reads it with a local clock. The buffer fill level is used to control the frequency of the local clock. The control is performed by continuously measuring the buffer fill level around its median position. If the fill level is greater than the median, the local service clock is assumed to be slow and the clock speed is increased. If the fill level is lower than the median, the clock is assumed to be fast and the clock speed is decreased. In order to minimize the effects of oscillations around the median, the fill level of the buffer may be maintained between two levels around the median. If the two levels are defined too close, then the frequency oscillation will still remain as a problem. If they are too far apart, then the reaction time may be too long to be effective. The oscillation problem gets even worse when the communication takes place among a group of users. If separate buffers are used for each connection, the problem is to monitor the filling levels at each buffer and take the appropriate action. If a single buffer is used, the filling level might oscillate rapidly. One solution is to assign one of the group members as the master and use its traffic to adjust the local clocks. This, however, requires a protocol to assign one of the group members as the master and procedures to transfer the master function to another group member when a failure occurs. Such problems that may arise in group communications with AAL 1 are not currently addressed in any standards organizations.

Another option is for the local clocks to be phase-locked to the network clock. In this case, there is no need to recover the source clock frequency at the destination end station, since the service clocks at these stations are all synchronized to the same network clock. Therefore, no AAL function is needed to transfer the local clock between the communicating entities. This approach would solve the problems associated with the adaptive clock method. However, it is not always practical and/or preferred to synchronize local clocks to the network clock. An example is the circuit emulation service, which, say, provides an ATM network connectivity between two T1 services in which the clock used by the T1 service may not be phase-locked to the network clock used by the ATM network.

The SRTS method provides a measure of information about the frequency difference between a locally available reference clock derived from the network and the source clock. To synchronize the sender and receiver to the same frequency, the method assumes that the same network reference clock is available to both entities. In a B-ISDN network based on synchronous optical network (SONET), the reference clock hierarchy is derived from

$$155.52 \text{ MHz} \times 2^{-k},\ k = 0, 1, \ldots 11$$

which can accommodate all service rates from 64 kHz to 155.52 MHz. In general, let:

f_n = network clock frequency;
f_{nx} = network-derived reference clock frequency, $f_{nx} = f_{n/x}$, where x is an integer;
f_s = source clock frequency;
N = period of residual time stamps (RTS) in seconds;
T_n = the nth period of the RTS in seconds;
$\pm e$ = tolerance of the source clock frequency in parts per million, $e = 200 \times 10^{-6}$ in SRTS;
M_n (M_{nom}, M_{max}, M_{min}) = number of f_{nx} cycles within the nth (nominal, maximum, minimum) RTS period.

During the *n*th period, measured by *N* source clocks, there are M_n network-derived clock cycles, as illustrated in Figure 2.15.

Let M_q denote the largest integer smaller than or equal to M_n. M_q is made up of a nominal part and a residual part. M_{nom} corresponds to the nominal part of the f_{nx} cycles in *T* sec and it is fixed for the service. The residual part conveys the frequency difference information (and the effect of the quantization) and thus can vary. Since the nominal part is constant, it is assumed that M_{nom} is available at the receiver. In this case, only the residual part of M_q needs to be transmitted from origin to destination.

A simple way to represent the residual part of M_q is the RTS method, whose generation is illustrated in Figure 2.16. The network clock frequency

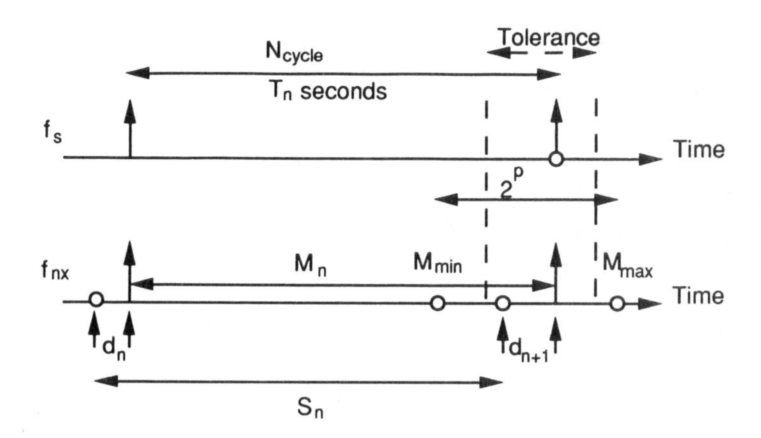

Figure 2.15 Residual time stamp concept.

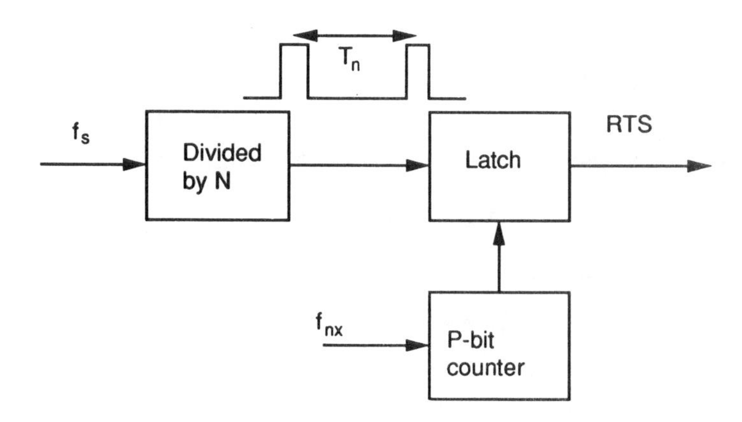

Figure 2.16 Generation of RTS.

(f_n) is divided by a constant (x) giving a network-derived reference clock frequency (f_{nx}). The derived network clock is used to clock a p-bit counter. The output of the counter is sampled at every N service clock cycles. This p-bit sample is the RTS. The RTS period is defined over a fixed number of SAR-PDU payloads, using the total number of bits in eight cells (with a total payload of 3,008 bits), i.e., $N = 3{,}008$.

For presentation purposes, let $f_s = 7.72$ MHz, $x = 16$, $f_n = 155.52$ MHz, $f_{nx} = 155.52/16 = 9.72$ MHz, $N = 3{,}008$, and $p = 4$ bits. Then,

$$T = 3{,}008/7.72 = 389.637 \times 10^{-6} \text{ seconds}$$

The expected number of derived network clock cycles (assumed to be known by both the transmitter and the receiver) is

$$M_{\text{nom}} = (389.637 \times 10^{-6})(9.72 \times 10^{6}) = 3{,}787.275 \text{ network cycles}$$

The counter is clocked at the derived network frequency over the interval needed to complete 3,008 cycles of the service clock. This would mean that the counter would increment 3,777 times; that is,

$$3{,}008\ (9.72 \times 10^{6})/(7.72 \times 10^{6}) = 3{,}777.488$$

Since the counter is 4 bits, it provides a modulo 16 operation. Hence, the RTS value at the end of 3,008 cycles of the service clock is 3,777 mod 16 = 1. Similarly, operations of the RTS process over 12 RTS periods are shown in Table 2.7.

Table 2.7
RTS Periods

RTS period	*Service Clock Cycles*	*Network Clock Cycles*	*RTS*
1	3,008	3,777	1
2	6,016	7,554	2
3	9,024	11,332	4
4	12,032	15,109	5
5	15,040	18,887	7
6	18,048	22,664	8
7	21,056	26,442	10
8	24,064	30,219	11
9	27,072	33,997	13
10	30,080	37,774	14
11	33,088	41,552	0
12	36,096	45,329	1

The RTS size is fixed at 4 bits. There is no field in the CS header to accommodate the 4-bit RTS. It is transmitted using the CSI bit of the SAR header in odd-sequence-number cells (i.e., 1, 3, 5, 7). Hence, eight ATM cells are required to pass one RTS value. The structure parameter in the AAL-UNIT-DATA-REQUEST and AAL-UNITDATA-INDICATION primitives is used to convey structure information between the AAL 1 and its user. The 47-byte payload used by the CS has two formats: P and non-P. In the non-P format, the entire payload is used for user information, with the first byte of the payload being the pointer field (this leaves 46 bytes in the cell payload for user data). The P format is used only when the sequence number is 0, 2, 4, or 6.

The 7 bits of the pointer field in the P format are used to indicate the offset, measured in bytes, between the end of the pointer field and the first start of the structured block in the 93-byte payload consisting of the (current) 46-byte payload and the 47 bytes of the next CS-PDUs.

Another function currently defined for AAL 1 is the correction method for bit errors and cell losses for unidirectional video services. This method combines FEC and octet interleaving, from which a particular CS-PDU is defined. In the transmitting CS, a special 4-byte code (i.e., Reed-Solomon code) is appended to 124 bytes of incoming data from the AAL user. The resulting 128-byte-long blocks are then forwarded to the byte interleaver. The interleaver is organized as a matrix of 128 columns and 47 rows. At the sender, the incoming 128-byte blocks are stored row by row; at the output, bytes are read out column by column. Each column forms a one-cell payload. The matrix has 128 × 47 = 6,016 bytes, corresponding to 128 SAR-PDU payloads. These 128 SAR-PDU payloads constitute one CS-PDU. The overhead of this method is 3.1% (i.e., 47 × 4/6,016) with an associated delay of 128 cell times.

For the synchronization of the CS-PDU, the CSI bit of the SAR-PDU header is set to 1 in the first 47-byte payload. Note that structured data transfer cannot be used with this mechanism. Within any CS-PDU matrix, this method can perform one of the following corrections:

- Four cell losses;
- Two cell losses and one errored byte in each row;
- Two errored bytes in each row if there is no cell loss.

2.3.2 AALs for VBR Services

AAL 1 is the only adaptation layer defined for CBR services, while all others are defined for VBR services. Variable-length packets, hereafter referred to as SDUs, are first received by the CS, which adds the CS header and CS trailer to the SDUs and passes it to the SAR layer to form the cell payloads. The SAR layer may append its own header and trailer to the CS-PDUs to form the SAR-

PDUs. The ATM layer adds the cell header and passes the cells to the physical layer for transmission. On the receive side, the processing takes place in reverse order. The cell headers are stripped at the ATM layer and the cell payload is passed to the SAR layer. SAR checks for transmission errors within each SAR-SDU, strips the SAR headers, reassembles the CS-PDU, and passes it to the CS layer. The CS layer constructs the CS-PDU, strips out the CS header and trailer, performs its functions, and delivers the PDU to the AAL user. The relationships between the ATM-AAL, ATM, and higher layers are illustrated in Figure 2.17.

Two modes of services are defined in the VBR SSCS sublayer: *message mode* and *streaming mode.* The message mode provides for the transport of either fixed-size or variable-length data units. A data unit across the AAL interface is passed as a single unit. In the case of small, fixed-size data units, an internal blocking function may be used at the transmitting AAL. This function collects a number of fixed-length data units into a single CS-PDU, thereby reducing the per-unit overhead associated with CS headers and trailers. At the receiver, the corresponding function is to deblock received CS-PDUs into fixed-

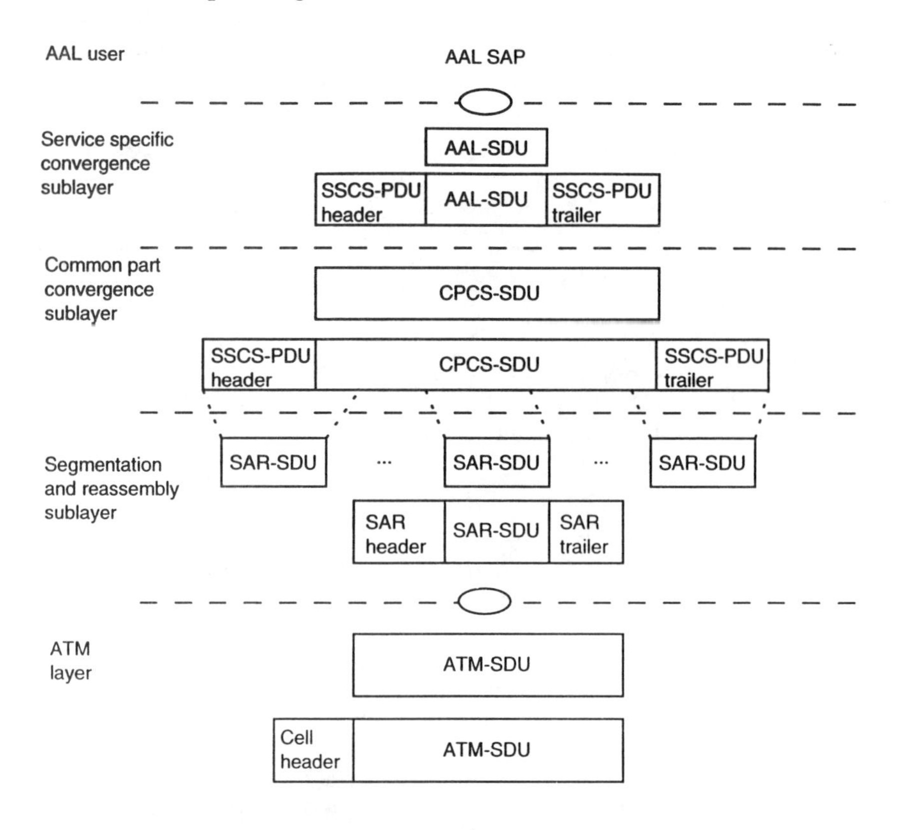

Figure 2.17 AAL functions in VBR services.

size data units. In the case of variable-length data units, the CS may include the segmentation function at the transmitting AAL to segment data units into smaller PDUs before processing at the CPCS. At the receiving end, the data units are reassembled to form the original data units. The blocking/deblocking and segmentation/reassembly functions are implemented in the service-specific coordination function (SSCF). If the SSCF is null, then data units passed across the AAL-SAP are mapped into one CPCS-PDU.

The streaming mode is used for variable-length data units to interleave their reception and transmission. Without this function, the complete data unit has to be received from the AAL user to be able to process it at the CS. When this function is enabled, the AAL user passes the data unit in one or more units and the transfer of these data units may occur separated in time. The streaming mode service includes an abort service by which the discarding of a partially transferred data unit may be requested. Similar to the message mode, SAR functions are included in the streaming mode. In addition, an internal pipelining function may be applied, providing a means by which the sending AAL entity initiates the transfer to the receiving entity before the complete data unit is received.

Both modes of services can offer two peer-to-peer operational procedures: assured operations and nonassured operations. Every data unit in the assured mode is delivered correctly to the receiver. This may require retransmission of missing or corrupted data units and the protocol for the two end entities to exchange messages. Flow control is mandatory in assured operations. In the case of nonassured operation, lost or corrupted SDUs are not corrected by retransmission. Both flow control and delivery of corrupted SDUs to the receiver are optional functions.

In VBR services, the CS-PDU size is not necessarily a multiple of the cell payload. Accordingly, the last segment of a CS-PDU or a single-segment CS-PDU may not fill the entire cell payload. This may necessitate the use of a *length indicator* (LI), which would indicate the number of useful bytes in the cell payload.

The SAR header for VBR services may also include an indication of the beginning, middle, or last segment of the CS-PDU or a single-segment PDU. Furthermore, the cell payload may be protected against bit errors by including a CRC field to the SAR trailer. Given this general framework, the details of three AAL protocols currently defined for VBR services are presented next.

2.3.2.1 AAL 2

AAL 2 is being developed for connection-oriented VBR services that require an end-to-end timing relationship. Typical examples of services that would use AAL 2 include VBR video, VBR audio, and VBR voice. The standardization of

AAL 2 has not yet been completed by the ITU-T. AAL 2 CPCS services are expected to include:

- Transfer of data units with a variable source rate;
- Transfer of timing information between source and destination;
- Indication of lost or errored information not recovered by AAL 2, if needed.

These services would require different AAL functions, such as:

- SAR of user information;
- Handling of cell delay variation;
- Handling of lost and misinserted cells;
- Source clock frequency recovery at the receiver;
- Recovery of source data structure at the receiver;
- Monitoring and handling of AAL protocol control information for bit errors;
- Monitoring of user information field for bit errors and possible corrective action.

2.3.2.2 AAL 3/4

AAL 3/4 is defined for connection-oriented and connectionless VBR services that do not require a timing relationship between the source and the destination. It has the basic functionality to support a connectionless network access (class D), as well as connection-oriented frame relay service (class C).

Combinations of service modes and operational procedures that may be used in AAL 3/4 SSCS are summarized in Table 2.8. Both modes of service may offer either assured or nonassured peer-to-peer operational procedures.

Based on the AAL framework illustrated in Figure 2.17, various primitives exchanged between the AAL user and the AAL, between the SSCS and the CPCS, and the CPCS and the ATM layer are summarized as follows.

Table 2.8
Combination of Service Mode and Internal Function

	Segmentation/Reassembly	*Blocking/Deblocking*	*Pipelining*
Message mode: long variable-size data units	Optional	Not applicable	Not applicable
Message mode: short fixed-size data units	Not applicable	Optional	Not applicable
Streaming	Optional	Not applicable	Optional

The primitives between the AAL user and the SSCS are service-specific, and they are not yet specified. If the SSCS is null, this sublayer would only provide a mapping between the AAL user and the CPCS identified as AAL-UNITDATA.request, AAL-UNITDATA. indication, AAL-U-Abort.request, AAL-U-Abort.indication, and AAL-P-Abort.indication, consistent with the naming convention used at AAL 3/4 SAP. The corresponding primitives between the SSCS (or AAL user) and the CPCS are referred to as *invoke* and *signal* (as opposed to *request* and indication, respectively), since there is no SAP between the two layers.

Table 2.9 summarizes the parameters of CPCS-UNITDATA.invoke and CPCS-UNITDATA.signal primitives.

Similarly, the primitives used for the abort service, CPCS-U-Abort.invoke, AAL-U-Abort.signal, and AAL-P-Abort.signal, are defined as follows. The invoke and signal primitives are used by the CPCS or the CPCS user to indicate that a partially delivered CPCS-SDU is to be discarded by instruction from its peer entity. These primitives do not have any parameters. CPCS-P-Abort.signal is used by the CPCS entity to signal to its user that a partially delivered CPCS-SDU is to be discarded due to an error in the CPCS. No parameters are defined. All three primitives are used only with the streaming mode.

The primitives used between the SAR and the CPCS layers are defined similarly to these primitives as follows: SAR-UNITDATA.invoke, SAR-UNIT-DATA.signal, SAR-U-Abort.invoke, SAR-U-Abort.signal, and SAR-P-Abort.signal.

Table 2.9
CPCS-UNITDATA.invoke and CPCS-UNITDATA.signal Primitive Parameters

Parameter	*Definition*	*MM*	*SM*
Interface data	specifies the data unit exchanged between the CPCS and SSCS	Y	Y
More	Specifies whether the communicated data unit is the beginning, continuation, or the end of the CPCS-SDU	N	Y
Maximum length	Indicates the maximum length of the CPCS-SDU (used only with the first invoke or signal primitive related to a CPCS-SDU)	N	Y
Reception status	Indicates that the data unit delivered (i.e., with the signal primitive) may be corrupted	Y	Y

Y = Used (either optional or mandatory).
N = Not used.
MM = Message mode.
SM = Streaming mode.

The transmitting AAL 3/4 SAR sublayer receives variable-length packets from the CPCS and generates SAR-PDUs containing up to 44 bytes of SAR-SDU data and 4 bytes of SAR header and trailer. The SAR-PDU structure is illustrated in Figure 2.18.

The segment type (ST) is a 2-bit field that indicates a SAR-PDU as containing the beginning of message (BOM), continuation of a message (COM), end of a message (EOM), or a single-segment message (SSM). The 4-bit sequence number (SN) field is used to number the SAR-PDUs based on modulo 16 (i.e., 0 to 15). The multiplexing identification (MID) field is 10 bits long and is used to allow multiplexing AAL connections into a single ATM connection. All SAR-PDUs of a particular AAL connection are assigned the same MID value. Using this field, it is possible to interleave and reassemble SAR-PDUs of different SAR-SDUs. A typical example of the use of the multiplexing feature of AAL 3/4 and the MID field is the connectionless service in ATM networks, where several CS-PDUs originally belonging to different sources may be transported using the same VPI/VCI. For example, consider the case where traffic generated from a LAN arrives at a border node (i.e., interworking unit) that receives the LAN packets and transfers them over the B-ISDN backbone to a connectionless server from where it is routed to another border node to its destination LAN. In this scenario, all packets originating at one LAN use the same connection from the origin border node to the connectionless server. Accordingly, the cells of different packets have the same cell header (i.e., VPI/VCI). These cells are demultiplexed at the server based on their MID values. Figure 2.19 illustrates multiplexing of AAL connections into the same ATM connection.

The 6-bit LI field is the number of bytes that belongs to the SAR-SDU. This value is equal to 44 when ST = BOM or COM. It is restricted to be between 4 and 44 for ST = EOM and 8 and 44 for ST = SSM. The value 63 is used with ST = EOM to indicate that the SAR-PDU is required to be aborted.

A CRC-10 that is carried in the 10-bit CRC field is used over the entire payload (excluding the CRC field) to detect bit errors.

The SAR sublayer functions are defined as follows.

- Preservation of SAR-SDU, which is accomplished by using the ST and the LI;

Cell header	ST	SN	MID	SAR-PDU payload (44bytes)	LI	CRC-10

Figure 2.18 AAL 3/4 SAR-PDU structure.

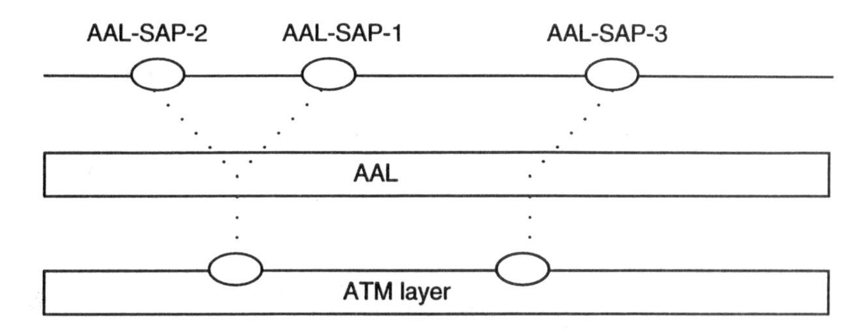

Figure 2.19 Relationship between AAL-SAP and ATM-SAP.

- Error detection and handling, which uses CRC-10/SN to provide the means to detect and handle bit errors as well as lost or gained (i.e., misinserted) SAR-PDUs (gained SAR-PDUs and the ones with bit errors are discarded);
- Multiplexing/demultiplexing, which allows multiple CPCS connections to be multiplexed on a single ATM layer connection using the MID field;
- Abort, which provides for the means to abort a partially transmitted SAR-SDU.

The AAL 3/4 CPCS receives fixed- or variable-length AAL user packets (assuming SSCS is null), forms CPCS-PDUs, and passes them to the SAR sublayer for further processing. The CPCS-PDU format for AAL 3/4 is illustrated in Figure 2.20.

The common part indicator (CPI) is a 1-byte field used to interpret subsequent functions in the CPCS-PDU header and trailer. This may include the counting units for the values specified in the buffer allocation size (BASize) and length fields. The use of this field for other values may be defined in the future. The beginning (Btag) and ending (Etag) tag fields allow the association of the CPCS header and trailer. The transmitting end inserts the same value in these two fields and changes the value for each successive CPCS-PDU. The receiving end matches the two values. The sequence of the values used in successive CPCS-PDUs has no significance. BASize is a 2-byte field that indicates the maximum buffer requirement to receive the CPCS-PDU. The CPCS-PDU is required to be a multiple of 4 bytes. The padding (PAD) field is 0 to 3

CPI	Btag	BA size	CPCS-PDU payload	PAD	AL	Etag	Length

Figure 2.20 CPCS-PDU format for AAL 3/4.

bytes long to achieve this 4-byte alignment. Similarly, the alignment (AL) field is a 1-byte field used to achieve a 4-byte alignment. Finally, the length field is used to encode the length of the CPCS-PDU payload field.

Various functions performed at the AAL 3/4 CPCS include the following:

- Message or streaming service mode;
- Preservation of CPCS-SDUs, which provides the delineation and transparency of CPCS-SDUs;
- Error detection and handling, which provides the detection and handling of CPCS-PDU corruption. This function uses the Btag and Etag as well as the length field to detect length mismatches and Btag/Etag mismatch.

2.3.2.3 AAL Type 5

Similar to AAL 3/4, AAL 5 is proposed for VBR services that do not require a timing relationship between the source and destination. As in other AALs, AAL 5 consists of SSCS, CPCS, and SAR sublayers. Different SSCS protocols for different types of services may be defined. SSCS may also be null, in which case it will only provide a primitive mapping of the equivalent primitives of the AAL user to CPCS, and vice versa. The primitives between the AAL 5 user and the SSCS are service-specific and are not yet specified. If SSCS is null, then it would only provide a mapping between the AAL user and the CPCS, and they are identified as AAL-UNITDATA.request, AAL-UNITDATA.indication, AAL-U-Abort.request, AAL-U-Abort.indication, and AAL-P-Abort.indication, consistent with the naming convention used at a SAP. The corresponding primitives between the SSCS (or AAL user) and the CPCS are invoke and signal (as opposed to request and indication, respectively) since there is no SAP between the two layers.

Table 2.10 summarizes the parameters of CPCS-UNITDATA.invoke and CPCS-UNITDATA.signal primitives.

Similarly, the primitives used for the abort service, CPCS-U-Abort.invoke, AAL-U-Abort.signal, and AAL-P-Abort.signal, are defined as follows. The invoke and signal primitives are used by the CPCS or the CPCS user to indicate that a partially delivered CPCS-SDU is to be discarded by instruction from its peer entity. These primitives do not have any parameters. CPCS-P-Abort.signal is used by the CPCS entity to signal to its user that a partially delivered CPCS-SDU is to be discarded due to an error in the CPCS. No parameters are defined. All three primitives are used only with the streaming mode.

The primitives used between the SAR and the CPCS layers are defined similarly to these primitives as follows: SAR-UNITDATA.invoke, SAR-UNITDATA.signal, SAR-U-Abort.invoke, SAR-U-Abort.signal, and SAR-P-Abort.signal.

Table 2.10
CPCS-UNITDATA.invoke and CPCS-UNITDATA.signal Primitive Parameters

Parameter	*Definition*	*MM*	*SM*
Interface data	Specifies the data unit exchanged between the CPCS and SSCS	Y	Y
More	Specifies whether the communicated data unit is the beginning, continuation, or the end of the CPCS-SDU	N	Y
CPCS loss priority (CPCS-LP)	Indicates the loss priority for the associated CPCS-SDU; the priority can take only one of the two values: high or low	Y	ffs
CPCS congestion indication (CPCS-CI)	Indicates whether the associated CPCS-SDU has experienced congestion in the network	Y	ffs
CPCS user-to-user indication (CPCS-UU)	Transferred transparently by the CPCS between peer CPCS users	Y	ffs
Reception status	Indicates that the data unit delivered may be corrupted	Y	ffs

Y = Used (either optional or mandatory).
N = Not used.
MM = Message mode.
SM = Streaming mode.
ffs = For further study.

The SAR sublayer functions are performed on an SAR-PDU basis. The SAR sublayer accepts variable-length SAR-SDUs that are integral multiples of 48 bytes from the CPCS and delivers SAR-PDUs containing 48 bytes of SAR-SDU data. Accordingly, there is no SAR header/trailer added to the SAR-PDU.

AAL 5 uses an SDU type indicator in the cell header PTI field, as shown in Table 2.11.

SDU type = 0 indicates that this is the continuation cell of a user frame, whereas SDU type = 1 indicates the end of a SAR-SDU (i.e., the last cell).

Table 2.11
AAL-5 Use of PTI Coding (AAL 5–related subset of Table 2.2)

PTI Coding	*Meaning*
000	User data cell, congestion not experienced, SDU type = 0
001	User data cell, congestion not experienced, SDU type = 1
010	User data cell, congestion experienced, SDU type = 0
011	User data cell, congestion experienced, SDU type = 1

Congestion indication is also defined at the PTI field (i.e., the second PTI bit). The loss priority of a SAR-SDU is transported using the CLP bit at the cell header.

Various functions performed at the AAL 5 SAR sublayer are defined as follows:

- Preservation of SAR-SDU, which provides an "end of SAR-SDU" indication;
- Handling of congestion information, which provides for the passing of congestion information between the layers above the SAR sublayer and the ATM layer in both directions;
- Handling of loss priority information, which provides for the passing of CLP information between the layers above the SAR sublayer and the ATM layer in both directions.

The CPCS-PDU format is shown in Figure 2.21.

The padding (Pad) field of 0 to 47 bytes is used so that CPCS-PDU multiple of 48 bytes. The 1- CPCS-UU field is used to transparently transfer CPCS user-to-user information. The CPI is a 1-byte field used to align the CPCS-PDU trailer to a multiple of 4 bytes. Other functions may be added in the future. The length field is a 2-byte field used to indicate the length of the CPCS-PDU payload field. Finally, CRC-32 is a 4-byte field used to protect the entire payload including the first 4 bytes of the CPCS-PDU trailer.

Various functions defined at the AAL 5 CPCS are summarized as follows:

- Preservation of CPCS-SDU, which provides for the delineation and transparency of CPCS-SDUs, using the SDU type indication;
- Preservation of CPCS user-to-user information, which provides for the transparent transfer of CPCS user-to-user information using the CPCS-UU field of the CPCS-PDU;
- Error detection and handling, which uses the length indicator and the CRC fields to determine whether a CPCS-SDU is corrupted or not;
- Abort, which provides the means to abort a partially transmitted CPCS-SDU;
- Padding, which provides for the 48-byte alignment of the CPCS-PDU;

User data	Pad (0-47 bytes)	CPCS-UU (1 byte)	Length field (2 bytes)	CRC-32

Figure 2.21 AAL type 5 CPCS-PDU format.

- Handling of congestion information, which provides for the passing of congestion information between the layers above the CPCS and the SAR layer in both directions;
- Handling of loss priority information, which provides for the passing of CLP information between the layers above the CPCS and the SAR layer in both directions.

2.3.2.4 AAL 3/4 versus AAL 5

Both AAL 3/4 and AAL 5 protocols are defined for VBR connection-oriented as well as connectionless applications that do not require an end-to-end timing relationship between the communicating entities. The natural question is, why are the two AALs defined?

AAL 3/4 was defined first. It supports multiplexing AAL connections into a single ATM connection using the MID field. Though this is a nice feature to have, it constitutes a large part of the SAR-PDU overhead (i.e., 10 bits). Another SAR layer overhead is the 10-bit CRC field. Arguments against the use of the CRC field are mostly based on the fact that errors in the cell payload would be detected by end station protocols (i.e., applications), and there is no need to have this in the cell payload.

Essentially, the effective payload of an ATM cell with AAL 3/4 is at best equal to 83% (44/53). Seeing that the last cell of a PDU may be filled only partially, the real utilization that can be achieved with AAL 3/4 is even lower than this value. The total overhead of an AAL 5 CPCS-PDU, on the other hand, is 8 bytes, and there is no SAR sublayer overhead. AAL 5 requires the additional overhead bits at the last cell of the PDU if the PDU is not a multiple of 48 bytes. Compared with the AAL 3/4, type 5 has the same effective payload usage for CS-PDU sizes of 88 bytes or less and smaller overhead for 88-byte or larger CS-PDUs. As the CPCS-PDU size increases, the effective utilization increases and it is less than or equal to at 90.5%, that is, the maximum effective utilization that can be achieved in an ATM network (48/53).

The main disadvantage of AAL 5 is that the multiplexing of AAL connections into a single ATM layer connection requires a special solution that can be provided only at the SSCS. An SSCS with this feature is not yet defined and feasible solutions appear to be rather complex to implement and limited in capability.

2.4 PHYSICAL LAYER

The physical layer transports ATM cells between two adjacent ATM layers. The ATM layer is independent of the physical layer and it is able to operate over a wide variety of physical link types. Currently available physical layer interfaces specified for ATM are listed in Table 2.12. These interfaces are described throughout this section.

The physical layer in ATM has more functionality than is typically associated with this layer. In legacy networks, the physical layer merely propagates bits across a medium. In ATM, the physical layer delivers cells (not bits) to the ATM layer, which requires determining the cell boundaries.

2.4.1 Structure of the ATM Physical Layer

Figure 2.22 illustrates the structure of the ATM physical layer. The physical layer provides the ATM layer with access to the transmission media. In the transmit direction, the ATM layer passes ATM cells (the 48-byte cell payload and the 5-byte header excluding the HEC value) to the physical layer. In the receive direction, the ATM layer receives 53-byte cells from the physical layer. Various functions performed at the physical layer are classified into two sublayers: transmission convergence (TC) and physical media dependent (PMD).

Table 2.12
Physical Layer Interfaces for ATM

	Transmission Rate (Mbps)	*Throughput (Mbps)*	*System*	*Medium*	*Campus/ WAN*
DS-1 (T-1)	1.544	1.536	PDH	Coax	Both
E-1	2.048	1.92	PDH	Coax	Both
DS-3 (T-3)	44.736	40.704	PDH	Coax	WAN
E-3	34.368	33.984	PDH	Coax	WAN
E-4	139.264	138.24	PDH	Coax	WAN
SDH STM-1 SONET STS-3c	155.52	149.76	SDH	Single-mode fiber	WAN
SDH STM-4c SONET STS-12c	622.08	599.04	SDH	Single-mode fiber	WAN
FDDI-PMD	100	100	Block coded	Multimode fiber/STP	Campus
Fiber channel	155.52	149.76	Block coded	Multimode fiber	Campus
Raw cells	155.52	155.52	Clear channel	Single-mode fiber	WAN
Raw cells	622.08	622.08	Clear channel	Single-mode fiber	WAN
Raw cells	25.6	25.6	Clear channel	UTP-3	Campus
Raw cells	51.84	49.536	SONET	UTP-3	Campus
STS 3-C	155.52	149.76	SONET	UTP-5	Campus

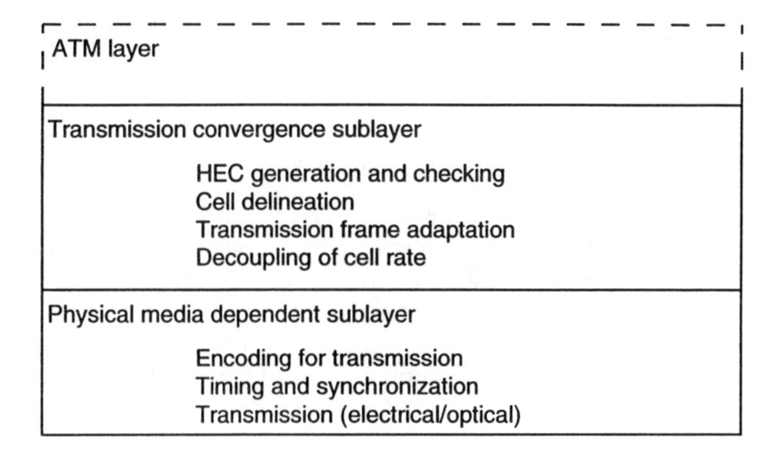

Figure 2.22 Structure of the ATM physical layer.

2.4.1.1 Transmission Convergence Sublayer

The TC sublayer constructs PMD payloads and generates the required protocol information for the physical layer. Various functions performed at the TC sublayer include HEC cell generation and verification, cell rate decoupling, and transmission frame generation and recovery. The PMD sublayer deals with the electrical or optical properties of the physical media.

Some physical layer interfaces use framed transmission structures. The principle of framed transmission was developed with time-division multiplexing systems in the context of transmitting PCM voice (discussed in detail in Chapter 3). In PCM voice, an 8-bit voice message is generated every 125 μs. Therefore, in framed transmission, blocks of bits are transmitted every 125 μs. The number of bits in a frame depends on the speed of the transmission link. For example, a T1 frame consists of 193 bits, whereas a SONET STS-3c frame consists of 19,440 bits.

A frame consists of a frame payload and overhead bits. The overhead bits are used to determine the start and end of a frame and to carry various signaling and maintenance information. The payload is used to carry user information.

Framed transmission is also referred to as *synchronous transmission,* since a frame is transmitted by the physical layer every 125 μs regardless of whether there is any user information to transmit. This may seem at first contrary to ATM, which is asynchronous, meaning that cells arrive at the physical layer at irregular times as determined by the nature of the applications. However, synchronous transmission at the physical layer does not impose any constraint on the asynchronous nature of ATM. Another transmission structure available for ATM is clear channel transmission. In this mechanism, there is no frame structure. Instead, cells in the physical media are transmitted asynchronously.

Each cell begins with a starting delimiter and ends 53 bytes later. There is no ending delimiter.

Transmission Frame Generation and Recovery

In frame-oriented transmission systems, the TC layer generates frames at the sender and recovers them from the bit stream at the receiver. The frame generation includes placing frame-related information and ATM cells into a well-defined frame structure. Similarly, the TC layer at the receiver recovers frames from which it recovers ATM cells and the frame-related information. The frame-related information is used for functions such as physical layer operations and maintenance and line testing.

HEC Cell Generation and Verification

The physical layer is passed a 53-byte cell from the ATM layer. At this time, the 53-byte cell is complete except for the HEC byte. This byte is computed by the TC sublayer and inserted into the HEC field of the cell header before it is passed to the physical media. At the receiver side, the physical layer performs error checking for the cell header integrity using the received HEC value. Only the cells with no detected errors in the cell header are passed to the ATM layer. Cells with detected errors at the cell header are discarded. The HEC used in ATM has the capability to detect and correct single-bit errors. It can also detect double-bit and some other combinations of bit errors. When an error is detected, however, it is not possible to know for sure whether there is a single-bit error or multiple-bit error. If the error correction capability is applied to an errored cell header, then it is possible that the cell header is still errored after the correction if there were originally multiple-bit errors.

Cell Rate Decoupling

Although there are exceptions to it (clear channel transmission), in general, a continuous bit stream is required at the physical medium (framed transmission). When a continuous stream of cells at the physical layer is required by the operational constraints of the transmitter and the receiver, then the question arises, what happens if there are not enough cells passed down from the ATM layer to the physical layer to fill up the pipe? To keep up the continuous bit stream flowing at the link, the TC sublayer inserts *idle* cells into the gaps of the user cell stream. These cells are discarded at the receiver TC sublayer; that is, they are not passed to the ATM layer. These idle cells are identified uniquely by four values in the cell header: VPI = 0, VCI = 0, PTI = 0, and CLP = 1.

Cell Delineation

There are no cells at the transmission link, just a sequence of bits (signals). The physical layer receives the bit stream and reconstructs the ATM cells before passing them up to the ATM layer. Cell delineation is the function of determining the cell boundaries (i.e., the beginning and the end of cells) from the received bit stream. I.432 [4] presents how the HEC field of the cell header can be used for this purpose. The process is illustrated in Figure 2.23. The receiver, at any given time, is in one of the following three states: hunt, presync, and sync. In the hunt state, the incoming bit stream is monitored to detect a 5-byte word with a correct CRC. That is, when the remainder of the 5-byte word divided by the CRC-8 used to determine the value of the HEC field is equal to zero, then it is assumed that this is the header of an ATM cell, thereby determining the cell boundary. Note that if the alignment is incorrect, the probability of a match will be 1 in 256 (i.e., a match in a 1-byte field). When a match occurs, the receiver moves to the presync state. In this state, δ consecutive matches are searched for. If found, then the receiver moves to the sync state. If a mismatch is found during the presync state, the receiver moves back to the hunt state. While in the sync state, synchronization is assumed to be lost when α consecutive mismatches for the 5-byte headers occur (i.e., the remainder is not equal to zero), which forces the receiver to move to the hunt state. While in the sync state, the HEC field is used to detect and correct single-bit errors or discard cells when multiple-bit errors are discarded. ITU-T recommends $\delta = 6$ and $\alpha = 7$, counts that make the cell delineation process robust against false delineation and false misalignment.

2.4.1.2 Physical Medium–Dependent Sublayer

So far, we have reviewed various functions that might be provided in the TC sublayer. Next we review different functions performed at the PMD sublayer.

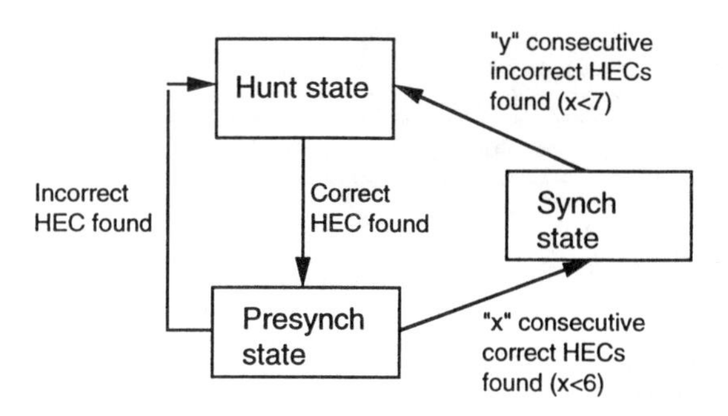

Figure 2.23 Cell delineation procedure.

This sublayer is much closer to the classical physical layer used in legacy networks.

The basic function of the PMD sublayer in the transmit direction is to take a stream of bits and transport it transparently across a link. In the receive direction, the PMD detects and recovers the arriving bit stream and passes it to the TC sublayer for recovering transmission frames (if framed transmission is used) and ATM cells. Figure 2.24 illustrates the relationship between the PMD and TC sublayers.

The two sublayers exchange data streams, requiring two sublayers to be synchronized to each other. Transmit and detect functions at the PMD are respectively the actual placement and recognition of the electrical or optical signals on a wire or a fiber. The timing function generates appropriate timing for transmitted signals and derives correct timing for received signals.

Another function performed at the PMD sublayer is line encoding/decoding. The PMD may operate on a bit-by-bit basis or with a group of bits at a time. The latter case is referred to as *block-coded transmission*. The basic idea behind the block-coded transmission is to treat a group of bits and translate each group into another bit pattern before transmitting it on the line. Two frequently used encoding techniques are 4B/5B and 8B/10B. In the former case, each group of 4 bits is coded into 5-bit groups for transmission, whereas 8-bit groups are coded into 10-bit groups in the latter.

Using either encoding technique requires the actual transmission rate to be increased by 25%. For example, the actual transmission rate of the FDDI

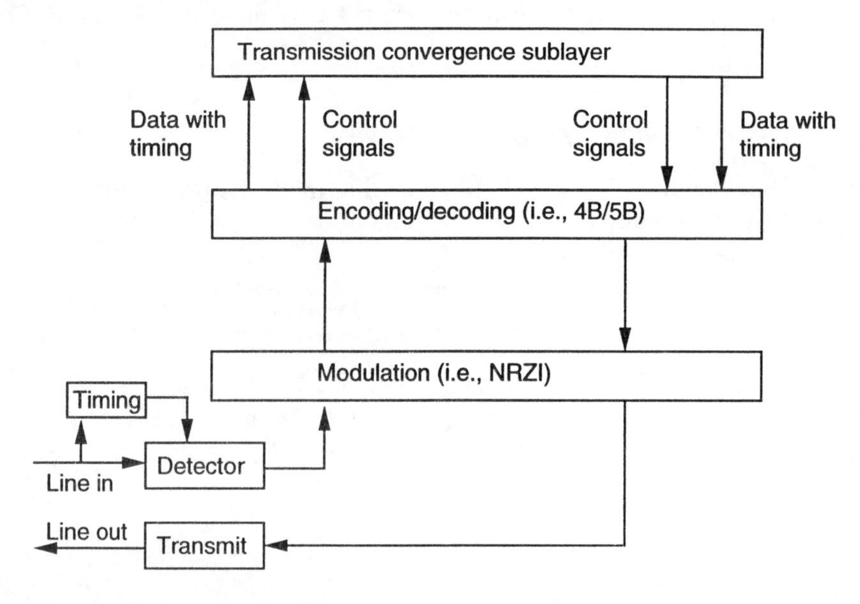

Figure 2.24 The relationship between the PMD and TC sublayers.

PMD is 125 Mbps, of which only 100 Mbps is used for data traffic. There are various reasons for this extra complexity:

- It is necessary to detect the bit boundaries from the received signal. If there are not enough transitions between bits (i.e., 1 to 0 and 0 to 1), then it becomes difficult to detect the bit boundaries. Clearly, the transitions between 0s and 1s are not controllable in user data. For example, the binary representation of the decimal number 15 is binary 1111, and that is what is passed to the PMD sublayer. A 4-bit group represents 16 different values. If these are transmitted as 5-bit groups, then there are 32 possible values from which 16 can be selected so that there are enough transitions to facilitate synchronization of a receive clock circuit.
- Some of the extra values that become available with the extra bit can be chosen to transfer control information between the two ends of a transmission link.
- Transmitting bits in groups makes it easier (and less costly) to pass the received bit streams to other circuits in the transmission system (and the other parts of the adapter) at a slower rate. For example, if 4B/5B encoding is used, then the bit stream received at 100 Mbps may be passed to the other parts of the adapter at the rate of 25, because the data is passed around in groups of 4 bits. Note, however, that block-coded transmission is not necessary to achieve this flexibility.

Table 2.13 illustrates 4B/5B encoding used with 100-Mbps PMD.

Table 2.13
4B/5B Encoding Used with 100 Mbps PMD

Symbol Type	*Symbol*	*Code Group*	*Meaning*
Line state symbols	I	11111	Idle
	H	00100	Halt
	Q	00000	Quiet
Starting delimiter (SD)	J	11000	First byte of SD
	K	10001	Second byte of SD
Control indicators	R	00111	Logical 0 (reset)
	S	11001	Logical 1 (set)
Ending delimiter	T	01101	Terminates data stream
Data symbol	0	11110	Binary 0000
	1	01001	Binary 0001'
	—	—	—
	9	10011	Binary 1001
	A	10110	Binary 1010
	—	—	—
	F	11101	Binary 1111

2.4.2 ATM Physical Layer Interfaces

Although ATM technology was originally developed for low-error-rate, high-speed fiber links, it is important to realize that fiber is not the only physical medium that can be used to take advantage of what ATM has to offer; it can use the nonfiber cabling in current networks. From the actual cable plant standpoint, ATM can run on a campus over different cables such as unshielded twisted pair—category 3 (UTP-3) and category 5 (UTP-5), shielded twisted pair (STP), and fiber. In a wide area, ATM can be deployed on coax copper and both multimode and single-mode fiber.

2.4.2.1 SONET/SDH ATM Interfaces

SONET was not designed with ATM in mind, and ATM has not been designed to run over SONET. SONET was designed originally by Bellcore so that public networks could efficiently support their current switching and transmission systems in optical-fiber backbones. It is now an international standard recommended by the ITU-T (referred to as SDH) and other standards bodies such as the American National Standards Institute (ANSI) and the European Telecommunications Standards Institute (ETSI).

Various SONET interfaces for ATM have been specified, including 51.84 Mbps (for campus use) and 155 and 622 Mbps (for WAN use). The basic structure of SONET is an 810-byte frame sent every 125 μs. Each byte of this frame provides a 64-kbps channel. This basic frame is referred to as Synchronous Transport Signal level 1 (STS-1). Accordingly, STS-1 operates at 51.84 Mbps.

Each SONET frame includes payload and frame overhead bytes. Conceptually, the STS-1 frame may be viewed as a 9- by 90-byte container, as illustrated in Figure 2.25. The first three columns of every row are referred to as the *SONET frame overhead* and they are used for various framing and OAM functions. The synchronous payload envelope is a 9- by 87-byte frame that floats within the physical frame structure. The payload is allowed to start anywhere within the physical frame (and span two physical frames). The start of a payload in a physical frame is pointed by corresponding SONET overhead bytes. Multiple STS-1 frames can be multiplexed together to form higher speed signals. An STS-n signal operates at n times the speed of STS-1 (i.e., $51.84 \times n$). The three levels defined for use in ATM networks are STS-1, STS-3, and STS-12.

ITU-T has defined the worldwide standard SDH. The basic frame in SDH is referred to as synchronous transport module level 1 (STM-1). STM-1 has the exact same format as STS-3c. Faster speeds in SDH are obtained again by multiplexing STM-1 frames into a larger frame. STM-n carries n times the payload of an STM-1 frame, which has the same structure as STS-3n.

SONET signals may be transported by either electrical or optical means. For example, the ATM Forum UTP-3 interface at 51 Mbps uses SONET framing

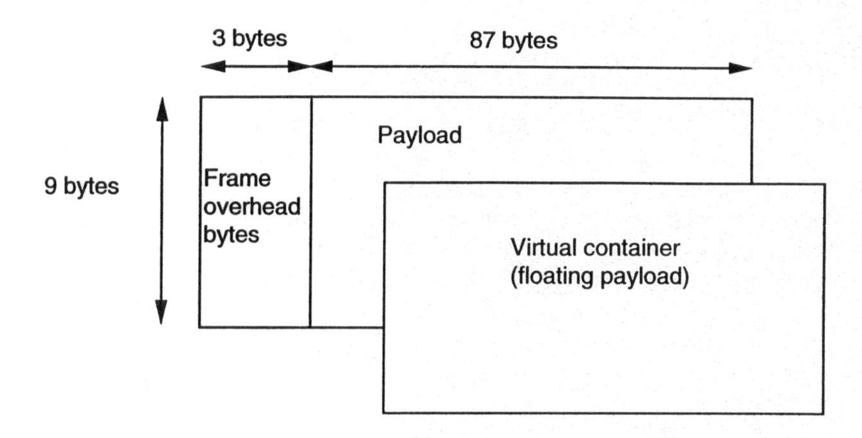

Figure 2.25 STS-1 frame structure and synchronous payload envelope.

and runs on copper, whereas the other STM-1 and STM-4c interfaces run on fiber. STS electrical signals, when transmitted over fiber, are converted to a corresponding optical signal called an *optical carrier* (OC). Parallel to the signaling hierarchy, the OC hierarchy starts at 51.84 Mbps, with OC-*n* denoting the OC rate of *n* times the basic transmission rate of 51.84 Mbps.

When used in ATM networks, cells in the synchronous payload are concatenated row by row without regard to row or cell boundaries. For example, an STM-1 frame is a 9- by 261-byte container. Excluding the path overhead column (9 bytes), cells are placed one after the other in a row. In the first row, then, there will be four ATM cells and the fifth cell will have 48 bytes. The second row will start with the remaining 5 bytes of the last cell of the first row and include four other ATM cells and 43 bytes of a cell, and so on. Since the payload size is not a multiple of ATM cell size, the last cell in the last row will have the rest of its payload in the next synchronous payload, which will be transmitted in the next physical frame.

Various functions defined in TC and PMD sublayers in ATM with SONET are listed in Table 2.14.

SONET overhead is divided into three layers: path, line, and section, as illustrated in Figure 2.26.

There is no overhead associated with the transportation of optical pulses through the fiber medium. The main function of the physical medium is to receive an STS-*n* signal stream, convert each bit into an optical pulse, and transmit the pulse across the fiber medium toward the far-end terminal.

The section layer transports the STS-*n* frame across the physical medium, using the physical layer for transport. Functions at this layer include framing, scrambling (if used), section overhead processing, and error monitoring. The overhead used at this layer is created and interpreted by the section terminating equipment.

Table 2.14
SONET ATM Physical Layer Processing

	HEC generate/verify Cell scramble Cell delineation	ATM
	Path signal identification	Path
TC	Frequency/pointer processing Multiplexing Scrambling	
	Framing	Line
PMD	Bit timing and coding	Section
	Physical medium	Photonic

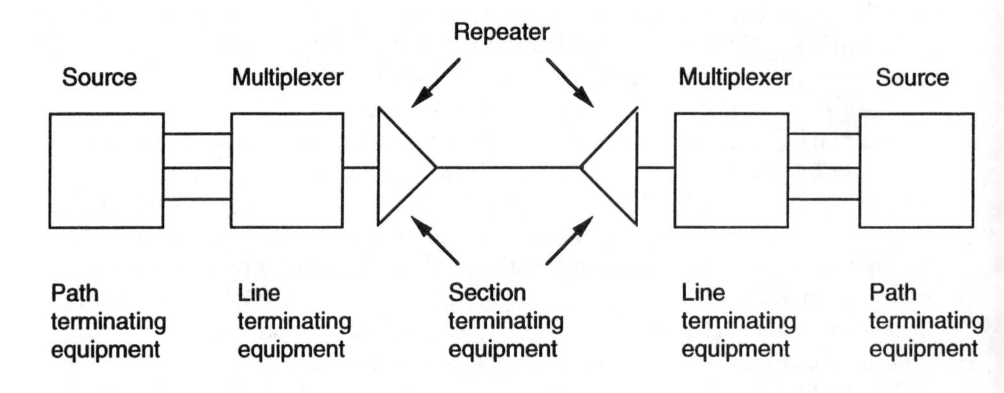

Figure 2.26 SONET layered architecture.

The line layer is responsible for transporting the payload and the line overhead to its peer at the far end. All lower layers (i.e., section and path layers) provide transport. This layer maps the payload and the line overhead into STS-*n* frames. The overhead used at this layer is created and interpreted by the line terminating equipment.

The path layer maps network services into the SONET payload format as supported by the line layer, and transports network services between two SONET multiplexing nodes. This communication between the path terminating equipment at each end is carried out via the path overhead. The path terminating equipment creates and interprets the overhead for this layer.

Various OAM flows defined at each layer is listed in Table 2.15.

Table 2.15
SONET OAM Error Detection

OAM Flow	*Error Detection*
F1: section/regenerator	Loss of signal or frame synchronization Degraded error performance
F2: line/digital section	Loss of signal or frame synchronization Degraded error performance
F3: path	Loss of cell delineation Degraded error performance Uncorrectable error in the cell header Degraded header error performance Payload pointer loss Failure of insertion and suppression of idle cells

2.4.2.2 PDH Interfaces

Plesiochronous (meaning nearly synchronous) digital hierarchy (PDH) was developed over the past 40 years for economic transmission of voice and data in public networks. All PDH interfaces use framed structures regardless of speed. Table 2.16 lists the currently defined PDH interfaces for ATM in three international systems.

In general, there are two ways to use the PDH interfaces for ATM cell transport: direct mapping and mapping base on the physical layer convergence protocol (PLCP). In direct mapping, the bytes within each frame are treated as a continuous stream of bytes. Frame boundaries are not heeded. Cell boundaries are found by cell delineation. PLCP is used to provide a form of rate synchronization. For example, in the DS-3 implementation, a floating payload frame is built within the DS-3 frame. The payload frame is synchronized to the 125-μs system

Table 2.16
PDH Interfaces Defined with ATM Mapping

North America		*Europe*		*Japan*	
Signal	*Transmission Rate (Mbps)*	*Signal*	*Transmission Rate (Mbps)*	*Signal*	*Transmission Rate (Mbps)*
DS-1	1.544	DS-1E	2.048	J-1	1.544
DS-3	44.736	DS-3E	34.368	J-2	6.312
		DS-4E	139.264		

clock, but the physical frame is not required to be aligned with the same clock. That is, there is no fixed relationship between the PLCP frame and the physical frame. The PLCP frame can start anywhere inside the DS-3 frame. However, the starting position, once established, will remain constant until a failure or discontinuation occurs. A DS-1 frame is 193 bits long. ATM cells are carried in the DS-1 payload using bits 2 to 193 by using direct mapping, as illustrated in Figure 2.27. Bit 1 (F) provides F3 OAM functions such as detection of loss of frame alignment and performance monitoring.

The DS-3 PLCP frame consists of 12 rows of ATM cells, each preceded by four octets of overhead as illustrated in Figure 2.28. Nibble (4 bits) stuffing is required after the 12th cell to fill the 125-μs PLCP frame. Although the PLCP frame is not aligned with the 44.736-Mbps physical frame, the bytes of the PLCP frame are nibble-aligned with the payload envelope.

Various physical layer functions performed in PDH interfaces for ATM are summarized in Table 2.17.

Similarly, the error detection (OAM) functions provided at PDH interfaces for ATM are summarized in Table 2.18.

2.4.2.3 155-Mbps Fiber Channel PMD Physical Interface

The 155-Mbps multimode fiber interface is a 27-cell frame. The first cell in each frame has a special format: a 5-byte delimiter and a 48-byte field reserved for OAM functions. Each frame has a 26-cell payload. Accordingly, the payload transmission rate is 149.76 Mbps, the same as the STS-1 payload. The cell boundaries are synchronous with respect to the frame structure. The frame format is illustrated in Figure 2.29.

The link encoding uses a 8B/10B code (i.e., 8 data bits are coded as 10 bits on the fiber). Each frame is not synchronized to the system 125-μs clock, but is generated every 73.61 μs (as opposed to 125 μs in every other framed

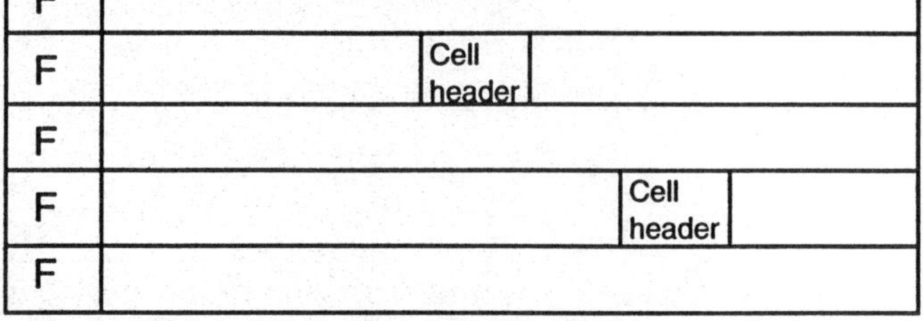

Figure 2.27 An example of direct mapping of cells onto a DS-1 frame.

PLCP framing (2 bytes)	POI (1 byte)	POH (1 byte)	PLCP payload (53 bytes)	
			First ATM cell	
			Eleventh ATM cell	13 or 14 nibbles

Figure 2.28 PLCP frame format to carry ATM cells.

Table 2.17
PDH Interfaces Physical Layer Processing

TC	Cell rate decoupling HEC generate/verify PLCP framing (if PLCP-based mapping is used) Cell delineation Path overhead utilization PLCP timing (if PLCP-based mapping is used)
PMD	Bit timing and line coding Physical medium

transmission). This particular choice of frame length allows the payload transmission rate of this interface to be aligned with the STM-1 payload transmission rate (i.e., 149.76 Mbps).

In order to propagate the system 125-μs clock, a special strobe character is inserted every 125-μs. This is possible due to the use of 8B/10B coding, which provides additional codes that are used for special functions (as in Table 2.13).

Table 2.18
OAM Flows in PDH Interfaces

OAM Flow	*Error Detection*
F1: section/regenerator	Loss of signal Loss of F1 physical layer OAM cell recognition Degraded error performance
F2: line/digital section	Null
F3: path	Loss of cell delineation Loss of F3 physical layer OAM cell recognition Uncorrectable error in the cell header Degraded header error performance

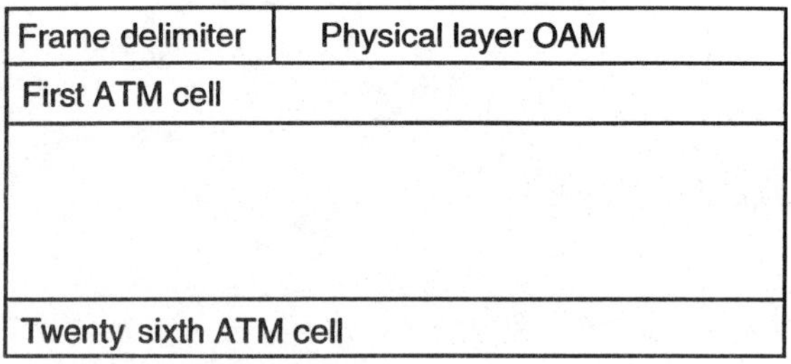

Frame delimiter	Physical layer OAM
First ATM cell	
Twenty sixth ATM cell	

Figure 2.29 155-Mbps fiber channel PMD physical interface.

The TC sublayer functions provided at this interface include cell delineation, HEC generation/verification, and 125-μs-clock recovery. The physical layer OAM information is used to provide transmission and reception of maintenance signals and low-level PM functions.

2.4.2.4 Cell Stream ATM Physical Layer Interfaces

In the cell stream physical layer interfaces, cells are sent asynchronously, and there is no frame structure as there is with PDH or SDH. If there is no cell to transmit, the line contains continuous idle codes. Cell boundaries are asynchronous; that is, a cell can begin anytime the line is idle. Each cell is preceded by a special bit sequence, referred to as the *start of a cell code.* Furthermore, a minimum of one idle code is transmitted on the link every half second.

The two interfaces of this type available for ATM are the 25.6-Mbps UTP-3 and the 100-Mbps multimode fiber interface.

The 25.6-Mbps physical layer interface uses the token ring PMD to support ATM on UTP-3. Token ring currently runs at 16 Mbps. Doubling the clock rate would provide a 32-Mbps interface. For ATM, a 4B/5B encoding is used, resulting in a 25.6-Mbps transmission rate. Various functions performed at the TC sublayer include HEC generation and verification, clock recovery, and cell delineation. The PMD sublayer provides bit timing and line coding as well as generating and receiving physical medium signals.

The 100-Mbps multimode fiber interface for ATM (referred to as TAXI) is based on the FDDI PMD standard and is intended to be used in private networks. Accordingly, there is no need to include physical layer OAM complexity in this interface, since these functions are mainly used in WANs.

2.5 HUMAN NOSE AND ATM

B-ISDN in general, and ATM in particular, is nothing like what we have experienced before and has the very ambitious goal of eventually becoming the standard means of communication all over the world. Since the initial concepts of ATM were introduced, a very large number of technical complaints and objections about the suitability of ATM as the transfer mode of communications networks have been raised.

One of the major complaints is that the cell size is too small to utilize network resources efficiently. For example, an AAL type 3/4 payload is 44 bytes, which results in at most 83% efficiency (i.e., 44/53). A link in a packet network can be provisioned to 80% to 85% utilization, after which the effective throughput of user applications decreases due to large delays and buffer overflow probabilities. Thus, the effective link utilization in an ATM network is about 66% to 70%. This is not necessarily the worst-case scenario, since it is possible for cells to be partially filled with VBR services. The ATM cell size decision was driven by the telephony industry's need to avoid the deployment of expensive echo canceler technology. Long delays and the need for echo cancelers would be inevitable with large cell sizes.

Another objection was the use of a fixed-size cell. It has been demonstrated in numerous forums that the use of variable packet lengths provides a performance superior to that of ATM. Arguments against variable-length packets focus on the cell size being appropriate for most current data applications (i.e. LAN traffic, terminal-to-host traffic). However, fixed-size packets simplify switch fabric design, buffer management, and efficient multiplexing. These concerns overwhelm the disadvantages, particularly for high-performance applications. Some other arguments include the complexity of effective bandwidth management, signaling, and the adaptation layers at high speeds. It is generally agreed that these issues need to be satisfactorily addressed before ATM networks can be widely deployed.

The following analogy from [5] between ATM and the human nose summarizes the complete story in a quite interesting way:

> The human nose, while humble in purpose, features a design that few engineers could take pride in. For instance, the large sinuses in the cheekbone drain from the top. Any self-respecting engineer would have placed the drain at the bottom. The nose is runny, unsanitary, and hangs upside down over the mouth. A better design would feature a nostril at each side of the head, improving sanitation and providing a more directional sense of smell. Despite these shortcomings, few of us spend any time in debate about how to advance the state of the art in nasal design. Odd as it may seem at first, important lessons about ATM can be drawn from the way we think about noses. Many objections can be raised about various design features of ATM. Indeed, many have been. But because the standards organizations have moved and are unlikely to undo what they have wrought, we are faced with the task of moving beyond complaint and technical argument to get on with the business of living with this odd thing called ATM.

As was the case with the human nose, we will invent tissues, handkerchiefs, and sprays necessary to make the most out of ATM networks. The challenge is learning how to take advantage of the capabilities the ATM technology offers.

References and Bibliography

[1] ITU-T Recommendation I.413, "B-ISDN User Network Interface," 1991.

[2] ITU-T Recommendation I.363, "B-ISDN ATM Adaptation Layer (AAL) Specification," 1991.

[3] ATM Forum ATM User Network Interface Specification, Version 3.1, 1994.

[4] ITU-T Recommendation I.432, "B-ISDN User Network Interface Specification," SG XVIII, Report R-34, 1990.

[5] Stevenson, D., "Electropolitical Correctness and High Speed Networking, or, Why ATM is Like a Nose," *ATM Networks,* I. Viniotis and R. O. Onvural, eds., New York: Plenum, 1993.

3 Source Characterization in ATM Networks

There are various issues that need to be resolved before ATM networks can become widely deployed. Designers are currently being faced with a wide range of problems such as implementing mechanisms for efficient congestion control, call admission, and routing to developing lightweight transport protocols. In order to design and develop network functions, it is necessary to comprehend the characteristics and the requirements of the traffic to be carried.

ATM networks are expected to support a diverse set of applications with a wide range of characteristics. Unfortunately, for the time being, there are no comprehensive measurements to permit designers to satisfactorily address the characteristics of various types of B-ISDN applications in a realistically accurate manner. Figure 3.1 illustrates the degree of understanding on the traffic characteristics of different types of applications.

The characteristics of voice sources have been studied for several decades in the context of telephone networks and are relatively well understood. A CBR video submits its bit stream to the network at a constant rate. As will be discussed later, the quality of CBR video services and the amount of bit rate required in the network increases as the rate data submitted to the network increases. The difficulty arises in choosing the (constant) bit rate to provide the desired service quality while minimizing the amount of bit rate used in the network. Although packet-switching networks have been around for about three decades, the source behavior of data sources is not well understood. Another view characterizing data sources is that it is well understood, but it is simply unpredictable. There is no doubt that several types of data sources are unpredictable. Given this fact, there is a great deal of interest in the development of models and techniques to support (unpredictable) data sources efficiently. In particular, there is no typical data application. Furthermore, for a given class of data services, there is no typical source behavior. Image and VBR video transmission over communications networks are relatively new research areas. The current knowledge of their source behavior is limited and based on different system implementations.

Source characterization at the macro level is defining the source traffic characteristics and its quality-of-service (QOS) requirements. The traffic charac-

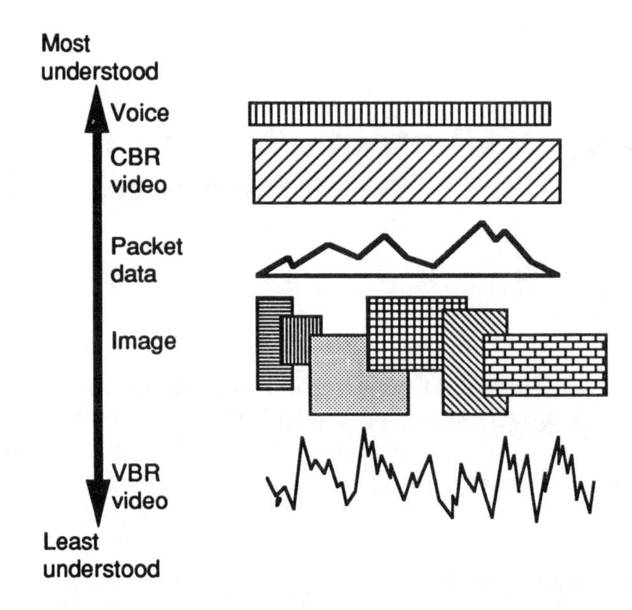

Figure 3.1 Degree of understanding on the traffic characteristics of various applications.

teristics of an application are the minimum set of parameters that a user can be expected to declare while providing the network management as much information as possible to effectively control network traffic and achieve high resource utilization. For example, a user can easily define its maximum cell generation rate as the speed of its interface. Based on this metric only, the network has no choice but to allocate resources to support this maximum rate. However, this may result in underutilization of network resources for sources with average cell generation rates that are significantly less than their maximum.

During the duration of a connection, the period at which a source generates traffic is referred to as an *active* period, whereas a *silent* period corresponds to the time between the active periods during which no traffic is generated. In CBR traffic, digital data generated by a source is presented to the network as a constant-rate bit stream, either by the use of smoothing buffers or by controlling the rate at which bits are generated. This type of traffic arises in services such as video source coding (producing CBR) or in phone conversations where silent periods are also transmitted. Knowing its maximum bit rate is sufficient to characterize a CBR source. On the other hand, traffic generation in a VBR source either alternates between the active and silent periods, or a continuous bit stream is generated at varying rates. Accordingly, characterization of VBR sources is a more complex task.

Peak rate is the maximum bit rate at which a source generates traffic during its active periods. ITU-T defines instantaneous peak cell rate and integrated peak cell rate to denote, respectively, the reciprocal of the minimum interarrival time of cells during the active period and the number of cells generated by a source measured over a predefined short interval T divided by T. There are two possible characterizations of average cell arrival rate defined by ITU-T: true average cell rate, which denotes the number of cells generated (measured) during the connection duration divided by the length of the duration, and average cell rate, which is the number of cells generated by a source measured over a long interval of time T divided by T.

VBR sources are referred to as *bursty* sources. However, the definition of this term is not unique in the literature. Some common uses include:

- The ratio of the peak bit rate to average bit rate;
- The average burst length, that is, the mean active period during which a source generates traffic at its peak rate;
- The squared coefficient of the variation of the interarrival times of cells C^2, where C^2 = Variance/(Expectation)^2.

Unless mentioned otherwise, following the ITU-T definition, the term *burstiness* is used throughout the book to correspond to the ratio of the peak-to-average traffic generation rate. Various parameters frequently used to characterize VBR sources include:

- R_p, peak cell generation rate;
- N, average number of cells in an active period;
- T_i, average time from the start of an active period to the start of the next active period;
- m, average cell rate;
- β, traffic burstiness.

Now let a^{-1} and s^{-1}, respectively, denote the average duration of active and silent periods. Let T be the minimum interarrival times of cells during the active period. Then we have

$$T_i = (a^{-1} + s^{-1}) \quad R_p = 1/T \quad m = \{a^{-1}/(a^{-1} + s^{-1})\} \quad \beta = p/m \quad N = a^{-1}/T \quad (3.1)$$

The burstiness of some B-ISDN applications and their peak rates are given in Figure 3.2.

QOS is the user's view of a service. Defining QOS in B-ISDNs is a difficult task due to:

- The many different types of users;
- The many different types of services;
- The subjective dependence on the user's view of the service.

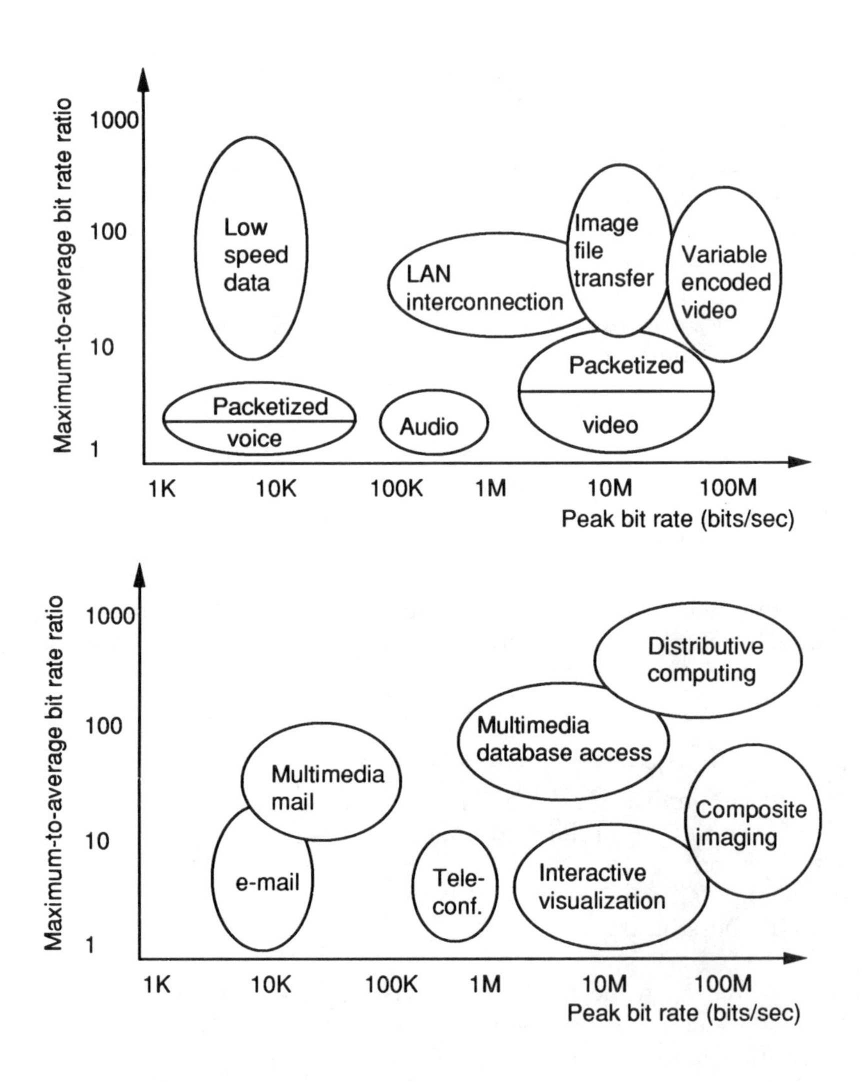

Figure 3.2 Burstiness of some B-ISDN services.

A definition of QOS is given in ITU-T Recommendation I.350 [101]: "Quality of Service is defined as the collective effect of service performances which determine the degree of satisfaction of a user of the specific service."

In connection-oriented networks, users first establish an end-to-end connection in the network prior to actual information transfer. Part of this process is the negotiation that takes place between the users and the network to set up a path connecting two or more nodes and connection parameters. At the end of the information transfer, the connection is released by one of the users. Some performance metrics of interest in connection-oriented networks are the time to establish and release connections and the probability that a connection request is rejected due to insufficient resource availability in the network. In packet-switching networks, packets generated by the sender are stored at intermediate nodes and switched from node to node until delivered to the end node. End-to-end delay characteristics, cell loss probability, and bit error rate are some of the parameters used to define service requirements of connections.

Viewed as connection-oriented packet-switching networks, the QOS metrics in ATM networks are classified into two categories: call control parameters and information transfer parameters. The current state of the art and the factors that determine the traffic generation rates of CBR and VBR services are discussed, respectively, in Sections 3.1 and 3.2, while call control and information transfer parameters of B-ISDN applications are reviewed in Section 3.3.

3.1 CBR SERVICES

CBR services generate traffic at a constant rate and can be simply described by their peak rates. The burstiness of a CBR source is equal to 1, and the source is active during the duration of the connection (or the silent periods are also transmitted at the peak rate). Typical examples of CBR services include voice, video, and audio with bit rate requirements given in Table 3.1.

CBR voice in ATM networks is transmitted with AAL 1 using the pulse code modulation (PCM) technique. PCM is a method for converting an analog signal to a digital format by representing the quantized amplitude samples of the original signal with binary code pulses, as illustrated in Figure 3.3.

Recommendation G.711 [104] specifies 64-kbps CBR voice. At this rate, the analog signal is sampled every 125 μs and the signal amplitude is quantized into 256 steps, giving PCM a primary frame length of 8 bits. Since 47 bytes of the 48-byte ATM cell payload is available with AAL 1, up to 5.875 ms of voice is transmitted with one cell. Current techniques for audio and video transmission rely on CBR coding mainly because existing data networks do not

Table 3.1
Bit Rate Requirements of Some CBR Services

Service	*Bit Rate (kbps)*
Telephony	64
Hi-fi stereo	1,400
Group III fax	14.4
Group IV fax	64
Proprietary fax	1,500

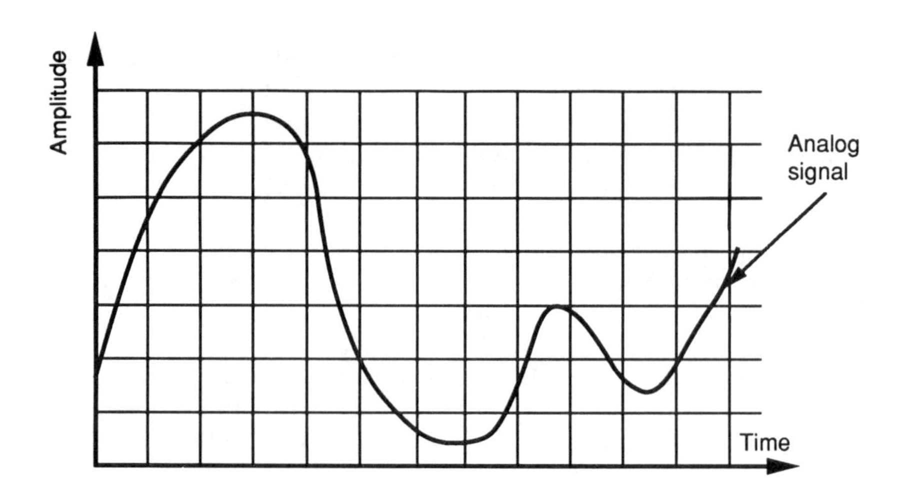

Figure 3.3 PCM of analog signal.

permit fine-grained bit rate allocation. The audio quality depends on the sample rate and sample resolution, and its bit rate requirement is

$$\text{Audio bit rate} = (\text{bits/sample})(\text{samples/sc}) \quad (3.2)$$

Conceptually, CBR audio is not different than voice. The main difference between the two applications is that audio requires more quantization levels sampled at higher rates. In particular, with a typical frequency range of 200 to 3,400 Hz, a sampling rate of 8 kHz provides a rather natural reproduction of telephone speech. If the audio signal is vocal and instrumental music instead of speech, then it has a frequency range of 10 to 20,000 Hz. In current compact disc (CD) technology, audio is stored using 16-bit PCM coding of 44.1-kHz sampled signals giving a bit rate of 700 kbps. In the case of stereo music, this bit rate is doubled to 1.4 Mbps. Once the emerging standards for audio coding are finalized, single-channel CD quality audio bit rate is expected to be reduced to 128 to 196 kbps. A high-level view of a CBR video system is illustrated in Figure 3.4. The rate at which a video bit stream is submitted to the network may be determined by the bit rate allocated in the network. Allocating CBR to a video source might cause quality degradation for scenes with a large information content and wasted bit rate for scenes with a small information content. This phenomenon is further explored later, in the context of VBR video. Since the information contents of images differ from one frame to another, the bit rate at the output of a video encoder varies over time. Hence, a smoothing buffer is necessary to regulate the bit rate of CBR video at the network interface. The buffer size is determined by the delay characteristics of applications. Since

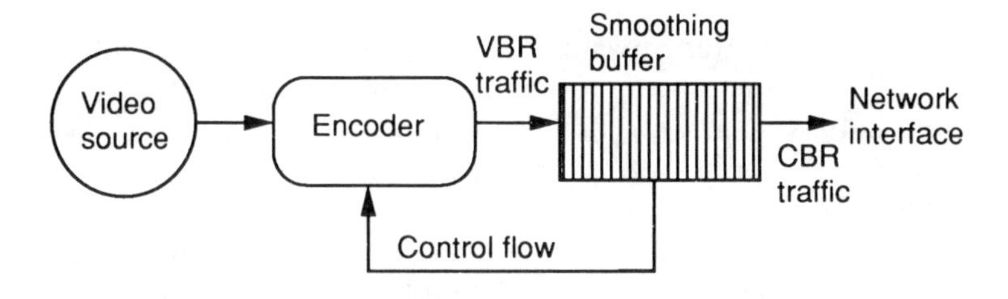

Figure 3.4 A high-level view of a CBR video system.

they can hold a finite amount of data, smoothing buffers are subject to overflows. However, video services are not tolerant to cell losses. To solve the overflow problem, it is necessary to reduce the bit generation rate at the encoder when the number of bits at the smoothing buffer reaches a threshold. Although this method solves the buffer overflow problem, reduced bit rate is achieved at the expense of increased quantization levels, causing degradation of the image quality.

CBR traffic is easy to manage in the network. A CBR is reserved for each CBR connection throughout its duration, regardless of whether the source is actively transmitting or in a silent state. This is an inefficient use of transmission capacity, as illustrated in Figure 3.5. Since the amount of information generated by most applications varies over time, it is possible to reserve less capacity in the network than the application's peak bit rate, thereby allowing more connections to be multiplexed and increasing the resource utilization. In initial deployments, a large portion of the traffic in ATM networks is expected to be CBR voice, video, and audio. In the future, designers will have a better understanding of the dynamics of VBR traffic and will be able to design efficient techniques to manage VBR traffic in the network, thereby achieving high resource utilization.

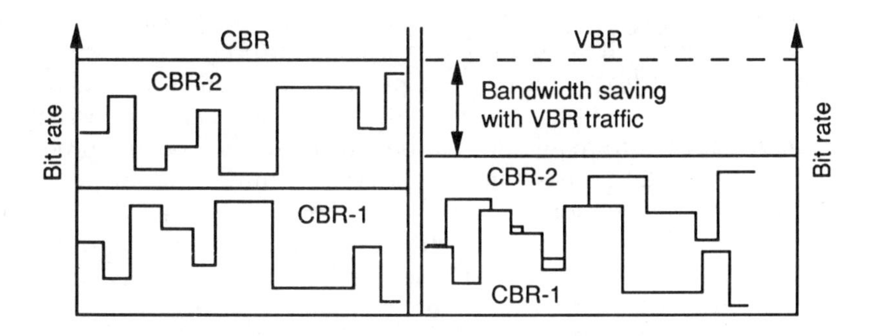

Figure 3.5 Transmission capacity usage for two CBR and VBR services.

High utilization of transmission links is particularly important in the wide area. With the introduction of fiber-optic transmission in the backbone, the question that naturally arises is how important it is to save the network bandwidth. As new multimedia services emerge, the demand for transmission capacity increases. A VCR-quality video distribution would require about 1.5 Mbps for about 1.5 hours. High-quality distance learning applications based on MPEG II (Motion Picture Experts Group) (I-frame-only encoding, as discussed later in this chapter) may require up to 20 Mbps in several transmission links over a wide area (i.e., long distances) to hundreds of different destinations (i.e., high schools in a state). Savings of 10% that might be possible with one such VBR video service would allow more than T1 link capacity without causing degradation of the service quality (VBR might in fact produce higher quality). As proved again and again in the past, there is no doubt that there will be new applications demanding higher bit rates that will chew up seemingly large transmission capacities currently viewed as unlimited.

3.2 VBR SERVICES

The traffic generated by networking applications generally either alternates between the active and silent periods or has a varying bit rate generated continuously. Furthermore, the peak-to-average bit rate of a VBR source is often much greater than 1. Presenting VBR traffic to the network as CBR traffic by means of buffering or, instead, artificially controlling its bit generation rate has the drawbacks of underutilization of network resources and QOS degradation. Although CBR service simplifies the network management and control tasks, it is more natural to provide VBR services to VBR sources and thereby provide better service and a framework to achieve higher resource utilization. For example, a voice source alternates between active and silent periods. Similarly, a video source (generally depending on the coding scheme used) generates a continuous bit stream at varying rates. Taking advantage of such source activities, VBR services result in multiplexing gains. In [57], a hardware test is performed to investigate the effect of multiplexing video sources. A video source measured to produce a 370-kbps average bit rate with a standard deviation of 63 kbps was observed to require 1.15 Mbps to achieve a blocking probability of 10^{-3}. When 5 and 25 of these video sources are multiplexed in two experiments, the standard deviation of the bit rate of the multiplexed traffic is measured to be equal to 76 and 27, respectively. With 25 sources multiplexed, it is observed that only 450 kbps for each source is required to achieve the same loss probability of 10^{-3}. This results in a 2.5:1 multiplexing gain in bit rate compared to carrying the same video sources in the CBR mode. Similar conclusions in multiplexing gains are also observed in [92,93].

3.2.1 Video Services

Video is presented to users as a series of frames in which the motion of the scene is reflected in small changes in sequentially displayed frames. Frames are displayed at the terminal at some constant rate, such as 24 or 30 frames/sec, enabling the eye to integrate the differences within the frame into a moving scene. Digital representation of video signals requires the transformation of a continuous image field into the discrete domain. This transformation is a mapping from the continuous image samples into a finite set of discrete amplitudes that span the intensity range of the image, similar to the PCM method used for voice. Each quantized amplitude is defined uniquely by a digital code word, referred to as a *pixel*. For example, monochrome images are generally uniformly quantized into 256 gray levels requiring 8 bits/pixel. On the other hand, full color images may generally require 24 bits/pixel and 8 bits/pixel, for example, for each red, blue, and green component. A color image can be represented in different color systems. Widely used color systems include R-G-B (red, green, and blue) in the computer industry, Y-U-V (Y is for luminance or brightness, and U and V are for color difference signals Y-R and Y-B, respectively) in the television industry, and C-M-Y-K (cyan, magenta, yellow, and black) in the printing industry. Within each color system, the constituent parts are called *components*.

The video encoder rate in CBR video is controlled by varying the precision of the quantizer. In particular, the quantizer step may be selected as a function of the instantaneous encoder buffer occupancy (compared with H.261 based systems as discussed next).

$$Q(t) = 2[32\, Be(t)/Bmax] + 2$$

where $[x]$ denotes the largest integer smaller than or equal to x, $Be(t)$ is the encoder buffer occupancy at time t, and $Bmax$ is the buffer size. $Q(t)$ is bounded below by 2 and above by 62. Based on the buffer occupancy, the quantizer step may be varied (linearly in the above example) to prevent buffer underflow/overflow as the encoder buffer fills and empties.

Typical consumer video cassette recorders (VCR) deliver an image resolution of 200 × 300 pixels. Similarly, the resolution in XGA and VGA displays are, respectively, equal to 1,024 × 768 pixels and 640 × 480 pixels. Figure 3.6 (cf. [79]) illustrates how various metrics affect the application bit rate of VBR video (and similarly for audio), where the raw bit rate requirement of a video service, given in (3.3), depends on the QOS, which further depends on the color, spatial, and motion resolutions.

$$\text{Visual bit rate} = (\text{bits/pixel})(\text{pixels/frame})(\text{frames/sec}) \tag{3.3}$$

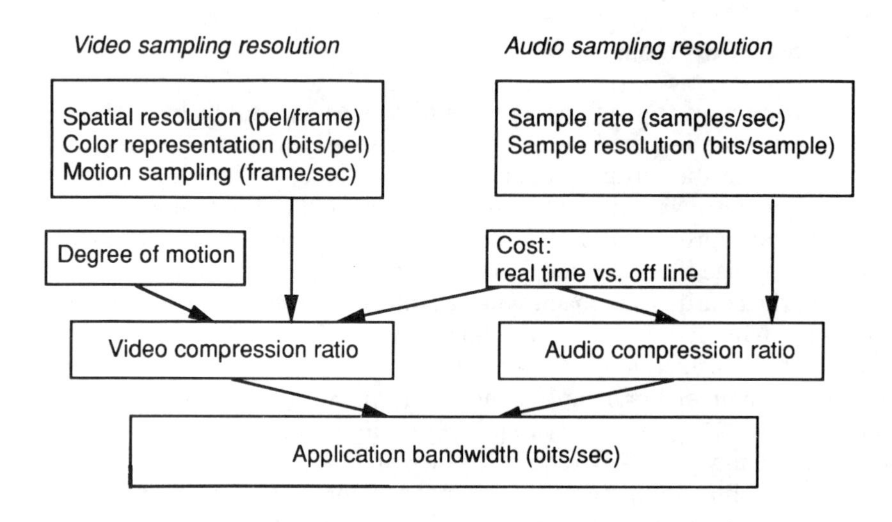

Figure 3.6 Bit rate requirements of video and audio applications.

Table 3.2 illustrates the raw bit rate requirements of a selected subset of video services (cf. [79]). Typically, the raw bit rate requirements of video applications vary anywhere from a few to several hundreds of megabits per second. Even with rapid advances in the technology enabling the deployment

Table 3.2
Bit Rate Requirements of Video Services

		Bit Rate (kbps)	
Service	*Example*	*Raw*	*Compressed*
	Real time (1/4 screen, low resolution) (128 × 120 pixels) (9 bits/pixel; 15 frames/sec)	2,074	64
	Real time (1/4 screen, high resolution) (128 × 240 pixels) (9 bits/pixel; 15 frames/sec)	4,147	384
	Real time (full screen, high resolution) (128 × 240 pixels; 9 bits/pixel; 30 frames/sec)	8,294	2,000
Video	Non-real-time, low-resolution server (352 × 240 pixels; 9 bits/pixel; 10 frames/sec)	7,603	384
	VCR-quality server (352 × 240 pixels; 24 bits/pixel; 30 frames/sec)	60,825	1,100
	Studio-quality server (640 × 480 pixels; 24 bits/pixel; 30 frames/sec)	221,184	4,000
	High-definition TV (HDTV) (1,125 lines; 24 bits/pixel; 30 frames/sec)	800,000	60,000–127,000

of links with rates of gigabits per second, the large bit rate requirements of these applications are still quite costly for service offerings and unfeasible for most networks scheduled to be deployed in the near future, unless acceptable service quality can be delivered at considerably reduced bit rates. Fortunately, the nature of visual applications leads to mechanisms that can be used to reduce their bit rate requirements. In particular, video sequences contain a significant amount of repetition, both within a frame and between frames. It is very likely that the pixel values corresponding to neighbor locations within a frame are close to each other, referred to as *spatial redundancy*. Similarly, subsequent frames have a large amount of data common between them, referred to as *temporal redundancy*.

It is possible to reduce the application bit rate by taking advantage of such redundancies. As an example, Intel's proprietary coding technique DVI [77] can reduce the amount of data transmitted to the network from a video source by a factor of 20 or more depending on the scene activity and its contents. A video source with a play-out rate of 30 frames/sec, a resolution of 512 × 480 pixels, and full color of 24 bits/pixel would generate 177-Mbps raw data. DVI may reduce the application bit rate by decreasing:

- The spatial redundancy to 256 × 240 pixels by alternate pixel stripping;
- The pixel depth from 24 to 9 bits;
- The image data by 10–15 to 1 using image correlation analysis.

This results in a 5.5- to 8.3-Mbps bit rate, a savings of about 20:1. The procedure of eliminating or, more realistically, reducing the redundancies in data streams is referred to as *compression*. Compression algorithms are classified as lossless and lossy [67]. In lossless coding, the original quantized sample values are recovered exactly, assuming no bit errors took place during the transmission. The main disadvantage of these algorithms is that they result in relatively small compression ratios and are mainly used for data compression. Lossy coding, on the other hand, can no longer be expected to produce the original sample values exactly, introducing the possibility of distortion in the images. However, they produce good-quality images with high compression ratios.

Generally, lossy coding can result in compression ratios of 10:1 to 50:1 for images and 50:1 to 200:1 for video. Lossless methods, however, may produce ratios of up to 3:1 and are used only for applications that are sensitive to losses such as data and medical images.

Lossy algorithms exploit various aspects of the human visual system. For example, the eye is much more receptive to fine detail in the luminance (or brightness) signal than in the chrominance (or color) signals. Accordingly, the luminance signals are often sampled at higher spatial resolution. For example, in broadcast-quality television, the digital resolution of the sampled luminance signal is 720 × 480, while for the color signals it may be only 360 × 240. The compressed representation of the luminance signal may be assigned more bits (i.e., higher dynamic range) than are the chrominance signals.

The eye is less sensitive to energy with high spatial frequency than to that with low spatial frequency. This deficiency is exploited by coding the high-frequency coefficients with fewer bits and the low ones with more bits.

The compression ratio used in encoding video frames may be a function of the available bit rate in the network, the resolution required, and the degree of redundancies in video bit streams. As an example, consider a visual application with a 90- to 100-Mbps raw bit rate requirement. If 1.544-Mbps T1 links were to be used between two such applications, it would be necessary to have a compression ratio of 60 to 1. The required degree of resolution differs in various applications. Digital x-rays or computer-aided design may require very high resolution, whereas much lower resolutions can be acceptable in video phones. Similarly, if there is very little change in subsequent frames, as in a video phone showing a person talking, then the information contents of frames decrease only if the differences between the frames are coded and transmitted.

Various compression techniques that have been developed to reduce the amount of data transmitted in the network are discussed next, divided into two categories: intraframe coding and interframe coding [17], corresponding respectively two classes of techniques developed to reduce the spatial and temporal redundancies.

3.2.1.1 Intraframe Coding Techniques

Intraframe coding techniques attempt to reduce spatial redundancy by taking into account the fact that the pixel values corresponding to neighbor locations within a frame are close to each other. In general, these methods are relatively simple to implement and do not have large storage requirements.

Predictive Coding

In this coding scheme, the spatial redundancy is determined from the neighbor pixels and the difference between the current and predictive pixel values is quantized and coded. The number of neighbor pixels used for prediction is usually less than four. Differential pulse code modulation (DPCM) is one of the earliest coding techniques used for voice compression. It is an extension of the PCM method in which only the quantized difference between the actual sample and its predicted value is binary-coded and transmitted. Depending on the voice quality desired, DPCM can reduce the application bit rate from 3–5 to 1. It is a simple technique, which leads to economical hardware implementations, and it produces good quality for still images and voice. The main disadvantage of this method is its sensitivity to bit errors in the transmission link and that other coding techniques produce similar or better quality images with higher compression ratios.

Transform Coding

The fundamental concept in this coding scheme is to decorrelate the image data via an orthogonal transformation into a more compact form, thereby reducing the number of bits used to represent images. Transform coding is a high-performance coding technique used for still-image compression.

The discrete cosine transform (DCT) is the most commonly used transform coder. In this technique, an image is divided into a number of disjoint $N \times N$ pixel blocks. The choice of N depends on the desired compression ratio and the desired quality of the image. As N increases, the compression ratio increases while the image quality decreases. The most frequently used block sizes are 8×8 and 16×16. DCT is applied to each block of data, producing a set of transformed coefficients $\{F(u, v): 0 < u, v < N\}$, where $F(0, 0)$ corresponds to the mean of the pixel values in a block. The coefficients are then thresholded and quantized. Finally, the quantized coefficients are scanned into a one-dimensional sequence whose nonzero amplitudes and run lengths of zeros are entropy-coded before being passed to the channel. Some variants of DCT coding have been proposed to include variable block sizes and adaptive schemes that allocate more or fewer bits to different blocks depending on the amount of activity. The interested reader may refer to [16,19,20,50,54,71] for the details on DCT coding.

DCT is chosen as the coding transform technique for various image coding standards such as the Joint Photographic Experts Group (JPEG), MPEG, and H.261. In general, 8–16:1 compression ratios can be achieved by DCT coding. However, DCT is rather complex to implement in hardware.

Subband Coding

In subband coding, the video signal is decomposed into several frequency subbands and dealt with by using different coding techniques according to the characteristics of each subband. An intraframe subband/DCT hybrid scheme is proposed in [33]. In this approach, each frame is decomposed into four subbands: low-low (LL), low-high (LH), high-low (HL), and high-high (HH), where (xy) denotes the frequency contents of the image in the horizontal (x) and vertical (y) axis. The LL subband includes the lowest frequency contents of images that constitute the most important part of the image data for reconstruction, excluding sharpness. Accordingly, coding distortion at the LL subband should be kept at a minimum. Higher subbands mainly contain the edge and high-frequency contents. A human visual system is not very sensitive to very high frequency data. Hence, some truncation at these subbands can be used without causing significant visual degradation and further reduces the application bit rate requirements. This hybrid coding scheme is applied to HDTV systems and has been observed to produce good-quality images at a 120-Mbps compressed bit rate compared to an 800-Mbps raw bit rate [33], resulting in about a 6.5:1 compression ratio.

Vector Quantization

All three coding algorithms discussed above achieve compression gains by taking advantage of the correlated structures of images to reduce the number of bits required to represent an image. In a vector-quantization coding scheme, small blocks of data are treated as vectors and matched (i.e., best match) from a predefined code book according to some distance measurement. The index of the matched entry in the code book is then sent to the receiver, which reconstructs the image through a simple table lookup. A review of code book design can be found in [36].

The quality of images reconstructed with this coding scheme depends on the size of the code book: the larger the book size, the higher the image quality is. However, the code book design complexity increases exponentially with its size. A typical vector quantizer breaks an image into $M \times N$ pixel blocks, where each pixel can take one of K possible values. The set of all possible image blocks has K^{mn} elements, where each block can be thought of as a vector of dimension MN. Compression ratios of 16–24:1 can be achieved using this coding scheme with moderate code book sizes [67].

Various vector-quantization schemes are proposed in the literature. Classified vector quantization [76] divides the input vectors into several classes, and each class is associated with its own code book. This reduces the size of each code book but requires class indexes to be included in the information transmitted, thereby increasing the amount of overhead included with each frame. Finite-state vector quantization [22] attempts to eliminate this overhead by obtaining the classification of classes from the neighbor image contents.

Vector quantization has a simple decoder structure. However, due to the relatively larger storage and higher processing power that are required compared to other schemes, this scheme is used mainly for applications with low bit rates, such as audio and low-speed data.

3.2.1.2 Interframe Coding Techniques

Intraframe coding is mainly used for still images. When time is involved (i.e., motion pictures), the information redundancy in subsequent frames is generally significant, and a major portion of the current frame can be determined from the previous frames. Hence, it is beneficial to take the correlation between subsequent frames into consideration to further reduce (more than can be achieved by spatial redundancy alone) the amount of bit rate required by visual applications. Interframe coding techniques attempt to reduce temporal redundancies in two steps: motion estimation and motion compensation.

Motion Estimation Algorithms

Motion estimation is the computation of relative displacement of a block of image data between the current and previous frames. If the location of a block in a frame can be predicted from the previous frame, then only the displacement vector can be coded and included in the transmission, thereby reducing the amount of information transmitted per frame.

Pal-Recursive Algorithm (PRA). PRA estimates motion on a pixel-by-pixel basis [68]. The intensity $I(x_1, x_2)$ of a pixel at location (x_1, x_2) at the current frame is the intensity of the pixel at location $(x_1 - d_1, x_2 - d_2)$, where $D = (d_1, d_2)$ is the displacement vector. That is,

$$I(x_1, x_2) = I(x_1 - d_1, x_2 - d_2)$$

for some (d_1, d_2). Hence, estimating the value of a pixel at the current frame is reduced to the estimation of the distance vector D. PRA estimates D by minimizing the squared value of the displaced frame difference (DFD) recursively for each moving element, where

$$\mathrm{DFD} = I(x_1, x_2) - I(x_1 - d_1, x_2 - d_2)$$

PRA can estimate the motion accurately and is shown to reduce the bit rate requirements by 30% to 60% (cf. [69,78]). However, the algorithm is computation intensive and cannot be used in real time.

Block Matching Algorithm (BMA). Assuming that all pixels within a block undergo uniform motion, BMA predicts the motion on a block-by-block basis. Accordingly, a frame in this scheme is divided into a number of $N \times N$ blocks. Then the motion estimation criterion is based on a measurement $M(i, j)$ of the difference between the frame and the displaced block in the previous frame within a given searching area $(N + 2L) \times (N + 2L)$, where L is the maximum displacement allowed. The best match is obtained by optimizing $M(i, j)$.

The block recursive matching algorithm proposed in [88] combines the advantages of PRA and BMA by jointly optimizing the measurement index $M(i, j)$ and the displacement vector D. Because the solution of the joint-optimization problem is computationally expensive, a recursive algorithm is developed to calculate the two metrics.

Motion Compensation Algorithms

Motion compensation (MC) algorithms use the output of motion estimation computation to reduce the amount of data needed to reconstruct the current frame from the previous frame by the use of indexes and differences between the blocks.

The most frequently used scheme to reduce the temporal redundancy is the motion-compensated predictive coding, where the prediction error (defined as the difference between the current block and the motion-compensated block from the previous frame) is coded. Any coding technique described above for intraframe coding can be used to code the prediction error.

In general, the performance of motion-compensated coding depends on the following factors [67]:

- The amount of purely transitional motion of objects in the video scene;
- The ability of the algorithm to estimate the motion accurately;
- The robustness of the displacement estimation algorithm.

The effectiveness of compression and MC algorithms for video applications is investigated in [72]. The setup includes two conference-type scenes: a high-activity scene showing a person moving and a low-activity scene where the person remains stationary. The picture size is 256 × 240 pixels, producing a raw bit stream of 22 Mbps at a play-out rate of 15 frames/sec. Three experiments are performed: DPCM, DPCM with MC, and DCT with MC. The compressed bit rates with each scheme are given in Table 3.3, which illustrates the bit rate savings possible and thereby the effectiveness of the use of the MC scheme and data compression.

3.2.1.3 Standardization in Video Services

Multimedia services combine two or more of the following services: text, graphics, audio, still images, and motion video services. These services are currently

Table 3.3
The Effectiveness of MC Together with DCT or DPCM

Algorithm	*DPCM*		*MC + DPCM*		*MC + DCT*	
Scene	*Active*	*Inactive*	*Active*	*Inactive*	*Active*	*Inactive*
Average (Mbps)	2.7	1.7	2	1.2	0.252	0.139
Peak (Mbps)	9.2	5.3	5	3.1	0.454	0.264

evolving around the JPEG standard for still-image delivery and MPEG for motion video. A high-level view of JPEG encoding and decoding is shown in Figure 3.7 (cf. [5]).

Recalling that within each color system (i.e., R-G-B, Y-U-V, or C-M-Y-K) the constituent parts are called *components,* each component of the source image (or 16-level gray-scale image) in the JPEG encoder and decoder is divided into 8 × 8 nonoverlapping pixel blocks that undergo a DCT, producing 64 frequency coefficients. The resulting 64 coefficients represent the frequency contents of the given block. Next, the DCT coefficients are quantized. The quantization step varies with frequency and component. The dependence on frequency reflects the fact that the high-frequency coefficients subjectively matter less than the low-frequency ones and may therefore be quantized with a larger step size. Each component may have its own quantization table. Up to four quantization tables are allowed in JPEG. Following the quantization, the coefficients are reordered into a one-dimensional array and the quantized coefficients are losslessly encoded.

For simplicity, a 4 × 4 block is used in Figure 3.7 to represent these coefficients. The two-dimensional array is then read into a vector in a zigzag fashion. The coefficients then go through a quantizer (in the example, divided

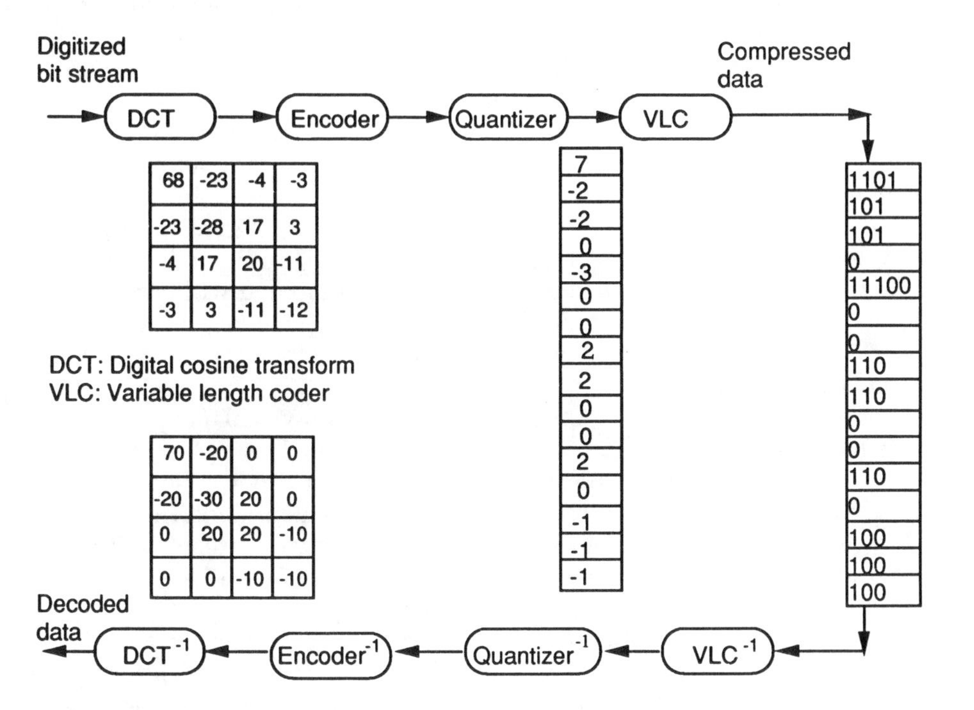

Figure 3.7 High-level view of JPEG encoder-decoder.

by 10) and are coded using the variable-length coder [47], producing the bit stream transmitted in the network.

At the receiver, the quantized coefficients are recovered from the bit stream using inverse variable-length coding and converted into a vector by inverse quantization. The data are then converted into a two-dimensional array via inverse encoding, which yields the image after the inverse DCT operation.

The MPEG coding algorithm was developed primarily for the storage of motion video on digital storage media. Recent applications of MPEG include a variety of video services from multimedia workstation to HDTV. The play-out rate in MPEG may vary from 24 to 60 frames/sec. The original MPEG standardization, referred to as MPEG-1, supports picture sizes of up to 352 × 240 with a target transmission rate of 1 to 1.5 Mbps. MPEG-1 was extended to MPEG-2, supporting a picture size of up to 704 × 480 with target transmission rates of 4 to 10 Mbps. The MPEG-2 standard consists of three data streams: MPEG-2 audio, MPEG-2 video, and MPEG-2 system.

MPEG-2 video specifies the coded bit stream for higher quality digital video. In addition to being compatible with MPEG-1, MPEG-2 supports increased image quality, various picture aspect ratios, and advanced features: MPEG-2 audio coding supports up to five full channels (i.e., left, right, center, and two surround channels). It also provides quality coding of mono and conventional stereo signals for bit rates at or below 64 kbps. The MPEG-2 system standard specifies how to combine multiple audio, video, and private data streams into a single multiplexed stream, as illustrated in Figure 3.8. It performs packetized stream control and synchronization and is designed to support a wide range of broadcast, interactive telecommunications, computing, and storage applications.

The systems layer defines two kinds of streams—program and transport streams—and processes the compressed video codec, audio codec, and data streams in two steps: the codec/data streams are combined with system-level information and packetized to produce packetized elementary systems (PES). Then the PESs are combined to form either a program stream (PS) or a transport stream (TS).

The PS supports the creation of a single audio-visual program, which could have multiple views and multichannel audio. It uses variable-length packets and is designed for transmission in error-free environments. The TS multiplexes a number of programs, composed of video, audio, and private data for transmission and storage. It performs packetized stream control and synchronization. It is designed for transmission in a lossy or noisy environment and uses a fixed-size 188-byte packet.

The MPEG video compression algorithm relies on two basic techniques: block-based MC for reduction in temporal redundancy and DCT for spatial redundancy. It is organized in a hierarchical format in order of spatial size. The basic coding units for the MPEG algorithm are 8 × 8 pixel blocks for spatial

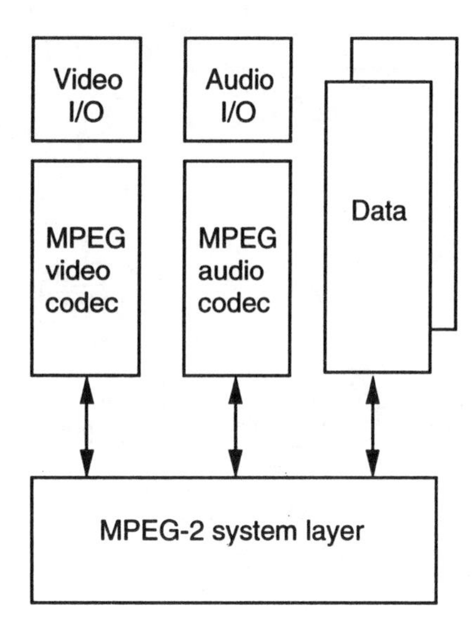

Figure 3.8 MPEG-2 components.

redundancy and 16 × 16 pixel macroblocks for MC. A group of macroblocks making up a horizontal segment of the image forms a slice of 512 × 16 pixels, and 30 slices are combined to form a picture (512 × 480 pixels).

MPEG produces three types of packets: intraframes (I), predicted frames (P), and interpolated frames (B). I frames are points for random access, refreshing the frame sequence, and preventing error propagation across frames, and they can tolerate only moderate compression. The first frame in a sequence is an I frame. P frames are coded with reference to a previous frame and are generally used as a reference for subsequent P frames. These predictively motion-compensated frames include motion vectors that represent the difference between the spatial location of the macroblock and that of its predictor. One drawback to using previous frames for MC is that regions in which new information is appearing cannot be adequately compensated for, since previous frames do not include related information. However, this information is most likely available at future frames. MPEG exploits this phenomenon by allowing backward-in-time prediction and interpolation by using previous as well as future frames. These frames are referred to as B frames. Accordingly, three types of MC may be applied to B frames: forward, backward, and interpolative. It is also possible to code these frames without any MC. The organization of frames in MPEG depends on the application parameters such as random accessibility and coding delay.

A high-level view of MPEG coding is illustrated in Figure 3.9 (cf. [5]). Similar to JPEG, each 8 × 8 pixel block of the incoming frame is encoded with

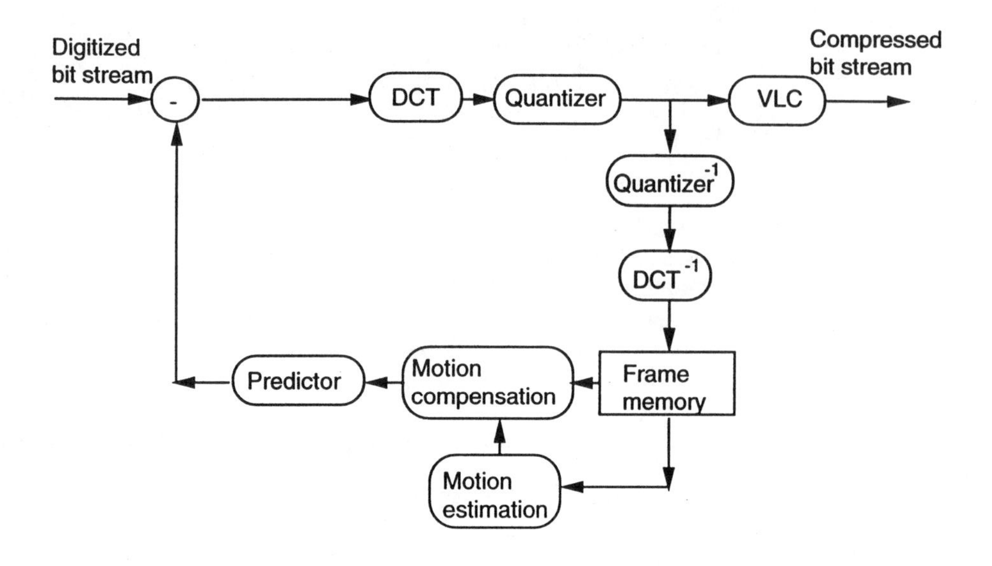

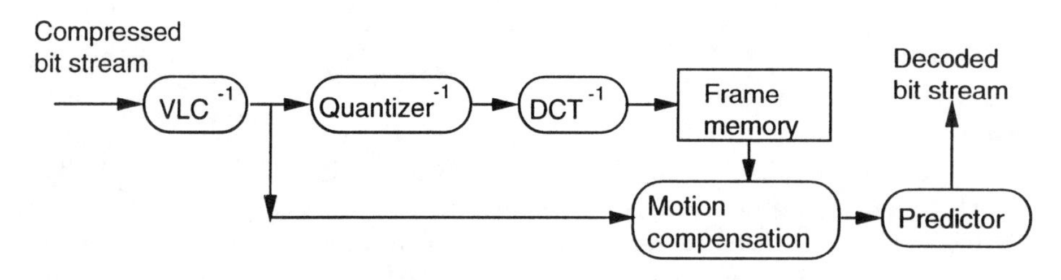

Figure 3.9 A high-level view of MPEG codec.

DCT and quantized. The output of the quantizer proceeds in two parallel paths: one to the variable-length coder and the other to quantization and inverse DCT, which yields a reconstructed block stored in the frame memory. Once the whole frame is reconstructed in the frame memory, interframe coding (i.e., motion estimation and compensation) is applied. The motion vector that represents the offset between the current block and a block in the previously reconstructed image that forms the best match is coded and transmitted. The predictor gives the motion-compensated block from the reconstructed frame. The difference between this and the original block is transform-coded, quantized, and variable-length-coded before being transmitted.

The I frames are compressed by taking the DCT of blocks of image data, quantizing the DCT coefficients, and variable-length-coding the quantized coefficients. P frames allow higher compression, whereas B frames allow highest compression ratios but require both a previous and a subsequent reference for prediction.

At the receiver, the bit stream is first decoded through inverse variable-length coding (VLC). The output of inverse VLC then passes through the inverse quantizer and the inverse DCT that yields the DCT coefficients. From these coefficients a block of data is reconstructed and stored in the frame memory. In the interframe mode, motion vectors which are used to provide the location of the predicted blocks are extracted from the VLC decoder.

The traffic characteristics of the MPEG coder are mainly affected by two parameters: B:I ratio, which determines the number of B frames between I frames, and the quantizer scale parameter that determines the video quality. As the B:I ratio increases, the bit rate decreases (since B frames allow high compression ratios), but the time to recover from errors increases. Similarly, large values of the quantizer scale decrease the bit rate at the expense of a reduction in the picture quality. The maximum bit rate of the MPEG-I coder is constrained to be less than 1.86 Mbps. The quality of video compressed with the MPEG-I scheme at rates of 1.2 Mbps has often been observed to be similar to that of VHS recording [34]. MPEG can reduce the raw bit rate twentyfold to fiftyfold (compared with its raw bit rate requirement), depending on the application and the desired picture quality.

ITU-T has recommended the H.261 video transmission protocol for use in real-time video delivery at bandwidth increments of 64 kbps (i.e., $P \times 64$ kbps, $P = 1, \ldots, 30$). It is mainly intended for video telephony and video teleconferencing applications in ISDN. A typical example of a real-time video service is distance learning, where a student at home may view a lecture taking place at a university. In this case, the student may require only remote video monitoring functionality; hence, the traffic flows mostly in one direction, from university to the house. On the other hand, video conferencing may require bidirectional communication and equipment at both locations. Another example of a real-time video application is videophone, which is essentially a scaled-down, limited-functionality residential video conferencing service.

Motion-compensated I-frame prediction and adaptive DCT coding are respectively used to reduce temporal and spatial redundancies. Two frame formats are defined: common intermediate format (CIF) and quarter-CIF (QCIF). The quality of video delivered increases proportionally to the amount of bit rate used. However, the image quality delivered by H.261 is perceived to be less than VCR quality at 1.544 Mbps (i.e., $P = 24$). Most equipment available today uses the H.261 protocol and requires a minimum of 128 kbps of bandwidth.

3.2.2 Voice

In a typical packet voice system, speech is digitized, coded, and packetized for transmission. PCM voice was briefly introduced in Section 3.1. To improve the transmission efficiency without reducing voice quality, PCM is extended to

DPCM, which is further extended to adaptive differential PCM (ADPCM), specified in ITU-T Recommendation G.721 [110].

A voice source alternates between talk spurts (active) and silent periods. CBR voice transmits silent periods as well as active periods, which is an inefficient way of using network resources. To achieve higher resource utilization, speech activity detection may be used at the VBR voice source so that voice packets are generated only when the source is active, thereby increasing the transmission efficiency. Transmitting cells generated only at active periods and using ADPCM coding, a compression ratio of better than 4:1, can be achieved without noticeable degradation in voice quality.

Voice connections require end-to-end timing between the source and the destination. Hence, AAL 2 may be used for VBR voice. At the 64-kbps peak bit rate, an ATM cell would carry 5.5 to 5.875 ms of voice (i.e., 44 to 47 one-byte voice frames). AAL 2 is not yet standardized. These numbers assume the use of 1 to 4 bytes of AAL overhead per cell.

ITU-T has formed an ad hoc group to investigate and define performance characteristics and objectives of selecting a single coding algorithm that would meet the requirements of various applications of speech coding, such as 16-kbps coding and VBR operation. Once finalized, the efforts of this group would further reduce the bit rate required for voice connections.

3.2.3 Data Applications

The term *data* is used for any application that uses coded text, that is, any application that is not voice, audio, video, or still image. Despite the fact that data networks have been operational for a number of decades, traffic characteristics of data sources are not well understood. The main difficulty arises due to the fact that there is no typical data connection. Large amounts of data are transmitted in a file transfer on a rather continuous basis during the duration of the connection, whereas only a few hundred bytes are generated by e-mail. Some applications require that end-to-end connections be established, while some others are best served in the connectionless mode. Interactive data services require bidirectional data exchange, while database updates may use only unidirectional connections. The problem of characterizing the source characteristics of data applications is further complicated by the fact that it is generally difficult to predict in advance the traffic characteristics of a connection, even if the particular application type is known. For example, in client-server computing, the amount of data exchanged and the source behavior may differ significantly from one application to another. Furthermore, data connections are not generally established between two users but between groups of users, as in the case of LAN interconnection. Keeping these difficulties in mind, the source characteristics of different data applications are discussed next.

3.2.3.1 LAN Interconnection

The fast growth in the number of personal computers (PC) and workstations and the rising need for interconnecting them with, for example, printers and servers in an office environment has led to the deployment of the first generation of LANs. As the need for interoffice communications rose, LANs were in turn interconnected via bridges and routers. It did not take long for users to start demanding scalable throughput for large volumes of data exchange between them and, for aggregate bit rate, requiring the design and deployment of high-bandwidth WANs. In addition, emerging multimedia applications that integrate voice, video, and data started to change the face of networking in local and wide areas due to their real-time and high-bit-rate requirements.

The concepts of B-ISDNs address all these requirements in a unique framework by providing scalable throughput and real-time transport capabilities, and facilitating the interworking between LAN and WAN technology. In order to provide LAN interconnection service, however, B-ISDNs must retain the features of connectionless networks. Yet ATM is a connection-oriented technique and end-to-end connections are required to be established in ATM networks before data transmission can start.

Furthermore, the ATM cell header provides a limited number of bytes for addressing (i.e., VPI and VCI fields), insufficient to be used as a global destination identifier. Hence, it is not possible to use ATM directly to provide LAN interconnection services. Various solutions proposed to offer connectionless services over ATM networks are discussed in detail in Chapter 9.

ITU-T has developed the use of connectionless servers (CLS) at ATM switching nodes (cf. [11,25,86,90]) to support connectionless service. The CLS approach consists of building a connectionless virtual-overlay network on top of ATM. The servers are directly attached to ATM switches and accessed via ATM connections. Preassigned VPs may be used for connections between servers and ATM switches as for any other terminal connected to an ATM switch. In addition, a set of VPs may be defined between ATM switches (that are connected to servers), which are used only by the connectionless traffic. Servers use the CLS network access protocol (CLNAP). ITU-T I.364 defines the basic framework for the selection of the QOS parameters and higher layer protocols such as routing, and addressing to transfer variable-length packets generated by a LAN station to one or more destinations without the need to establishing end-to-end connections.

The ATM Forum has defined a LAN emulation service to provide interoperability among applications running over legacy LANs and ATM networks. The basic principle of the LAN emulation framework is to hide the connection-oriented nature of ATM from the applications and emulate connectionless service. In this approach, the medium access control (MAC) layer is replaced by the LAN emulation layer that allows to run legacy LAN applications over ATM without requiring any changes to the applications.

Leaving the discussion on how connectionless service for LAN interconnection can be implemented in B-ISDNs until Chapter 9, we will next present the traffic characteristics of LAN traffic, based on various studies that have been reported in the literature. Figure 3.10 (cf. [43]) illustrates the typical packet-length distributions (information field length only) in Ethernet.

Packets in Ethernet tend to be of three different sizes, a consequence of three different application classes. Short packets are transmitted during terminal-to-host communications, whereas applications based on the network file system (NFS) protocol generate short packets in one direction followed by medium-sized packets in the other direction. The maximum packet size in Ethernet is 1,512 bytes, used mostly during file transfer applications. The Ethernet peak rate is 10 Mbps. Ethernet traffic can be modeled by an on/off process. The active period distribution is given in Figure 3.10, where the average packet size is 876.6 bytes, or 0.7 ms. The distributions of active and silent periods are not exponential.

Figure 3.11 illustrates a typical packet-length distribution on a token ring. These particular data were collected at IBM, Research Triangle Park [21]. Based on these measurements, token ring traffic, similar to Ethernet, is observed to behave as an on/off process. The average active period (which may include one or more packets) is 205 bytes with a squared coefficient of variation of 2.4. The peak rate of token ring traffic is either 4 or 16 Mbps.

Although the peak rates are known, the average bit rate at the IWUs at the boundary between LANs and an ATM network depends on the environment. The average duration and its distribution can be obtained from the packet-length distributions given above, but require some statistical knowledge on the number of back-to-back packets transmitted in an active period. The average

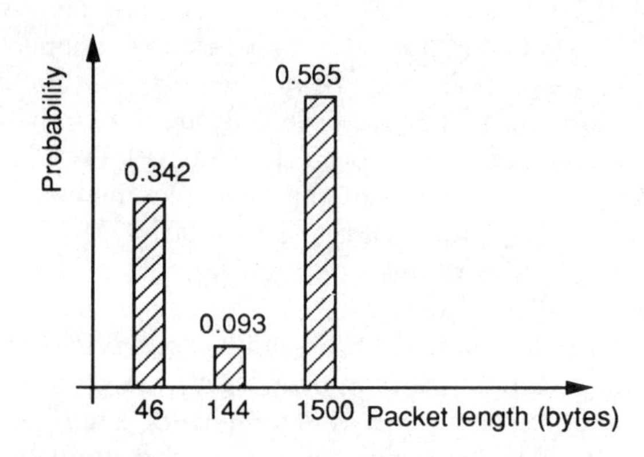

Figure 3.10 Ethernet packet-length distribution.

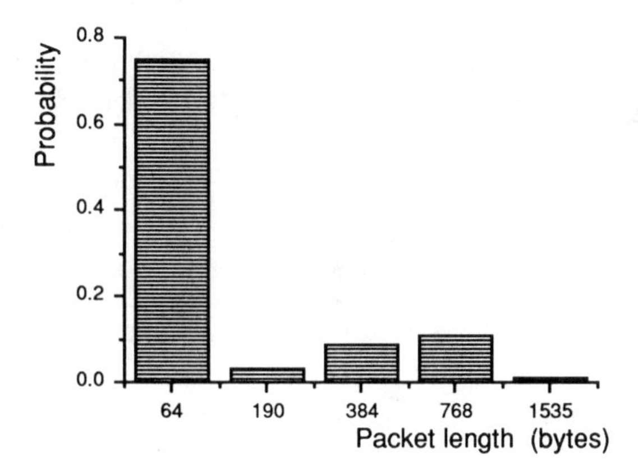

Figure 3.11 Token ring packet-length distribution.

duration of the silent period may then be determined if the average bit rate and the average duration of the active period are known. Accurate characterization of LAN traffic should also include the distribution of the silent period. Furthermore, as discussed above, most large- and medium-sized packets are sent in response to short request packets, as in NFS. Detailed studies are needed to investigate the existence and the performance implications of such correlations and the distribution of silent periods in order to accurately characterize LAN interconnection traffic in B-ISDNs. It is generally agreed that LAN traffic is unpredictable and cannot be characterized accurately. There are various proposals currently being considered in the standardization bodies toward specifying a connectionless traffic class in ATM and developing procedures to support this type of traffic in the network. For example, connectionless application in an ATM network may specify its peak and minimum cell rate requirements, and there might be mechanisms to allow the traffic to vary between these two limits (for details, refer to Chapter 9).

3.2.3.2 Interactive Communication

Consider a workstation connected to a communication system. In an interactive system, the workstation transmits a service request to some other location in the network. Sometime later, the workstation receives a response message and initiates its think period during which no communications take place. A pictorial illustration of the process is illustrated in Figure 3.12 (cf. [79]).

The throughput requirements of interactive applications depend on the lengths of the request and response messages and the duration of the interaction period, which is equal to the sum of the average system response time and the

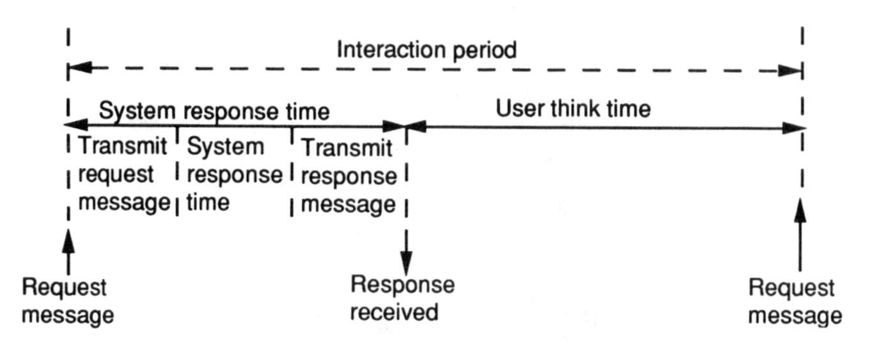

Figure 3.12 High-level view of interactive traffic.

user think time. Some typical applications and their average bit rate requirements are given in Table 3.4 [79].

In a database read application, the user is considered to be reading one page of information, whereas in database retrieval the user is considered to be browsing through a number of pages looking for a particular page. The typical 1,240-byte message length relates to performance benchmarks with IBM 3270 displays when they are used for database read, retrieval, and browsing [79]. The PC file server application represents the communication of files from a common alphanumeric data base to PC users, requested at human speeds. Also, these measurements have demonstrated that 80% of all file accesses are made to files less than 10,000 bytes long.

Users of data communications require high bit rates, not so much because they have a very large volume of data to transmit, but because they require rapid response times, particularly for interactive applications. One exception to this is high-performance computing in which a complex problem requiring very large processing times with large data requirements is solved on a set of supercomputers connected via a high-speed communications network.

Table 3.4
Average Bit Rate Requirements of Some Interactive Applications

Application	*Message Length (bytes)*	*Interaction Period (sec)*	*Throughput Demand (Bps)*
Database read	1,240	30	41
Database retrieval	1,240	9	138
Database browse	1,240	3	413
Shared PC file server	12,000	20	600

3.2.3.3 High-Performance Computing

The main challenge addressed in high-performance computing is to get much higher performance from multiple computers connected by high-bit-rate communication links than is possible on any one system. Assuming that algorithms can be devised to hide communication latency and applications can be rewritten to run in a distributed fashion, the time to solve a large and complex problem may be reduced in proportion to the number of computers that participate in the computation. Some problems that can potentially benefit from high-performance computing include chemical reaction dynamics, geophysics, combinatorial optimization, and global climate modeling. These applications are mainly characterized by their huge data and processing requirements.

In distributive computing, a task is divided into a number of pieces, and each piece is loaded into a separate computer. Subtasks then run simultaneously. During the computation, computers need to exchange or share information, thereby aggregating partial results obtained by each piece. The communications network provides transmission facilities over which the computers communicate with each other. If communication between the computers is slow compared to the speed of the computers or cannot be overlapped with computation, then each computer will be idle while waiting for data and will be underutilized. Packet size is an important factor in determining the communication throughput. In the experiments performed on the Cray supercomputer model D input/output (I/O) system in [84], it was observed that the performance of the I/O channel rated at 800 Mbps is equal to 280 Mbps with 16K packets. The packet size is required to be 256K to saturate the link. The latter is equivalent to a burst of approximately 5,600 cells.

To illustrate the high-bit-rate demands of high-performance computing applications, we next discuss the traffic measurements for the dynamic radiation therapy planning application [46]. Data in this system flow from a Cray supercomputer to a network terminal interface over a HIPPI link. Collected statistics are converted to simulate the traffic on an STS-12 SONET frame. Let the traffic rate denote the total amount of ATM traffic arriving over a fixed interval of N slots divided by the duration of the interval. When the traffic rate is averaged over 8,000 slots, the peak rate with $N = 200$ is observed to be about 145 Mbps, whereas it is equal to 600 Mbps (the effective STS-12 bit rate excluding SONET overhead). The traffic, as expected, alternates between active and silent periods. There are only three different active-period durations observed corresponding to HIPPI packets of lengths 24 bytes, 2264 bytes, and 16K, as illustrated in Figure 3.13.

The mean number of ATM cells is measured to be 372.9, with a squared coefficient of variation of 0.1555. Therefore, statistically the active period has little variation and can be modeled by an Erlang distribution with 6 phases or

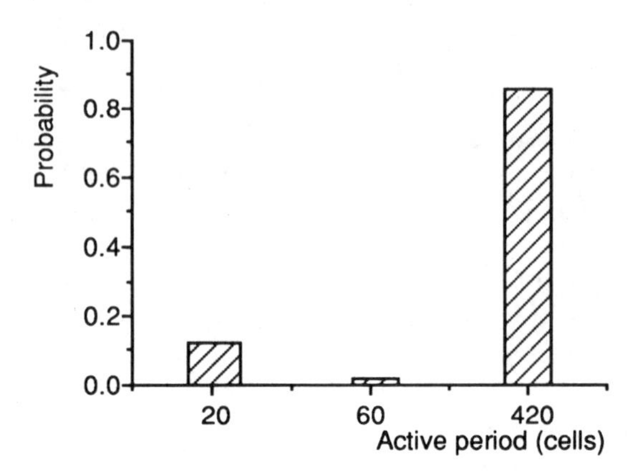

Figure 3.13 Active period durations in a radiation therapy planning environment.

by constant distribution. The silent period, on the other hand, is observed to have a large squared coefficient of variation of 83.

These early results indicate that neither the interrupted Poisson process (IPP) nor the interrupted Bernoulli process (IBP) is a likely candidate to accurately model the source behavior of data sources. One alternative is to assume that these two Markovian models produce good results and use them without any change. Alternatively, new models may be developed for data traffic where the duration of idle or silent periods is not exponential (or geometric). However, these models can be expected to be much more complex than IPP and IBP and may have a limited use (such as ARPA for video source modeling).

3.2.4 Multimedia Applications

The term *multimedia* is used to refer to the representation, storage, retrieval, and dissemination of machine-processable information expressed in multiple media, such as text, voice, video, graphics, image, audio, and video. Examples of multimedia applications include teleconferencing, entertainment video, medical imaging, and advertising.

Multimedia applications differ from unimedia applications in various ways. Various multimedia applications involve a group of users and require point-to-multipoint or multipoint-to-multipoint connections. In general, they impose strict real-time performance requirements on the network. While employing different types of media, most multimedia applications generate large amounts of bit streams. Table 3.5 illustrates the bit rate requirements of some multimedia applications as given in [79].

Various challenges to providing multimedia services in B-ISDNs include the following:

Table 3.5
Bit Rate Requirements of Some Multimedia Applications

Multimedia Application	*Bit Rate*
Voice-annotated text	32.3 kbps
Voice 32 kbps	
One page text per 30 sec (1.24 x (8/30) kbps)	
Voice-annotated office image (e.g., insurance claim processing)	37.3 kbps
Voice 32 kbps	
One form per 90 sec (60 x (8/90) kbps)	
Voice-annotated high-resolution image (e.g., medical diagnosis)	85.3 kbps
Voice 32 kbps	
One image per 90 sec* [(2,000 × 2,000 × 12/(10 × 90 × 1,000) kbps]	
CD-quality sound and office system (e.g., library systems)	392 kbps
Audio (384 kbps)	
One image per 60 sec* [(2,000 × 2,000 × 12/(10 × 60 × 1,000) kbps]	
Complex Teleconference	325 kbps
Voice (32 kbps)	
1.240K text page per 60 sec (1.24 × (8/60) kbps)	
35K graphics per 60 sec (35 x (8/60) kbps)	
Two high-resolution 60K images per 60 sec	
Two low-quality video windows (128 kbps)	
Video distribution system	1,484 kbps
CD-quality sound (1,100 kbps)	
VCR-quality video (384 kbps)	

*10:1 compression ratio; 2,000 × 2,000 resolution; 12 bits/pixel.

- Multiple call setup;
- Real-time control of integrated real-time services such as video, audio and voice;
- Synchronization between a group of users;
- Dynamic allocation of network resources that take into consideration the different service requirements of each application integrated in the service;
- Efficient multipoint transfer;
- Synchronization of different applications, such as between video and audio.

Managing communication between a group of users is much more complex than point-to-point connections. Call setup between more than two users can be performed either by a sequence of point-to-point connections or as a multiparty connection. The latter is preferred, since it would result in quicker connection establishment, but it is more complex to design. The problem of synchronization in real-time applications has been studied for years and various solutions have been proposed for point-to-point connections. The extension of such techniques to more than two communicating entities is not a trivial task. A bit stream generated by a source is distributed to a group of users. Efficient handling of such traffic requires efficient distribution capabilities in the network. Most switch fabrics proposed for ATM have limited multicasting capabilities. Temporal relationships among various applications required in multimedia services must be maintained by the network. In an asynchronous network, synchronizing different types of applications to each other is a fairly complex task due to random network delays.

Traffic models of various applications that are put together in multimedia services are presented throughout this section. The extension of these models to characterize the integrated environment of a multimedia service is not a trivial task and has not yet been satisfactorily addressed in the literature.

3.3 QOS METRICS IN ATM NETWORKS

ATM networks are connection-oriented packet-switching networks. Accordingly, the QOS metrics in ATM networks can be categorized into two classes. The first class includes the call control parameters associated with connection-oriented networks, and the second class is the set of information transfer parameters defined in packet networks.

3.3.1 Call Control Parameters

There are mainly three control metrics of interest in connection-oriented networks: connection setup delay, connection release delay, and connection acceptance probability. The connection setup delay is the time interval between the call setup message transfer and the call setup acknowledge message transfer, excluding the called user's response time. This metric is not an ATM-specific parameter and is mainly determined by the processing delays at various signal transfer points in the network. ITU-T Recommendation I.352 [102] defines the provisional values for the connection setup delay in 64-kbps ISDN exchanges for the longest reference connection of 27,500 km. Assuming that the processing delays at ATM exchanges are similar to the processing delays in ISDN exchanges, the mean setup delay should be less than 4,500 ms, with 95% of the delay values being less than 8,350 ms.

The connection release delay is the time interval between the call release message transfer and the receipt of the call release acknowledge message. The processing delays at the signal transfer points do not contribute to the value of this metric, which is also independent of ATM. Similar to the connection setup delay values, the connection release delay values for ISDN defined in [102] can be used for ATM networks. In this case, the mean release delay in ATM networks should be less than 300 ms, with 95% of the delay values being less than 850 ms.

The connection acceptance probability is the proportion of accepted calls over a long period of time, which is also referred to as the *blocking probability* in telephony networks. This is one of the most important performance metrics used in allocating the network resources (i.e., link speeds, network topology, and switch capacities). Techniques used for network dimensioning have been developed over several decades and have been used satisfactorily for both circuit-switched (i.e., telephony) and packet-switched networks. However, with the introduction of B-ISDN, there is a new uncertainty with regard to future traffic and service characteristics, which cannot be predicted accurately at this phase of the evolution. Nevertheless, attempts have been made to estimate the holding times and call attempt rates of various B-ISDN applications for initial planning. With λ and $1/\mu$ respectively denoting the call attempt rate and the average call holding time (i.e., duration of time the connection is active), Table 3.6 illustrates an early estimation of a traffic reference model for a B-ISDN [32].

A similar study was reported in [9] on the number of busy hour call attempts (BHCA) in a metropolitan area, as illustrated in Table 3.7. The BHCAs in this table are defined as (λ call attempts/sec) $\times$ 3,600 sec/hour.

3.3.2 Information Transfer Parameters

There is a set of information transfer parameters requested by applications and used by the network to determine whether or not the connection can be admitted, finding a path that can support its service requirements, and so forth. Such parameters in ATM networks include cell information field bit error ratio (BER), cell loss ratio (CLR), cell insertion ratio (CIR), end-to-end cell transfer delay (CTD), cell delay variation (CDV), and skew. Not all of these parameters may be used by every application. Instead, the set of parameters defined should provide enough flexibility to transform application requirements to ATM-specific metrics. For example, an application's service requirement may be error-free for 5 minutes. The definition of error-free, however, may differ from one application to another. A cell that arrives at its destination after an end-to-end delay of more than 20 ms may be considered in error for voice services, whereas delay may not be a part of error-free definition for a data service. The key issues in characterizing the set of information transfer parameters of connections include:

- How to express parameters in a form that is meaningful to applications;
- How to relate these to networkwide mechanisms;
- How to deal with the range of implementation-specific mechanisms that affect individual nodes' contributions to the overall (i.e., end-to-end) QOS of the connection without specification of node architecture or special features;
- Tradeoff of complexity for precision and accuracy.

Table 3.6
B-ISDN Traffic Reference Model

User Class	*Application*	*Peak Rate (bps)*	*Burstiness*	*Burst Length (sec)*	*Holding Time (sec)*	*Call Arrival Rate (calls/sec)*
Analog	Telephony	64K	1	CBR	100	0.0008
Narrowband	Telephony	64K	1	CBR	100	0.001
residence	Document retrieval	64K	1	CBR	300	0.0001
Narrowband	Telephony	64K	1	CBR	100	0.0019
business	Document retrieval	64K	1	CBR	300	0.00016
	Text	64K	1	CBR	8	0.000163
	Fax	64K	1	CBR	20	0.000175
	Data on demand	64K	1	CBR	60	0.000966
	File transfer	64K	1	CBR	2	0.00005
Narrowband	Telephony	64K	1	CBR	100	0.0218
private automatic	Document retrieval	64K	1	CBR	300	0.00243
branch exchange	Text	64K	1	CBR	8	0.0045
	Fax	64K	1	CBR	20	0.00255
	Data on demand	64K	1	CBR	60	0.01333
	File transfer	64K	1	CBR	2	0.00075
Broadband	Telephony	64K	1	100	100	0.001
residence	Video telephony	10M	5	1	100	0.0002
	Document retrieval	64K	200	0.25	300	0.000166
	Video retrieval	10M	5	10	540	0.000055
Broadband	Telephony	64K	1	100	100	0.004
business	Video telephony	10M	5	1	100	0.0002
	Document retrieval	64K	200	0.25	300	0.000833
	Video retrieval	10M	5	10	180	0.000555
	Color fax	2M	1	3	3	0.00333
	Data on demand	64K	200	40 ms	30	0.00666
	File transfer	2M	1	1	1	0.003
Broadband private	Telephony	64K	1	100	100	0.045
automatic branch	Video telephony	10M	5	1	100	0.001
exchange	Document retrieval	64K	200	0.25	300	0.001666
	Video retrieval	10M	5	10	180	0.002222
	Color fax	2M	1	3	3	0.003333
	Data on demand	64K	200	40 ms	30	0.02
	File transfer	2M	1	1	1	0.003
Broadband service	Document retrieval	64K	1	300	300	0.22466
center	Narrowband document retrieval	64K	200	0.25	300	0.011
	Broadband video retrieval	10M	5	1	480	0.004854

Burstiness = Peak bit rate/average bit rate.

Table 3.7
Number of Busy Hour Call Attempts: a Traffic Model for a B-ISDN Scenario in a Metropolitan Area

Teleservices	*Residential 20,000 Users*	*Small Business 4,700 Users*	*Medium Business 150 Users*	*Large Business 50 Users*
Telephony	2.1	5.0	40	200
Videophone	0.65	1.5	10	50
Motion videotex	0.2	1.0	6	20
Video retrieval	0.6	0.5	3	10
Telefax	—	2.3	20	100
Videotex	0.23	1.2	7	20
Teletex	—	1.1	6	10
Color fax	—	0.7	8.4	42
Interactive data	0.2	6.0	50	200
Low-speed file transfer	0.43	1.0	10	50
High-speed file transfer	—	3.0	20	100
CAD/CAM	—	60	10	20
TV	0.8	0.6	0.6	0.6
HDTV	0.6	0.5	0.5	0.5
Hi-fi distribution	0.7	0.9	0.9	0.9

3.3.2.1 Bit Error Ratio

BER is defined as the ratio of the bit errors in the information field to the total number of bits transmitted in the information field. This is not an ATM-specific metric in that it depends mainly on the transmission system. Compared to existing networks, BER in ATM is expected to be much smaller due to the introduction of fiber technology into the transmission medium. Table 3.8 pre-

Table 3.8
Recommended BER Values for Some VBR B-ISDN Applications

Application	*Bit Rate*	*BER**	*BER†*
Videophone	2 Mbps	3×10^{-11}	1.3×10^{-6}
Video conference	5 Mbps	10^{-11}	1.8×10^{-6}
TV distribution	20–50 Mbps	3×10^{-13}	6×10^{-7}
MPEG1	1.5 Mbps	4×10^{-11}	2.5×10^{-6}
MPEG2	10 Mbps	6×10^{-12}	1.5×10^{-6}

*Without error handling in AAL.
†Single-bit error correction on cell basis and additional cell loss correction in AAL.

sents the early recommendations for the bit error rates of various B-ISDN services (cf. [102]).

Since there is no communication system capable of providing error-free communication channels, bit errors are also expected to occur in ATM networks. However, as the BER characteristics of transmission system improve, together with the small ATM payload, link-by-link error protection for corrupted cells due to bit errors is omitted without significant impact. Assuming that bit errors occur randomly, the probability that there is no bit error in the 48-byte (384 bits) ATM cell payload on a link is $(1 - \text{BER})^{384}$. For example, with BER = 10^{-6}, this probability is equal to 0.99617, which increases to 0.999996 with BER = 10^{-9}.

3.3.2.2 Cell Loss Ratio

CLR is the ratio of the number of lost cells to the total number of cells sent by a user within a specified time interval. CLR is an ATM-specific metric and has a significant impact on the QOS provided to users. All AALs, except AAL 5, include cell sequence numbers to detect lost cells at the receiver. CLR objectives for some B-ISDN services based on the ITU-T IVS Baseline Document [106] are shown in Table 3.9. Cells are lost in the network for two reasons: buffer overflows and bit errors in the cell header which can be detected but cannot be corrected.

Due to the randomness of the traffic, it is possible that a cell arriving at a switching node may find the buffer full and be lost. Although this probability can, at least in theory, be controlled to negligible values, there is always a nonzero probability of the buffers overflowing in ATM networks, regardless of

Table 3.9
CLR Objectives for Various B-ISDN Applications

Application	*Bit Rate*	*CLR**	*CLR†*
Videophone‡	64 kbps–2 Mbps	10^{-8}	8×10^{-6}
Videophone§	2 Mbps	10^{-8}	8×10^{-6}
Video conference§	5 Mbps	4×10^{-9}	5×10^{-6}
TV distribution§	20–50 Mbps	10^{-10}	8×10^{-7}
MPEG1§	1.5 Mbps	10^{-8}	9.5×10^{-6}
MPEG2§	10 Mbps	2×10^{-9}	4×10^{-6}

*Without error handling in AAL.
†Single-bit error correction on cell basis and additional cell loss correction in AAL.
‡CBR service.
§VBR service; bit rates are average values for VBR services.

buffer sizes. It is also possible for the bit pattern in the cell header to change during the transmission. The ATM cell header includes the 8-bit HEC field, which provides single-bit error correction and a low-probability corrupted-cell delivery. The HEC field of the header is used at each node the cell passes through to ensure header integrity. There are four possible outcomes from this function. If there is no error in the header or there is an error that cannot be detected, the cell is sent to the next node along its path. If there is an error detected and it is possible to correct it, the cell is sent to the next node after the error is corrected. If an error is detected but the error cannot be corrected, the cell is discarded.

The effect of cell losses and actions taken on lost cells in ATM networks differs for different types of services. For example, video services are not tolerant of cell losses in which a lost cell may result in the loss of parts of a video frame, possibly resulting in the transmission integrity no longer being ensured. This, in turn, may cause service to be interrupted. Similarly, data services require low CLR values, whereas voice traffic may tolerate moderate cell losses.

For a given CLR, the average time between cell losses (ATBCL) is given in (3.4) and a sample set of values is given in Table 3.10.

$$\text{ATBCL} = \frac{53 \times 8 \times 10^{-7}}{\text{Encoding rate (Mbps)} \times \text{CLR}} \text{ sec} \tag{3.4}$$

Since cell losses occur more frequently as the encoding rate increases, extra countermeasures are required to be designed and taken in ATM networks, particularly for loss-sensitive applications. Cell loss requirements of applications are average figures and provide only partial information. For example, the average time between cell losses with a CLR of 10^{-6} is equal to 0.274 sec on a 155-Mbps link. The impact of having two consecutive cell losses in every 0.548 sec, on the average, has a more significant impact on the image quality compared to one cell loss in every 0.274 sec. Hence, we are better off if cell losses are distributed uniformly during the duration of a video service. On the contrary, we would like to have all cells following a lost cell to be lost in a data service, since the whole frame may need to be retransmitted by the source when it does not arrive at the receiver correctly.

Table 3.10
Average Time Between Cell Losses with CLR = 10^{-6}

Encoding rate	64 kbps	1.544 Mbps	45 Mbps	155 Mbps	620 Mbps
ATBCL	662.5 sec	27.5 sec	0.942 sec	0.274 sec	0.068 sec

3.3.2.3 Cell Insertion Ratio

As discussed earlier, it is possible that an error occurred at the header may not be detected by header error checking. In this case, if an undetected change in the bit pattern of a cell header corresponds to the address of another connection, then the cell is misrouted to a wrong destination. CIR is then defined as the ratio of cells delivered to a wrong destination to the total number of cells sent. This type of error is much more difficult to deal with than cell losses.

Cells arriving unexpectedly may cause loss of terminal synchronization for some types of services. In addition, misrouted cells increase the traffic flow on links other than links used by the connection they belong to. If those other links are already highly used, then they may not have enough bit rate to carry these misrouted cells, causing degradation of service to their existing traffic. Table 3.11 lists the CIR values of some B-ISDN applications as defined by the Research on Advanced Communication in Europe (RACE) consortium 1022 (Technology for ATM).

3.3.2.4 Cell Transfer Delay

Cell transfer delay between two points in the network is defined as the elapsed time from which the first bit of a cell leaves the first observation point to the time the last bit of the cell passes the second observation point. The two points are the two network interfaces at each end node if the considered metric is the end-to-end delay. Various factors that determine the cell transfer delay in ATM networks are reviewed next.

Coding Delay

Coding delay is the time required to convert a nondigital signal to digital bit patterns. For example, an analog signal generated by a voice, audio, or video source is sampled and digitized for transmission in the network. This delay

Table 3.11
Recommended CIR Values for Some B-ISDN Applications

Service	*CIR*
Telephony	10^{-6}
Data transmission	10^{-6}
Distributive computing	10^{-6}
Hi-fi sound	10^{-7}
Remote process control	10^{-6}

depends on the coding algorithm and the software/hardware used to do the coding. Coding delay may also occur in applications that do not require signal transformation. For example, large volumes of data may be compressed before being transmitted, resulting in coding delays.

Packetization Delay

This type of delay incurs while accumulating the required number of bits to form an ATM cell, which depends on the type of adaptation layer used and the source bit rate. For example, 64-kbps PCM voice transmitted using AAL 1 would introduce $47 \times 8/64$-ms packetization delay, whereas it takes only $44 \times 8/10$ μs to fill a cell in a 10-Mbps MPEG-II video transmission using AAL 2, assuming AAL 2 payload is 44 bytes (i.e., 4-byte AAL overhead).

Propagation Delay

Propagation delay occurs due to the speed of the light in the transmission medium and depends on the distance between the source and the destination. Typical values of propagation delay in different media are given in Table 3.12.

Transmission Delay

Before any processing can be done on it, all 424 bits of a cell have to arrive from the transmission link. This delay depends on the speed of the link and becomes negligible as the transmission speeds increase. For example, a cell on a T3 link would take about 9.2 μs to transmit, which reduces to 2.73 μs on an OC-3 link.

Switching Delay

The switching time is the total delay it takes for a cell to traverse the switch. It depends on the internal switch speed and the amount of overhead added to

Table 3.12
Propagation Delays in Different Media

Transmission Medium	*Propagation Delay*
Coax cable	4 μs/km
Optical-fiber cable	5 μs/km
Submarine coax cable	6 μs/km
Satellite (14,000-km altitude)	110 ms
Satellite (36,000-km altitude)	360 ms

the cell for routing within the switch. For example, a Banyan network with 256 input (output) ports built up using 2 × 2 switching elements would require 1-byte overhead per cell for self-routing within the switch. In a typical ATM switch, the total time it takes to switch a cell consists of a table lookup to determine the output port and the time it takes to traverse the switch from the input to the output port. The latter, in addition to the speed of the internal links, depends on the contention resolution technique used in the switch fabric.

Queuing Delay

ATM switches may have buffers at the input ports, within the switch, at the output ports, or a combination of input, internal, and output buffering (discussed in Chapter 5). Output buffering is used to resolve output contention where more than one cell arrives at an output port during the transmission time of one cell. Internal contention may occur within the switch if cells arriving from different input ports attempt to use the same switch elements simultaneously within the switch fabric. Internal buffers are used to resolve this type of contention. Finally, input buffers may be used to smooth out the effect of head-of-line blocking, which occurs when the cell at the head of the input port cannot be switched due to unavailability of resources within the switch fabric.

Reassembly Delay

For various applications, several cells of a frame are collected at the receiver before they are passed to the application. For example, a number of voice cells may be collected before the voice frames are started, to be played out to provide a continuous 64-kbps constant rate of service. Similarly, a CS protocol data unit for a data application may be composed of a number of cells and may require all of its cells to arrive before they can be forwarded to the network or transport layer.

Table 3.13 illustrates average end-to-end delay values of some B-ISDN services, as well as other performance metrics of interest as defined by RACE consortium 1022.

3.3.2.5 Cell Delay Variation (Jitter)

The end-to-end delay of the ith cell is given as $D + W_i$, where D is a constant that includes the propagation and transmission (plus the switching) delays, and W_i is the random delay component that arises out of buffering within the network. The interarrival times of cells at the receiver is given by

$$(D + W_{i+1}) - (D + W_i) = \delta \tag{3.5}$$

Table 3.13
Some Service Attributes for B-ISDN Applications

Service	*BER*	*CLR*	*Delay (ms)*
Telephony	10^{-7}	10^{-3}	
Without echo cancelers			<25
With echo cancelers			<500
Data transmission	10^{-7}	10^{-6}	1,000
Distributive computing	10^{-7}	10^{-6}	50
Hi-fi sound	10^{-5}	10^{-7}	1,000
Remote process control	10^{-5}	10^{-3}	1,000

Ideally, interarrival times of cells at the receiver are equal to the interexit times of cells, which is the case if $W_{i+1} = W_i$. However, due to randomness in the network, the W_i is a random variable and is not constant.

Various definitions of CDV used in the literature include:

- Variance of the transmission delay of a connection; that is, $E\{(W_i - E[W_i])^2\}$.
- Difference between the values of the transit delay of the cells of a connection: $W_{i+1} - W_i$; that is, $\Pr\{W_{i+1} - W_i > w\}$.

  Instant delay variation from the mean; that is, $\Pr\{W_{i+1} - E[W_i] > w\}$.

The following two definitions of CDV follow from related ITU-T recommendations and ATM Forum specifications: 1-point CDV and 2-point CDV.

The 1-point CDV for cell k, y_k, describes the variability in the pattern of cell arrival events observed at a single measurement point with reference to the connection's peak rate (i.e., $1/T$). Accordingly, the difference between the cell's reference arrival time (c_k) and the actual arrival time (a_k) at a measurement point (i.e., $y_k = c_k - a_k$). The reference arrival time is defined as follows:

$$c_0 = a_0 = 0$$

$$c_{k+1} = \begin{array}{l} c_k + T \text{ if } c_k \geq a_k \\ a_k + T \text{ otherwise} \end{array}$$

Positive values of the 1-point CDV correspond to cell clumping, whereas negative values correspond to gaps in the cell stream. The reference time defined above eliminates the effect of gaps and provides a measurement of cell clumping.

The 2-point CDV for cell k (v_k) describes variability in the pattern of cell arrival events observed at the output of a connection portion (MP_2) with reference to the pattern of the corresponding events observed at the input to the connection portion (MP_1). That is, v_k is the difference between the absolute cell transfer delay of cell k (x_k) between the two MPs and a defined cell transfer delay d between MP_1 and MP_2: $v_k = x_k - d$.

Table 3.14 presents the delay and end-to-end delay variation requirements of various video services (cf. [79]).

Table 3.14
Delay and Delay Variation Objectives for Two-Way Session Audio and Video Services

Application	*Delay (ms)*	*Jitter (ms)*
64-kbps video conference	300	130
1.5-Mbps MPEG NTSC video	5	6.5
20-Mbps HDTV video	0.8	1
16-kbps compressed voice	30	130
256-kbps MPEG voice	7	9.1

Jitter can be controlled at the receiver at the expense of larger buffers and increased delays. In particular, it is possible to temporarily store the arriving cells in a jitter removal buffer so that the departure rates of cells from the buffer is close to the interexit times of cells at the receiver, as illustrated in Figure 3.14. In order to remove the jitter completely, the buffer size at the decoder is determined from the maximum delay that a cell may incur in the network. However, the delay requirements of applications are required to be considered as well in determining the buffer size. In particular, it may not always be possible to delay cells to compensate for the maximum network delay, in which case it is preferable to drop cells delayed more than an acceptable value rather than attempting to handle the CDV.

3.3.2.6 Skew

Skew is defined as the difference in the presentation times of two related objects (i.e., video stream and audio stream). Coarse skew represents gross delays between an image and accompanying voice, whereas fine skew represents the

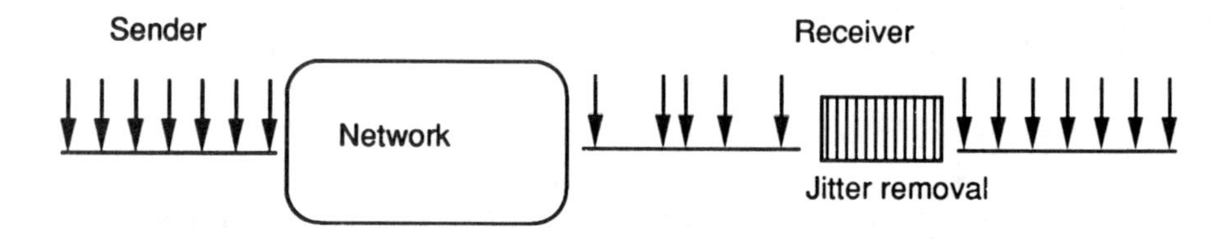

Figure 3.14 Jitter handling with buffering.

time delays between lip motion and voice. Skew objectives of various multimedia applications are given in Table 3.15 (cf. [79]).

Table 3.15
Skew Objectives in Multimedia Applications

Application	*Skew Objective*
Audio + text or still image (one-way session)	Coarse skew < 1 sec
Audio + video (multipoint-to-multipoint sessions)	Coarse skew < 200 ms Fine skew: audio in advance of video < 20 ms video in advance of audio < 120 ms
Complex teleconferencing Audio + video + still image + text	Coarse skew < 200 ms Fine skew: audio in advance of video < 20 ms video in advance of audio < 120 ms

3.3.3 Information Transfer Metrics of Various B-ISDN Applications

In this section, we discuss the QOS specifics of different types of B-ISDN applications.

3.3.3.1 Voice

A lost cell in CBR voice includes 47 bytes of voice information. With 64-kbps PCM coding in which the frame length is 1 byte, this is equivalent to 5.785 ms of voice. The effect of a lost cell, then, is a click heard at the receiver. To minimize the effect of lost voice cells, the block-dropping technique may be used. In this scheme, two cells are used to transmit 94 bytes of voice frames. The most significant bits of the 94 bytes are collected into one cell and transmitted with a high CLP, whereas the least significant bits are transmitted in another cell with a low priority. The low-priority cells can then be discarded in the network during congestion. At the receiver, these bits can be partially recovered, resulting in a slight loss of voice quality.

Delay in echo-free voice services cannot be detected if only one person is speaking at a time. Round-trip delays of 600 ms to 1.2 sec remain almost unnoticed in a telephone conversation (cf. [15,31]). However, these relatively large delay figures affect the overall conversational dynamics (e.g., more simultaneous talks and longer silent periods). In a typical conversation between two subscribers over a public network, 2-wire and 4-wire lines are respectively used in subscribers' loops and in the network. Different portions of the network are coupled together by differential transformers, often called *hybrids.* All significant echoes in a telephone network arise due to the impedance mismatch at the hybrids. The speech signal, say from user A to user B, is transmitted across hybrid A, the network, and hybrid B to user B. Part of the signal reflects back from hybrid B to user A so that, after the propagation delay, user A listens to a weakened part of his or her own speech, referred to as *talker's echo.* Similarly, some of the talker's echo is reflected from hybrid A back to user B and is referred to as *listener's echo.* As the end-to-end delays increase, echoes are heard as distinct signals and become more noticeable. The end-to-end delay for echo-free conversation, recommended by ITU-T, is limited to 25 ms. Beyond this limit, the echo perception cannot be suppressed and the use of echo cancelers is required.

In the case of ATM networks, coding and packetization delays are expected to constitute a large portion of the allowable 25-ms delay, requiring the use of echo cancelers even for connections over relatively short distances. A straightforward solution is to fill each cell only partially. For example, instead of 47 bytes, only a few bytes can be transmitted with each cell, thereby reducing the packetization delays. However, this would cause inefficient use of network resources. Another alternative is to fill a cell with voice frames of different sources. For example, with 2 voice frames carried in a cell from each source, a cell payload would carry traffic from 23 voice sources. Although this scheme solves the packetization delay problem and uses the cell payload efficiently, it also increases the complexity of the receiver design.

The jitter objective for voice services is about 130 ms and can be relatively easily removed by using buffers, as discussed previously.

3.3.3.2 Data

Data applications in general are intolerant of errors and data integrity (i.e., correct and insequence delivery) between the communication ends is required. Most data services other than connectionless data service require retransmission of data frames to guarantee correct delivery to the receiver. Data services take place between humans, humans and machines, or machines. When human interaction is involved, data applications are intolerant of large delays, whereas services between machines can tolerate moderately large delay values. For example, a user accessing a teller machine to do a banking transaction demands a response time of a few seconds, whereas account updates that take place between different information centers of a bank do not impose a strict delay constraint. Delay jitter is not a big concern in data services, since no synchronization or constant play-out rates are required at the receiver.

3.3.3.3 Video

Unlike voice, a primary video frame contains a large amount of information and is transmitted over a number of ATM cells. Losing a cell in the middle of a frame may cause the phase alignment to be lost for the rest of the frame and, depending on the synchronization scheme used, a cell lost may corrupt a large part of the frame. In general, an encoded video frame consists of a number of macro frames. The number and the contents of macro frames depend on the coding technique used. In this framework, the amount of distortion caused by a lost cell depends on the type of information included in the macro frame that the lost cell belongs to: a short flush, a distorted number of lines, or a soft cloud that spreads over several frames and diminishes gradually.

As far as the reconstruction of frames at the receiver is concerned, macro frames of high-resolution layers are not as important as macro frames of low-resolution layers. Hence, it is possible to transmit cells of different macro frames at different CLPs in the network. Since the ATM cell header defines only one priority bit, the cells can only be prioritized as high- and low-priority cells. Synchronization signals and the bit stream necessary to maintain minimum video quality may then be transmitted using high-priority CLP, whereas portions of the frame used to improve the image quality beyond the level provided by high-priority cells alone can be transmitted with low priority.

Cell losses occur in ATM networks in bursts. That is, given that a cell is lost, the probability that the consecutive cell is lost is higher than the average cell loss probability. Even though it may be possible to compensate for the loss of one cell in, say, every second at the receiver, the procedure is much more complicated if two consecutive cells are lost in every 2 sec. As discussed previously, the average cell loss rate required by video applications may not be meaningful if special care is not given in their calculation. For example, let us consider a framework that can compensate for, with an acceptable degradation in the image quality, cell loss rates of 10^{-5}. It might be necessary to request stricter service from the network, say, a cell loss rate of 10^{-7}, in order to minimize the effect of consecutive cell losses. Unlike data, cell loss compensation by retransmitting full frames (or parts of a frame) is not a viable alternative in ATM networks due to strict delay constraints.

Reducing the effect of a lost cell on the image quality by reconstruction depends on its contents. For example, in DPCM for voice coding, lost cells can be filled with all zeros, concealing the error partially. Recalling that VBR video uses AAL 2, the SAR headers at each cell include the sequence number, making it a relatively easy task to detect a lost cell. Once detected, there are mainly two approaches to error recovery [52]. Error control coding offers perfect recovery from errors until the number of errors exceeds the limit of the error code [61]. Use of this technique to recover from errors increases the overhead transmitted per frame and it is processing intensive. In error concealment by visual redundancy, perfect recovery may not be possible. With this scheme, the exact location of a lost cell within the frame, referred to as *erasures,* is approximately concealed by spatial and temporal interpolation or statistical image reconstruction methods [53,23].

Clock frequencies at different nodes in ATM networks may not be the same, causing the receiver to expect data at a faster or slower rate than being transmitted at the source, as illustrated in Figure 3.15 (cf. [51]). In the case of a faster clock at the receiver, cells can be considered lost by the receiver although they are arriving within acceptable delay limits. Similarly, a slower clock at the receiver causes the allowance of much more time than needed for arriving cells, which in turn increases the chances of buffer overflow. This problem can, to some extent, be solved by monitoring the input buffer level or using the time information transmitted by the sender and adjusting the clock. However, the problem is complicated by the fact that cells carrying time stamps may be lost.

Cell delay jitter complicates the decoder synchronization. It makes it difficult to derive a stable clock out of cell arrivals in CBR services in ATM networks because the jitter component is superposed on cell arrival times. In VBR coding, it is necessary to use time stamps where the sender transmits clock information embedded in the generated cell stream. Increasing the delay (up to levels that do not degrade the QOS) of cells to minimize the delay jitter is preferable to unpredictable arrival times of cells to the frame buffers. This maximal value of delay may be tolerable for some video services and may not be for others. In particular, transmission delays in one-way sessions (i.e., broadcast services) are not as important as they are in multiway sessions (i.e., video conferencing). In the latter, long delays impede the information exchange, and cell losses are preferred in order to achieve lower delays.

3.3.3.4 Multimedia Services

The service requirements of individual applications integrated in a multimedia service have already been discussed. Multimedia applications impose a new

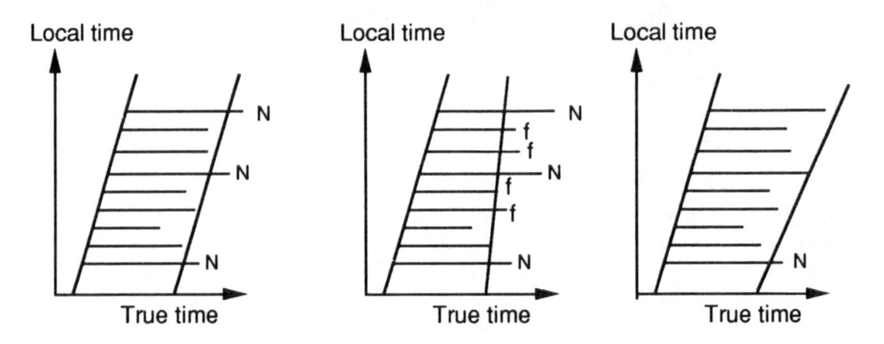

Figure 3.15 The effect of faster or slower clock at the receiver compared to sender clock.

requirement for synchronization among various information types, which can range from coarse synchronization such as sequencing the transmission of various objects (e.g., image followed by audio, followed by data, followed by image) to a more precise, fine synchronization such as synchronization of voice to the speaker's lip motion. For example, synchronization of voice and video cells (i.e., lip synchronization) appears to be acceptable for video-to-voice lags of –90 to 120 ms.

Multimedia services often take place between a group of users and require multipoint-to-multipoint connections. Although the problems are solved for connections that take place between two users, the clock recovery and synchronization among a group of users remain open issues. The delay requirements in distribution services in which the communication is bidirectional are relatively stricter than they are in unidirectional services.

Figure 3.16 summarizes the CDV and cell loss requirements of various applications that may be integrated in a multimedia service (cf. [95]). The simplest approach from a traffic management point of view is to make no differentiation between the service requirements of individual applications at the transport layer and assign the most stringent service requirement to all. However, this approach would require a larger bit rate in the network than is necessary and would waste transmission capacity.

Another approach is to establish different connections for each application according to its service requirements. This would increase the complexity of network management and application design. Furthermore, the jitter and skewing with this scheme are more difficult to handle. Alternatively, the complexity of this scheme can be decreased by appropriately partitioning the set of service demands of individual connections into a manageable number of classes, where

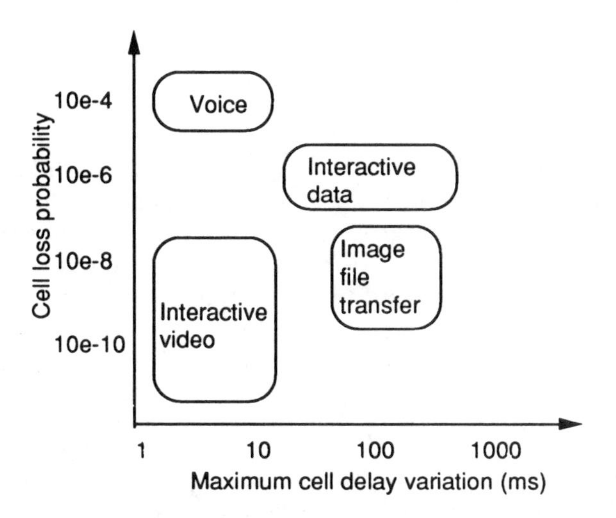

Figure 3.16 Performance requirements of various multimedia service component applications.

each class includes one or more types of applications with similar characteristics.

3.4 TRAFFIC MODELS

In this section, we will present different traffic models proposed for video and voice.

3.4.1 Video Traffic Models

Video frames are generated at a constant rate defined by the play-out rate. Since the amount of data transmitted per frame varies due to intraframe and interframe coding, video applications generate traffic in a continuous manner at varying rates. Typically, frames of high-activity scenes and scene changes contain large amounts of data followed by frames that contain less data, and these subsequent frames can be predicted from the previous ones. Video is a relatively new service in communications networks, and its traffic characteristics are not well understood. The choice of the bit rate of a CBR video service (assuming that it can be varied at the encoder) is a relatively new research area. Video service is also quite different from voice or data in that the bit streams of the latter exhibit various types of correlations between consecutive frames. The understanding of such correlations is important in the analysis of transmission buffers in intermediate switches and in the design of various network control services, as discussed in Chapter 4.

In mathematical terms, the correlation coefficient for lag t, $C(t)$, is defined as follows:

$$C(t) = \frac{E\{[x(n) - E(x(n))]\,[x(n+t) - E(x(n))]\}}{\mathrm{Var}\{x(n)\}} \tag{3.6}$$

where $x(n)$ is the amount of information generated at the nth frame, $E(x(n))$ is the expected value of $x(n)$, and $\mathrm{Var}\{x(n)\}$ is the variance of $x(n)$.

The numerator $E\{[x(n) - E(x(n))]\,[x(n+t) - E(x(n))]\}$ is called the *covariance* of $x(n)$ and $x(n+t)$. $C(t)$ measures the degree of association between the amount of data transmitted in the nth and $(n+t)$th frames. We note that -1 $C(t)$ 1. If $C(t) > 0$, then $x(n)$ and $x(n+t)$ are said to be positively correlated. In this case, $x(n+t)$ increases (decreases) with increasing (decreasing) values of $x(n)$. Similarly, if $C(t) < 0$, then $x(n)$ and $x(n+t)$ are said to be negatively correlated. Then $x(n+t)$ increases (decreases) with decreasing (increasing) values of $x(n)$. The two random variables are said to be uncorrelated if $C(t) = 0$. We note that the correlation coefficient is a measure of the degree of linearity between two random variables. A value of $C(t)$ close to 1 or -1 indicates a high degree of linearity, whereas if $C(t)$ is equal (or close) to zero, it *only* indicates the absence of a linear relationship between the two random variables and does *not* preclude the possibility of some type of nonlinear relationship. For presentation purposes, the activity levels at video applications can be classified into two categories: uniform and nonuniform activity levels.

3.4.1.1 Uniform-Activity-Level Video Scenes

In applications with uniform-activity-level scenes, the change in the information content of consecutive frames is not significant. A typical application of this type is a video telephone, where the screen shows a person talking. In general, correlations in video services with uniform activity levels last for a short duration and decay exponentially with respect to time.

Continuous-State Autoregressive Model

Let $\lambda(n)$ denote the bit rate of the nth frame. The first-order autoregressive model proposed in [65] estimates the bit rate at the nth frame from the bit rate at the $(n - 1)$st frame to be

$$\lambda(n) = a\lambda(n - 1) + b\omega(n) \tag{3.7}$$

where a and b are constants and $\omega(n)$ is a Gaussian random variable with a mean m. The mean $E(\lambda)$ and the autocovariance of the bit rate $C(n)$ are equal to

$$E(\lambda) = bm/(1 - a); \quad C(n) = b^2a^n/(1 - a^2) \tag{3.8}$$

which can be used to determine the two unknown variables a and b from the measured values of the mean and autocovariance. The model is found to be quite accurate compared with the actual measurements [65]. However, it is not suitable for use in numerical/analytical studies of queuing models, but is used mostly in simulation studies.

Discrete-State, Continuous-Time Markov Models

An analytically tractable model of a video source with uniform activity levels is developed in [65]. Let P and L respectively denote the maximum and minimum bit rate generated by a video source. Furthermore, let the possible bit rates between P and L be uniformly quantized into M steps, with a constant step size $A = (P - L)/M$. Referring to each quantization step as a state, we have state i corresponding to the bit rate $\lambda(i)$, where

$$\lambda(i) = i^*A \quad 0 \le i \le M \tag{3.9}$$

Then the actual bit rate of the source is $\lambda(i) + L$. The transition rates between the states of the system are given as follows:

$$R(j, k) = \begin{cases} (M - i)\alpha & \text{if } j = i \text{ and } k = i + 1;\ 0 \le i \le M - 1 \\ i\beta & \text{if } j = i \text{ and } k = i - 1;\ 1 \le i \le M \\ 0 & \text{otherwise} \end{cases} \tag{3.10}$$

This choice of rates is based on the observation that if a video source is in a high-activity phase, then it is more likely to transit to a lower activity phase than to a higher activity phase. Similarly, if the source is in a low-activity phase, then it is more likely to transit to a higher activity phase than to a lower activity phase.

M is chosen arbitrarily. Larger values of M would result in more granularity in the quantization of bit rates, thereby increasing the accuracy of the model. However, as M becomes larger, the number of states and the time complexities of models used to analyze such sources would increase. Hence, there is a tradeoff between the model accuracy and its solution complexity.

Let π_i denote the steady-state probability that the Markov process is in state i, $0 \le i \le M$. π_i is the solution of the following set of equations:

$$(M - i)\alpha\pi_i = (i + 1)\beta\pi_{i+1} \quad i = 0, \ldots, M - 1 \tag{3.11}$$

$$\sum_{i=0}^{M} \pi_i = 1$$

It is easy to verify by substitution that π_i is given as follows:

$$\pi_i = \frac{M!}{i!(M - i)!}\{\alpha/(\alpha + \beta)\}^i\{\beta/(\alpha + \beta)\}^{M-i}, \quad i = 0, \ldots, M \tag{3.12}$$

The unknowns A, α, and β are estimated by matching the mean $E(\lambda)$, variance $\mathrm{Var}(\lambda)$, and autocovariance function $C(\tau)$ of the Markov process, obtained from the measured data using the following equations.

$$E(\lambda) = \sum_{i=0}^{M} \pi_i\lambda(i) = MAp \tag{3.13a}$$

where $p = \alpha/(\alpha + \beta)$.

$$\begin{aligned}\mathrm{Var}(\lambda) &= E(\lambda^2) - \{E(\lambda)\}^2 \\ &= MA^2p(1 - p)\end{aligned} \tag{3.13b}$$

where $E(\lambda^2)$ is the second moment of the bit rate

$$C(\tau) = \text{Var}(\lambda)\exp\{-(\alpha + \beta)\tau\} \tag{3.13c}$$

Discrete State, Discrete-Time Markov Models

Similar to the discrete-state, continuous-time Markov model, the discrete-state, discrete-time Markov model is defined over the states 0 to M, where state i corresponds to quantized bit rate level $\lambda(i)$. In this model, the time is discretized into fixed-length slots, with the slot length being equal to frame generation intervals (e.g., 1/30 sec). At the end of each slot, the process moves from state i to either state $I + 1$, with probability α_i, or state $I - 1$, with probability β_i, or stays at state i, with probability γ_i, such that $a_i + \beta_i + \gamma_i = 1$. State transition probabilities are then

$$p(j, k) = \begin{matrix} \alpha_i & \text{if } j = i \text{ and } k = i + 1;\ 0 \le i \le M - 1 \\ \beta_i & \text{if } j = i \text{ and } k = i - 1;\ 1 \le i \le M \\ \gamma_i & \text{otherwise} \end{matrix} \tag{3.14}$$

The transition probabilities in the discrete model are required to have similar ordering to that of transition rates in the continuous-time model to simulate the transitions from higher activity scenes to lower activity scenes and vice versa.

$$\begin{matrix} \alpha_i \ge \alpha_{i+1} & 0 \le i \le M - 1 \\ \beta_i \ge \beta_{i-1} & 1 \le i \le M \end{matrix} \tag{3.15}$$

M is chosen arbitrarily. For the values of α_i and β_i, various sets of choices have been proposed in [44,45]:

$$\begin{aligned} &\text{Set I: } \alpha_i = c(1 - i/M);\ \beta_i = ci/M;\ \gamma_i = 1 - c \\ &\text{Set II: } \alpha_i = c_1(1 - i/M);\ \beta_i = c_2 i/M;\ \gamma_i = 1 - c_1 + (c_1 - c_2)i/M \\ &\text{Set III: } \alpha_i = c_1(i/M)^2;\ \beta_i = c_2 i^2;\ \gamma_i = 1 - c_1(1 - M)^2 - c_2 i^2 \end{aligned} \tag{3.16}$$

where c, c_1, and c_2 are constants.

The first set corresponds to a case where the bit rate of a video application has a symmetrically shaped bit rate distribution. The other two sets of values provide a more general framework for shaping the bit rate distribution by appropriately selecting the values of c_1 and c_2.

Both Markovian models capture the fast-decaying short-term correlations in the bit rates of consecutive frames, which lasts on the order of a few hundred milliseconds. These models are not accurate for modeling video services with nonuniform activity levels, because of the following two factors.

- Transitions from state i are allowed only to its neighbor states $i + 1$ and $i - 1$, which take only small changes in the bit rates into consideration.
- Long-term correlations in scene changes that last on the order of few seconds cannot be captured with these models.

Next we present the models for nonuniform-activity-level video scenes.

3.4.1.2 Nonuniform-Activity-Level Video Scenes

In motion video, in addition to the short-term fast-decaying correlations, there is a long-term slow-decaying correlation in the amount of information generated per frame that occurs at times of scene changes. The following model is an extension of the two Markovian models proposed for uniform-activity-level scenes to simulate sudden changes in the bit rates and long-term correlation in nonuniform-activity-level scenes.

Two-Dimensional Continuous-Time Markov Model

The two-dimensional continuous-time Markov model proposed in [81] is a generalization of the model developed in [65] for uniform-activity-level scenes. Each dimension of the model can be viewed as the one-dimensional Markov chain discussed above. In two dimensions, it is now possible to model the small bit rate fluctuations in consecutive frames to include jumps to the higher or lower bit rates, thereby modeling the correlation at scene changes. Let A_h and A_l respectively denote the high-rate and low-rate quantization step sizes. In general, $N_1 + 1$ low-rate and $N_2 + 1$ high-rate levels are defined where state (i, j) corresponds to bit rate $(iA_h + jA_l)$, $0 \le i \le N_2$, $0 \le j \le N_1$. According to the validation tests reported in [65], a single video source can be accurately characterized with $N_1 = 1$, whereas N_2 is chosen arbitrarily but is large enough to scan all likely bit rates. As discussed previously in the context of one-dimensional Markovian models, both the accuracy and the complexity of the model increases as N_1 increases. Figure 3.17 is a pictorial representation of the two-dimensional Markov process for a single source.

The transition rates between the states of the two-dimensional Markov chain are defined as follows:

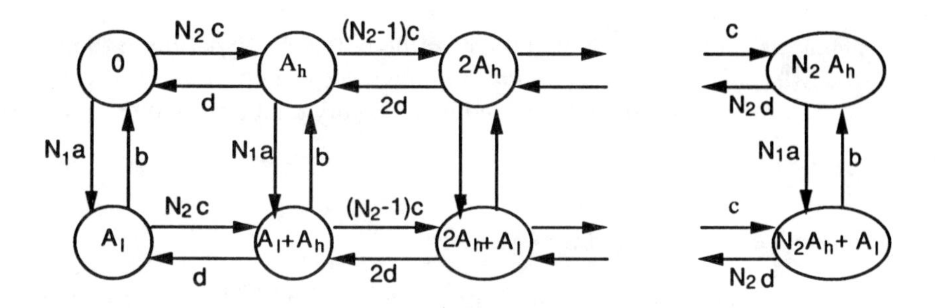

Figure 3.17 Two-dimensional continuous-time Markov model.

$$R\{(k_1, m_1), (k_2, m_2)\} = \begin{cases} (N_1 - j)a & \text{if } k_1 = k_2 = i \text{ and } m_1 = j;\ m_2 = j + 1; \\ & 0 \le i \le N_2;\ 0 \le j \le N_1 - 1 \\ jb & \text{if } k_1 = k_2 = i \text{ and } m_1 = j;\ m_2 = j - 1; \\ & 0 \le i \le N_2;\ 1 \le j \le N_1 \\ (N_2 - i)c & \text{if } k_1 = i;\ k_2 = i + 1 \text{ and } m_1 = m_2 = j; \\ & 0 \le i \le N_2 - 1;\ 0 \le j \le N_1 \\ id & \text{if } k_1 = i;\ k_2 = i - 1 \text{ and } m_1 = m_2 = j; \\ & 1 \le i \le N_2;\ 0 \le j \le N_1 \end{cases} \tag{3.17}$$

With $N_1 = 1$, the parameters used to characterize the Markov process in terms of source parameters are

$1/d$ = average time spent in the high activity level;
$1/c$ = average time spent in the low activity level;
q = fraction of time spent in the high activity level, $q = c/(c + d)$;
γ = ratio of the average data rate in the high activity level to that in the low activity level;
$C(0)$ = conditional variance;
$C(\tau)$ = conditional covariance function;
λ = the overall mean bit rate

Then

$$\begin{aligned} C(0) &= N_1 p(1 - p)A_1^2, \text{ where } p = a/(a + b) \\ C(\tau) &= C(0)\exp\{-(a + b)\tau\} \\ \gamma &= \{N_1 pA_1 + A_h\}/N_1 pA_1 \\ \lambda &= N_1 pA_1 + qA_h \end{aligned} \tag{3.18}$$

From the actual data, a set of statistics can be collected to estimate the values of q, $1/d$, $C(0)$, $C(\tau)$, γ, and λ. Using the above set of equations, the values of A_l, A_h, a, and b can then be determined, completely characterizing the Markov process.

Autoregressive Moving Average Process

In addition to the two types of correlations modeled with the two-dimensional Markov chains, the covariance in the bit streams of video sources is observed to further recorrelate at lags. The significance of this recorrelation is summarized as follows [40]:

- The autocovariance function does not vanish for lags $k \neq 0$, even when a large number of video codecs are superposed. Hence, the assumption that the superposition of a large number of sources behaves like a Poisson is not valid in this environment.
- The effect of correlation of the arrival process depends on the utilization of the queue. With utilization levels of up to 61%, its effect is observed to be negligible, whereas its effect on queuing behavior becomes significant at 76% utilization. At higher loads, it is further observed that a slight change in autocovariance causes significant changes in the behavior of performance metrics of interest (i.e., delay and loss probability).

The autoregressive moving average process (ARMA) model is proposed to take into account this recorrelation in addition to capturing the two types of correlations that occur in nonuniform-activity-level scenes. Let $\{X_n\}$ be a sequence of random variables representing the video cell arrival process. Assuming that

$$E[X_n] = E[X_0] \text{ and Covariance}(X_{n+k}, X_n) = \text{Covariance}(X_k, X_0) = R(k)$$

the autocovariance function $R(k)$ depends only on lag k. Characterizing the arrival statistics of ATM cells using the ARMA model is a three-step procedure [40]:

1. Measurement step. The long-term mean, variance, and autocovariance are estimated from the cell output of the coder.
2. Parameter estimation. The ARMA process consists of a finite order filter, a recursive filter, and memoryless nonlinearity. The parameters of the mathematical model of ARMA are estimated from the measured data.
3. Transfer function. This produces the cell interarrival sequence by taking white noise into consideration.

The ARMA arrival processes are used in Monte Carlo simulations to estimate the probability distribution function of the queuing delay and the mean and variance of the interdeparture time seen by an arriving cell. However, these models cannot be used in the numerical and analytical analysis of queues.

3.4.2 VBR Voice Source Models

An active period of a voice source corresponds to a talk spurt, whereas a silent period corresponds to durations during which the speaker pauses to breathe, gather thoughts, or listen to the other speaker. The silent periods constitute 60% to 65% of the transmission time of voice calls in each direction. More specifically, the average active and silent periods are measured to be respectively equal to 352 and 650 ms (cf. [82]). Furthermore, in a normal conversation, the active period fits the exponential distribution reasonably well, while the duration of the silent periods is less well approximated by the exponential distribution (cf. [14]). Nevertheless, the most frequently used models of voice sources in the literature assume that the duration of both the active and silent periods are exponentially distributed. The accuracy of this model has been validated on a single queue where a number of voice sources are multiplexed. It has been found that the accuracy of the two-state model with exponentially distributed durations is fairly good if more than 25 voice sources are multiplexed, whereas the model is less suitable if the number of sources multiplexed is less than 10.

3.4.2.1 Interrupted Poisson Process

The IPP is a Poisson process that is alternatively turned on for an exponentially distributed period of time (active period) and turned off for another independent exponentially distributed period of time (silent period). During the active period, the interarrival times of packets are exponentially distributed (i.e., in a Poisson manner), while no packets are generated during the silent period. All processes are assumed to be mutually independent.

Let $1/\sigma_A$, $1/\sigma_S$, and λ respectively denote the average duration of the active and silent periods and the packet-generation rate during the active period. Then the probability distribution function of the active period $\Pr\{X \le t\}$, the silent period $\Pr\{Y \le t\}$, and the interarrival times of packets during the active period $\Pr\{Z \le t\}$ are

$$\Pr\{X \le t\} = 1 - e^{-\sigma_A t} \quad \Pr\{Y \le t\} = 1 - e^{-\sigma_S t} \quad \Pr\{Z \le t\} = 1 - e^{-\lambda t} \tag{3.19}$$

Furthermore, let π_A and π_S respectively denote that the two-state Markovian on/off process is in active and silent periods. Then

$$\pi_A = \sigma_S/(\sigma_A + \sigma_S) \quad \pi_S = \sigma_A/(\sigma_A + \sigma_S) \tag{3.20}$$

Now let $\phi(s)$ denote the Laplace transform of the interarrival times of packets and consider the on/off process just after a packet arrival. The time until the next event is the minimum(X, Z). The event is an arrival with probability $\lambda/(\lambda + \sigma_A)$ and it is a change from an active state to a silent state with probability $\sigma_A/(\lambda + \sigma_A)$. We note that minimum$(X, Z)$ is exponentially distributed with rate $(\lambda + \sigma_A)$. At the end of the silent period, the process repeats itself. Hence we have

$$\phi(s) = \frac{\lambda}{\lambda + \sigma_A}\frac{\lambda + \sigma_A}{\lambda + \sigma_A + s} + \frac{\lambda_A}{\lambda + \sigma_A}\frac{\sigma_S}{\sigma_S + s}\frac{\lambda}{\lambda + \sigma_A}\left(\frac{\lambda + \sigma_A}{\lambda + \sigma_A + s}\right)^2 \tag{3.21}$$

$$+ \left(\frac{\sigma_A}{\lambda + \sigma_A}\right)^2\left(\frac{\sigma_S}{\sigma_S + s}\right)^2\frac{\lambda}{\lambda + \sigma_A}\left(\frac{\lambda + \sigma_A}{\lambda + \sigma_A + s}\right)^3 + \ldots$$

After some manipulation,

$$\phi(s) = \frac{\lambda(\sigma_S + s)}{s^2 + (\lambda + \sigma_S + \sigma_A)s + \sigma_S\lambda} \tag{3.22}$$

The first moment of the interarrival times of packets $E(X)$ and the squared coefficient variation c^2 are then

$$E(X) = \frac{\sigma_A + \sigma_S}{\lambda\sigma_S}; \quad c^2 = 1 + \frac{2\lambda\sigma_A}{(\sigma_A + \sigma_S)^2} \tag{3.23}$$

An IPP with parameters (s_A, σ_S, λ) is equivalent to a hyperexponential distribution with parameters (p_1, μ_1, p_2, μ_2), where

$$\begin{aligned}
\mu_1 &= 0.5\{(\lambda + \sigma_A + \sigma_S) + [(\lambda + \sigma_A + \sigma_S)^2 - 4\lambda\sigma_A]^{1/2}\} \\
\mu_2 &= 0.5\{(\lambda + \sigma_A + \sigma_S) - [(\lambda + \sigma_A + \sigma_S)^2 - 4\lambda\sigma_A]^{1/2}\} \\
p_1 &= (\lambda - \mu_2)/(\mu_1 - \mu_2) \\
p_2 &= 1 - p_1
\end{aligned} \tag{3.24a}$$

Similarly, the parameters of the equivalent IPP, $(\sigma_A, \sigma_S, \lambda)$, for a hyperexponential distribution with parameters (p_1, μ_1, p_2, μ_2), are

$$\begin{aligned}
\lambda &= p_1\mu_1 + p_2\mu_2 \\
\sigma_A &= p_1p_2(\mu_1 - \mu_2)^2/\lambda \\
\sigma_S &= \mu_1\mu_2/\lambda
\end{aligned} \tag{3.24b}$$

We are now going to address the following problem. Given the first three moments of interarrival times of packets, $E(X)$, $E(X^2)$, and $E(X^3)$, what are the parameters of the hyperexponential distribution (equivalent IPP) that has the same first three moments? Provided that

$$E(X)E(X^3) > 1.5\{E(X^2)\}^2 \tag{3.25}$$

the parameters (p_1, μ_1, p_2, μ_2) are uniquely determined from the following equations.

$$\begin{aligned} u &= \{6E(X)E(X^2) - 2E(X^3)\}/[3\{E(X^2)\}^2 - 2E(X)E(X^3)] \\ v &= \{12\{E(X)\}^2 - 6E(X^2)\}/[3\{E(X^2)\}^2 - 2E(X)E(X^3)] \\ \mu_1, \mu_2 &= 0.5[u \pm (u^2 - 4v)^{1/2}] \\ p_2 &= \frac{\mu_1\mu_2}{\mu_1 - \mu_2}\left(E(X) - \frac{1}{\mu_1}\right) \\ p_1 &= 1 - p_2 \end{aligned} \tag{3.26}$$

If equation (3.25) does not hold, then there are many other ways of selecting the parameters of the hyperexponential distribution. The following set is suggested in [66] in the context of approximate analysis of queuing networks:

$$\mu_1 = 2/E(X) \quad \mu_2 = 1/\{E(X)c^2\} \quad p_2 = 0.5\mu_1/\{(\mu_1 - \mu_2)c^2\} \tag{3.27}$$

In these equations, the squared coefficient of variation is assumed to be greater than or equal to 1, which from (3.23) is always the case. Note that if (3.27) is used to obtain the parameters of the hyperexponential distribution (matching the first two moments only), then there is no one-to-one correspondence between its parameters and the equivalent IPP. In this case, one of the parameters of IPP can be chosen arbitrarily. For example, the parameter λ can be chosen so that $1/\lambda$ is the cell transmission time in the medium.

3.4.2.2 Interrupted Bernoulli Process

The IBP is the discretized version of the IPP. Time is slotted, with the slot length being equal to the cell time in the medium. A slot is either in an active or silent state. A slot in an active state contains a cell with probability α and no cell with probability $1 - \alpha$, while no cells arrive in a silent state. Given that the slot is in an active state (regardless of whether or not the slot contains a cell), the next slot is also in the active state with probability $(1 - p)$. Similarly, given that the slot is in the silent state, the next slot is also in the silent state with probability q, and it changes to an active state with probability $(1 - q)$.

Accordingly, both the active period $\Pr(X = x\}$ and the silent period $\Pr\{Y = y\}$ are geometrically distributed. That is,

$$\Pr(X = x\} = (1 - p)p^{x-1} \quad \Pr(Y = y\} = (1 - q)q^{y-1} \quad x, y \geq 1 \tag{3.28}$$

with respective average duration times of $1/(1 - p)$ and $1/(1 - q)$.

Similar to the continuous-time version, the solution to the steady-state probabilities of being in active, π_A, and silent, π_S, states gives

$$\pi_A = (1 - q)/(2 - p - q) \quad \pi_S = (1 - p)/(2 - p - q) \tag{3.29}$$

The mean, $E(X)$, squared coefficient of variation c^2 and the third-moment $E(X^3)$ of the interarrival times of cells of an IBP can be obtained from the z transform $\phi(z)$ of the IBP and are defined in terms of its parameters (p, q, α) as follows:

$$\begin{aligned}
\phi(z) &= \frac{z\alpha[p + z(1 - p - q)]}{(1 - \alpha)(p + q - 1)z^2 - [q + p(1 - \alpha)]z + 1} \\
E(X) &= (2 - p - q)/[\alpha(1 - q)] \\
c^2 &= (1 - \alpha) + \alpha\frac{(1 - p)(p + q)}{(2 - p - q)^2} \\
E(X^3) &= 6\left\{\frac{2(1 - \alpha)}{(1 - q)\rho^2} + \frac{q}{(1 - q)^2\rho} - \frac{2(1 - \alpha)}{\alpha(1 - q)\rho} - \frac{q}{\alpha(1 - q)^2}\right. \\
&\quad \left. + \frac{(1 - \alpha)^2}{\rho^3}\right\} + \frac{3c^2 + 3 - 2\rho}{\pi^2}
\end{aligned} \tag{3.30}$$

where $\rho = 1/E(X)$ is the probability that a slot contains a cell, defined as the source utilization. Assuming that a slot in an active period always contains a cell (i.e., $\alpha = 1$), then (3.30) reduces to

$$E(X) = \frac{2 - p - q}{(1 - q)} \quad c^2 = \frac{(p + q)(1 - p)}{(2 - p - q)^2} \quad E(X^3) = \frac{6(1 - \rho)q}{\rho(1 - q)^2} + \frac{3c^2 + 3 - 2\rho}{p^2} \tag{3.31}$$

Given the values of $E(X)$, c^2, and $E(X^3)$, the parameters of an IBP that match these metrics cannot be obtained in a closed form due to the complexity of (3.30) and (3.31). The following algorithm proposed in [74] provides a good estimation for the parameters (p, q, α). Let $|a|$ denote the absolute value of a.

Step 0: Let $\alpha = \rho$; min = 100,000.
Step 1: Set $\alpha = \alpha + \epsilon$. If $\alpha \geq 1$, then STOP.
Step 2: If the following two conditions do not hold, then go to step 1. Condition 1: $(c^2 - 1)\alpha + 3\alpha\rho - 2\rho^2 > 0$; condition 2: $(c^2 - 1)\alpha - 3\alpha\rho + 2\rho^2 + 2\alpha^2 > 0$.

Step 3: Calculate p and q as follows:

$$p = \frac{\alpha c^2 + 3\alpha\rho - 2\rho^2 - \alpha}{\alpha c^2 - \alpha\rho + 2\alpha^2 - \alpha} \qquad q = \frac{\alpha c^2 - 3\alpha\rho + 2\alpha^2 - \alpha + 2\rho^2}{\alpha c^2 - \alpha\rho + 2\alpha^2 - \alpha} \tag{3.32}$$

Step 4: Calculate the third moment, m_3, of the IBP with parameters (p, q, α) from (3.31). If $|m_3 - E(X^3)| < \epsilon$, then STOP. If min $> |m_3 - E(X^3)|$, then save the parameters (p, q, α) and set min $= |m_3 - E(X^3)|$, and then go to step 1.

The algorithm searches for the minimum difference between the third moment of the distribution and the IBP with estimated parameters. If the third moment is not used, then it is necessary to set the value of one of the three parameters arbitrarily and calculate the other two from the equations for the first two moments. In particular, if we set $\alpha = 1$, then (3.31) may be used to estimate the values of p and q. In this case, we have

$$p = 1 - \frac{2(1-\rho)}{c^2(1-\rho)^{-1} + 1}; \quad q = 1 - \frac{(1-\rho)\rho}{(1-\rho)}$$

References and Bibliography

[1] Aartsen, G., et al., "Error Resilience of a Video Codec for Low Bit Rates," *Proc. ICASSP*, New York, 1988, pp. 1312–1315.

[2] Akiyama, T., T. Takahashi, and K. Takahashi, "Adaptive Three Dimensional Transform Coding for Moving Picture," *Proc. 1990 Picture Coding Symp.*, Cambridge, MA, 1990, pp. 8.2.1–8.2.2.

[3] Anagnostou, M. E., et al., "Quality of Service Requirements in ATM Based B-ISDNs," *Computer Communications*, Vol. 14-4, 1991, pp. 197–204.

[4] Andrade, J., W. Burakowski, and M. Villen-Altamirano, "Characterization of Cell Traffic Generated by an ATM Source," *ITC-13*, 1991, pp. 545–550.

[5] Ang, P. H., P. A. Ruetz, and D. Auld, "Video Compression Makes Big Gains," *IEEE Spectrum*, October 1991, pp. 16–19.

[6] Baker, R. L., and R. M. Gray, "Image Compression Using Non-adaptive Special Vector Quantization," *Proc. 16th Asilomar Conf. Circuit and Systems*, 1982, pp. 35–61.

[7] Banarjee, S., H. Feder, T. Kilm, and D. Sparrell, "32 kbit/s ADPCM Algorithm and Line Format Standard for US Networks," *GLOBECOM '85*, 1985, pp. 1143–1147.

[8] Bell, T., and K. Pawlikowski, "The Effect of Data Compression on Packet Sizes in Data Communication Systems," *ITC-13*, 1991, pp. 551–556.

[9] Bermejo, L., P. Parmentier, and G. H. Petit, "Service Characteristics and Traffic Models in a Broadband ISDN," *Electrical Communication*, Vol. 64-2/3, 1990, pp. 132–138.

[10] Bierling, M., "Displacement Estimation by Hierarchial Block Matching," *Proc. 3rd SPIE Symp. Visual Comm.*, Cambridge, MA, 1988.

[11] Biocca, A., et al., "Architectural Issues in the Interoperability between MANs and the ATM Network," *Proc. XIII ISS*, Stockholm, 1990.

[12] Blake, S. L., and T. L. Mitchell, "Issues in Video Transmission over Broadband ATM Networks," *Proc. 3rd Symp. on Comm. Signal Processing Expert Systems and ASIC VLSI Design*, Greensboro, NC: IEEE Press, 1992, pp. 420–444.

[13] Boiocchi, G., et al., "ATM Connectionless Server: Performance Evaluation," in *Proc. Modeling and Performance Evaluation of ATM Technology*, Perros, Pujolle, and Takahashi, eds., Amsterdam: North Holland, 1993.

[14] Brady, P. T., "A Model for Generating On/Off Speech Patterns in Two Way Conversations," *BSTJ,* Vol. 48, 1969, pp. 2445–2472.

[15] Brady, P. T., "Effects of Transmisison Delay on Conversational Behavior on Echo Free Telephone Circuits," *BSTJ,* 1968, pp. 73–91.

[16] Chen, C.-T., "Adaptive Transform Coding via Quadtree-Based Variable Block Size DCT," *IEEE Proc.* ICASSP, Glasgow, Scotland, 1989, pp. 1858–1857.

[17] Chen, C.-T., and T. R. Hsing, "Digital Coding Techniques for Visual Communications," *J. Visual Comm. and Image Representation,* Vol. 2-1, 1991, pp. 1–16.

[18] Chen, C.-T., and D. J. Le Gall, "A K-th Order Adaptive Transform Coding Algorithm for Image Data Compression," *SPIE Proc. Appl. of Dig. Image Process, XII,* San Diego, Vol. 1153, 1989, pp. 7–18.

[19] Chen, W.-H., and W. K. Pratt, "Scene Adapter Coder," *IEEE Trans. Comm.,* Vol. COM-32, 1984, pp. 225–232.

[20] Chen, W.-H., and C. H. Smith, "Adaptive Coding of Monochrome and Color Images," *IEEE Trans. Comm,* Vol. COM-25, 1977, 1285–1292.

[21] Chimento, P., personal conversation.

[22] Chiu, C.-Y., C.-T. Chen, and D. J. Le Gall, "Quadtree-Structured Finite State Vector Quantization for Still Images," *Proc. 1989 Conf. Information Sci. and Systems,* Baltimore, 1989, pp. 412–420.

[23] Cochennec, J.-Y., et al., "Asynchronous Time Division Networks: Terminal Synchronization for Video and Sound Signals," *GLOBECOM '85,* New Orleans, 1985, pp. 791–794.

[24] Cosmas, J. P., and A. Odinma-Okafor, "Characterization of Variable Rate Video Codecs in ATM to a Geometrically Modulated Deterministic Process Model," *ITC-13,* 1991, pp. 773–780.

[25] Crocetti, P., et al., "ATM Based SMDS for LANs/MANs Interconnection," *Proc. XIV ISS,* Yokohama, 1992.

[26] Crochiere, R. E., and L. R. Rabiner, *Multirate Digital Signal Processing,* New York: Prentice-Hall, 1983.

[27] Dagnino, R. M. R., M. R. K. Khansari, and A. Leon-Garcia, "Prediction of Bit Rate Sequences of Encoded Video Signals," *IEEE JSAC,* Vol. 9, No. 3, April 1991, pp. 305–314.

[28] Daigle, J., and J. Langeford, "Models for Analysis of Packet Voice Communications Systems," *IEEE JSAC,* September 1986.

[29] Ding, W., "A Unified Correlated Input Process Model for Telecommunication Networks," *ITC-13,* 1991, pp. 539–544.

[30] Eckberg, A. E., "Generalized Peakedness of Teletraffic Processes," *ITC-10,* 1983.

[31] Emling, J. W., and D. Mitchell, "The Effects of Time Delay and Echoes on Telephone Conversations," *BSTJ,* 1963, pp. 2869–2891.

[32] Galassi, G., G. Rigolio, and L. Verri, "Resource Management and Dimensioning in ATM Networks," *IEEE Network Mag.,* May 1990, pp. 8–17.

[33] Le Gall, D. J., H. Gaggioni, and C.-T. Chen, "Transmission of HDTV Signals under 140 Mbps Using a Subband Decomposition and the DCT," *Proc. 2nd Int. Workshop on Signal Proc. of HDTV,* L'Aquita, Italy, 1988.

[34] Le Gall, D. J., "MPEG: A Video Compression Standard for Multimedia Applications," *Comm. ACM,* Vol. 34-4, 1991, pp. 46–58.

[35] Ghanbari, M., "Two-Layer Coding of Video Signals for VBR Networks," *IEEE JSAC,* Vol. 7, No. 5, June 1989, pp. 771–781.

[36] Gray, R. M., Vector Quantization, *IEEE ASSP. Mag.,* April 1984, pp. 4–29.

[37] Gruber, J. G., "A Comparison of Measured and Calculated Speech Temporal Parameters Relevant to Speech Activity Detection," *IEEE Trans. Comm.,* Vol. COM-30, No. 4, April 1982, pp. 728–738.

[38] Gruber, J. G., and N. H. Le, "Performance Requirements for Integrated Voice and Data Networks," *IEEE JSAC,* December 1983, pp. 981–1005.

[39] Gruber, J. G., "Delay Related Issues in Integrated Voice and Data Networks," *IEEE JSAC,* June 1981, pp. 786–800.
[40] Grunenfelder, R., J. P. Cosmas, S. Manthorpe, and A. Odinma-Okafor, "Characterization of Video Codecs as Autoregressive Moving Average Processes and Related Queueing System Performance," *IEEE JSAC,* Vol. 9, No. 3, April 1991, pp. 284–293.
[41] Grunenfelder, R., J. P. Cosmas, S. Manthorpe, and A. Odinma-Okafor, "Measurement and ARMA Model of Video Codecs in an ATM Network," *ITC-13,* 1991, pp. 981–985.
[42] Guillemin, F., and J. W. Roberts, "Jitter and Bandwidth Enforcement," *GLOBECOM '91,* 1991, pp. 261–265.
[43] Gusella, R., "A Measurement Study of Diskless Workstation Traffic on an Ethernet," *IEEE Trans. on Comm.,* Vol. 39-9, 1990.
[44] Huang, S., "Source Modelling for Packet Video," *ICC '88,* 1988, pp. 1262–1267.
[45] Huang, S., "Modeling and Analysis for Packet Video," *GLOBECOM '89,* 1989, pp. 881–885.
[46] Holtsinger, D. S., "Analysis of VISTAnet Traffic Measurements," manuscript.
[47] Huffman, D. A., "A Method for the Construction of Minimum Redundancy Codes," *Proc. IRE,* 1952, pp. 1098–1101.
[48] Jain, A. K., *Fundamentals of Image Processing,* New York: Prentice-Hall, 1989.
[49] Jayant, N. S., V. B. Lawrence, and D. P. Prezas, "Coding of Speech and Wideband Audio," *AT&T Tech. J.,* Vol. 69-5, 1990, pp. 25–41.
[50] Kaneko, M., Y. Hatori, and A. Koike, "Improvement of Transform Coding Algorithm for Motion-Compensated Interframe Prediction Errors—DCT/SQ Coding," *IEEE JSAC,* Vol. 5, 1987, pp. 1068–1078.
[51] Karlsson, G., and M. Vetterli, "Packet Video and Its Integration into the Network Architecture," *IEEE JSAC,* Vol. 7-5, 1989, pp. 739–751.
[52] Karlsson, G., and M. Vetterli, "Subband Coding of Video Signals for Packet Switched Networks," *Proc. SPIE Conf. Visual Comm. Image Process.* II, Vol. 845, Cambridge, MA, 1987, pp. 446–456.
[53] Karlsson, G., and M. Vetterli, "Subband Coding of Video for Packet Networks," *Opt. Eng.,* Vol. 27, 1988, pp. 574–586.
[54] Kato, Y., N. Mukawa, and S. Okuba, "A Motion Picture Cosing Algorithm Using Adaptive DCT Encosing Based on Coefficient Power Distribution Classification," *IEEE JSAC,* Vol. 5, 1987, pp. 1090–1099.
[55] Kawashima, K., and H. Saito, "Teletraffic Issues in ATM Networks," *Comp. Networks and ISDN Sys.,* Vol. 20, 1990, pp. 369–375.
[56] Kishimoto, R., Y. Ogata, and F. Inumaru, "Generation Interval Distribution Characteristics of Packetized Variable Rate Video Coding Data Streams in an ATM Network," *IEEE JSAC,* Vol. 7, No. 5, June 1989, pp. 833–841.
[57] Kishino, F., K. Manabe, Y. Hayashi, and H. Yasuda, "Variable Bit-Rate Coding of Video Signals for ATM Networks," *IEEE JSAC,* Vol. 7, No. 5, June 1989, pp. 801–806.
[58] Koga, T., K. Niwa, Y. Iijima, and I. Iinuma, "A Low Bit Rate Motion Video Coder for Videoconference," *Opt. Eng.,* Vol. 26-7, 1987, pp. 590–595.
[59] Lazar, A. A., G. Pacifici, and J. S. White, "Real-Time Traffic Measurements on MAGNET 11," *ICC '90,* 1990, pp. 1191–1196.
[60] Lee, D.-S., K.-H. Tzou, and S.-Q. Li, "Control Analysis of Video Packet Loss in ATM Networks," *SPIE Visual Communications and Image Processing '90,* Vol. 1360, 1990, pp. 1232–1243.
[61] Lee, P. J., "Forward Error Correction Coding for Packet Loss Protection," presented at *First Int. Packet Video Workshop,* Columbia Univ., New York, May 1987.
[62] Li, S.-Q., "Study of Packet Loss in Packet Voice Systems," *IEEE Trans. Comm.,* XX.
[63] Liou, M. L., "Visual Telephony as an ISDN Application," *IEEE Communications,* February 1990, pp. 30–38.

[64] Madsen, H., and B. F. Nielsen, "The Use of Phase Type Distributions for Modeling Packet-Switched Traffic," *ITC-13,* 1991, pp. 593–599.
[65] Maglaris, B., D. Anastassiou, P. Sen, G. Karlsson, and J. Robins, "Performance Models of Statistical Multiplexing in Packet Video Communications," *IEEE Trans. Comm.,* Vol. 36, 1988, pp. 834–844.
[66] Marie, R., "An Approximate Analytical Method for General Queueing Networks," *IEEE Trans. Soft. Eng.,* Vol. 5, No. 5, 1979, pp. 530–538.
[67] Mitchell, T. L., and S. L. Blake, "Techniques for Video Transport at the B-ISDN Basic Rate," *CCSP Tech. Rep.* TR 92/8, Dept. ECE, North Carolina State University, 1992.
[68] Netravali, A. N., and J. D. Robbins, "Motion Compensated TV Coding: Part I," *BSTJ,* Vol. 58-3, 1979, pp. 631–670.
[69] Netravali, A. N., and J. D. Robbins, "Motion Compensated Coding: Some New Results," *BSTJ,* Vol. 59-9, pp. 1735–1745.
[70] Nikolaidis, I., and I. F. Akyildiz, "Source Characterization and Statistical Multiplexing in ATM Networks," Tech. Rep. GIT-CC-92/24, *Georgia Tech.*
[71] Nill, N., "A Visual Model Weighted Cosine Transform for Image Compression and Quality Assesment," *IEEE Trans. Comm.,* Vol. COM-33, 1985, pp. 551–557.
[72] Nomura, M., T. Fujii, and N. Ohta, "Basic Characteristics of Variable Rate Video Coding in ATM Environment," *IEEE JSAC,* Vol. 7, No. 5, June 1989, pp. 752–760.
[73] Ohta, N., M. Nomura, and T. Fujii, "Variable Rate Video Coding Using Motion Compensated DCT for Asynchronous Transfer Mode Networks," *ICC '88,* 1988, pp. 1257–1261.
[74] Park, D., and H. G. Perros, "Departure Process of IBP/Geo/1/K Queue and Analysis of Tandem Configuration," *Tech. Rep., Computer Science Dept.,* North Carolina State University, 1992.
[75] Puri, A., "Multiframe Conditional Motion Compensated Interpolation and Coding," *Proc. 1990 Picture Coding Symp.,* Cambridge, MA, 1990, pp. 8.3.1–8.3.2.
[76] Ramamurthi, B., and A. Gersho, "Clasified Vector Quantization of Images," *IEEE Trans. Comm.,* Vol. COM-34, 1986, pp. 1105–1115.
[77] Ripley, G. D., "DVI—A Digital Multimedia Technology," *Comm. ACM,* Vol. 32-7, 1989, pp. 811–822.
[78] Robbins, J. D., and A. N. Netravali, "Spatial Subsampling in Motion Compensated TV Coder," *BSTJ,* Vol. 61, 1982, pp. 1895–1917.
[79] Russell, J., "Multimedia Networking Performance Requirements," *ATM Networks,* I. Viniotis and R. O. Onvural, eds., New York: Plenum, 1993, pp. 187–198.
[80] Saito, H., "Optimal Control of Variable Rate Coding with Incomplete Observation in Integrated Voice/Data Packet Networks," *European J. Op. Res.,* Vol. 51, 1991, pp. 47–64.
[81] Sen, P., B. Maglaris, N. Rikli, and D. Anastassiou, "Models for Packet Switching of Variable-Bit-Rate Video Sources," *IEEE JSAC,* Vol. 7, No. 5, June 1989, pp. 865–869; INFOCOM '90, 1990, pp. 124–131.
[82] Sriram, K., and W. Whitt, "Characterizing Superposition Arrival Processes in Packet Multiplexers for Voice and Data," *IEEE JSAC,* Vol. 4-6, 1986, pp. 833–846.
[83] Sriram, K., R. S. McKinney, and M. H. Sherif, "Voice Packetization and Compression in Broadband ATM Networks," *IEEE JSAC,* Vol. 9, No. 3, April 1991, pp. 294–304.
[84] Stevenson, D., "Electropolitical Correctness and High Speed Networking, or, Why ATM is like a Nose," *ATM Networks,* I. Viniotis and R. O. Onvural, eds., New York: Plenum, 1993, pp. 15–20.
[85] Sutherland, J., and L. Litteral, "Residential Video Services," *IEEE Comm. Mag.,* Vol. 30-7, 1992, pp. 36–41.
[86] Tirtaatmadja, E., and R. A. Palmer, "The Application of Virtual Paths to the Interconnection of IEEE 802.6 Metropolitan Area Networks," *Proc. XIII ISS,* Stockholm, 1990.
[87] Tubaro, S., "A Two Layer Video Coding Scheme for ATM Networks," *Signal Processing: Image Communications,* Vol. 3, 1991, pp. 129–141.

[88] Tzou, K.-H., T. R. Hsing, and N. A. Daly, "Block-Recursive Matching Algorithm (BRMA) for Displacement Estimation of Video Images," *IEEE Proc. ICASSP,* Florida, 1985, pp. 359–362.
[89] Vaisey, D. J., and A. Gersho, "Variable Block Size Image Coding," *Proc. ICASSP,* Dallas, April 1987, pp. 25.1.1–25.1.4.
[90] Van Landegem, T., and P. Peschi, "Managing a Connectionless Virtual Overlay Network on Top of ATM, *ICC '91,* Denver, 1991.
[91] Verbiest, W., and M. Duponcheel, "Video Coding in an ATD Environment," *Proc. 3rd Int. Conf. New Syst. Services Telecomm.,* Liege, Belgium, 1986, pp. 249–253.
[92] Verbiest, W., L. Pinnoo, and B. Voeten, "The Impact of ATM Concept on Video Coding," *IEEE JSAC,* Vol. 6, No. 9, December 1988, pp. 1623–1632.
[93] Verbiest, W., and L. Pinnoo, "A Variable Rate Video Codec for Asynchronous Transfer Mode Networks," *IEEE JSAC,* Vol. 7, No. 5, June 1989, pp. 761–770.
[94] Weinstein, C. J., "Fractional Speech Loss and Talker Activity Model for TASI and for Packet-Switched Speech," *IEEE Trans. Comm.,* Vol. 26, 1978, pp. 1253–1257.
[95] Woodruff, G. M., and R. Kositpaiboon, "Multimedia Traffic Management Principles for Guaranteed ATM Network Performance," *IEEE JSAC,* Vol. 8, 1990, pp. 437–445.
[96] Yasuda, Y., H. Yasuda, N. Ohta, and F. Kishino, "Packet Video Transmission through ATM Networks," *GLOBECOM '89,* 1989, pp. 876–880.
[97] Yatsuzuka, Y., "Highly Sensitive Speech Detector and High Speed Voiceband Data Discriminator in DSI-ADPCM Systems," *IEEE Trans. Comm.,* Vol. COM-30, No. 4, April 1982, pp. 738–750.
[98] Yoshida, J., and R. Doherty, "JVC, C-Cube Team for High Quality MPEG Extension," *Electronic Eng. Times,* March 1991, pp. 4.
[99] "Broadband Service Principles," ETSI Recommendation, 1.2xx (draft).
[100] ITU-T SG XV, Draft Revision of Recommendation H.261, Document 572.
[101] ITU-T Recommendation I.350, 1988.
[102] ITU-T Recommendation I.352, "Network Performance Objectives for Connection Processing Delays in an ISDN," Volume III, Fascicle III.8, Blue Book, 1988.
[103] ITU-T Recommendation I.364, "Support of Broadband Connectionless Data Services on B-ISDN," Geneva, 1992.
[104] ITU-T Recommendation G.711, "Pulse Code Modulation (PCM) of Voice Frequencies," ITU-T Red Book, Vol. III, Jasc III.3, VIIIth plenary assembly, Spain, 8–19 October 1984.
[105] ITU-T Annex 1 to Report of Working Party XVIII/2—"Report of the work of ad hoc group on 32 kbit/s ADPCM," COMM XVIII-R 28-E, 1984.
[106] ITU-T IVS Baseline Document, SG XVIII/8, June 1992, Geneva.
[107] ISO/IEC/JTC1/SC2/WG8, "Adaptive Discrete Cosine Transform Coding Scheme for Still Image Processing," Document N640 R1.
[108] ISO/MPEG, MPEG Video Simulation Model 2, Document 90/041.
[109] RTT Belgium, WD10 ATD Information Field Size, ITU-T SG XVIII Meeting, Seoul, South Korea, 25 January–5 February 1988.
[110] ITU-T Recommendation G.721, "Differential Pulse Code Modulation (DPCM) of Voice Frequencies."
[111] Bisdikian, C., "Performance Analysis of Single-Stage Output Buffer Packet Switches with Independent Batch Arrivals," to appear in *Computer Networks and ISDN Systems.*
[112] Bisdikian, C., W. Matragi, and K. Sohraby, "A Framework for Jitter Analysis in Cell Based Multiplexers," to appear in *Performance Evaluation.*
[113] Bisdikian, C., and B. Patel, "On the Cell Burst Size at the Output of a 'Leaky Bucket' Source Traffic Shaper in an ATM Network," *28th Annual Conf. on Information Sciences and Systems CISS '94,* Princeton University, 16–17 March 1994.
[114] Bisdikian, C., W. Matragi, and K. Sohraby, "A Study of the Jitter in ATM Multiplexers," in *High Speed Networks and Their Performance,* H. G. Perros and Y. Viniotis, eds., Amsterdam: North Holland, 1994, pp. 219–236.

[115] Bisdikian, C., W. Matragi, and K. Sohraby, "A Study of the Jitter in ATM Multiplexers," in *High Speed Networks and Their Performance,* H. G. Perros and Y. Viniotis, eds., Amsterdam: North Holland, 1994, pp. 219–235.

[116] Bisdikian, C., W. Matragi, and K. Sohraby, "Jitter Analysis in ATM Multiplexers," *Telecommunications Systems Conference: Modelling and Analysis,* Nashville, 28 February–3 March 1993, pp. 489–490.

[117] Lynch, J., "ATM Multimedia System Structure," manuscript, IBM Research Triangle Park, 1994.

[118] Matragi, W., C. Bisdikian, and K. Sohraby, "Jitter Calculus in ATM Networks: Single Node Case," *IEEE INFOCOM '94,* Toronto, 12–16 June 1994, pp. 232–241.

[119] Matragi, W., K. Sohraby, and C. Bisdikian, "Jitter Calculus in ATM Networks: Multiple Node Case," *IEEE INFOCOM '94,* Toronto, 12–16 June 1994, pp. 242–251.

[120] Matragi, W., C. Bisdikian, and K. Sohraby, "On the Jitter and Delay Analysis in ATM Multiplexers," *SUPERCOMM/ICC '94,* New Orleans, 1–5 May 1994, pp. 738–744.

[121] Matragi, W., C. Bisdikian, and K. Sohraby, "Jitter Calculus in ATM Networks: Single Node Case," *INFOCOM '94,* Toronto, 12–16 June 1994.

[122] Matragi, W., K. Sohraby, and C. Bisdikian, "Jitter Calculus in ATM Networks: Multiple Nodes Case," *INFOCOM '94,* Toronto, June 12–16 1994.

[123] Matragi, W., C. Bisdikian, and K. Sohraby, "On the Jitter and Delay Analysis in ATM Multiplexers," *SUPERCOMM/ICC '94,* New Orleans, 1–5 May 1994.

[124] Pancha, P., and M. El Zarki, "MPEG Coding for Variable Bit Rate Video Transmission," *IEEE Comm. Mag.,* May 1994, pp. 54–66.

[125] Patel, B. V., and C. C. Bisdikian, "On the Dimensioning of Leaky Bucket and Its Effects on the Burst Size Distribution in ATM Networks," *5th Maryland Workshop on Very High Speed Networks,* University of Maryland, 14–16 March 1994.

[126] Patel, B. V., and C. C. Bisdikian, "On the Performance Behavior of ATM End-Stations," *IEEE INFOCOM '95,* Boston, 2–6 April 1995.

[127] Patel, B. V., and C. C. Bisdikian, "Performance Characteristics of End-Stations in an ATM Network as Viewed by Applications and Networks," *IEEE 19th Conf. Local Computer Networks,* Minneapolis, 2–5 October 1994, pp. 316–323.

Traffic Management in ATM Networks 4

Traffic management in communications networks deals with the controlled use of network resources to prevent the network from becoming a bottleneck. In particular, when network resources are allocated to more connections or traffic than they can effectively support, network performance for users degrades (i.e., buffers start to overflow and delays increase beyond acceptable levels). Therefore, it is necessary to manage resource utilization and control the traffic so that the network can operate at acceptable levels even at times when the offered load to the network exceeds its capacity.

In circuit-switched networks, each connection is allocated a fixed amount of bandwidth, and a constant data rate in the network is provided to communicating entities throughout the duration of the connection. For example, in a telephone network each connection requires a 64-kbps channel. If a channel between the caller and the called party is available, the connection is established. Otherwise, it is rejected. This simple call admission procedure is sufficient to control congestion in circuit-switched networks, since the dedicated bandwidth is always available for a connection and there is no contention for network resources once a channel is allocated. ISDNs use a separate packet-switched signaling network (see Chapter 7). The signaling network in these systems (like any other packet-switching network) is subject to congestion, as discussed next.

Traffic control is a much more complex task in packet-switched networks due to the random nature of the traffic arrival pattern and contention for network resources. In simple terms, in these networks there is a queue associated with each link at every switching node in the network. As the arrival rate at a link approaches its transmission rate, the queue length grows dramatically. As a node becomes congested, buffers start to overflow (i.e., packets that arrive at a time the buffer is full are discarded). Dropped packets are eventually retransmitted by an upstream node (or by the source), causing the traffic load to further increase. As the number of retransmissions increases, more nodes become congested and more packets are dropped. Eventually, the network can reach a catastrophic state in which most of the packets in the network are retransmissions. Accordingly, the transmission links in packet-switching networks may not be utilized at more than 80% to 85%. However, due to the random nature of the traffic, provisioning links at such utilization levels does not solve the congestion control problem. The possibility of setting the resource utilization at lower levels, say 55% to 60%, so that simple admission control schemes may be sufficient to prevent congestion is not an acceptable alternative, since the network provider wants to use its resources as much as possible while providing its users with acceptable levels of service. Then, the challenge in packet switching is to develop a framework that maximizes the utilization of network resources while controlling the traffic flow in the network so that the temporary periods of overload that occur due to the stochastic nature of the traffic do not turn into sustained periods of congestion, which causes the network performance to degrade dramatically.

In general, two types of traffic control are used in packet-switched networks, namely, flow control and congestion control. Flow control is concerned with the regulation of the rate at which the sender transmits packets to match the rate at which the destination station receives data, so that it is not overwhelmed. Typically, flow control in current networks is exercised with various types of window techniques, which are discussed later in the context of ATM networks. Flow control between the source and the destination does not help much toward reducing the possibility of congestion within the network. In particular, it is necessary to maintain the number of packets within the network below the level at which network performance starts degrading dramatically. To minimize the effects of congestion, each node in the network can regulate the traffic flow on its input links by forcing them to slow down (or stop) their transmission as the possibility of congestion increases, for example, when the number of packets at its buffer reaches some threshold level. This, in turn, might cause the buffers at the upstream nodes to start to fill up, forcing their upstream nodes to slow down. Eventually, sources realize that the network is becoming congested and regulate their own traffic, allowing the network resources to recover from the congested state. Although methods developed as variations of this concept work well in networks with slow links, they do not scale well to ATM networks.

In this chapter, various congestion control techniques proposed for ATM networks are introduced and their effectiveness and performance models are discussed.

4.1 WHAT IS SO DIFFERENT IN ATM NETWORKS?

The success or failure of ATM networks depend on the development of an effective congestion control framework. Despite the experience gained from circuit-switching networks for about a century and from packet-switched networks for several decades, congestion control in ATM networks remains an unresolved issue. Some aspects of ATM networks that complicate the control problem include the following:

1. Various B-ISDN VBR sources generate traffic at significantly different rates. Terminal-to-terminal communication may generate traffic of only a few kilobits per second, whereas an HDTV application will use several megabits per second of network bandwidth. Furthermore, bit generation rates of some applications often have time-varying characteristics.
2. A single source may generate multiple types of traffic (i.e., voice, data, image) with different characteristics.
3. In addition to the performance metrics of call blocking and packet loss probabilities in current networks, ATM networks have to deal with cell delay variation, maximum delay, and skewness.
4. Different services have different types of QOS requirements at considerably varying levels. Statistical multiplexing of VBR traffic into the same medium, together with CBR traffic, complicates the predictability of network performance for both types of applications. Real-time services are particularly vulnerable to performance degradation.
5. Traffic characteristics of various types of services are not well understood.
6. As the transmission speeds increase, the ratio of call duration to cell transmission time increases, adding a new dimension to the problem. Due to large bandwidth propagation delay with high-speed links, there are very large numbers of cells in transit at any time in the network. Furthermore, large propagation delays compared with the transmission times give rise to long periods between the onset of congestion and its detection by the network control elements.
7. High transmission speeds limit the available time to do on-the-fly processing at the intermediate nodes.

Considering the high transmission speeds, the set of control algorithms to be used should be as simple as possible (to the extent it is feasible) to allow hardware implementations. QOS parameters agreed upon at the call setup phase should be consistently provided to each connection, regardless of the traffic characteristics of other connections in the network and the source behaviors. Source characterization remains an open problem for several types of B-ISDN applications. Furthermore, it may not always be possible for a source to accurately define its traffic characteristics. Hence, control techniques should not only use simple source characterizations that facilitate users developing an intuitive understanding, but they should also be robust and minimize the consequences of inaccurate source specifications. A method that artificially constrains the characteristics of user traffic streams is unacceptable in this environment. For example, reducing the peak rates of connections by delaying cells before they are submitted to the network may not be acceptable for delay-sensitive applications, although doing so would smooth the traffic in the network, thereby making it easier to manage. A poorly designed congestion control framework would cause underutilization of network resources. The problem is further complicated by the fact that ATM networks must not only support applications that we know of today, but also emerging multimedia services.

Congestion control mechanisms proposed for ATM networks are classified into two categories: *preventive control* and *reactive control*. Preventive congestion control techniques attempt to prevent congestion by taking appropriate actions before they actually occur. However, it is generally agreed that preventive control techniques are not sufficient to eliminate congestion problems in ATM networks and that when congestion occurs it is necessary to react to the problem. Proposed reactive control techniques initiate the recovery from a congested state. In a reactive scheme, the network is monitored for congestion. When congestion is detected, sources are requested to slow down or stop transmission for a while until the congestion is cleared. The main problem with the reactive scheme is the large propagation delay bandwidth product in ATM networks, which introduces the possibility that by the time a source receives a notification it may be too late to react.

Each class of control is applicable at different time frames. In particular, reactive schemes must operate at time frames greater than the propagation delay, whereas preventive techniques are designed to be effective across all time frames. Figure 4.1 illustrates the current state of the art in control techniques proposed for ATM networks and the time frames that are most effective.

4.2 RESOURCE PROVISIONING

Resource provisioning is an important traffic management function for existing networks. Its major role is to provide an acceptable level of connection blocking performance. The network topology, the number of links and their bandwidths, and the number of switching and access nodes are all determined by some understanding of traffic requirements. As time goes on, the number of users, the amount of traffic generated, and types of applications used will change. Accordingly, a network design based on a particular set of assumptions may no longer have the correct set of resources to support its users once the assumptions start changing. As this happens, new links are added, existing links are replaced with their higher bandwidth counterparts, new nodes are added, and so forth. Similar network engineering activities will also take place for B-ISDNs. In addition, ATM networks have unique resource provisioning requirements. VPs in ATM networks are semipermanent connections with deterministic, preassigned bandwidths. Let us consider a case in which a particular VP between two ATM switches is underutilized. Since, by definition, its bandwidth cannot be shared by other VPs along the links the VP is mapped onto, the amount of traffic that can be carried on these links is artificially restricted. A short-term solution may be to reduce its bandwidth. Since this type of VP management is in response to short-term traffic fluctuations, it is necessarily suboptimal. As

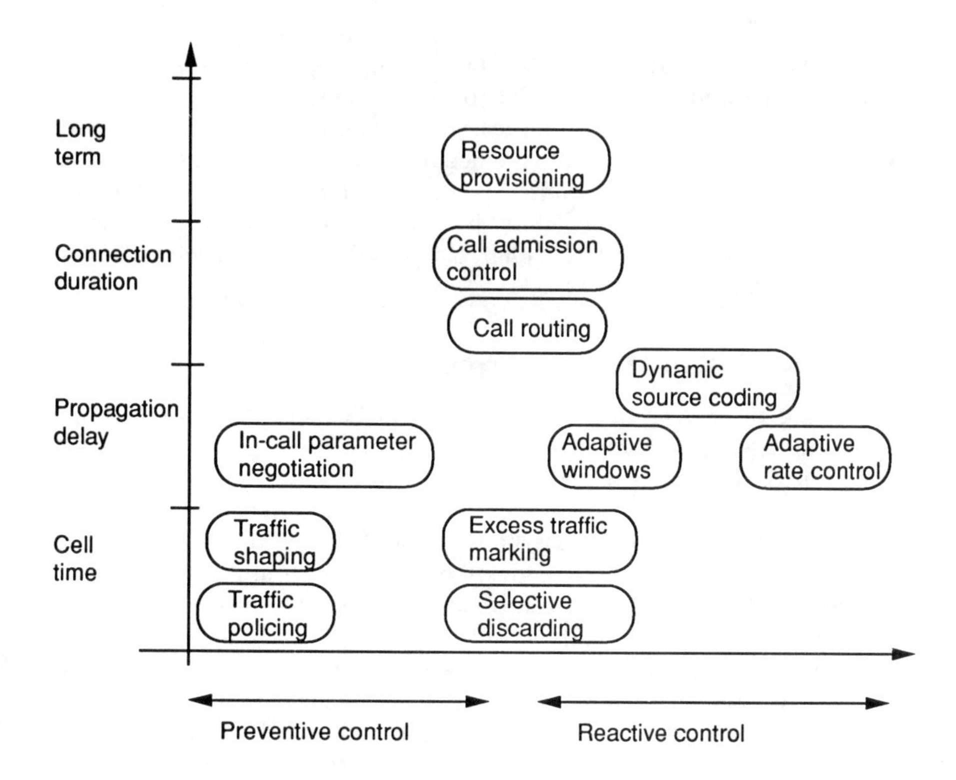

Figure 4.1 Traffic control options in ATM networks.

time passes and the traffic characteristics in the network changes, it would become necessary to redefine the VPs in the network and reassign their bandwidth. This type of VP management is based on networkwide information and will achieve near-optimal solutions under stable traffic conditions. The problem of defining the set of VPs and their parameters for a given network topology has not yet been satisfactorily addressed in the literature.

4.3 CALL ADMISSION CONTROL

When a new connection request is received at the network, the call admission procedure is executed to decide whether to accept or reject the call. A call is accepted if the network has enough resources to provide the QOS requirements of the connection request without affecting the QOS provided to existing connections. Accordingly, there are two questions that need to be answered:

- How can the amount of bandwidth required by a new connection be determined?
- How can we make sure that the service levels required by existing connections are not affected when multiplexed together with this new connection?

Any technique designed in response to these two questions should function in real time and should attempt to maximize the utilization of network resources. The first step is to determine the set of parameters required to describe the source activity adequately for the accurate prediction of the performance metrics of interest in the network. As discussed in Chapter 3, this largely remains an open issue. Nevertheless, in what follows, we assume that there are analytically tractable models to characterize VBR sources, such as multistate Markovian models.

4.3.1 Superposition of Arrival Streams

For modeling purposes, an ATM network can be viewed as a collection of queues connected in a manner determined by the network topology. A server corresponds to a transmission link. There is a finite buffer associated with each server to temporarily store incoming cells when cells arrive at a faster rate than they can be transmitted. This may be the case, for example, when more than one connection is active simultaneously on a link with multiple connections.

In theory, it is possible to develop a Markovian model of a communications network and solve it numerically with the new connection to determine if the network can accommodate it or not. In practice, such a queuing model consists of tens and hundreds of queues with traffic generated from thousands of sources. It is impossible to solve such queuing models numerically in real time, even for small networks (due to large storage and processing requirements). A well-known approach in queuing theory to analyze large queuing networks is to decompose the network into individual queues and analyze each queue in isolation. This method, although approximate for most cases of interest, has been used satisfactorily to obtain the performance metrics of various real systems.

To solve a queue in isolation requires the characterization of its arrival streams, effective service process, and buffer capacity so that the behavior of the queue in isolation is approximately the same as it is within the network. Let us for a moment consider the original queuing network at a time a cell departs from queue i and attempts to enter queue j. If there is space available at queue j, then the cell joins the queue. If, on the other hand, there is no space available, the cell is dropped and cleared from the system. Hence, the service process at node i is not affected by the state of its downstream nodes, assuming that intermediate nodes do not retransmit cells or exchange messages with their neighbors, which is expected to be the case in ATM networks. Similarly, the behavior of a downstream queue (i.e., its buffers and service process) is not affected by the state of its upstream nodes. Hence, the service process and the buffer capacity of a queue in isolation is the same as it is in the original network. The arrival process at a queue, on the other hand, is the superposition of the departure processes from its upstream queues and arrivals from outside the network (i.e., traffic is generated by sources that are directly attached to the node).

Although the complexity of the problem is reduced significantly with this decomposition approach, the problem itself is not yet solved. Let us for a moment assume that the stochastic behavior of the cell arrival process of a connection at an intermediate node in the network is exactly the same as it is at the source, the characterization of which we assume we know. Then transmission link i in the network can be modeled as a single queue with N arrival streams, a finite buffer K, and constant service time (being equal to 53 $\times$ 8/link speed), where N is the total number of connections currently multiplexed on link i. The analysis of such single-queuing systems quickly becomes infeasible as the number of arrival streams multiplexed to the queue increases. For example, let us consider a case where each source is modeled as a two-state Markovian model. Then obtaining, for example, the loss probability at this queue would require the solution of $2^N(K + 1)$ linear equations. A queue with a 100-cell buffer would then require the solution of 11,264 equations with only 10 sources, which cannot be performed in real time. A typical approach to overcome this difficulty is to reduce the number of arrival streams by superposing all the component arrival processes into a small number of streams, thereby reducing the dimensionality of the problem.

To obtain its performance metrics when a new connection request arrives, a queue can be analyzed with two arrival streams: one corresponding to that of the new connection and another one to represent the superposition of existing arrival streams on that link. Depending on how the superposed traffic is modeled (often approximately), the dimensionality of the problem may be reduced, thereby making it feasible to obtain a solution. It may also be possible to further reduce the dimensionality of the problem by superposing all arrivals, including the new connection, into a single stream.

Various methods proposed for modeling superposed traffic are presented in Appendix A.

4.3.2 Bandwidth Allocation

Bandwidth allocation deals with determining the amount of bandwidth required by a connection for the network to provide the required QOS. There are two alternative approaches for bandwidth allocation: deterministic multiplexing and statistical multiplexing. In deterministic multiplexing, each connection is allocated its peak bandwidth, causing large amounts of bandwidth to be wasted for bursty connections, particularly for those with large peak-to-average bit rate ratios. Deterministic multiplexing can eliminate cell level congestion almost totally. However, there is still a nonzero probability that buffers will overflow and cells will be lost with this scheme. In particular, consider a transmission link with a K cell buffer and N connections such that the sum of the peak rates of all connections is less than or equal to the provisioned link bandwidth. If $N \leq K$, then the probability of a cell loss is equal to zero. However, if $N > K$, then it is possible for more than K sources to become active and simultaneously generate cells. Only K of N arriving cells can be placed in the buffer and the rest are lost. Although this probability is negligible for most practical applications today, it is nevertheless not equal to zero.

Deterministic multiplexing goes against the philosophy of ATM framework, since it does not take advantage of the multiplexing capability of ATM and restricts the utilization of network resources. An alternative method is statistical multiplexing. In this scheme, the amount of bandwidth allocated in the network to a VBR source (hereafter referred to as the *statistical bandwidth* of a connection) is less than its peak, but necessarily greater than its average bit rate. Then the sum of peak rates of connections multiplexed onto a link can be greater than the link bandwidth as long as the sum of their statistical bandwidths is less than or equal to the provisioned link bandwidth.

The bandwidth efficiency due to statistical multiplexing increases as the statistical bandwidths of connections get closer to their average bit rates and decreases as they approach their peak bit rates. In general, though, statistical multiplexing allows more connections to be multiplexed in the network than deterministic multiplexing, thereby allowing better utilization of network resources.

The main difficulty in calculating the value of the statistical bandwidth of a connection is twofold:

- Guaranteeing the QOS requirements of individual connections;
- Assuring that the QOSs provided to existing connections do not degrade to unacceptable levels when multiplexed together with the new connection.

Accordingly, the statistical bandwidth of a connection does not only depend on its own stochastic characteristics, but it also depends on the characteristics of existing connections in the network. The problem of determining the bandwidth requirements of connections and aggregated traffic at intermediate nodes in the network by taking into consideration the effects of multiplexing largely remains an open issue.

In general, efficiency gain due to statistical multiplexing is a factor of various connection characteristics and network parameters, as illustrated in Figure 4.2. As the number of sources multiplexed increases, it is highly likely that the aggregated traffic would be less bursty at the downstream nodes in the network than the individual connections at access nodes. If this is the case, then the statistical bandwidth of the aggregated traffic approaches their average bit rates. Accordingly, more connections can be supported in the network.

However, it is not necessarily the case that the individual connection behavior becomes smoother along the path of the connection. In particular,

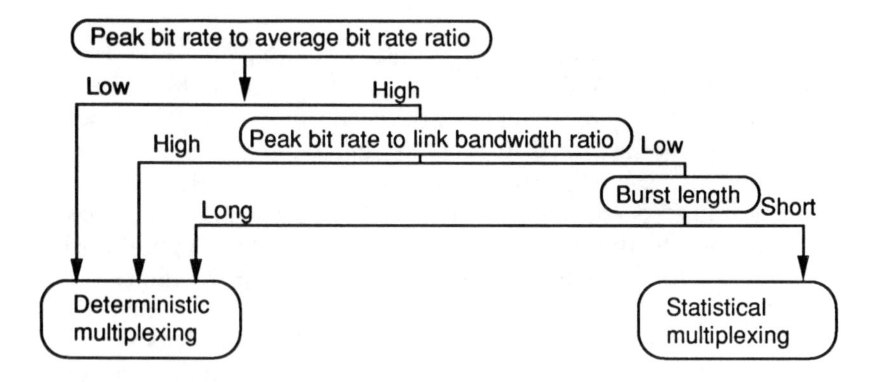

Figure 4.2 Deterministic multiplexing versus statistical multiplexing.

when high-bit-rate sources are multiplexed together with low-bit-rate sources, the smoothing effect of low-bit-rate sources to high-bit-rate ones can be negligible, whereas high-bit-rate sources may have a dominant effect on low-bit-rate sources. Depending on how bursty the high-bit-rate sources are, the efficiency gain due to statistical multiplexing may or may not be significant.

The burst length of connections is another major factor that determines the efficiency gain due to statistical multiplexing. For presentation purposes, in Table 4.1, let us consider two bursty connections with the same peak and average bit rates.

The first type has long active and silent periods, whereas the two periods are much shorter in the second type. Both are modeled as IBPs in which the source utilizations are fixed at 0.5 and there is a cell arrival at every slot during the busy period. Furthermore, the average durations of the active and silent periods are two and four slots for connection type 2 and type 1, respectively. Using c^2 as the definition of burstiness, connection type 1 is burstier than type 2 and accordingly uses more network resources if both connections require the same grade of service (i.e., cell loss probability).

Table 4.1
Traffic Parameters

Parameter	*Connection 1*	*Connection 2*
α	1	1
p_1	0.5	0.75
q_1	0.5	0.75
c^2	1.5	0.5

Another major factor in achieving bandwidth efficiency is the ratio of the connection's peak bit rate to the link bandwidth. In packet-switching networks, an arriving packet is fully received from the incoming link before its transmission starts at the outgoing link. Consider a packet that is 200 μs at the incoming link. If it requires the same transmission time at the outgoing link, that is, both links have the same speed, then it will take 400 μs from the time its first bit arrives at the node to the time its last bit is transmitted, assuming that the packet arrives at a time when there is no other packet at the node. Excluding ATM cell overhead, if the packet arrives in consecutive cells, then this time would go down to 220 μs, assuming it consists of 10 cells. This is due to the overlapping of receive and transmit operations at the incoming and outgoing links. In particular, if the packet arrives as consecutive cells, then the transmission time of the first cell at the outgoing link overlaps with the receiving time of the second cell from the incoming link, the third cell overlaps with the second, and so on. Accordingly, the total time now equals the packet arrival time plus the transmission time of the first cell.

Extending this example to a more realistic scenario, let us consider the case in which there is contention for the outgoing link (i.e., multiple connections from different incoming links are active simultaneously) and the cell header overhead is included in the transmission times. The total time a packet spends at a node is less with ATM if a number of cells of the packet have been transmitted before its last cell arrives from the incoming link. Then ATM is a very efficient technique in its use of buffers: a large number of connections may be multiplexed and the buffer requirements at the intermediate nodes are small. This efficiency is expected to be the case in B-ISDNs, where the backbone bandwidth is expected to be much larger than the user interface speeds.

4.3.3 Call Admission Algorithms

In this section, we discuss various call admission algorithms proposed in the literature. Their mathematical derivations and the details are given in appendixes at the end of the book. Although these techniques may be used in early deployments of ATM networks, they should be viewed as intermediate solutions to the problem. In particular, these algorithms either allocate more bandwidth in the network than required to provide the QOS requirements for connections, thereby causing underutilization of network resources, or they may underestimate the bandwidth requirements of connections, necessitating additional safeguard mechanisms to be used in conjunction with them, such as provisioning links to low utilization levels.

4.3.3.1 Gaussian Approximation

In this method, each connection is characterized by its average bit rate m_i and standard deviation σ_i. With n multiplexed connections, the problem is determining the total bandwidth required by n connections c_0 so that the probability that the instantaneous aggregate bit rate exceeding c_0 is less than a given value ϵ.

Let A be a random variable denoting the aggregate bit rate of n multiplexed connections. Then the problem is to determine the value of c_0 such that $\Pr\{A > c_0\} \leq \epsilon$. Assuming that the aggregate bit rate distribution is Gaussian, the derivation of c_0 is given in Appendix C. For completeness, only the formulas are given here. Let m and σ^2 respectively denote the mean and standard deviation of the aggregate (i.e., superposed) traffic. Then $c_0 \approx m + \alpha\sigma$. The parameter α is the inverse of the Gaussian distribution with one possible value given by $\alpha = \sqrt{2 \ln(1/\epsilon) - \ln 2\pi}$. Using this framework, the steps of the call admission procedure are given as follows:

1. Upon receiving a new connection request with parameters (m_{n+1}, σ_{n+1}) on a link with aggregated traffic parameters m and σ:
2. Calculate the new mean, $m' = m + m_{n+1}$ and standard deviation $\sigma' = \sqrt{\sigma^2 + \sigma_{n+1}^2}$.
3. Let $c_0 = m' + \alpha\sigma'$.
4. If $c_0 \leq C$, the provisioned link bandwidth, then accept; otherwise reject the connection.

Despite its simplicity, the Gaussian approximation method has a limited use because:

- The aggregate bit rate is Gaussian if the stationary distributions of individual connections are themselves Gaussian. It is a realistic assumption only if a large number of connections with similar parameters (i.e., m_i and σ_i) are multiplexed together. It is particularly inaccurate when the standard deviations of connections differ rather significantly, which is often the case in ATM networks unless only connections with similar characteristics are multiplexed together.
- All connections are treated as if they have the same cell loss requirements, although in reality they may differ significantly.
- The method does not fully explore the amount of statistical gain that can be achieved because it does not take into consideration the buffer size.

4.3.3.2 Fast Buffer Reservation

In this scheme, a bursty source is characterized by a two-state Markovian model. When a source becomes active, a prespecified number of buffer slots in the link buffer is reserved for the duration of its active period. At the end of the active period all reserved slots are released. This process is repeated throughout the duration of the connection. The technique developed in [143] uses marked cells to specify the transitions between active and silent periods. Let:

B = number of buffer slots available at the link buffer;
B_i = number of buffer slots to be reserved for connection i;
b_i = number of slots currently in use;
s_i = state of connection i (active or idle);

The operation of the fast buffer reservation scheme is summarized as follows:

1. Decrease b_i by 1 every time a cell belonging to connection i is transmitted from the buffer.
2. If the connection is in a silent state and:
   - a. If the received cell specifies the beginning of an active period and:
     - (1) If $B_i > B$, then discard the cell;
     - (2) If B_i B then:
       - (a) Set up the timer;
       - (b) Change s_i from silent to active;
       - (c) $B = B - B_i$;
       - (d) If $b_i < B_i$, then:
         - (i) $b_i = b_i + 1$;
         - (ii) Place the cell in the buffer as an unmarked cell.
       - (e) If $b_i = B_i$, then place the cell in the buffer as a marked cell.
   - b. If the received cell does not specify the beginning of an active period, then discard the cell.
3. If the connection is in an active state, then:
   - a. If the received cell does not specify the end of an active period, then:
     - (1) Reset the timer;
     - (2) If $b_i < B_i$, then:
       - (a) $b_i = b_i + 1$;
       - (b) Place the cell in the buffer as an unmarked cell.
     - (3) If $b_i = B_i$, then place the cell in the buffer as a marked cell.
   - b. If the received cell specifies the end of an active period or the timer expires, then:
     - (1) change s_i from silent to active;
     - (2) $B = B + B_i$.

A timer is used to force the eventual return to the silent state and to ensure reserved slots are released. The buffer slots are reserved only upon reception of a cell, referred to as the *start-of-the-burst cell,* which specifies the beginning of an active period. If the required number of buffer slots is not available, then all cells of the connection are discarded until another start-of-the-burst cell is received.

For loss-sensitive applications, a lost cell forces the complete or a major portion of the burst to be eventually retransmitted by the sender. In this case, it is preferable to drop the whole burst instead of attempting to transmit parts of it. By doing so, the resource contention at the downstream nodes is reduced. Bursts of such applications are sent by a start-of-the-burst marked cell followed by *middle* cells, ending with an *end-of-the-burst* marked cell. With this scheme, if the buffer space requested by the first cell of the burst is not available, then all the cells of the burst are dropped.

The fast-buffer reservation scheme can handle loss-tolerant traffic relatively easily as well. In this case, each cell of the burst except the last one is transmitted as a start-of-the-burst cell. Then, if a cell is lost due to unavailability of buffer space, the consecutive cells still have a chance to enter the buffer. In addition to the three types of marked cells specifying the start, end, and middle of a burst, *loner* marked cells are defined [143] to be used for single cell bursts or to specify low-priority cells.

The definition of the four types of cells requires the use of two bits at the ATM cell header. For example, the least significant 2 bits of the VCI field may be used for marking. However, this is a nonstandard use of the VCI field and introduces the problem of compatibility between interconnected networks. Another approach proposed in [143] uses the CLP bit as follows. When in a silent state, a cell with the CLP bit set is treated as a loner, whereas a cell with the CLP bit cleared corresponds to the start-of-the-burst marked cell. Similarly, when in an active state, a cell with the CLP bit set is treated as an end-of-the-burst marked cell, whereas a cell with the CLP bit cleared corresponds to a middle cell. With this scheme, low-priority cells cannot be sent within a burst. Furthermore, the network cannot filter out clipped bursts, which results in decreased transmission efficiency.

To decide if a new connection can be multiplexed with the existing connections on a link, the probability of requiring more buffer slots than are available is calculated and referred to as the *excess demand probability*. If the excess demand probability is greater than a predefined value, the new connection is rejected; otherwise it is accepted. Let:

λ_i = peak bit rate of connection i;
μ_i = average bit rate of connection i;
x_i = random variable representing the number of buffer slots needed by connection i. x_i is either equal to 0 (in a silent state) or B_i (in an active state).

$$p_i = \Pr\{x_i = B_i\} = \mu_i/\lambda_i \text{ and } \Pr\{x_i = 0\} = 1 - \mu_i/\lambda_i$$

Using the peak-to-link-rate ratio, the number of buffer slots required by an active source is assumed to be equal to

$$B_i = \lceil L\lambda_i/R \rceil \tag{4.1}$$

where $\lceil z \rceil$ denotes the smallest integer greater than or equal to z, L is the total number of buffer slots at the link buffer, and R is the link rate. Consider now a link carrying n connections with buffer demands $x_1, \ldots, x_n$. The total buffer demand X is the sum of n random variables; that is,

$$X = \sum_{i=1}^{n} x_i$$

The excess demand probability can be calculated in real time recursively using the derivations given in Appendix D. However, certain anomalies are observed to occur in this framework [143]. Consider two connections, both of which require all the buffer slots when active and consider $p_1 = 0.9$, $p_2 = 0.01$. The excess buffer demand probability with the two connections is equal to 0.009. However, when the second connection becomes active, there is no buffer space available with probability 0.9, meaning that only 10% of its bursts will succeed. If the excess demand probability is greater than 0.009, then the second connection will be accepted, which is clearly unacceptable.

This problem is addressed using the probability that the number of buffer slots requested will not be available when source i transmits its burst instead of the excess demand probability. This probability is bounded above by the probability $\Pr\{X - x_i > L - B_i\}$, referred to as the contention probability of connection i.

$$\Pr\{X = L\} = \Pr\{X - x_i > L - B_i\} \leq (1/p_i)\Pr\{X > L\} \tag{4.2}$$

Hence, as long as p_i is not too small, the excess demand probability is not too much larger than the contention probability. However, the two differ significantly as p_i decreases. When p_i is small, tight bounds on the contention probability are obtained with some additional computation [143]:

$$\begin{aligned} \Pr\{X - x_i > j\} = \{1/(1 - p_i)\}\sum_{h=0}^{k-1}\{-p_i/(1 - p_i)\}^h \Pr\{X > j - hB_i\} \\ + \{-p_i/(1 - p_i\}^k \Pr\{X > j - hB_i\} \end{aligned} \tag{4.3}$$

where k is chosen such that $\{p_i/(1 - p_i)\}^k$ is sufficiently small. The call admission procedure in this framework is then based on the excess demand probability when p_i is small or on the contention probability if p_i is large. That is:

- If p_i is small then $p = \Pr\{X > L\}$; otherwise $p = \Pr\{X - x_i > j\}$.
- If $p < \epsilon$ for a given ϵ, then accept the call; otherwise reject it.

4.3.3.3 Equivalent Capacity

Similar to the other models, a connection in this method is characterized by a two state model in which the flow of bits is generated at a peak rate during the active period whereas no bits are generated during the silent period. Unlike the other techniques discussed so far, this model is based on a flow model. In general, let Z be a random variable denoting the bit rate of a source feeding a link with a finite buffer and a transmission speed A. Then the dynamics of the queue in the flow model are defined as follows:

1. If $Z = z < A$ and:
   a. The buffer is empty, then it remains empty;
   b. The buffer is not empty, then its content decreases at a constant rate of $A - z$.
2. If $Z = z = A$, then the buffer content does not change.
3. If $Z = z > A$ and:
   a. The buffer is not full, then the buffer content increases at a constant rate of $z - A$;
   b. The buffer is full, then the cells are lost at a constant rate of $z - A$.

Let:

R = peak bit rate of the connection;
b = average duration of the active period;
ρ = source utilization (i.e., probability that the source is in an active state);
μ = transition rate out of active state (i.e., $\mu = 1/b$);
λ = transition rate out of the silent state (i.e., $\lambda = \rho/\{b(1 - \rho)\}$);
c = link speed;
X = buffer size.

Assuming each source is characterized as a two-state on/off model and the duration at each state is exponentially distributed and independent of each other, the amount of bandwidth required by a connection is estimated in [55,56] as follows.

The amount of bandwidth required by a connection (referred to as the equivalent capacity) is obtained in isolation as the answer to the following question. If a connection with parameters (R, m, b) is input to a link with buffer capacity X, what should be the transmission rate for this link to achieve a desired buffer overflow probability ϵ?

Equation (4.4) gives the equivalent capacity c of a source. The interested reader may refer to Appendix B for the derivation of c.

$$c = R\frac{y - X + \sqrt{(y - X)^2 + 4X\rho y}}{2y} \tag{4.4}$$

with $y = \alpha b(1 - \rho)R$. In this framework, the total bandwidth of n multiplexed connections is equal to the sum of the equivalent capacities of individual connections c_i; that is,

$$C = \sum_{i=1}^{n} c_i$$

However, C significantly overestimates the required bandwidth for the aggregate traffic, since the interaction between individual connections is not taken into consideration. To capture the effect of multiplexing, the Gaussian approximation is used together with the equivalent capacities. In particular, the total bandwidth required for the aggregate traffic of n connections C is given by

$$C = \min\left\{m + \alpha'\ \sigma,\ \sum_{i=1}^{n} c_i\right\} \tag{4.5}$$

Note that the mean m_i and the variance σ_i^2 of the connection bit rate with the above flow model are respectively equal to $R_i b_i$ and $m_i(R_i - m_i)$ and that

$$m = \sum_{i=1}^{n} m_i,\ \sigma^2 = \sum_{i=1}^{n} \sigma_i^2$$

Using (4.5), the steps of the call admission procedure are given as follows:

1. Given the parameters $(m_{n+1}, R_{n+1}, b_{n+1})$ of a new connection and the current values of the aggregated traffic statistics m, σ, and $\Sigma_{i=1}^{n} c_i$:
   a. Calculate c_{n+1} from (4.4).
   b. Calculate the new values of $m' = m + m_{n+1}$ and $\sigma^{2'} = \sigma^2 + m_{n+1}(R_{n+1} - m_{n+1})$.
2. Use (4.5) to calculate the new value of C':
   a. If C' is less than the link bandwidth, then accept the connection; otherwise reject it.

4.3.3.4 Flow Approximation to Cell Loss Rate

Let us now consider a VBR source alternating between active and silent periods characterized only by its peak p_i and average m_i bit rates. No assumptions on the distributions of the two periods are required, except that the probability of source i being active (idle) is equal to m_i/p_i $(1 - m_i/p_i)$. Accordingly, this method requires a minimal amount of information to characterize a source while taking into consideration the burstiness of the traffic.

Consider a link with N independent sources, a transmission rate of C cells/sec, and a buffer size of M. Let r_i and R be the random variables denoting, respectively, the cell arrival distribution of connection i and the aggregate traffic from N connections; that is,

$$R = \sum_{i=1}^{N} r_i$$

Given this parameterization, a continuous cell stream with rate R arrives at the queue and departs at rate C in the flow model of the system under consideration. With q denoting the number of buffer slots available, the rate at which cells are lost, L, is given as follows:

$$L = \sum_{X>C} (X - C) \Pr(R = X)\{1 - \Pr(q > 0 \mid R = X)\} \tag{4.6}$$

That is, if $X > C$ and the buffer is full, the rate at which cells are lost is equal to $(X - C)$. The derivation of what follows is given in Appendix E.

A supremum of L with respect to the burst length is given by

$$OF = \sum_{X>C} (X - C)\Pr(R = X) \tag{4.7}$$

Using (4.7), an upper bound on the cell loss probability PV is equal to

$$PV = OF / \sum_{i=1}^{N} m_i \tag{4.8}$$

An approximation to PV is given by

$$PV = \frac{\prod_{n=1}^{N} (m_i/p_i)\{e^{-s^* p_n} - 1\} + 1}{s^* e^{s^* C} \sum_{i=1}^{N} m_i} \tag{4.9}$$

with s^* obtained as the root of

$$\sum_{i=1}^{N} \frac{m_i e^{sp_i}}{(m_i/p_i)\{e^{sp_i} - 1\} + 1} - \frac{1}{s} - C = 0 \tag{4.10}$$

PV can now be used to make call admission decisions similar to the other techniques discussed above. There are two drawbacks of this approach. First, it does not use the buffer size in decision making, thereby restricting the amount of statistical gains that can be achieved. Secondly, the procedure does not distinguish between the cell loss requirements of individual connections. Cell loss probability for each connection PV_n can be obtained by modifying (4.9) to obtain OF_n as follows (cf. [45]):

$$OF_n = \sum_{X>C} (X - C) \Pr(R = X) p_n / X, \qquad n = 1, \ldots, N \tag{4.11}$$

Then the call admission procedure can compare the calculated value of PV_n for a given current load on a link with the QOS requirement of the connection and make the decision to admit or reject.

4.3.3.5 Nonparametric Approach

The nonparametric approach [157] is based on the peak and average cell rates of connections and does not require any knowledge of the distribution of the arrival process. Consider a link with a transmission capacity of C cells/sec. A link in this context can be a VP with a preassigned bandwidth. A slot length at this link is defined as the time to transmit one cell (i.e., $1/C$). P_{CLR} is the cell loss ratio the link is provisioned for. Furthermore, let R_i' and a_i' denote the peak and average cell rate of connection i. In this approach, the transmission buffer is monitored and observed at every r slots, referred to as the observation period. The new parameters R_i and a_i are defined as $R = \lceil rR_i'/C \rceil$ and $a_i = \lceil ra_i'/C \rceil$ denoting the maximum and average number of cells that arrive during the observation period r, respectively.

The call admission procedure is based on an estimate of the blocking probability when there are n multiplexed connections onto the link. In particular, let $U(n, r)$ denote this estimate. If $U(n, r)$ is conservative (i.e., $P_{CLR} \cdot \leq U(n, r)$), the call admission procedure is based on first calculating $U(n + 1, r)$, which is the CLR after the new connection request is superposed together with n already established connections. If $U(n + 1, r)$ is greater than the desired loss ratio, then the new connection is accepted; otherwise it is rejected.

A recursive scheme to estimate $U(n, r)$ is proposed in [157]. Let $A = \Sigma_{i=1}^{n} a_i$ be the average number of cells arriving from all n multiplexed connections. Then

$$U(n, r) = S(r + 1)/A$$

where $S(r + 1)$ with new connection $n + 1$ is calculated from $S(j)$, i = 0, . . . , r, and the parameters of the new connection are calculated as

$$\left\{1 - (a_{n+1}/R_{n+1})\right\} S(r + 1) = \begin{cases} (a_{n+1}/R_{n+1}) S(r + 1 - R_{n+1}) & r + 1 \geq R_{n+1} \\ (a_{n+1}/R_{n+1})[(R_{n+1} - r + 1) + S(0) & r + 1 < R_{n+1} \end{cases}$$

Then $U(n + 1, r)$, used in the call admission decision, is given by $S(r + 1)/(A + a_{n+1})$. The derivation of $S(r + 1)$ is given in Appendix F.

4.3.3.6 Heavy Traffic Approximation

Heavy traffic approximation [158] is based on the asymptotic behavior of the tail of the queue length distribution in an infinite capacity queue with constant service times and a Markovian cell arrival process governed by a probability matrix $P(z)$. $P\{\text{queue length} > i\} = p$ can be used to make call admission decisions. In particular, the arrival process is general enough to model the superposition of N sources into a queue. If the probability p with the new connection included in the superposed process is less than or equal to the desired loss ratio, then the new connection is accepted; otherwise it is rejected.

The probability p is estimated as $P\{\text{queue length} > i\} = \alpha(1/z^*)^i$, which requires the estimation of the values of z^* and α.

For an on/off source with parameters R denoting its peak rate, m denoting its average rate, and b denoting the average burst length, the constant α is set to the total utilization at the queue with N multiplexed connections; that is,

$$\alpha = \sum_{i=1}^{N} m_i / R_i$$

Assuming that the on and off periods are exponentially distributed, z^* is approximated as

$$z^* = 1 + \frac{1 - \alpha}{\sum_{i=1}^{N} m_i (1 - \rho_i)^2 b_i}$$

with $\rho = m/R$. Furthermore, the approximation is extended in [159] to sources with arbitrary on and off period distributions. In this case,

$$z^* = 1 + \frac{2(1 - \alpha)}{\sum_{i=1}^{N} m_i (1 - \rho_i)^2 b_i [c_i^2(\text{on}) + c_i^2(\text{off})}$$

where $c_i^2(\text{on})$ and $c_i^2(\text{off})$ are the squared coefficient of variation of the on and off periods, respectively.

4.3.3.7 A Comparative Study of Various Call Admission Algorithms

This section is based on a comparative study of various call admission algorithms reported in [160]. The three algorithms considered are the equivalent capacity method, the heavy traffic approximation, and the nonparametric approach.

Based on the experiments reported in [160], the equivalent capacity method achieves the highest level of statistical multiplexing while guaranteeing the desired application CLR at the link.

The heavy traffic approximation becomes effective as the ratio of the buffer size to the burst length increases. If the buffer size is small, then it coincides with the peak bandwidth allocation (i.e., deterministic allocation). It appears that the nonparametric approach is less sensitive to buffer size. In particular, increasing the buffer size a hundredfold may not result in a significant increase in the amount of statistical multiplexing that can be achieved with the nonparametric approach. As the buffer size goes to infinity (which is not possible in practice), both the equivalent capacity and the heavy traffic approximation converges to the same level of statistical multiplexing.

The next set of experiments reported in [160] investigates the effect of CLR on the amount of statistical multiplexing that can be achieved with each method while all other parameters (i.e., buffer size and source traffic parameters) are kept the same. It was observed that, of the three methods, the nonparametric approach is the most sensitive to CLR, followed by the heavy traffic approximation and the equivalent capacity. As expected, the amount of statistical multiplexing increases as the CLR requirement of the source increases in all three methods. Relatively, in each method, the amount of statistical multiplexing increases the most with the nonparametric approach and the least with the equivalent capacity method.

4.4 TRAFFIC SHAPING

The amount of bandwidth reserved for a connection is bounded below by its average bit rate and above by its peak bit rate. For most VBR sources, cells are generated at the peak rate during the active period, while no cells are transmitted during the silent period. Taking advantage of this source behavior, it is possible to reduce the peak rate by buffering cells before they enter the network so that the departure rate from the queue is less than the peak arrival rates of cells (note that the departure rate has to be greater than the average bit rate; otherwise the queue would become unstable). Shaping the peak rate of connections can be done at the source equipment or at the network access point.

To illustrate the effectiveness of shaping, consider a scenario in which packets are generated by sources attached to a 16-Mbps token ring. The average packet size is 1,000K, and traffic utilization at the ring is 10%. Then, assuming that a burst consists of one packet, the average duration of the active and silent periods are respectively equal to 0.5 and 4.5 ms. Furthermore, the average bit rate is 1.6 Mbps, giving a peak-to-average bit ratio of 10. As an example, if we use the equivalent capacity method (i.e., (4.4)), the reserved bandwidth in the network is equal to 15.8 Mbps. Let us now consider the same scenario after the traffic is shaped before entering the network so that its peak rate is now equal to 4 Mbps. Then the duration of the active and silent periods are respectively equal to 2 and 3 ms and the peak-to-average bit ratio becomes 2.5. With the substitution of these values, (4.4) now yields an equivalent bandwidth of 2.8 Mbps, more than fivefold savings in bandwidth.

Hence, shaping the traffic before it enters the network can result in significant bandwidth savings, particularly for sources with high burstiness. However, this is achieved at the expense of delaying the user traffic, which cannot be tolerated by delay-sensitive traffic. Accordingly, the amount of shaping (i.e., reduction at the peak rate) is limited by the amount of delay that can be tolerated.

4.5 TRAFFIC POLICING

Call admission is not sufficient to prevent congestion, the main reason being that users may not stay within the connection parameters negotiated at the call setup phase, because:

- Users may not know or they may underestimate the connection requirements;
- User equipment may malfunction;
- Users may deliberately underestimate their bandwidth requirements in order, for example, to pay less;
- Users may deliberately try to crush the network.

Accordingly, the network must ensure that sources stay within their connection parameters negotiated at the connection setup phase. This function, referred to as *traffic policing* or *usage parameter control* (UPC), resides at the access point to the network (i.e., at the UNI). A policing function should detect a nonconforming source as quickly as possible and take appropriate actions to minimize the potentially harmful effect of the excess traffic. This should be achieved transparently to conforming users, in that the traffic generated by such sources should not be artificially delayed at the interface.

Actions that a policing function can take when a source is detected to be nonconforming include:

- Dropping violating cells;
- Delaying violating cells in a queue so that the departure from the queue conforms with the contract;
- Marking violating cells differently than the cells that stay within the negotiated parameters and transmitting them so that the network can treat them differently when congestion arises;
- Adaptively controlling the traffic by informing the source when it starts to violate its contract.

The set of traffic parameters to be controlled is the set of parameters used to characterize a source. Since the latter remains an open issue, the former is necessarily unresolved. Various policing mechanisms have been proposed in the literature. In most of these schemes, the controlled source parameters include the peak and average bit rates and the length of active periods.

4.5.1 Leaky Bucket

The leaky bucket scheme was first introduced in [2]. Since then, a number of its variants have been proposed. The basic idea behind this approach is that a cell, before entering the network, must obtain a *token* from the token pool. An arriving cell will consume one token and immediately depart from the leaky bucket if there is at least one available in the token pool. Tokens are generated at a constant rate and placed in a token pool. There is an upper bound on the number of tokens that can be waiting in the pool and tokens arriving at a time the token pool is full are discarded. The size of the token pool imposes an upper bound on the burst length and determines the number of cells that can be transmitted back to back, controlling the burst length. The maximum number of cells that can exit the leaky bucket is greater than the pool size, since while cells arrive and consume tokens, new tokens are generated and placed at the pool.

The two types of enforcement action that can be taken with the leaky bucket scheme and whether or not there is a user buffer give rise to four different versions of the leaky bucket scheme, as illustrated in Figure 4.3. In scheme 1(a), an arriving cell is dropped if it arrives when there is no token waiting in the token pool. Since tokens are generated at a constant rate, this scheme can be used to control either the peak or the average cell transmission rate (but not both).

Instead of being dropped, arriving cells in schemes 2(b) and 4(d) are placed in a buffer if they arrive when there is no token available at the pool. If there is a cell waiting in the queue when a token arrives, then the first cell in the queue will immediately consume the arriving token and exit the leaky bucket. Hence, there cannot be any tokens in the token pool and cells in the buffer simultaneously. The buffer size is finite and cells arriving at a time the buffer is full are discarded. In the buffered version, the operation of leaky bucket is no longer transparent to the user due to delays introduced by buffering. The token pool size can be set to a large value to reduce the buffering delays, but this would cause large bursts to enter the network, limiting the effectiveness of the method, and would cause the cell loss probabilities to increase within the network.

An alternative to discarding cells before they enter the network is to allow them to enter and discard them within the network at the congested nodes. The main reasoning behind doing this is to increase the resource utilization in the network. Furthermore, due to the statistical nature of user traffic, this approach provides a safeguard mechanism that penalizes sources for not knowing their traffic characteristics more accurately or for transmitting excess traffic

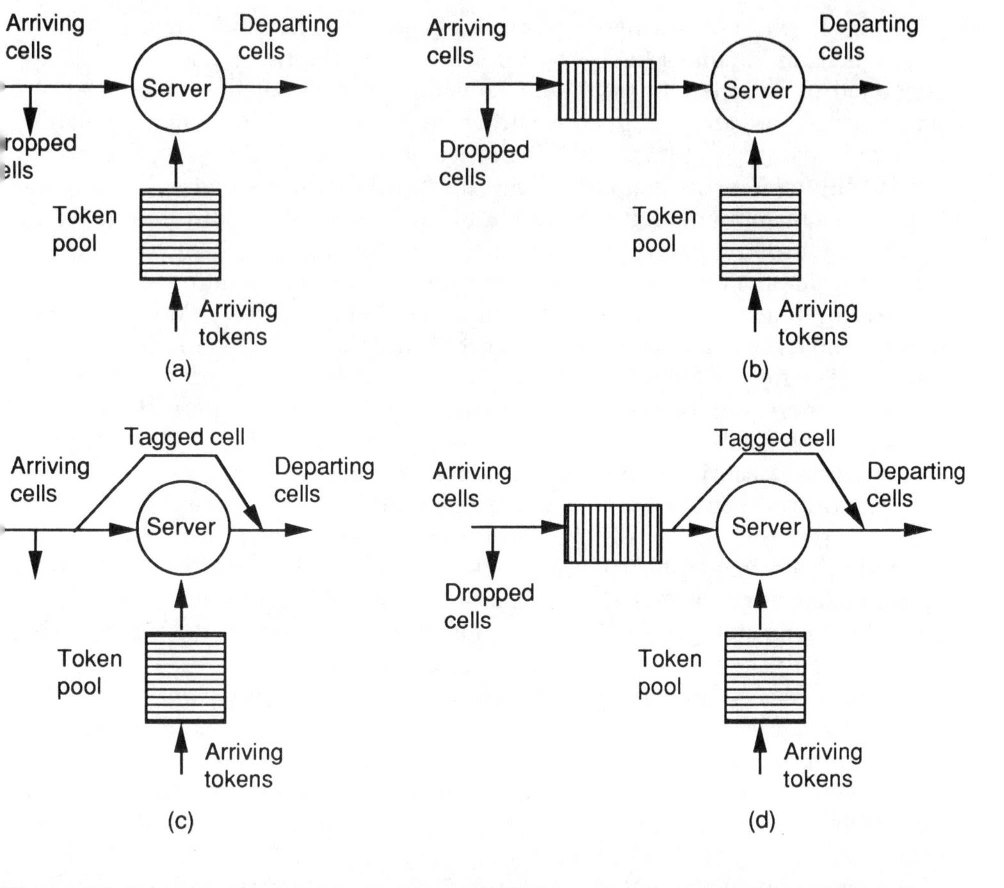

Figure 4.3 Four leaky bucket schemes: (a) scheme 1; (b) scheme 2; (c) scheme 3; (d) scheme 4.

in a short period of time. Schemes 1 and 3, or 2 and 4, in Figure 4.3 are essentially the same, except that schemes 2 and 4 allow cells of nonconforming sources to enter the network. It is necessary to guarantee that the traffic of conforming sources is not negatively affected due to this excess traffic. The ATM cell header includes a reserved CLP bit that can be used for this purpose. As long as the source is observed to be staying within its connection parameters, cells are transmitted to the network with their CLP bits unset. As soon as the policing function detects that a source is nonconforming, the CLP bit is set before cells are transmitted. This is repeated for each cell until the source is observed to be conforming again. This framework is sufficient for the network to treat these cells differently. In particular, a network node discards the cells with CLP bits set before they discard cells with CLP bits unset when congestion occurs. Two discarding mechanisms are discussed in Section 4.6.

The leaky bucket scheme proposed in [8] uses red and green tokens in implementing scheme 4 (similarly for scheme 2). The two types of tokens are generated at different rates. A cell consumes only a single color token. In particular, if the number of cells waiting in the queue is less than a threshold, then cells use green tokens to exit the leaky bucket, whereas red tokens are used if the queue size is greater than the threshold. Cells that exit the leaky bucket by consuming a red cell have a lower loss priority within the network and are discarded before the cells that use green tokens. This scheme introduces two more parameters: the threshold and the red token generation rate.

Leaky bucket can be used to police both VPs and VCs. For example, a carrier who leases VPs for private networks is not interested in policing individual VCs multiplexed onto these VPs or in their source characteristics. The only metric of interest to the carrier is the peak aggregate bit rate offered to the VPs, which can be easily policed at the access nodes in the network. Scheme 1 or 3 of Figure 4.3 can be employed for this purpose, depending on whether the excess traffic will be allowed to enter the network after being marked. To police the peak rate only, it is sufficient to set the token generation rate to the VP bandwidth. For example, policing a 10-Mbps VP would require one token to be generated at every 42.4 μs. Furthermore, a small token pool that can hold a few tokens is enough to compensate for small variations in the aggregate bit rate.

The selection of leaky bucket parameters is much more complex when it is used to police individual connections (i.e., VCs). In this case, the source characterization includes at least the peak and the average bit rates of sources. As we will illustrate shortly, a single leaky bucket cannot be used to police both metrics simultaneously. As with VP policing, the peak bit rate of a connection can be easily policed by employing a leaky bucket for each connection. The average bit rate is not so easy to police. In particular, setting the token generation rate to the average cell generation rate in scheme 1 or 3 is not sufficient, because sources generate traffic at their peak rates during their active periods and do not generate any traffic during their silent periods. A large number of cells in this setting would either be dropped or marked, regardless of whether the source conforms to its connection parameters or not.

Let us now recall the following two definitions of average measurements defined in Chapter 3: *true average cell rate* denotes the number of cells generated during the connection duration divided by the length of the duration, and, *average cell rate* is the number of cells generated by a source measured over a "long" interval of time T divided by T. Policing the true average cell rate is useless, because by the time it is determined that a source does not conform to its true average cell rate, the connection is terminated and the network is already flooded with excessive cells. The difficulty in policing the average cell rate is to define the length of time T to collect statistics. The details of how leaky bucket is used to police the peak and average cell rates are described in

Chapter 6, and a method to analyze the leaky bucket scheme is discussed in Appendix H.

Effectiveness of the Leaky Bucket Scheme

As discussed previously, policing the peak rate of a connection using a leaky bucket is relatively straightforward. Accordingly, hereafter, we only focus on the effectiveness of the leaky bucket scheme with regard to policing the burstiness of the connection.

In order for its operation to be transparent to conforming sources, it is generally agreed that the cell loss probability at the leaky bucket should be less than 10^{-10} regardless of the actual cell loss requirements of applications, since the latter is an end-to-end performance metric requested from the network.

The cell loss probability at a leaky bucket depends on the value of $C + K$ rather than the individual values of C and K (cf. Appendix H). Hence, dimensioning a leaky bucket requires the values of the two parameters $C + K$ and N. Once $C + K$ is determined, the individual values of C and K can be determined by considering the delay requirement of the application. Buffered leaky bucket introduces a nonzero waiting time, even for conforming sources. The waiting time decreases as the token generation time and the token pool size increase (and buffer size decreases). However, as the token pool size increases, the probability of transmitting longer bursts to the network increases, which may cause performance degradation for the aggregated traffic in the network.

A detailed study was reported in [70] on the effectiveness of the leaky bucket scheme. The definition of burstiness used is the squared coefficient of variation c^2 of the interarrival times of cells. An unbuffered leaky bucket is dimensioned for a Bernoulli source with $\rho = 0.1$ cells/slot and $c^2 = 0.9$, so that the cell loss probability is less than 10^{-10}.

Two sets of experimentation are reported. In the first set of experiments, the mean arrival rate is kept constant, whereas c^2 of the cell interarrival time is varied to simulate a source that conforms to the negotiated average bit rate but violates the burstiness part of the contract. Note that if appropriate network resources are not reserved in the network, a source that submits cells into the network with a higher c^2 than the negotiated value would not only receive degraded performance from the network, but would also cause higher cell losses for the traffic of conforming sources. In the second set of experiments, both the average cell rate and its squared coefficient of variation are varied. Their conclusions include:

- A leaky bucket with a small token pool size and high token arrival rate may be used to effectively police the c^2 of the cell interarrival time. However, the performance of leaky bucket in policing the c^2 of the cell interarrival time degrades as the token pool size increases.

- Policing the average cell interarrival time while being transparent to conforming sources requires the use of large token pool sizes. However, in this case, cell losses are insufficient to maintain the average cell departure rate at levels less than or equal to that of the average cell interarrival time.
- A leaky bucket cannot be configured to police both the average and squared coefficient variation of cell interarrival times.

A similar set of experiments is repeated with the buffered leaky bucket. Although the above conclusions still hold, buffered leaky bucket is observed to have a better performance in policing the c^2 of the cell traffic when a source violates both the average and c^2 of the cell traffic. A bufferless dual leaky bucket is proposed in [71]. In this scheme, there are two token pools with different token pool sizes and generation rates. The first token pool is kept small with a large token generation rate to police the c^2 of the source traffic, while the second token pool is configured to police the average cell rate with a large token pool size. Note that the cell departure rate from the leaky bucket is bounded above by the token generation rate. Hence, the token generation rate at the second pool is required to be set higher (but close) to the average cell rate. An arriving cell is required to consume one token from each token pool. If a token is not available in either one of the pools, then the arriving cell is lost. A lost cell consumes a token even if the other token pool is empty. Dimensioning a bufferless dual leaky bucket requires four parameters to be determined: the size and the generation rate for each of the token pools. In [71] these parameters are calculated by treating each token pool in isolation.

4.5.2 Window-Based Techniques

In addition to the leaky bucket scheme, various window-based policing mechanisms have been proposed to regulate traffic flow submitted to the network. In this section we discuss four of them: the jumping window, triggered jumping window, moving window, and exponentially weighted moving-average.

4.5.2.1 Jumping Window

The jumping window (JW) scheme imposes an upper bound m on the number of cells accepted from a source during a fixed time interval T, referred to as the window. The new interval starts immediately after the end of the preceding interval. Once m cells are received and transported, all consecutively arriving cells are dropped and are not allowed to enter the network. As with the leaky bucket scheme, these excessive cells can be marked to have low loss priority using the CLP bit and may be accepted, although this is an implementation issue.

With the JW mechanism, the rate at which cells offered to the network, λ_p, is equal to m/T. Hence, there are two independent variables, m and T, and one equation $\lambda_p = m/T$. With $m = 1$ and $T = N$ (i.e., token generation time), the JW scheme becomes equivalent to the leaky bucket scheme. In general, the peak or average cell submission rate to the network can be policed relatively easily by choosing m, say, arbitrarily. Then $T = m/\lambda_p$, where λ_p is the rate to be controlled. We note that the JW scheme cannot be used to control both the average and peak rates simultaneously.

The choice of m cannot be done totally arbitrarily. For example, a JW with parameters $m = 5$ and $T = 10$ would give the same controlled rate as the JW with parameters $m = 10$ and $T = 20$. The difference is the number of cells submitted to the network during a window. As T increases, it takes longer to detect that the controlled rate is exceeded. However, as T decreases, the JW scheme may become nontransparent to users that stay within the negotiated parameters. For example, consider a source that generates deterministically 10 cells during its active period and stays inactive for a 30-cell transmission time. Then assuming for presentation purposes that the window starts simultaneously with the active period, a JW with $m = 1$ and $T = 4$ would cause up to 3 out of every 4 cells to be dropped, although the source does not violate its contract.

In order to find the probability pv that a JW will drop the cells of a conforming source, let $x_j(T)$ be a random variable denoting the number of cells generated by a source during an interval of time T. In general, assuming that the interarrival times of cells is a renewal process, the Laplace transform of the counting process $\phi^n(s)$ of having n arrivals during a period of time t is given as follows, where $\phi(s)$ is the Laplace transform of the interarrival times of cells and λ is the mean cell generation rate of the source (cf. [122]).

$$\phi(s) = (\lambda/s^2)[\phi^{n-1}(s) - 2\ \phi^n(s) + \phi^{n+1}(s)] \quad n \geq 1 \tag{4.12}$$

Then pv is given by the ratio of the number of dropped cells by the JW scheme to the total number of cells generated; that is,

$$pv = \frac{\sum_{i=1}^{\infty} i x_{m+i}(T)}{\sum_{i=1}^{\infty} i x_i(T)} \tag{4.13}$$

where the denominator of (4.13) is equal to λT (cf. [66]) and the numerator is given by

$$\lambda T - N + \sum_{i=0}^{m-1} (N - i) x_i(T)$$

If the source is modeled by a two-phase Markovian model with exponentially distributed active and silent periods, given the Laplace transform $\phi(s)$ of the interarrival times of cells, (4.13) is equivalent to (cf. [122])

$$pv = \frac{\zeta^{-1}\{\phi^m(s)/s^2\}_{t=T}}{T} \tag{4.14}$$

where $\zeta^{-1}\{f(s)\}$ denotes the inverse Laplace transform of the function $f(s)$. Equation (4.14) can be used to obtain the violation probability of a JW with parameters m and T and the interarrival time distribution of cells. This probability should be less than 10^{-9} in order for the policing mechanism to be transparent to conforming users.

4.5.2.2 Triggered Jumping Window Mechanism

In the JW scheme, the window is not synchronized with the source activity. As its name indicates, the window in the triggered jumping window (TJW) mechanism is triggered by the first arriving cell (of the busy period). Hence, consecutive windows in TJW are not necessarily consecutive in time, but are triggered by the arriving cells.

The probability of violation in TJW can be obtained in a similar way to that of the JW mechanism. The main difference is that in a JW, the beginning of a window does not correspond to an arrival, whereas in a TJW the window starts with an arrival. In particular, the only difference with the above framework is the probability distribution of the number of arrivals during a period of time t, which in this case is (cf. [122])

$$\phi(s) = (1/s^2)[\phi^{n-1}(s) - \phi^n(s)], \quad n \geq 1 \tag{4.15}$$

The violation probability pv can now be calculated from (4.13) after (4.12) is replaced by (4.15).

4.5.2.3 Moving Window

In the moving window (MW) scheme, as with the JW mechanism, the maximum number of cells allowed during a predefined interval of time is constant. The main difference between the two mechanisms is that, in MW, each cell is remembered for exactly one window (i.e., T time units). Therefore, the MW scheme can be interpreted as a window that is steadily moving along the time axis.

To obtain the violation probability, the scheme can be modeled as an m-server queue with service time being equal to the duration of the window. A cell arriving when there are m cells in the system is assumed to be lost. The cell loss probability is then equal to the violation probability (cf. [122]).

4.5.2.4 Exponentially Weighted Moving Average

The exponentially weighted moving-average (EWMA) mechanism operates in a similar way to the JW scheme. The window size T is constant and a new window is triggered immediately after the preceding one ends. The difference between the two is that, in EWMA, the number of cells m_i accepted during the ith window varies from one window to the next. In particular, m_i is an exponentially weighted sum of the number of accepted cells, x_{i-1}, in the preceding window and the mean number of cells m:

$$m_i = \frac{m - \delta S_{i-1}}{1 - \delta} \qquad 0 \leq \delta < 1 \tag{4.16}$$

with $S_{i-1} = (1 - \delta)\, x_{i-1} + S_{i-2}$.

Equation (4.16) can be equivalently written as

$$m_i = \frac{m - (1 - d)(dx_{i-1} + \ldots + d^{i-1}x_1) - d^{i+1}S_0}{1 - d} \tag{4.17}$$

where S_0 is the initial value of the EWMA measurement.

Factor δ controls the flexibility of the mechanism with respect to the burstiness of the traffic. If $\delta = 0$, then m_i is constant and EWMA reduces to the JW mechanism. Variable source behavior is allowed with $\delta > 0$.

No analytical expression or methodology has been reported to analyze the performance or to determine the values of T and δ of the EWMA scheme.

4.5.2.5 Effectiveness of Window-Based Schemes

It is generally agreed that the leaky bucket performs better than window-based schemes, and is also recommended by ITU-T and the ATM Forum as the policing mechanism of choice. In reality, neither type satisfactorily solves the problem of policing the source behavior, in which the metrics of interest include the peak and average bit rates, burst duration, and even the distribution of the active and silent periods; all of these parameters are needed to characterize the traffic in the network accurately. Some of the problems identified include:

- The average cell rate is required to be estimated from a small number of samples, which may lead to large errors and hence incorrect policing decisions.

- The time to react and take corrective actions increases as the window size and the sampling interval increases in that excessive cells may already be submitted to the network by the time the source is detected to be violating its negotiated contract.
- Since there is some uncertainty in the characterization of source parameters, the parameters of policing mechanisms are not accurately chosen, limiting the effectiveness of the mechanisms.

In a final comparison, the leaky bucket and the EWMA appear to be the most effective of all mechanisms discussed. Others cannot cope well with short-term fluctuations in the cell streams [122].

4.6 SELECTIVE DISCARDING

As discussed previously, policing schemes are designed to protect the network and conforming users from the excess traffic generated by nonconforming users. However, to compensate for the uncertainties due to the statistical nature of user traffic and to allow higher resource utilizations, cells of nonconforming users may be admitted to the network after they are marked. This is based on the assumption that these cells can be dropped within the network if necessary when congestion occurs. Furthermore, users may choose to prioritize their cells before they are transmitted to the network. For example, a voice source may place the most significant bits of its frames into one cell and the least significant bits in another cell. Since the effect of losing cells of the latter type is less significant than the former type, different loss priorities can be assigned appropriately to the two types of cells. The same framework may be employed in other types of applications, such as video, audio, and still images.

The ATM cell header format allows the use of two priority levels (i.e., 1 bit). Hence, a cell is either a high- or a low-priority cell. Then the buffer space at the intermediate nodes can be used by incoming cells according to their priorities. Two discarding mechanisms have received considerable attention in the literature: push-out and threshold.

4.6.1 Push-out

In this mechanism, both low- and high-priority cells are admitted to the network as long as there is a space available at the intermediate switch buffer. If a low-priority cell arrives at a time the buffer is full, it is discarded. If a high-priority cell arrives during the time the buffer is full, it will be discarded only if there is no low-priority cell waiting in the queue. However, if there is at least one low-priority cell at the buffer, the high-priority cell replaces the low-priority cell and enters the queue. The only exception to the replacement of low-priority cells by high-priority ones is if a low-priority cell currently being transmitted cannot be replaced.

The main drawback of this mechanism is its implementation complexity. While high-priority cells are replacing low-priority cells, it is still necessary to ensure that the sequence of cells is preserved. Hence, the buffer can no longer be a first in, first out (FIFO). Therefore, there is a considerable overhead in managing the buffer, that is, in keeping track of where the low-priority cells are stored, as well as the sequence of both high- and low-priority cells and overall sequence of cells.

4.6.2 Threshold

In this case, a threshold value less than the buffer capacity is used to regulate the buffer occupancy between the high- and low-priority cells. Both types of cells are admitted to the queue as long as the total number of cells waiting in the queue is less than or equal to the threshold. Once the number of cells in the queue exceeds the threshold, all low-priority cells are discarded until the queue size falls below the threshold value. High-priority cells continue to enter the queue as long as there is a space available at the queue.

The main problem in this approach is determining the value of the threshold. If it is set to a very low value, low-priority cells at the buffer may be unnecessarily discarded (i.e., allowing more low-priority cells would not affect the QOS given to high-priority cells), thereby restricting the effectiveness of marking cells. If, on the other hand, the threshold is set to a large value, then the performance of high priority cells may degrade, since there may not be enough space left to accommodate them. Furthermore, although the threshold value depends on the characteristics (e.g., load, burstiness, correlation) of both types of cell streams, more importance is given to the characteristics of high-priority cells to guarantee their QOS requirements. Therefore, it may be necessary to adjust the threshold value as the traffic characteristics at the buffer change. Unlike the push-out mechanism, however, the threshold mechanism can be implemented in FIFOs without much complexity.

Both mechanisms produce rather similar cell loss behavior for both types of cells if the threshold value is set accurately according to the traffic load and its characteristics. Threshold is preferred over push-out due to its lower implementation simplicity.

4.7 REACTIVE CONGESTION CONTROL MECHANISMS

Momentary periods of cell losses occur within the network when a large number of sources start generating traffic simultaneously. Although preventive techniques reduce the buffer overflow probabilities, it is not possible to totally eliminate momentary periods of cell losses in the network because of the statistical nature of the traffic. Lost cells result in retransmissions by the source nodes, thereby increasing network traffic and turning the momentary buffer overflows to sustained periods of cell losses. This in turn may cause a catastrophe, reducing the effective network throughput eventually to zero. Furthermore, most preventive techniques allocate more resources than are actually required, thereby restricting the efficient use of network resources. Finally, preventive schemes often require accurate source characterization. Therefore, in addition to their preventive counterparts, reactive control mechanisms are necessary to monitor the congestion level in the network, notify sources when congestion is detected, and take action based on the congestion information. The main objective of a reactive scheme is to prevent momentary periods of overload from turning into sustained periods of cell losses. Reactive control mechanisms have been satisfactorily used in low-speed packet-switched networks. However, the time to react to congestion increases proportionally to the propagation delay bandwidth product, decreasing their effectiveness. As the link speeds increase, most cells in the network are in transit, not at switch buffers. For example, the number of cells in transit on a 5,000-km OC-3 link is more than 9,000, assuming 5-μs/km propagation delay. The buffer sizes required at the intermediate nodes to store large numbers of cells in transit momentarily during a congestion at the downstream nodes may be prohibitively expensive or may not be feasible at all, considering the delay requirements of applications. More likely, cell buffers at the intermediate nodes are expected to be on the order of a few hundred cells, mainly to limit the end-to-end delays and the cell delay variation. Furthermore, by the time the sources are informed of the congestion in the network, it may be too late to react effectively. Moreover, by the time sources are ready to react, there may no longer be a congestion to react to.

Accordingly, reactive control mechanisms are not as effective in ATM networks as they are in low-speed packet-switching networks. The effectiveness of a reactive mechanism in this environment mainly depends on the connection duration, the burst length, and the distance involved between the two communicating entities. Finally, in ATM networks, it may not be easy to identify which sources are causing the congestion. Hence, most reactive schemes require a number of sources to throttle their traffic submission. This introduces the issue of fairness.

The design and implementation of reactive control schemes in ATM networks is still an open issue. Nevertheless, they are required as safeguard mechanisms and may potentially be used to increase resource utilizations beyond what can be achieved by preventive schemes alone. Various reactive control mechanisms are discussed next.

4.7.1 End-Node Notification Techniques

Once congestion is detected in an intermediate network node, the end nodes need to be notified to be able to react. There are three techniques proposed for congestion notification in ATM networks.

4.7.1.1 Estimation by the End Nodes

In the scheme proposed in [59], a source sends time-stamped *probe cells* along the connection route periodically to measure the response time of cells between the source and the destination. These cells are treated by the network as regular cells and have meaning only at the receiver. Since there is no field in the header, the receiver is required to differentiate between data and control cells by looking into the cell payload. The probe cells are used at the end node to estimate the one-way delay. When the destination node detects congestion along the route based on this estimate, it notifies the source to adjust its rate.

The main disadvantage of this technique is that it requires control cells to be created at the origin and to be processed at the receiver. Although the amount of bandwidth required for the control cells is not significant, it is nevertheless an extra traffic burden to the network. With thousands of connections per access node, the processing burden can be significant. This technique has not been adopted by the B-ISDN standardization bodies and is left as an option for network providers.

4.7.1.2 Explicit Backward Congestion Notification

In the explicit backward congestion notification (EBCN) scheme, each node in the network monitors the queue occupancies of its trunks. When the queue size of a trunk reaches a predefined threshold value, a special cell is prepared and sent to the source nodes of all connections that pass through the congested node. This method minimizes the time it takes to notify a source that there is a congested node along its path. Since a special cell is used for this purpose, the 48-byte payload (minus overhead used to specify that the cell is carrying a congestion indication) of the cell can be used to include a variety of information about the congested node. Accordingly, sources can react to congestion along their path effectively. This scheme is used in current low-speed networks as a back-pressure mechanism. In particular, a congested node may notify its upstream nodes to throttle its traffic when a congestion state is about to be entered. The upstream node then notifies its upstream nodes to slow down or stop sending traffic if it reaches a congestion state. Continuing in this manner, sources may eventually be forced to control their traffic submission to the network. Despite these advantages, EBCN is not accepted by the standardization committees as a viable approach for congestion indication in ATM networks. This is mainly due to the high-speed transmission links, which limit the effectiveness of the method and, more importantly, the use of special cells, which imposes a considerable processing burden on intermediate nodes. Furthermore, if the same VCI value is not used in both directions, then a congested intermediate node has to establish a connection to each source before the congestion notification cell can be transmitted. Because of the latencies and the implementation complexities involved, the EBCN may be used in some designs as an exception measure.

4.7.1.3 Explicit Forward Congestion Notification

As with the EBCN scheme, each node in the explicit forward congestion notification (EFCN) scheme monitors the queue occupancies of its trunks. When the queue size of a trunk reaches a predefined threshold value, all cells passing through that trunk are marked until the congestion period ends, that is, the queue occupancy falls below another predefined threshold value. Once the EFCN bit of a cell is set at a node, it cannot be modified by any other node along the path to the receiver. A reserved bit in the cell header that may be used for this purpose is defined by ITU-T. A cell received at a destination node with the EFCN bit set indicates that there is a congested node along the path of this connection. In this scheme, receivers have no information about which nodes in the network are congested and there is no indication that the particular connection the marked cell belongs to is nonconforming to the negotiated connection parameters.

Due to large propagation delays relative to the transmission times of cells, by the time a marked cell arrives at a receiver, it is possible that the congested nodes that marked the cell are no longer congested. Nevertheless, the source has no means of knowing that the congestion is cleared and has to slow down as required. Hence, receivers do not react to congestion indication very quickly. Instead, statistics are collected so that there can be an accurate determination of whether there is a momentary or sustained congestion along the path. If it is decided that the latter is the case, then the receiver sends a notification back to the source.

4.7.2 Adaptive Rate Control

In adaptive rate control, the rate at which traffic is submitted to the network is varied by a source depending on the congestion status information available to it. In an adaptive rate control mechanism based on the EFCN scheme proposed in [97], the source traffic passes through a variable-rate server. The rate of the server is controlled by the feedback information from the destination node. Initially, the server starts submitting traffic at a nominal rate and slowly increases the transmission rate to the peak rate of the source. When a congestion notification is received from the destination node, the access rate to the network is reduced to the nominal rate and is not increased for a "backoff" period. The congestion threshold values in the network are observed to be fairly low in order for this scheme to be effective. To decide whether the received marked cells at the destination node indicate a momentary or sustained congestion in the network, they are placed at the queue that discards the cells at a predetermined constant rate. When the number of cells in this queue reaches a threshold value, a notification is sent to the source. Tuning the parameters of this system is not trivial. In particular, the effectiveness of this reactive scheme depends on the choice of threshold values used both at the intermediate nodes and at the receivers, nominal transmission rate and the rate at which the transmission rate is increased from its nominal rate to the peak bit rate.

4.7.3 Incall Parameter Negotiation

Bandwidth allocation is well suited to VBR applications with connection holding times long enough to justify the overhead involved to reserve resources in the network. The services for these applications include call establishment and termination steps which, in addition to some other functions, reserve and release bandwidth in the network. On the other hand, there are other types of applications that would require transmission of only one burst (or a few bursts) of data. Typical applications that may fall into this category are LAN interconnection, e-mail, and small-file transfer services. The time to establish and terminate connections for these types of applications can be much longer than the actual time it takes to transfer the bursts. The problem of supporting such applications in B-ISDNs while minimizing the amount of wasted network resources is a complex task and remains an unresolved issue. In this section, we only focus on the congestion-related aspects of the problem, particularly bandwidth allocation.

Incall parameter negotiation techniques are proposed to minimize the call setup overhead. Since ATM networks are connection-oriented, end-to-end connections are required to exist before the data transmission can start. The challenge is then to find the most effective means of supporting these applications within the ATM framework. One alternative is to establish VPs between the corresponding end nodes, thereby defining a virtual network that provides interconnectivity between the users of such services. Each VP with a reserved constant bandwidth in this framework originates at an end node that has complete control over the VP. It is then a relatively easy task to manage the traffic on such VPs by transmitting, delaying, or rejecting burst transfers, depending on the current load in the VP. The main disadvantage of this approach is the inefficient use of network resources, since these VPs cannot be used by others. This inefficiency can be reduced somewhat by the use of switchable VPs as opposed to single VPs connecting the end nodes.

Nevertheless, this approach illustrates that there is a feasible yet undesirable solution to the problem. Alternative efficient fast bandwidth reservation techniques are proposed in [29,30,31,50]. During the call setup, first a path from the source to destination is determined. Then a special cell is sent to the nodes along the path to reserve the peak bandwidth of connection. Each node can either accept, deny, or grant less bandwidth than requested. If the connection request is granted, then the burst is transmitted. The reserved bandwidth is released at a node along the path immediately after the last cell of the burst leaves the node. With this approach, the connection termination phase is eliminated and the establishment phase overhead is reduced. Furthermore, the buffer requirements of connections at intermediate nodes are minimal, since the bandwidth reserved is equal to their peak requirements. This in turn may cause CBR and VBR connections that use some of the links along the path to be rejected until the reserved bandwidth is released, restricting the resource utilization. Alternatively, the reservation of a predefined number of buffers may be requested at the intermediate nodes, allowing better utilization of transmission bandwidth at the expense of using extra buffers. As with bandwidth reservation, each node in this scheme may accept, reject, or grant fewer buffers than requested.

4.7.4 Dynamic Source Coding

When a congestion indication is received by a source, the rate at which traffic is submitted to the network may be reduced (or stopped) temporarily. For delay-insensitive traffic, cells may be delayed in the source buffers. For real-time delay-sensitive traffic, buffering is not a desired alternative. As discussed in Chapter 3, however, the source may either reduce the traffic submission rate or may mark cells that are not so essential, which may be reconstructed from other cells at the receiver with a low CLP before they are transmitted. For example, voice frames can be transmitted with the most significant 4 bits in one cell and the least significant 4 bits in another. The CLP bit of the latter is set before they are transmitted to provide the option of discarding these cells before the others during congestion. In the case of video services, the quantization step may be increased to reduce the amount of information generated per frame. Like voice frames, video frames with different levels of significance may be transmitted with different CLPs. For example, the I frames in MPEG coding include synchronization information and have a higher loss priority than P frames, which carry the difference from a reference frame.

An EFCN scheme may be used to inform sources of a congested network state. As discussed previously, the effectiveness of this scheme, and therefore dynamic source coding, is constrained by the propagation delay bandwidth product, in that by the time a source is aware of congestion along its path, it may be too late to react.

4.8 SUMMARY

An effective congestion control framework in ATM networks that allows high resource utilizations will greatly influence the ability of B-ISDNs to compete and be a universally accepted solution. Although a great deal of expertise has been developed over the years from circuit- and packet-switching networks, the congestion control in this new environment has a completely new face. Protocols for congestion control at intermediate nodes must be simple to be able to keep up with the small cell transmission times at high-speed links. The large bandwidth propagation delay products in ATM networks have introduced new challenges to the congestion control problem. Simple reactive control mechanisms are no longer effective in this new environment due to long feedback delays relative to transmission speeds, and large numbers of cells are in transit, making it a relatively difficult task to react to congestion. Therefore, the effectiveness of reactive schemes is limited by the duration of feedback delays and requires very large buffers. The former is a physical constraint imposed by the speed of light. The latter can be used for delay-insensitive traffic, whereas the delays imposed by large buffers are not acceptable for real-time traffic.

Preventive techniques, on the other hand, have had limited success in solving the problem. They are often sensitive to the parameters of the source traffic, which itself remains an open issue. Even with accurate source characterization, the proposed techniques often restrict utilization of network resources. Furthermore, due to the random nature of the traffic and the inaccuracies introduced, it is not possible to eliminate the congestion control problem with the use of only preventive control techniques while achieving high resource utilizations. The problem is further complicated due to the existence of different applications with varying QOS requirements.

In summary, the congestion control framework in ATM networks remains an open issue that needs to be satisfactorily addressed.

References and Bibliography

[1] Ahmadi, H., R. Guerin, and K. Sohraby, "Analysis of Leaky Bucket Access Control Mechanism with Batch Arrival Process," *Globecom '90.*

[2] Akhtar, S. "Congestion Control in a Fast Packet Switching Network," MS Thesis, Washington University, 1987.

[3] Albin, S. L., "Approximating a Point Process by a Renewal Process II: Superposition Arrival Processes to Queues," *Op. Res.,* Vol. 32, No. 5, September-October 1984, pp. 1133–1162.

[4] Andrade, J., W. Burakowski, and M. Villen-Altamirano, "Characterization of Cell traffic Generated by an ATM Source," *ITC-13,* 1991, pp. 545–550.

[5] Anick, D., D. Mitra, and M. M. Sondhi, "Stochastic Theory of a Data-Handling System with Multiple Sources," *BSTJ,* Vol. 61, No. 8, October 1982, pp. 1871–1894.

[6] Awater and G. A., and F. C. Schoute, "Optimal Queuing Poilicies for Fast Packet Switching of Mixed Traffic," *IEEE JSAC,* Vol. 9, No. 3, April 1991, pp. 458–467.

[7] Baiochi, A., et al., "Loss Performance Analysis of an ATM Multiplexer Loaded with High Speed ON-OFF Sources," *IEEE JSAC,* Vol. 9, No. 3, April 1991, pp. 388–393.

[8] Bala, K., I. Cidon, and K. Sohraby, "Congestion Control for High Speed Packet Switched Networks," *INFOCOM '90,* 1990, pp. 520–526.

[9] Berger, A. W., "Determination of Load-Service Curves for Distributed Switching Systems: Probabilistic Analysis of Overload-Control Schemes," *ITC-13,* 1991, pp. 435–440.

[10] Berger, A. W., "Performance Analysis of a Rate Control Where Tokens and Jobs Queue," *IEEE JSAC,* Vol. 9, No. 2, 1991, pp. 165–170.

[11] Berger, A. W., "Overload Control Using Rate Control Throttle: Selecting Token Bank Capacity for Robustness to Arrival Rates," *IEEE Trans. Autom. Control,* Vol. 36, No. 2, February 1991, pp. 216–219.

[12] Berger, A. W., et al., "Performance Characterizations of Traffic Monitoring and Associated Control Mechanisms for Broadband 'Packet' networks," *GLOBECOM '90,* pp. 350–354.

[13] Bhattacharya, P. P., and A. Ephremides, "Optimal Scheduling with Strict Deadlines," manuscript.

[14] Blondia, C., "The N/G/1 Finite Capacity Queue," *Stochastic Models,* Vol. 5, No. 2, 1989.

[15] Borgonovo, F., and L. Fratta, "Policing Procedures: Implications, Definitions and Proposals," *ITC-13,* 1991, pp. 859–866.

[16] Boyer, P. F., "Congestion Control for the ATM," *7th Int. Teletrafflc Congress Seminar,* Morristown, NJ, 9–11 October 1990.

[17] Briem, U., T. H. Theimer, and H. Kroner, "A General Discrete-Time Queuing Model: Analysis and Applications," *ITC-13,* 1991, pp. 13–19.

[18] Butto, M., E.Cavallero, and A. Tonietti, "Effectiveness of the 'Leaky Bucket' Policing Mechanism in ATM Networks," *IEEE JSAC,* Vol. 9, No. 3, April 1991, pp. 335–342.

[19] Castelli, P., E. Cavallero, and A. Tonietti, "Policing and Call Admission Problems in ATM Networks," *ITC-13,* 1991, pp. 847–852.

[20] Chao, H. J., "Design of Leaky Bucket Access Control Schemes in ATM Networks," *ICC '91,* 1991, pp. 180–187.

[21] Chen, M., "A Services Features Oriented Congestion Control Scheme for a Broadband Network," *ICC '91,* 1991, pp. 105–109.

[22] Chen, T. M., J. Walrand, and D. G. Messerschmitt, "Dynamic Priority Protocols for Packet Voice," *IEEE JSAC,* Vol. 7, No. 5, June 1989, pp. 632–643.

[23] Chipalkatti, R., J. F. Kurose, and D. Towsley, "Scheduling Policies for Real-Time and Non-Real-Time Traffic in a Statistical Multiplexer," *INFOCOM '89,* 1989, pp. 774–783.

[24] Daigle, J. N., and J. D. Langford, "Queuing Analysis of a Packet Voice Communication System," *INFOCOM '85,* 1985, pp. 18–26.

[25] Daigle, N., and J. D. Langford, "Models for Analysis of Packet Voice Communications Systems," *IEEE JSAC,* Vol. SAC-4, No. 6, September 1986, pp. 847–855.

[26] Ding, W., "A Unified Correlated Input Process Model for Telecommunication Networks," *ITC-13,* 1991, pp. 539–544.

[27] Dittmann, L., and S. B. Jacobsen, "Statistical Multiplexing of Identical Bursty Sources in an ATM Network," *GLOBECOM '88,* 1988, pp. 1293–1297.

[28] Dittmann, L., S. B. Jacobsen, and K. Moth, "Flow Enforcement Algorithms for ATM Networks," *IEEE JSAC,* Vol. 9, No. 3, April 1991, pp. 343–350.

[29] Doshi, B., and H. Heffes, "Overload Performance of an Adaptive, Buffer-Window Allocation Scheme for a Class of High Speed Networks," *ITC-13,* 1991, pp. 441–446.

[30] Doshi, B., and S. Dravida, "Congestion Controls for Bursty Data in High Speed Wide Area Networks: In Call Parameter Negotiations," *ITC Seventh Specialist Seminar on Broadband Tech.,* 1990.

[31] Doshi, B., S. Dravida, P. Johri, and G. Ramamurthy, "Memory, Bandwidth, Processing and Fairness Considerations in Real Time Congestion Controls for Broadband Networks," *ITC-13,* 1991, pp. 153–159.

[32] Doshi, B. T., et al., "Retransmission Protocols and Flow Controls on High Speed Packet Networks," *ITC-13,* 1991, pp. 309–314.

[33] Dron, L. G., G. Ramamurthy, and B. Sengupta, "Delay Analysis of Continuous Bit-Rate Traffic Over an ATM Network," *IEEE JSAC,* Vol. 9, No. 3, April 1991, pp. 402–407.

[34] Dziong, Z., K.-Q. Liao, and L. Mason, "Flow Control Models for Multi-Service Networks with Delayed Call Set Up," *INFOCOM '90,* 1990, pp. 39–46.

[35] Dziong, Z., K.-Q. Liao, L. Mason, and N. Tetrault, "Bandwidth Management in ATM Networks," *ITC-13,* 1991, pp. 821–827.

[36] Eckberg, A. E., "The Single Server Queue with Periodic Arrival Process and Deterministic Service Times," *IEEE Trans. Comm.,* Vol. COM-27, No. 3, March 1979, pp. 556–562.

[37] Eckberg, A. E., D. T. Luan, and D. M. Lucantoni, "Bandwidth Management: A Congestion Control Strategy for Broadband Packet Networks—Characterizing the Throughput Burstiness Filter," manuscript.

[38] Eckberg, A. E., D. T. Luan, and D. M. Lucantoni, "Meeting the Challenge: Congestion and Flow Control Strategies for Broadband Information Transport," *GLOBECOM '89,* 1989, pp. 1769–1773.

[39] Elwalid, A. I., D. Mitra, and T. E. Stern, "Statistical Multiplexing of Markov Modulated Sources: Theory and Computational Algorithms," *ITC-13,* 1991, pp. 495–500.

[40] Fendick, K., V. Saksena, and W. Whitt, "Dependence in Packet Queues: a Multi-Class Batch Poisson Model," *ITC-12,* 1989, pp. 1450–1454.

[41] Filipiak, J., "Structured Systems Analysis Methodology for Design of an ATM Network Architecture," *IEEE JSAC,* Vol. 7, No. 8, October 1989, pp. 1263–1273.

[42] Fisher, W., and K. S. Meier-Hellstern, *The MMPP Cookbook,* INRS Telecommunications, Verdun, Quebec, and AT&T Bell Labs., Holmdel, NJ, 1990.

[43] Forys, L. J., and D. E. Smith, "Servicing Bursty Systems," *Comp. Networks and ISDN Sys.,* Vol. 20, 1990, pp. 171–177.

[44] Gallager, R. G., *Information Theory and Reliable Communication,* John Wiley and Sons, 1968.

[45] Gallassi, G., G. Rigolio, and L. Fratta, "ATM: Bandwidth Assignment and Bandwidth Enforcement Policies," *GLOBECOM '89,* 1989, pp. 1788–1793.

[46] Garcia, J., and 0. Casals, "Priorities in ATM Networks," manuscript.

[47] Golestani, S. J., "Congestion-Free Transmission of Real-Time Traffic in Packet Networks," *INFOCOM '90,* 1990, pp. 527–536.

[48] Gerla, M., J. A. S. Monteiro, and R. Pazos, "Topology Design and Bandwidth Allocation in ATM Nets," *IEEE JSAC,* Vol. 7, No. 8, October 1989, pp. 1253–1262.

[49] Gersht, A., and K. J. Lee, "A Congestion Control Framework for ATM Networks," *INFOCOM '89,* 1989, pp. 701–710.

[50] Gilbert, H., O. Aboul-Magd, and V. Phung, "Developing a Cohesive Traffic Management Strategy for ATM Networks," *IEEE Comm. Mag.,* Vol. 29, No. 10, 1991, pp. 26–45.

[51] Golestani, S. J., "Congestion Free Transmission of Real Time Traffic in Packet Networks," *INFOCOM '90,* 1990, pp. 527–536.

[52] Gravey, A., and G. Hebuterne, "Mixing Time and Loss Priorities in a Single Server Queue," *ITC-13,* 1991, pp. 147–152.

[53] Gusella, R., "Characterizing the Variability of Arrival Processes with Indexes of Dispersion," *IEEE JSAC,* Vol. 9, No. 2, February 1991, pp. 203–212.

[54] Guerin, R., H. Ahmadi, and M. Naghshineh, "Equivalent Capacity and Its Application to Bandwidth Allocation in High-Speed Networks," *IEEE JSAC,* Vol. 9, 1991, pp. 968–981.

[55] Gun, L., and R. Guerin, "A Unified Approach to Bandwidth Allocation and Access Control in Fast Packet Switching Networks," *INFOCOM '92,* 1-12, Italy, 1992 pp. 1–12.

[56] Gun, L., and R. Guerin, "A Framework for Bandwidth Management and Congestion Control in High Speed Networks," to appear in *Computer Networks and ISDN.*

[57] Gun, L., and R. Guerin, "An Approximation Method for Capturing Complex Traffic Behavior in High Speed Networks," to appear in *Performance Evaluation.*

[58] Gun, L., and R. Guerin, "Effectiveness of Dynamic Bandwidth Management Mechanisms in ATM Networks," *INFOCOM '93,* San Fransisco, 1993, pp. 358–367.

[59] Haas, Z., and J. H. Winters, "Congestion Control by Adaptive Admission," *INFOCOM '91,* 1991, pp. 560–569.
[60] Hac, A., "Congestion Control and Switch Buffer Allocation in High-Speed Networks," *INFOCOM '91,* 1991, pp. 314–322.
[61] Hahne, E. L., C. R. Kalmanek, and S. P. Morgan, "Fairness and Congestion Control on a Large ATM Data Network with Dynamically Adjustable Windows," *ITC-13,* 1991, pp. 867–872.
[62] Hashida, 0., Y. Takahashi, and S. Shimogawa, "Switched Batch Bernoulli Process (SBBP) Queue with Application to Statistical Multiplexer Performance," *IEEE JSAC,* Vol. 9, No. 3, April 1991, pp. 394–401.
[63] Hashida, 0., and Y. Takahashi, "A Discrete-Time Priority Queue with Switched Batch Bernoulli Process Inputs and Constant Service Time," *ITC-13,* 1991, pp. 521–526.
[64] Haskell, B. G., "Buffer and Channel Sharing by Several Interframe Picturephone Coders," *BSTJ,* Vol. 51, No. 1, January 1972, pp. 261–289.
[65] Heffes, H., "A Class of Data Traffic Processes—Covariance Function Characterization and Related Queuing Results," *BSTJ,* Vol. 59, No. 6, July–August 1980, pp. 897–929.
[66] Heffes, H., and D. Lucantoni, "A Markov Modulated Characterization of Packetized Voice and Data Traffic and Related Statistical Multiplexer Performance," *IEEE JSAC,* Vol. SAC-4, No. 6, September 1986, pp. 856–868.
[67] Heyman, D. P., "A Performance Model of the Credit Manager Algorithm," manuscript.
[68] Hirano, M., and N. Watanabe, "Traffic Characterization and a Congestion Control Scheme for an ATM Network," *Int. J. Digital and Analog Comm. Sys.,* Vol. 3, 1990, pp. 211–217.
[69] Hirano, M., and N. Watanabe, "Characteristics of a Cell Multiplexer for Bursty ATM Traffic," *ICC '89,* 1989, pp. 399–403.
[70] Holtsinger, D. S., "Performance Analysis of Leaky Bucket Policing Mechanisms," Ph.D. thesis, Department of Electrical and Computer Engineering, NC State University, 1992.
[71] Holtsinger, D. S., "Design and Analysis of the Dual Leaky Bucket Policing Mechanism for ATM networks," manuscript.
[72] Hong, D., T. Suda, and J. J. Bae, "Survey of Techniques for Prevention and Control of Congestion in an ATM Network," *ICC '91,* 1991, pp. 204–210.
[73] Hoshida, O., Y. Takahashi, and S. Shimogawa, "Switched Batch Bernoulli Process (SBBP) and the Discrete Time SBBP/G/1 Queue with Application to Statistical Multiplexer Performance," *IEEE JSAC,* Vol. 3, 1991.
[74] Huang, S., "Source Modelling for Packet Video," ICC '88, 1988, pp. 1262–1267.
[75] Huang, S., "Modeling and Analysis for Packet Video," *GLOBECOM '89,* 1989, pp. 881–885.
[76] Hughes, D. A., H. S. Bradlow, and G. Anido, "Congestion Control in an ATM Network," *ITC-13,* 1991, pp. 835–840.
[77] Hui, J. Y., "Resource Allocation for Broadband Networks," *IEEE JSAC,* Vol. 6, No. 9, December 1989, pp. 1598–1608.
[78] Ide, I., "Superposition of Interrupted Poisson Processes and Its Application to Packetized Voice Multiplexers," *ITC-12,* 1988, pp. 3.1B.2.1–3.1B.2.7.
[79] Jalali, A., and L. G. Mason, "Open Loop Schemes for Network Congestion Control," *ICC '91,* 1991, pp. 199–203.
[80] Joos, P., and W. Verbiest, "A Statistical Bandwidth Allocation and Usage Monitoring Algorithm for ATM Networks," *ICC '89,* 1989, pp. 415–422.
[81] Jou, Y. F., A. A. Nilsson, and F. Lai, "Analysis of a Finite Capacity Polling System under Bursty and Correlated Arrivals," *Modeling and Performance Evaluation of ATM Technology,* H. Perros, G. Pujolle, Y. Takahashi, eds., Amsterdam: North Holland, 1993.
[82] Kamitake, T., and T. Suda, "Evaluation of an Admission Control Scheme for an ATM Network Considering Fluctuations in Cell Loss Rate," *GLOBECOM '89,* 1989, pp. 1774–1780.
[83] Karlsson, J. M., H. G. Perros, and I. Viniotis, "Adaptive Polling Schemes for an ATM Bus," *ICC '91,* 1991, pp. 403–407.

[84] Kim, B. G., "Characterization of Arrival Statistics of Multiplexed Voice Packets," *IEEE JSAC,* Vol. SAC-1, No. 6, December 1983, pp. 1133–1139.

[85] Kroner, H., "Comparative Performance Study of Space Priority Mechanisms for ATM Networks," *Proc. IEEE Infocom,* 1990.

[86] Kroner, H., P. J. Kuhn, and G. Willmann, "Performance Comparison of Resource Sharing Strategies Between Lost-Call-Cleared and Reservation Traffic," *ITC-13,* 1991, pp. 639–645.

[87] Kuczura, A., "The Interrupted Poisson Process as an Overflow Process," *BSTJ,* Vol. 52, No. 3, March 1973, pp. 437–448.

[88] Le Boudec, J., "An Efficient Solution Method for Markov Models of ATM Links with Loss Priorities," *IEEE JSAC,* Vol. 9, No. 3, April 1991, pp. 408–417.

[89] Le Boudec, J., "A Generalization of Matrix Geometric Solutions for Markov Models," *IBM Research Report,* RZ 1903, 1989.

[90] Leland, W. E., "Window-Based Congestion Management in Broadband ATM Networks: The Performance of Three Access-Control Policies," *GLOBECOM '89,* 1989, pp. 1794–1800.

[91] Li, S.-Y.-R., "Algorithms for Flow Control and Call Set-Up in Multi-Hop Broadband ISDN," *INFOCOM '90,* 1990, pp. 889–895.

[92] Li, S.-Q., "A General Solution Technique for Discrete Queuing Analysis of Multimedia Traffic on ATM," *IEEE Trans. Comm.,* Vol. 39, No. 7, July 1991, pp. 1115–1132.

[93] Liao, K. Q., and L. G. Mason, "A Discrete-Time Single Server Queue With a Two Level Modulated Input and Its Applications," *GLOBECOM '89,* 1989, pp. 913–918.

[94] Lim, Y., and J. Kobza, "Analysis of a Delay Dependent Priority Discipline in a Multiclass Traffic Packet Switching Node," *INFOCOM '88,* 1988, pp. 889–898.

[95] Louvion, J. R., J. Boyer, and J. B. Gravereaux, "Statistical Multiplexing of VBR Sources in ATM Networks," *3rd IEEE CAMAD Workshop,* 1990.

[96] Lucantoni, D. M., and S. P. Parekh, "Selective Cell Discard Mechanisms for a B-ISDN Congestion Control Architecture," *Proc. ITC-7 Seminar,* New Jersey, 1990.

[97] Macrucki, B., "Explicit Forward Congestion Notification in ATM Networks," *Proc. TRICOMM '92,* H. G. Perros, ed., New York: Plenum, 1992.

[98] Madsen, H., and B. F. Nielsen, "The Use of Phase Type Distributions for Modeling Packet-Switched Traffic," *ITC-13,* 1991, pp. 593–599.

[99] Magd, 0. A., H. Gilbert, and M. Wernik, "Flow and Congestion Control for Broadband Packet Networks," *ITC-13,* 1991, pp. 853–858.

[100] Mitra, D., "Optimal Design of Windows for High Speed Data Networks," *INFOCOM '90,* 1990, pp. 1156–1163.

[101] Miyao, Y., "A Call Admission Control Scheme in ATM Networks," *ICC '91,* 1991, pp. 391–396.

[102] Murase, T., H. Suzuki, and T. Takeuchi, "A Call Admission Control for ATM Networks Based on Individual Multiplexed Traffic Characteristics," *ICC '91,* 1991, pp. 193–198.

[103] Murata, M., Y. Oie, T. Suda, and H. Miyahara, "Analysis of a Discrete-Time Single-Server Queue with Bursty Inputs for Traffic Control in ATM Networks," *GLOBECOM '89,* 1989, pp. 1781–1787.

[104] Nagarajan, R., J. F. Kurose, and D. Towsley, "Approximation Techniques for Computing Packet Loss in Finite-Buffered Voice Multiplexers," *IEEE JSAC,* Vol. 9, No. 3, 1991, pp. 368–377.

[105] Neuts, M. F., "A Versatile Markovian Point Process," *J. Appl. Prob.,* Vol. 16, 1979, pp. 764–779.

[106] Neuts, M. F., *Matrix Geometric Solutions in Stochastic Models,* Baltimore: Johns Hopkins University Press, 1981.

[107] Norgaard, K., "Evaluation of Output Traffic from an ATM Node, *ITC-13,* 1991, pp. 533–537.

[108] Norros, I., J. W. Roberts, A. Simonian, and J. T. Virtamo, "The Superposition of Variable Bit Rate Sources in an ATM Multiplexer," *IEEE JSAC,* Vol. 9, No. 3, April 1991, pp. 378–387.

[109] Ohba, Y., M. Murata, and H. Miyahara, "Analysis of Interdeparture Processes for Bursty Traffic in ATM Networks," *IEEE JSAC,* Vol. 9, No. 3, April 1991, pp. 468–476.

[110] Ohta, M., "Fluid Model for a Traffic Congestion Prediction," *ITC-13,* 1991, pp. 665–670.

[111] Park, D., H. G. Perros, and H. Yamashita, "Approximate Analysis of Discrete Time Tandem Queuing Networks with Bursty and Correlated Input Traffic and Customer Loss," Tech. Rep., Computer Science Dept., North Carolina State University, 1992.

[112] Park, D., and H. G. Perros, "m-MMBP Characterization of the Departure Process of an m-MMBP/Geo/1/K queue," Tech. Rep., Computer Science Dept., North Carolina State University, 1993.

[113] Perros, H. G., *High Speed Networks,* New York: Plenum, 1992.

[114] Perros, H. G., and R. O. Onvural, "On the Superposition of Arrival Processes for Voice and Data," *Proc. 4th Int. Conf. Data Comm. Sys. and Their Performance,* Barcelona, 1990, pp. 341–357.

[115] Petr, D. W., and V. S. Frost, "Optimal Threshold Based Discarding for Queue Overload Control," *Proc. ITC-7 Seminar,* New Jersey, 1990.

[116] Ramamurthy, G., and B. Sengupta, "Delay Analysis of a Packet Voice Multiplexer by the ΣD/D/1 Queue," *IEEE Trans. Comm.,* Vol. 39, No. 7, July 1991, pp. 1107–1114.

[117] Ramamurthy, G., and R. S. Dighe, "A Multidimensional Framework for Congestion Control in B-ISDN," *IEEE JSAC,* Vol. 9, No. 9, 1991, pp. 1440–1451.

[118] Ramamurthy, G., and R. S. Dighe, "Distributed Source Control: A Network Access Control for Integrated Broadband Packet Networks," *INFOCOM '90,* 1990, pp. 896–907.

[119] Ramaswami, V., "The N/G/1 Queue and Its Detailed Analysis," *Adv. Appl. Prob.,* Vol. 12, 1980, pp. 222–261.

[120] Ramaswami, V., M. Rumsewicz, W. Willinger, and T Eliazov, "Comparison of Some traffic Models for ATM Performance Studies," *ITC-13,* 1991, pp. 7–12.

[121] Rasmussen, C., J. H. Sorensen, K. S. Kvols, and S. B. Jacobsen, "Source-Independent Call Acceptance Procedures in ATM Networks," *IEEE JSAC,* Vol. 9, No. 3, April 1991, pp. 351–358.

[122] Rathgeb, E. P., "Modeling and Performance Comparison of Policing Mechanisms for ATM Networks," *IEEE JSAC,* Vol. 9, No. 3, April 1991, pp. 325–334.

[123] Rigolio, G., and L. Fratta, "Input Rate Regulation and Bandwidth Assignment in ATM Networks: an Integrated Approach," *ITC-13,* 1991, pp. 141–146.

[124] Rossiter, M. H., "The Switched Poisson Process and the SPP/G/1 Queue," *ITC-12,* 1988, pp. 3.1 B.3.1–3.1 B.3.7.

[125] Sabourin, T., G. Fiche, and M. Ligeour, "Overload Control in a Distributed System," *ITC-13,* 1991, pp. 421–427.

[126] Saez, L. B., and G. H. Petit, "Bandwidth Resource Dimensioning in ATM Networks: a Theoretical Approach and Some Study Cases," *ITC-13,* 1991, pp. 929–934.

[127] Saito, H., "The Departure Process of an N/G/1 Queue," *Performance Evaluation,* Vol. 11, 1990, pp. 241–251.

[128] Saito, H., M. Kawarasaki, and H. Yamada, "An Analysis of Statistical Multiplexing in an ATM Transport Network," *IEEE JSAC,* Vol. 9, No. 3, April 1991, pp. 359–367.

[129] Saito, H., "Optimal Control of Variable Rate Coding with Incomplete Observation in Integrated Voice/Data Packet Networks," *European J. Op. Res.,* Vol. 51, 1991, pp. 47–64.

[130] Saito, H., "Optimal Queuing Discipline for Real-Time Traffic at ATM Switching Nodes," *IEEE Trans. Comm.,* Vol. 38, No. 12, December 1990, pp. 2131–2136.

[131] Schormans, J., J. Pitts, and E. Scharf, "Time Priorities in ATM Switches," *ITC-13,* 1991, pp. 527–532.

[132] Schulzrinne, H., J. F. Kurose, and D. Towsley, "Congestion Control for Real-Time Traffic in High-Speed Networks," *INFOCOM '90,* 1990, pp. 543–550.

[133] Sidi, M., W.-Z. Liu, I. Cidon, and I. Gopal, "Congestion Control Through Input Rate Regulation," *GLOBECOM '89,* 1989, pp. 49.2.1–49.2.5.

[134] Sole, J., J. Domingo, and J. Garcia, "Modelling the Bursty Characteristics of ATM Cell Streams," manuscript.

[135] Sriram, K., P. K. Varshney, and J. G. Shantikumar, "Discrete-Time Analysis of Integrated Voice/Data Multiplexers With and Without Speech Activity Detectors," *IEEE JSAC,* Vol. SAC-1, No. 6, December 1983, pp. 1124–1132.
[136] Sriram, K., and W. Whitt, "Characterizing Superposition Arrival Processes in Packet Multiplexers for Voice and Data," *IEEE JSAC,* Vol. SAC-4, No. 6, September 1 986, pp. 833–846.
[137] Sriram, K., R. S. McKinney, and M. H. Sherif, "Voice Packetization and Compression in Broadband ATM Networks," *IEEE JSAC,* Vol. 9, No. 3, April 1991, pp. 294–304.
[138] Sriram, K., and D. M. Lucantoni, "Traffic Smoothing Effects of Bit Dropping in a Packet Voice Multiplexer," *INFOCOM '88,* 1988, pp. 759–770.
[139] Stern, T. E., "A Queuing Analysis of Packet Voice," *GLOBECOM '83,* 1983, pp. 71–76.
[140] Stewart, W. J., "On the Use of Numerical Methods for ATM Models," in *Modeling and Performance Evaluation of ATM Technology,* H. Perros, G. Pujolle, Y. Takahashi, eds., Amsterdam: North Holland, 1993.
[141] Sykas, E. D., K. M. Vlakos, I. S. Venieris, and E. N. Protonotarios, "Simulative Analysis of Optimal Resource Allocation and Routing in IBCN's," *IEEE JSAC,* Vol. 9, No. 3, April 1991, pp. 486–492.
[142] Tucker, R. C. F., "Accurate Method for Analysis of a Packet Speech Multiplexer with Limited Delay," *IEEE Trans. Comm.,* Vol. 36, No. 4, April 1988, pp. 479–483.
[143] Turner, J. S., "Managing Bandwidth in ATM Networks with Bursty Traffic," *IEEE Network,* 1992, pp. 50–58.
[144] Tseng, K. H., and M. T. Hsiao, "Admission Control of Voice/Data Integration in an ATM Network," *ICC '91,* 1991, pp. 188–192.
[145] Unteregelsbacher, E., and H. T. Mouftah, "PDF Based Congestion Control in ATM Networks," *ICC '91,* 1991, pp. 211–215.
[146] Verbiest, W., L. Pinnoo, and B. Voeten, "The Impact of ATM Concept on Video Coding," *IEEE JSAC,* Vol. 6, No. 9, December 1988, pp. 1623–1632.
[147] Virtamo, J. T., and J. W. Roberts, "Evaluating Buffer Requirements in an ATM Multiplexer," *GLOBECOM '89,* 1989, pp. 1473–1477.
[148] Wang, Q., and V. S. Frost, "Efficient Estimation of Cell Blocking Probability for ATM Systems," *ICC '91,* 1991, pp. 385–390.
[149] Whitt, W., "Approximating a Point Process by a Renewal Process, 1: Two Basic Methods," *Op. Res.,* Vol. 30, No. 1, January-February 1982, pp. 125–147.
[150] Whitt, W., "The Queuing Network Analyzer," *BSTJ,* Vol. 62, No. 9, November 1983, pp. 2779–2815.
[151] Whitt, W., "Performance of the Queuing Network Analyzer," *BSTJ,* Vol. 62, No. 9, November 1983, pp. 2817–2843.
[152] Wong, A. K., "On Buffer Dimensioning and Static Rate-Control in Broadband ATM Networks," *ICC '91,* 1991, pp. 280–286.
[153] Yamada, H., and S. Sumita, "A Traffic Measurement Method and Its Application for Cell Loss Probability Estimation in ATM Networks," *IEEE JSAC,* Vol. 9, No. 3, April 1991, pp. 315–324.
[154] Yamashita, H., H. G. Perros, and S. W. Hong, "Performance Modeling of a Shared Buffer ATM Switch Architecture," *ITC-13,* 1991, pp. 993–998.
[155] Yamashita, H., and R. Onvural, "On Packet Loss in ATM Networks," manuscript.
[156] Ziegler, C., and D. L. Schilling, "Waiting Times at Very Fast, Constant Service Time Merger Nodes," *IEEE Trans. Comm.,* Vol. COM-32, No. 2, February 1984, pp. 189–194.
[157] Saito, H., "Call Admission Control in an ATM Network Using Upper Bound on Cell Loss Probability," *IEEE Trans. Comm.,* Vol. 40, 1992, pp. 1512–1521.
[158] Sohraby, K., "Heavy Traffic Multiplexing Behavior of Highly Bursty Heterogenous Sources and Their Admission Control in High Speed Networks," *Proc. IEEE Infocom,* 1992, pp. 1518–1523.

[159] Sohraby, K., "On the Theory of General On-Off Sources with Applications in High Speed Networks," *Proc. IEEE Infocom,* 1993, pp. 401–410.

[160] Elsayed, K. M., and H. G. Perros, "Call Admission Control in High Speed Networks," Tech. Rept., Computer Science Dept., North Carolina State University, 1994.

[161] Elwalid, A. I., and D. Mitra, "Effective Bandwidth of General Markovian Traffic Sources and Admission Control of High Speed Networks," *IEEE Trans. Networking,* 1993, pp. 329–343.

5 ATM Switching

Conceptually, ATM networks are packet-switching networks in that each ATM cell in the network is transmitted independently. ATM is connection-oriented in that end-to-end connections are established before the traffic transfer can start flowing. In simple terms, an ATM switching node transports cells from the incoming links to the outgoing links based on the information stored in its routing table using the routing label at the cell header. In particular, a connection setup task mainly performs two functions at each switching node: (1) for each connection, it defines a unique set of connection identifiers used at incoming and outgoing links, and (2) it sets up routing tables at each switching node, providing an association between the incoming and outgoing connection identifiers for each connection. VPI and VCI are the two connection identifiers used in ATM cells. In order to uniquely identify each connection, VPIs are uniquely defined at each link and VCIs are uniquely defined at each VP. Hence, a connection identifier is a triplet (link identifier, VPI, VCI).

The first step in establishing end-to-end connections is to determine a path from source to destination. At the end of this step, the sequence of links that will be used by the connections and their identifiers will be known. Let us now review how VPI and VCI values may be set up at the switches along the end-to-end path.

In the case of pure VC switching, a sequence of messages is exchanged between every neighbor node along the path to set up a routing table entry associated with the connection. The entry basically maps the incoming triplet to the outgoing triplet. For presentation purposes, let us consider three neighbor nodes $i - 1$, i, and $i + 1$. The routing table entry at node i may be set as follows. Node $i - 1$ sends a message to node i that includes information on the incoming and outgoing link identifiers. Node i assigns a VPI/VCI to the connection and creates the incoming triplet part of the identifier. Node i then sends two messages: one to node $i - 1$ defining the VPI/VCI value used at its incoming link (which corresponds to an outgoing link at node $i - 1$) and another one to node $i + 1$ that includes information on its incoming and outgoing links. Node $i + 1$ replies back to node i, which includes the required information to set up the outgoing triplet part of the table entry at node i. Once the routing tables at all nodes along the path are set, cells start to follow. At each node, the VPI/VCI, together with the identification of the link cells are coming from, is used for each arriving cell to determine the outgoing link and the new VCI.

In the case of VP switching, the procedure is similar to that of VP/VC switching, with the main difference being the number of routing tables set per connection. In particular, there is no distinction between a VP and a VC if the VP is defined over a single link. VP switching is meaningful when VPs are defined across more than one link. Furthermore, VPs are semipermanent connections in that routing tables for each VP are preset by network management functions. The routing table entries in this case maps the incoming link identifier and VPI to the outgoing link identifier and VPI. Hence, no setup messages need to be exchanged between the neighbor nodes that belong to a VP. In particular, the only nodes involved in message exchanging are the two end nodes of the VP where the VP originates and ends. The connection setup procedure is the same as it is in VP/VC switching after each VP is treated as one link. The routing table entry at the node where one VP ends and another one starts maps the incoming triplet to the outgoing triplet.

In either type of switching, the routing label at the cell header is read, a table lookup is performed to determine the outgoing link and the new routing label used at that link, the routing label is updated to its new value, and the cell is switched from the incoming link to the outgoing link.

In this chapter we discuss the switching requirements of ATM networks, present a general view of various switch architectures proposed in the literature, and discuss different techniques used to analyze their performance metrics.

5.1 PRELIMINARIES

An $N \times N$ switch can be viewed as a black box with N inputs and N outputs that transports cells from any incoming link to any outgoing link. Incoming links are connected to a switch fabric through *input ports.* After the cell header is processed to determine its outgoing link, it is passed to the switch fabric to be delivered to its outgoing link. The interface between the switch fabric and the outgoing link is referred to as the *output port.* When more than one cell attempts to access an output port simultaneously, then a phenomenon called *output conflict* occurs. When this happens, only one of the contending cells can be read out by the output port. Other cells may either be stored in a buffer until they can be read out or dropped. In *output buffering,* these cells are stored between the switch fabric and the output port. In *input buffering,* storage is provided between the incoming link and the input port, where cells that are blocked due to output contention are kept until they can be delivered to the output ports. If the input buffer is a FIFO, a blocked cell at the head of the queue blocks all other cells that may be waiting. In particular, *head of line* (HOL) blocking occurs when some of the cells that might be waiting in the queue, which could otherwise be switched to their destination ports, are forced to wait for the HOL cell to be transmitted first. Finally, *internal blocking* may occur if more than one cell contends for the same resource simultaneously within the switch fabric. Similar to the case of output contention, all but one of the contending cells may be temporarily stored at buffers either within the switch fabric (referred to as *internal buffering*) or at input buffers. Excluding the transmission link overhead for the simplicity of presentation (i.e., SONET frame overhead or control cells in the block-coded scheme), cells can potentially arrive at the switch from an incoming link every (53 × 8/link speed) sec. On OC-3 links, this time is 2.74 μs, which is reduced to about 0.7 μs on an OC-12 link. This corresponds respectively to switching rates of 365,000 cells/sec and 1,466,000 cells/sec for each incoming link. A relatively small 16 × 16 switch with 10 OC-3 and 6 OC-12 input links is then required to switch more than 12 million cells per second.

Different applications require different types of connections in ATM networks. For example, a voice service may take place between two end points, whereas more than two end nodes may participate in teleconferencing. In general, two types of connections are established across a switch:

1. Point-to-point connections, where the connection is established between two entities. Most current services fall into this category.
2. Point-to-multipoint connections, where the cell stream generated by a source node is distributed to two or more nodes. A typical example of this type of connection is video distribution in which a video server serves multiple destination nodes.

In terms of switching requirements, a cell is either switched point-to-point from an incoming link to an outgoing link or point-to-multipoint from one link to a number of outgoing links. The latter is referred to as *multicasting*, in which only a subset of outgoing links receive the switched cell, or *broadcasting*, in which an incoming cell is transported to every outgoing link at the switch.

With the advances in the transmission technology providing the bandwidth required to support B-ISDN applications, the challenge is now to design switches that can deal with the several hundreds of millions of cells arriving at a node every second and can support different types of connection requirements of applications.

Various ATM switches with different architectures have been proposed in the literature. Following the classification in [69], switch fabrics proposed for ATM networks are classified into three categories:

- Shared-medium architecture;
- Shared-memory architecture; Space-division architecture.

Each architecture is discussed in detail in Section 5.2.

5.2 SHARED-MEDIUM ARCHITECTURES

In a shared-medium architecture, incoming cells are multiplexed into a common medium, typically a bus or a ring. The medium speed, in general, is greater than or equal to the sum of the transmission rates of incoming links attached to it. Then a small FIFO that has a capacity to hold only a few cells is sufficient to store incoming cells until they can access the medium. Output contention cannot occur in this architecture, since two or more cells cannot arrive at an output port simultaneously. However, the arrival rate of cells at a particular outgoing link may exceed the link bandwidth for a short period of time, and output buffers are used to store cells that arrive at a faster rate than they can be served.

Each output port is assigned a unique address. Once the outgoing link of an arriving cell is determined, the output port address is added to each cell before it is passed to the medium. This address is decoded by each output port interface at the medium and filtered to determine whether to copy the incoming cell or not. Cells addressed to a particular output port are copied to corresponding output buffers to be read out by the transmission link.

Shared-medium architectures naturally support multicast and broadcast and perform well when the medium speed is greater than the sum of the transmission rates of incoming links attached to the switch. As the number of links attached to the switch and their speeds increase, running shared media at very high rates may no longer be technologically feasible and the medium speed becomes the bottleneck. Accordingly, shared-medium architectures do not scale well and can support a relatively small number of ports. Alternatively, shared medium can be used as the switching elements of large switches in which these units are connected to others according to some topology. Three examples of shared-medium architectures proposed for ATM switching are discussed next.

5.2.1 ATM Output Buffer Modular Switch

The ATM output buffer modular (ATOM) switch architecture was developed by NEC [64]. A switching element referred to as an ATOM chip consists of a bus and output buffers. The number of links that can be attached on a chip is limited by the bus speed. The architecture achieves scalability by interconnecting a number of chips together through serial links, as illustrated in Figure 5.1.

A virtual circuit within the switch is established for each ATM connection, and routing tables are set at each ATOM chip along this internal virtual circuit. Then the switch itself becomes a virtual circuit network and cells are routed within the switch using the internal virtual circuit identifiers encapsulated to each cell at the input ports. Multicasting is achieved relatively easily by using

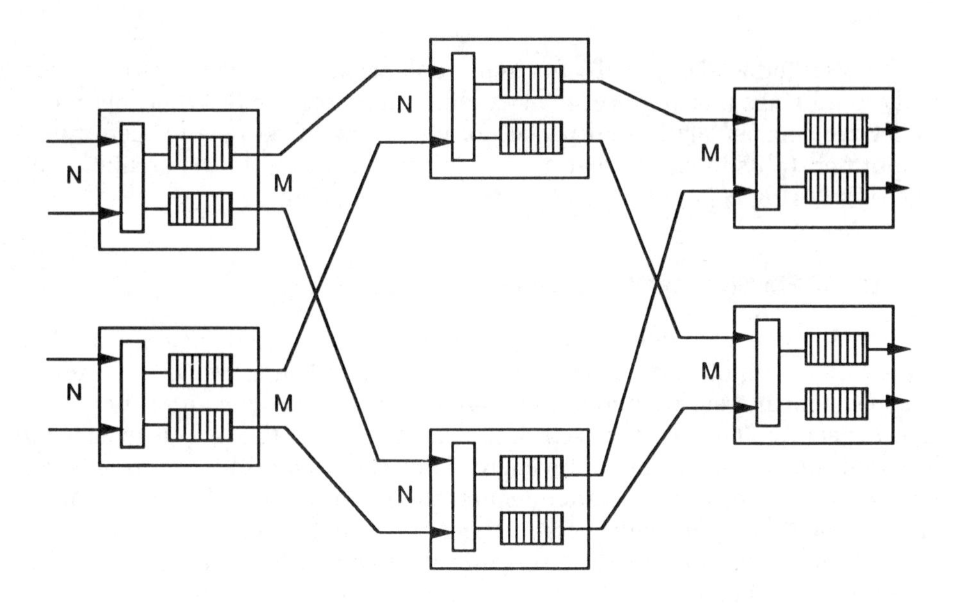

Figure 5.1 ATOM switch architecture.

a broadcast module. For cells to be multicast, the internal virtual circuit is an index to a broadcast table.

5.2.2 Packetized Automated Routing Integrated System

The packetized automated routing integrated system (PARIS) is an experimental high-speed packet-switching system that was designed to transport voice, video, and data all in packetized form [22]. The architecture was originally developed for variable-length packets, but it supports ATM cells without any extra overhead. The switch is a bus with a bandwidth greater than the aggregate capacity of all incoming links. Both input and output buffers are used. A round-robin type of exhaustive service policy, in which each input buffer is served in a cyclical manner, is used to arbitrate the access to the bus. Small input buffers that can hold four of the largest packets are enough to make the switch non-blocking. Output buffers are dimensioned such that packet loss probability due to a momentary overload situation is less than the desired QOS requirements.

5.2.3 Synchronous Composite Packet Switching

The synchronous composite packet-switching (SCPS) architecture is proposed for integrated circuit- and packet-switching functions. The architecture consists of switching modules interconnected via multiple rings [66]. Each switching module is interfaced externally to circuit- and packet-switched lines and internally to all rings that perform the switching function. An SCPS header and a trailer are encapsulated to all packet- or circuit-switched data to identify the output port (plus to carry other control information used in the architecture). Access to a ring is based on time-division multiplexing.

5.3 SHARED-MEMORY ARCHITECTURES

A shared-memory switch consists of a single dual-ported memory module shared by all input and output ports. Incoming cells are multiplexed into a single stream and are written to the shared memory. The memory is organized into logical queues, one for each output port. Cells at the output queues are also multiplexed into a single stream, read out, demultiplexed, and transmitted on the output lines. In this architecture, the main bottleneck is the memory access time to support both incoming and outgoing traffic.

The memory can logically be organized into either *full sharing* or *complete partitioning*. In the former, the entire memory is shared by all output ports. An arriving cell is dropped only when the memory is full. In the latter, an upper bound is imposed on the number of cells waiting in the queue of each output port, and a cell is dropped if this limit is reached at a particular queue, even though there is space available in the memory. Full sharing provides a better cell loss probability than complete partitioning by using the memory more efficiently (i.e., if there is a space, then the cell is accepted). However, this scheme may not be fair at times when a burst of cells arrives at a particular output port, reducing the space available and eventually causing service degradation at other ports.

Two shared-memory architectures are presented next.

5.3.1 The Prelude Switch

The Prelude switch was developed by CNET [7]. In the prototype, each packet is 15 bytes with a 1-byte header. The number of input and output ports is equal to the number of bytes in a packet (i.e., 16). In order to achieve high memory speeds, 16 memory modules are used such that each byte of a packet is stored in a different memory module. Although the discussion is based on 16-byte packets, for simplicity of presentation, Figures 5.2 and 5.3 illustrate the related concepts, using 4-byte packets as an example.

Incoming cells are fed into a phase alignment and supermultiplexing stage, which performs frame alignment by locking the byte rate and the phase of each cell onto the local central clock. This operation is accomplished through the clock adaptation queues in such a way that packets are extracted from the phase alignment stage so that they are shifted by 1 byte from one link to the next.

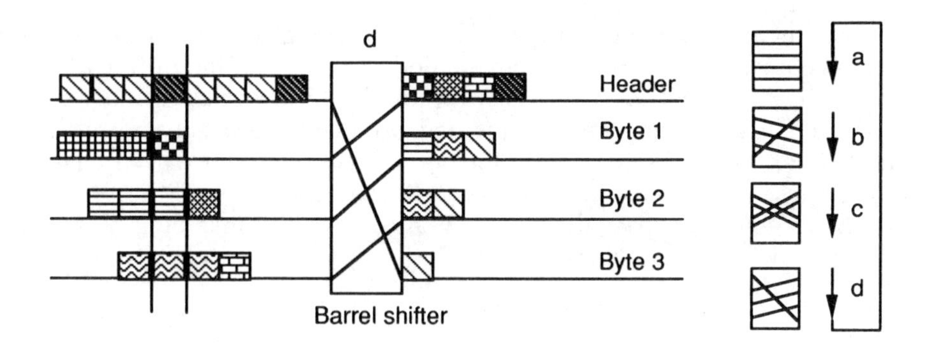

Figure 5.2 Parallel-diagonal multiplexing using a rotative space-division switch.

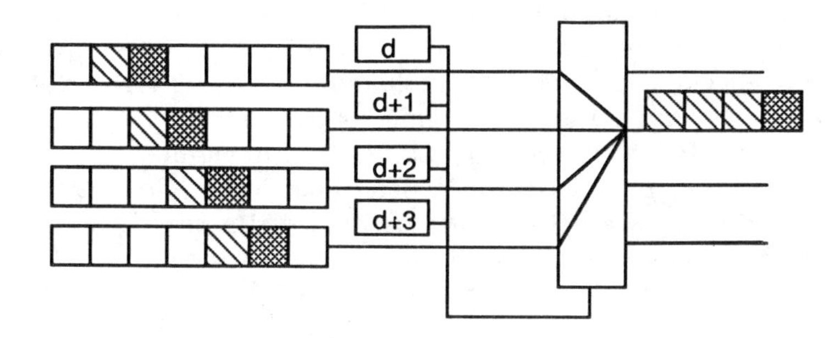

Figure 5.3 Extraction of packets from memory.

These diagonally aligned packets are next fed to a rotative space-division switch, a *barrel shifter,* which assumes 16 different switching patterns. This concept is illustrated in Figure 5.2.

The rotative cycle is synchronized with the arriving packets in such a way that the header of each packet ends up on the first output of the switch and the subsequent 15 bytes are switched sequentially to the remaining 15 output lines, each byte being offset in time with respect to the previous 1×1 byte.

The controller processes the headers, determining the outgoing link and the new information to put in the header. After being processed, the new header is stored in an empty location at the first memory bank and the following bytes of the same packet are stored in the other memory banks diagonally, incrementing the address register by 1 for each subsequent byte. The address where the header is stored is then placed on the queue for the corresponding outgoing link.

On the output side of the switch, cells are extracted via an output barrel shifter that reverses the process of the input barrel shifter to reassemble the cells, as illustrated in Figure 5.3. In particular, the controller delivers the address of a packet to be read out by the corresponding outgoing link during a 1-byte period. The rest of the packet is easily retrieved from the other memory banks. In particular, packets being retrieved are fed to a rotative switch that reconstructs packets in exactly the same way as the one used at the input. The crucial step is to supply the address of the header to be extracted so that it is consistent with the stage of the space-division switch in its cycle, thereby guaranteeing the proper switching of the packet.

5.3.2 Hitachi's Shared-Buffer Switch

In this switch architecture, the memory is fully shared by the output ports. Output queues are formed using linked lists and multicasting is supported by writing multiple copies of a cell to the shared memory. A high-level view of the switch architecture is shown in Figure 5.4 (cf. [44]).

Serial-to-parallel (S/P) modules perform serial-to-parallel data conversion. A header conversion module (HD CNV) determines which link the cell will be put in. In addition, there are three types of circuitry: the *switching chip*, which contains memory, a multiplexer, and a demultiplexer; the *control chip*, which contains read and write registers, one pair for each buffer; and *input address buffers*, which keep track of the status of unused memory locations.

After a cell passes through the header conversion stage, a write address register is accessed to get a memory location for the arriving cell, the cell is written to the memory, and the memory address is queued to the corresponding outgoing link. Simultaneously, a new empty buffer address, if one is available, is supplied from input address buffers to the control unit. Similarly, at each time slot, one packet buffer from each linked list is identified by the control unit read registers, retrieved, and transmitted by the outgoing link.

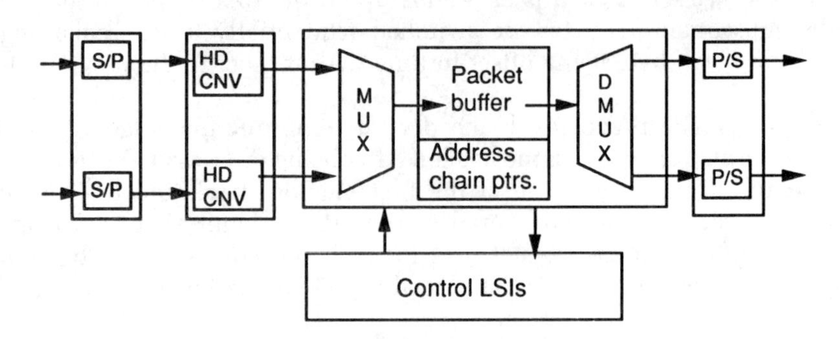

Figure 5.4 Hitachi's shared-buffer switch.

Simultaneously, the read registers and the contents of idle address buffers are updated. In order to meet the speed requirements, multiple chips in parallel are used.

5.3.3 IBM's Switch

The switching element of IBM's switch is a shared-memory switch with 16 input and 16 output ports. Multiple switching elements can be used to expand the number of ports, to increase the port speeds, and/or to increase overall switch performance by allowing extra buffering.

Conceptually, the structure of IBM's switching element looks like the shared-buffer architecture illustrated in Figure 5.4. Each port operates at speeds of up to 400 Mbps. The packet buffer can hold up to 128 cells shared by all output ports. The packet buffer is accessed in parallel. That is, a particular byte of each buffer is accessed in the same cycle.

Within the control section, there is a table that controls the order of bytes stored in the packet buffer. This operation is mainly used to reorder the bytes of the routing header so that they are in the correct order at the inputs of the next-stage switching element.

Associated with each port, there are 16 input routers and 16 output routers. At the input side, data are routed from an input port to its position in the allocated packet buffer 1 byte at a time in parallel. Output routers are used to read cells out from the packet buffer to output ports.

When a cell is received, it is allocated a free buffer by the control section. A routing vector that specifies the list of switching elements and their corresponding port identifiers in each switching element is appended to the cell. This is referred to as *source routing* in the switch fabric, since the routing vector is determined by the initial switching element in the switch fabric using routing tables at the switch. After a cell is received, the control logic analyzes the first byte of the current routing vector and places the address of the packet buffer holding the cell into the corresponding output queue for its destination port. Each output router obtains the packet buffer address of the first cell in its queue and reads out the cell.

The multicast and broadcast operations are performed using special routing tags that do not correspond to any port identifiers. To perform a multicast, the routing tag of the cell corresponds to a list of output ports. A single copy of the cell is kept in the packet buffer and the address of the buffer is copied to all corresponding output queues.

The switching element is designed to allow the use of multiple chips to expand the number of ports, to increase the port speeds, or to increase overall switch performance by allowing extra buffering.

The input and output ports of a switching element may be connected to a number of switching elements so that the whole system behaves as if it is a single switching element. When a packet buffer in a switching element is full, the next switching element is used to accept incoming cells. To keep the cell sequence integrity, output routers can read cells out from a switching element while the input routers are placing incoming cells in another switching element.

One way to increase the number of ports (in multiples of 16) is to use the principles of space-division architecture discussed in Section 5.4, with each switching element being the shared-buffer switch.

The port speeds may be increased by either connecting two or more switching elements to concatenated input and output ports or by grouping ports of a single switching element together such that a new port can support speeds equaling the sum of individual port speeds grouped together.

Let us first consider connecting two switching elements with concatenated input and output ports. Parallel data paths of 8 bits between the adapters and the switching element are in this case doubled to 16 bits. The first byte is sent to the first switching element, whereas the second byte is sent to the second switching element. The operations of the two switching elements are locked, so the two operate synchronously. Each switching element receives the same routing tag and operates independently (but in a synchronous manner) to transmit cells from incoming ports to the output ports. This scheme increases the speeds of all ports in the switching element.

Alternatively, the total number of ports in a switching element may be reduced by grouping a number of them together. This would provide a few ports with higher speeds supported in a switching element. For example, if two ports are grouped together, the resulting port could support 800 Mbps, while the remaining 14 ports could operate at 400 Mbps each. The combined ports are logically viewed as a single port, while the internal operations at the switching element do not change.

5.4 SPACE-DIVISION ARCHITECTURES

There are two major drawbacks to shared-medium and shared-memory switch architectures. First, multiplexing is required at the input side and demultiplexing at the output side of the switch, restricting the scalability of the switch to support a large number of ports. Second, buffer management and control functions are often centralized, which increases the complexity of the switching node.

In space-division switching, multiple cells from different input ports can be transferred concurrently on multiple links. Each cell transfer requires the establishment of a dedicated physical path through the switch from incoming to outgoing links. These switches also allow the control to be distributed within the switch, thereby reducing its design complexity.

The basic building block in a space-division switch is a crosspoint that can be enabled or disabled by a control unit. As shown in Figure 5.5, each

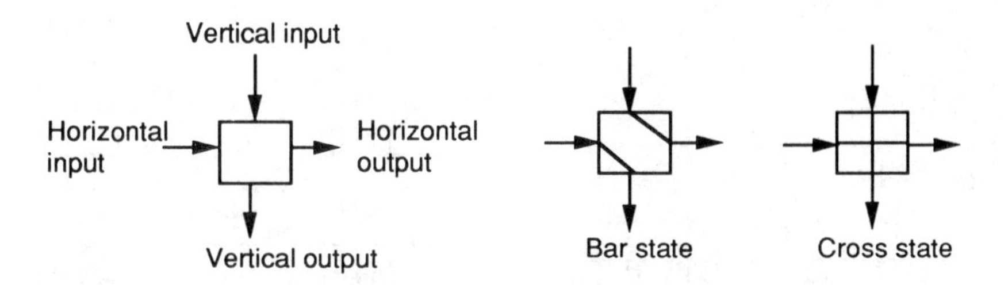

Figure 5.5 A crosspoint and its states.

crosspoint has two input ports and two output ports and allows concurrent activation of two separate paths.

Output contention in a crosspoint occurs when the two input lines simultaneously request connection to the same output line. If this occurs, one of the two contending cells is granted access to the output port, whereas the other may be either dropped or buffered temporarily until the port becomes available and the cell is granted access.

When buffers are used, they may be placed at the input ports or within the crosspoint. In either case, the buffer size is finite and the use of buffers does not totally solve the problem of output contention. In particular, it is possible that the buffers will become full, causing cells to be dropped due to the lack of space to store incoming cells.

Figure 5.6 illustrates an 8×8 *crossbar switch*, where each square corresponds to a crosspoint. In general, N^2 crosspoints are used in an $N \times N$ crossbar switch. Connection from any input port i to output port j is accomplished by engaging the corresponding crosspoint (i, j) in the $N \times N$ matrix. As long as cells at each input port are destined for different output ports, the crossbar switch allows N connections to be simultaneously established, thereby resulting in concurrent delivery of N cells. Accordingly, it is an internally nonblocking switch. The main disadvantage of crossbar switches is that the switch complex-

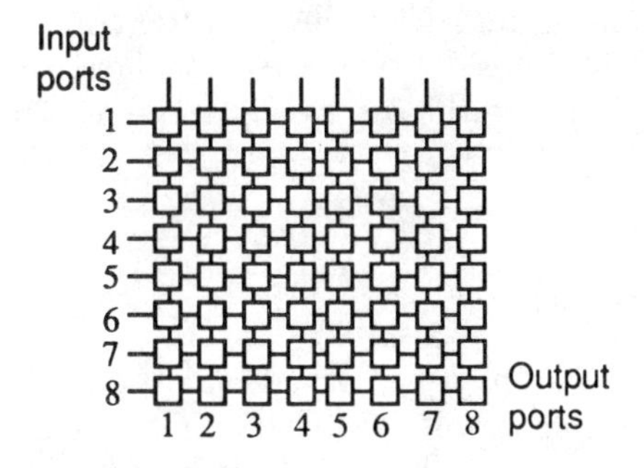

Figure 5.6 An 8×8 crossbar switch.

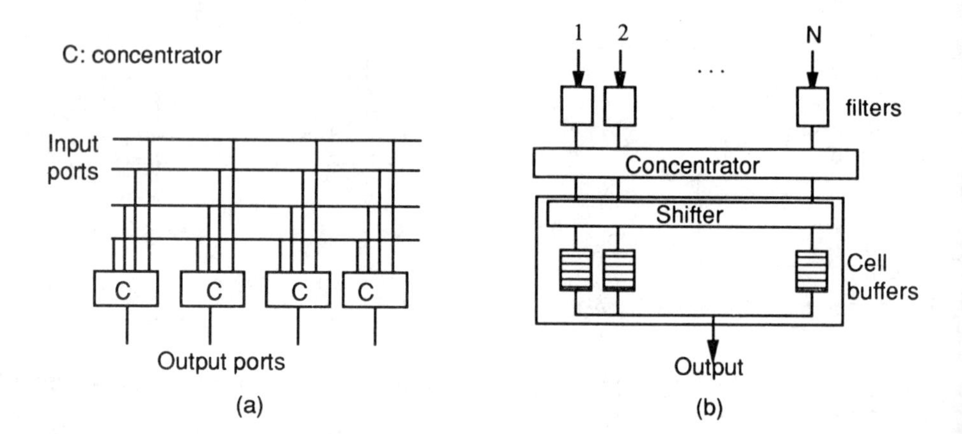

Figure 5.7 Knockout switch architecture: (a) 4 × 4 knockout switch; (b) concentrator/buffer. C = concentrator.

ity grows with N^2, of which at most N are used at any given time. Furthermore, there is a unique path between any input and output port and the loss of a crosspoint would prevent the connection between the two ports involved.

A crossbar switch with concentration, known as a *knockout switch*, was developed by AT&T [26]. The architecture, illustrated in Figure 5.7(a), is based on a quasicrossbar architecture. Each input drives a bus connected to all output concentrators/buffers, as shown in Figure 5.7(b). The concentrator reduces the number of inputs to the output buffer from N to L ($L \ll N$). A shared-buffer design, which works as a FIFO for the L buffers, precedes the concentrator. Since L is smaller than N, there can be cell losses in the concentrator.

The concentrator works in accordance with the *knockout tournament* principle. Of the up to N cells that enter the first level of the concentrator, only one winner takes the first spot. The remaining cells enter the second level of competition and one winner takes the second spot. This procedure continues until L winners are determined. Each level of the concentrator is made up of knockout elements that choose between two packets; the winner goes to the next level of competition, and the loser goes to the next stage to compete with other losers.

5.4.1 Banyan Networks

Unlike crossbar switches, which are based on a matrix topology, Banyan networks are based on the tree topologies. A 1 × 8 multistage binary tree is shown in Figure 5.8. In Banyan networks, each input port is the root of a tree that branches over a number of intermediate switching elements, with output ports

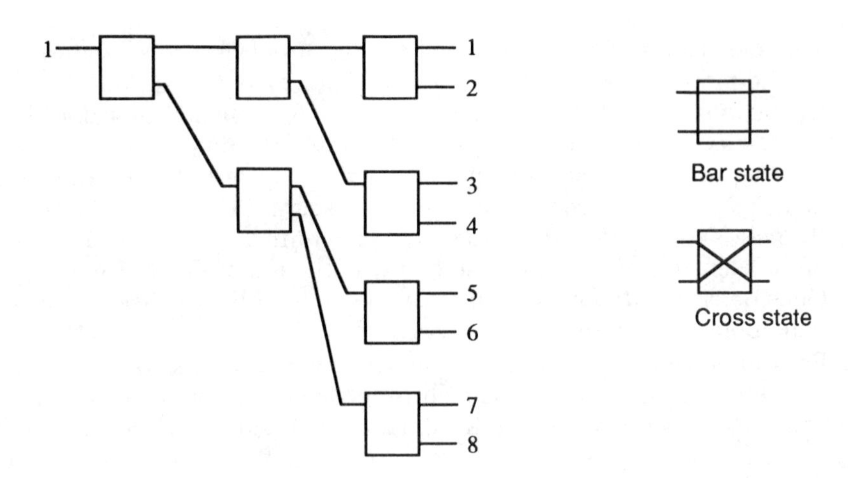

Figure 5.8 A 1 × 8 multistage binary tree.

as its leaves. Then the forest of N trees that share all the links and switching elements except the roots forms an $N \times N$ Banyan switching network. There are different ways of forming such trees, resulting in different topologies of Banyan networks. Figure 5.9 illustrates two architectures.

Regardless of the particular form they take, all Banyan networks with $n \times n$ switching elements have the following properties:

1. Each switching element is an $n \times n$ crossbar switch, and the switch consists of $\log_n N$ stages. Then the total number of crosspoints in a Banyan network is equal to $n^2(N/n) \log_n N$, which is less than N^2, the total number of crosspoints in an $N \times N$ crossbar switch.

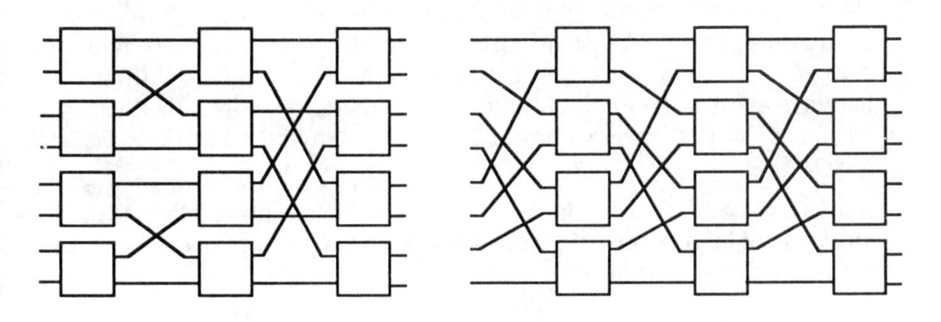

Figure 5.9 Example topologies of Banyan networks.

2. There exists a unique path connecting an input port to an output port. Furthermore, Banyan networks are self-routing switches. The path from an incoming port to an output port can be established in a distributed manner as follows. Let the output port of a switching element be numbered from top to bottom, starting at 0 and ending at $n - 1$. The routing vector is $(r_1, r_2, \ldots, r_k)$, where k is the number of stages and r_j is the port number the cell is routed to at the switching element in the jth stage. The values of the elements of a routing vector are a function of the destination port. Once determined, the switching element at the jth stage uses r_j to decide which output port the cell is destined for.
3. Banyan networks are internally blocking, which occurs when more than one cell attempts to use the same link between two stages. Let us consider an $N \times N$ Banyan network built from 2×2 switching elements. An output of a switching element at stage k, $k \leq 0.5 \log_2 N$, can be reached from 2^k different input ports. For $k > 0.5 \log_2 N$, the number of output ports that can be reached from the output of a switching element is less than 2^k. The maximum I/O connections on an internal link are reached at stage $k = 0.5 \log_2 N$ and are equal to $2^{0.5 \log_2 N}$ (i.e., $\sqrt{N}$).

   Internal blocking causes throughput degradation, which increases proportionally with the number of ports in the network. The three approaches used to reduce the effect of internal blocking are internal speedup, internal/input buffering, and sorter networks that precede the switch fabric. These are discussed in detail later in this section.
4. In an internally nonblocking switch, any input port can be connected to any output port provided that the connections to distinct output ports are requested. Similarly, up to N concurrent connections can be established in an $N \times N$ Banyan network. Unlike crossbar switches, however, due to internal blocking, not all $N!$ permutations of concurrent connections can be realized. Since the total number of switching elements in the network is equal to $0.5N \log_2 N$ and each switching element can be either in a bar state or in a cross state, the total number of different states in the network is equal to $2^{0.5}N \log^{2N}$. Considering the fact that each state corresponds to N distinct, internally nonconflicting paths and that there is a unique path between every I/O port combination, the total number of realizable permutations of N concurrent connections in a Banyan network is equal to $2^{0.5N \log_2 N}$ (i.e., $\sqrt{N^N}$).
5. Banyan networks are scalable. Their modular structure allows the construction of larger networks from smaller ones without the need to modify the algorithms used for their operation.

5.4.1.1 Blocking in Banyan Networks

Let us consider an $N \times N$ Banyan network built of 2×2 switching elements with internal links running at the same speed as the fastest incoming link (note that the internal speed is in fact greater than this value to accommodate the self-routing routing header used internally at the switch). Recalling that at most N cells can be contending for the same output port of a switching element, a Banyan network becomes internally nonblocking if the internal links run N times faster than the fastest incoming link. For a 16×16 switch with 155-Mbps OC-3 links, the internal link speeds should be more than 600 Mbps, which increases to more than 1.2 Gbps in a 64×64 switch. As the number of ports and the link speeds increase, the required internal speeds for nonblocking becomes technologically difficult to achieve and complicates switch design. Typical speedup factors expected in ATM switches range from 2 to 4.

In addition to internal blocking, as with other internally nonblocking space-division switches, output contention occurs in Banyan networks when more than one cell (in a switching element at the last stage) contends for the same output port.

To reduce the amount of throughput degradation caused by internal blocking and output contention, buffers may be provided at the input ports, output ports, or internally at the switching elements, respectively referred to as input buffering, output buffering, and internal buffering.

5.4.1.2 Input Buffering

Buffers at the input ports may be used against both internal blocking and output contention. In general, an input buffer is implemented as a FIFO. A cell that cannot be switched to its output port during the current cycle occupies the HOL position at the FIFO and retries during consecutive cycles. In this case, cells behind the HOL cell are forced to wait in the FIFO until they reach the HOL position in the queue. Due to HOL blocking, a Banyan network with input buffering is not *work conservative*. In particular, during a cycle, a path from an input port to the output port of non-HOL cells waiting in the queue can be established without causing any internal or output blocking of any other HOL cell at other ports. However, input buffering does not allow such cells to be switched during the current cycle, thereby forcing some internal links to be idle even though there are cells waiting in the system that can use these links.

5.4.1.3 Output Buffering

Due to output contention, an internally nonblocking switch can still block at the output ports. If output buffering is not used, only one of the contending cells can go through the output port. Depending on the scheme used, other contending cells may be dropped (i.e., lost) or stored either at input or internal buffers. With output buffering, all cells contending for the same output port are stored at the output ports until they can be read out by the corresponding transmission link. Hence, output buffering results in a better switch throughput than input buffering, since only one of the contending cells from different input ports can be delivered to the output in the latter.

However, output buffering requires outputs of all switching elements at the last stage of a Banyan network to be capable of simultaneously connecting to all output ports. Furthermore, the output buffer must be able to receive and manage cells arriving simultaneously. This is not an easy task, particularly as the link (and therefore the switching) speeds increase; for example, an 8 × 8 switch with OC-3 links would require more than 1.2 Gbps of memory bandwidth. Hence, the scalability of Banyan networks is constrained when output buffering is used.

5.4.1.4 Internal Buffering

Buffers at each internal link may be used to temporarily store cells contending for the same output port of a switching element. The buffers can be placed at the input or output ports, or may be a shared memory, as shown in Figure 5.10 for a 2 × 2 switching element. As discussed previously, input buffering causes HOL blocking. Output buffering, on the other hand, requires the memory bandwidth to be equal to twice (for 2 × 2 switching elements) the speed of the internal links. It also requires much more real estate than the part of the chip that performs the switching function. In both cases, buffers are essentially a couple of FIFOs.

Shared memory eliminates HOL blocking and does not require the additional connectivity complexity introduced by output buffering. However, shared memory can no longer be implemented as a FIFO and requires buffer management logic to be included in the switching element, which is not necessarily less complex than implementing output buffering, particularly at the chip level.

The use of internal buffers, although it reduces the probability of internal blocking, has various drawbacks. Since the number of cells arriving at the switching elements simultaneously can be as large as N, large buffers may be

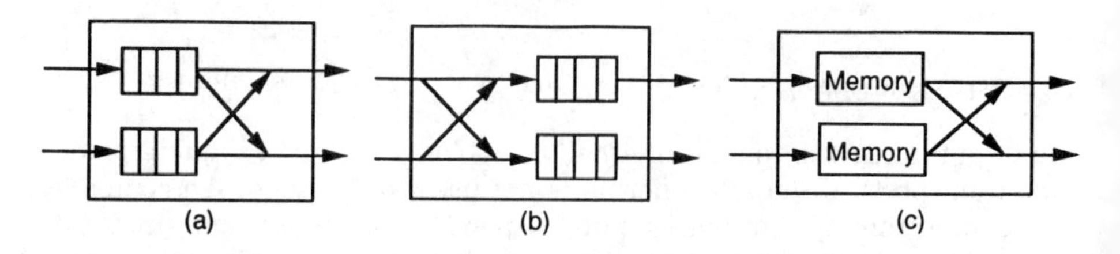

Figure 5.10 Three types of internal buffering: (a) input buffering; (b) output buffering; (c) shared memory.

required, further increasing the chip complexity. Internal buffers also introduce random delays within the switch fabric, causing undesired cell delay variation in the network. Furthermore, internal buffers are most effective when the traffic is uniformly distributed at the output ports and when the cell stream in an incoming link is not correlated in any link. Neither of these conditions, in general, holds in ATM networks. Although cells belonging to different frames are interleaved, the cell stream on a link is highly likely to be correlated in that, given that a cell is switched to a particular output port, the probability that the consecutive cells on the link will be destined for the same output port is greater than the unconditional probability that they will go to that particular port. Next we discuss an approach that eliminates the need for internal buffering by essentially eliminating the internal blocking problem in Banyan networks.

5.4.1.5 Batcher Sorting Network

Another solution to the internal blocking problem in Banyan networks is the use of sorter networks preceding the switch. If we recall that there are various permutations of N concurrent I/O port connections realizable in Banyan networks, the question addressed with this technique is how to sort incoming cells in such a way that when they access the input ports of the switch, no two cells share an internal link, thereby making Banyan networks internally nonblocking.

Let l_j denote the output port address of the HOL cell at the jth input port. Then, assuming that they are all distinct output port addresses, the list $(l_1, l_2, \ldots, l_N)$ can be sorted to a list $(k_1, k_2, \ldots, k_N)$ so that the sorted list corresponds to one of such realizable permutations, where k_j denotes the output port address of the cell at the jth input port. Furthermore, it is possible to do this sorting using space-division networks. The Batcher sorting network is the least complex among all the sorting algorithms known with respect to the total number of crosspoints used to sort N cells.

Two building blocks are used in a Batcher sorter: the *up sorter*, a 2×2 crosspoint that sorts two numbers in descending order and *down sorter*, in ascending order. Accordingly, two numbers are sorted in ascending and descending order, respectively, by the down sorter and the up sorter.

The Batcher sorter then operates as follows. Two pairs of numbers are first sorted by using two 2×2 sorters. Next, the two sorted lists of two numbers are sorted by using a 4×4 sorter. Two sorted lists of four numbers are then sorted by using an 8×8 sorter, and so on. The resulting sorting network consists of multistages of Banyan sorting networks of increasing size. The successive stages of sorters are connected by links in an omega network pattern, as shown in Figure 5.9. Figure 5.11 illustrates an 8×8 Batcher sorter connected to a Banyan network. Each $n \times n$ sorter consists of either up or down sorters. Furthermore, when two lists of numbers are to be sorted and merged, one list is produced by using up-sorter switching elements and the other by using down-

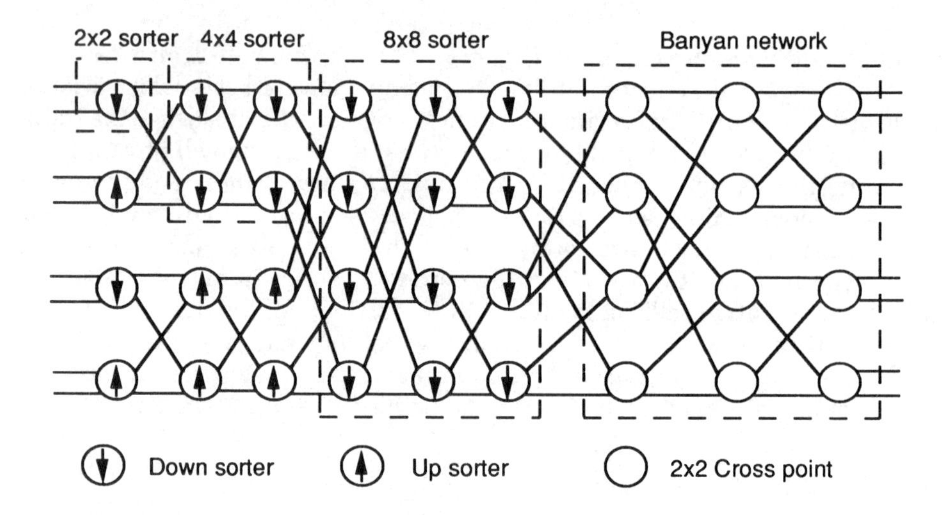

Figure 5.11 The Batcher sorting network.

sorter switching elements. In order to sort 2^k elements, k stages of sorters are needed, where the stage i sorter itself consists of i stages of 2×2 sorters. The total number of stages is then equal to $k(k + 1)/2$. With $N/2$ sorters in each stage, the total number of 2×2 sorters used in a Batcher sorter is equal to $0.25N \log_2 N (\log_2 N + 1)$.

A Banyan network preceded by a Batcher sorter network becomes internally nonblocking if all cells are destined for different output ports. The Batcher sorting network can also be used to resolve output conflicts. In particular, the sorting network has the property that when self-routing addresses are sorted, duplicated output requests appear adjacent to each other in the sorted order. In this case, it is relatively easy to check whether or not cells at the adjacent output ports of the sorting network have the same output port address and to allow only one of the contending cells to proceed, thereby eliminating output contention.

5.4.1.6 Copy Networks

Another drawback of Banyan networks is the lack of support for multicasting. In particular, Banyan networks do not have the facilities to multicast an incoming cell to a number of outgoing links. A self-routing copy network to support multicasting is proposed in [46]. The rest of this section on copy networks is based on the framework developed in [46,33].

Let us consider a switching network consisting of a compact superconcentrator and a copy distribution network preceding a routing network as illustrated in Figure 5.12. All links connecting different modules are numbered from top to bottom starting with 0. c_i denotes the number of copies to be made of the cell at the incoming link i; a_i denotes an indicator bit for link i; $a_i = 0$ if $c_i = 0$ and $a_i = 1$ if $1 \le c_i$; and b_i denotes the output link number of the concentrator to which the cell at incoming link i is routed.

The concentrator moves the cell at the incoming link i to its output port b_i.

$$b_i = \sum_{k=0}^{i-1} a_k$$

Then the copy distribution network takes this cell from its input, b_i, and makes c_i copies while transporting these cells to its output links m_i to M_i, where

$$m_i = \sum_{k=0}^{i-1} c_k;\; M_i = \sum_{k=0}^{i} c_k - 1$$

The concentrator and the copy distribution networks can be made self-addressing by using Banyan networks. In particular, the two networks are implemented in [46] using the folded network of Figure 5.9 with reversed input and output lines, as illustrated in Figure 5.13. The self-addresses that need to be provided are b_i in the concentrator and the interval addresses m_i and M_i in the copy distribution network. The address

$$b_i = \sum_{k=0}^{i-1} a_k$$

can be determined relatively easily. The interval address, represented by the two bounds, can be used for self-routing in the copy distribution network by the following interval splitting algorithm.

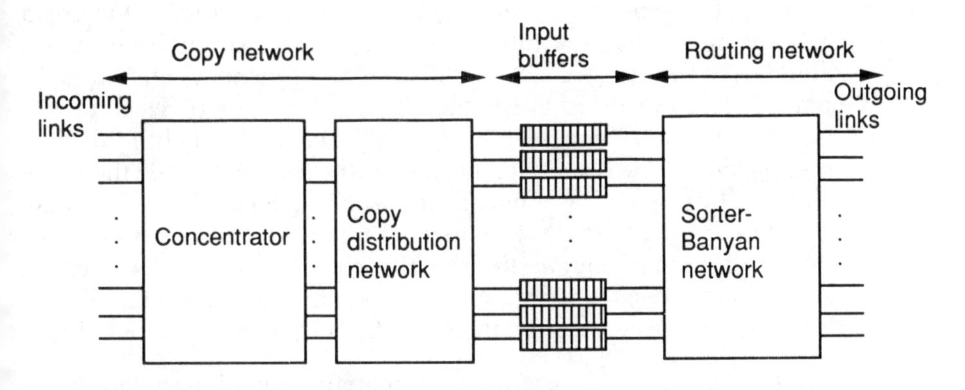

Figure 5.12 A Banyan network that facilitates multicasting.

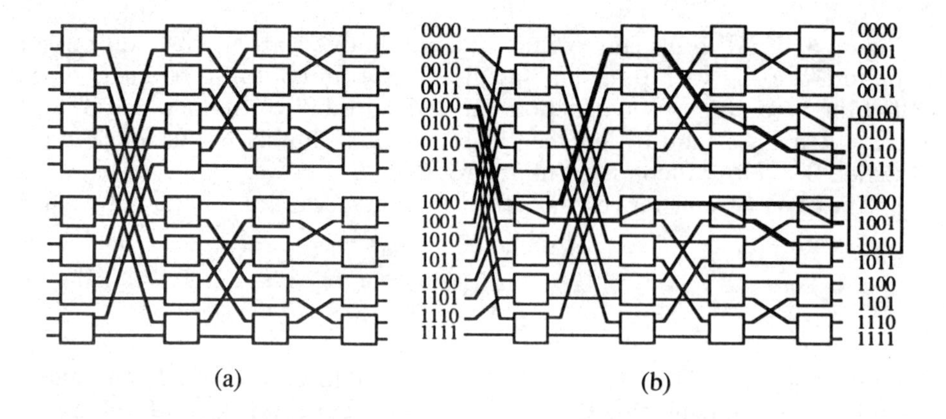

Figure 5.13 Concentrator and copy networks.

Interval Splitting Algorithm

1. If the kth bits of the two bounds are both 0s, then pass the interval from the current switching element at the kth stage to the next switching element at the $(k + 1)$st stage using the *upper* outgoing link. The values of the two bounds are the same in the next stage.
2. If the kth bits of the two bounds are both 1s then, pass the interval from the current switching element at the kth stage to the next node at the $(k + 1)$st stage using the *lower* outgoing link. The values of the two bounds are the same in the next stage.
3. If one of the kth bits of the two bounds is 0 and the other is 1, then:
   a. Split the interval into two sets according to whether the kth bit is 0 or 1.
   b. Pass the interval associated with 0 from the current switching element in stage k to the next switching element in stage $k + 1$ using the upper outgoing link. The new interval at stage $k + 1$ has the same lower bound as at the kth stage. The upper bound of the new interval is the same as the lower bound after the bits $k + 1$ to n are replaced by 1s.
   c. Pass the interval associated with 1 from the current switching element in stage k to the next switching element in stage $k + 1$ using the lower outgoing link. The new interval at stage $k + 1$ has the same upper bound as at the kth stage. The lower bound of the new interval is the same as the upper bound after the bits $k + 1$ to n are replaced by 0s. (At this point, two copies of the cell are made; one is transmitted from the upper outgoing link and the other from the lower outgoing link.)

All input and output addresses are represented in their binary form as an n-bit binary number. At the k-th stage of the copy distribution network, we

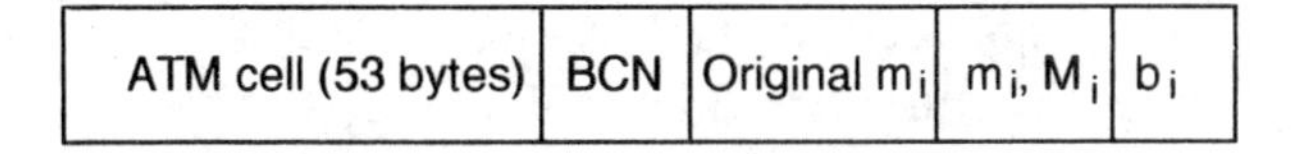

Figure 5.14 Packet header format for the copy network.

look at the k-th bit counted from left to right. Hence, at the k-th stage, the splitting (and the copying) decision for each cell is made by comparing only the kth bit of the lower bound of the interval. If they are different, then the new value of the lower or upper bound of the interval can be generated using a simple logic.

In Figure 5.13, $b = 0100$, $m = 0101$, and $M = 1010$. At stage 1, the two bit values are different. Hence, a copy is made and two cells are transmitted to the next stage with $m = 0101$ and $M = 0111$ at the upper outgoing link, and $m = 1000$ and $M = 1010$ at the lower outgoing link. At stage 2, no splitting is done, since the second bits of respective bounds are the same at each cell. At the third stage, both cells have different values at the third bit position and both are split. From top to bottom, the new values of interval bounds are equal to ($m = 0101$, $M = 0101$), ($m = 0110$, $M = 0111$), ($m = 1000$, $M = 1001$), and ($m = 1010$, $M = 1010$). At the fourth stage, the second and the third cells from the top are further split, and six cells are delivered at addresses 0101 to 1010.

After the copies are made by the copy distribution network, point-to-point addresses are needed to switch cells from the incoming links of the sorter/Banyan network to its output ports. Two identifiers are needed for this purpose: one to uniquely identify the original cell and another one to identify the copies. The two values used in [46] are referred to as the *broadcast channel number* (BCN) and the *copy index*. The copy index e_i is given by $y_i - m_i$, where y_i is the output port number at the last stage of the copy distribution network and m_i is the original lower bound given by

$$\sum_{k=0}^{i-1} c_k$$

Given the BCN and m_i, the point-to-point self-routing address of cells at the sorting/Banyan network can then be obtained from an address translation table residing at the switching node. The address of the cell at the input port of the concentrator has the form illustrated in Figure 5.14.

5.4.2 Nonblocking Space-Division Architectures

At the expense of losing the self-routing property, space-division switch architectures can be made internally nonblocking to realize all $N!$ combinations without the use of sorting networks. Let us consider a three-stage space-division switch. N input lines are distributed over N/n groups of n lines. Without loss of generality, N/n is, for presentation purposes, assumed to be an integer number. There are N/n switching elements at the first and the third stages and each group of lines is connected to one first-stage switching element. Similarly, each third-stage switching element has n outgoing links connected to it. There are k switching elements at the second stage, with k to be determined. Each switching element at the first stage is connected to all k switching elements at the second stage. Similarly, each switching element at the second stage is connected to all third-stage switching elements. Hence, the switching elements at the first, second, and third stages are $n \times k$, $(N/n) \times (N/n)$, and $k \times n$ crossbar switches, respectively. This switch architecture is shown in Figure 5.15.

A connection from an input port to an output port is established by finding a path from the corresponding first-stage switching element to the third-stage switching element through a second-stage switching element. Hence, as long as there is at least one path that connects the first-stage switching element to the corresponding third-stage switching element, the cell at the input port can be switched to its output port if there is no output contention. Given N, the question is how to choose the values of k and n so that this switch architecture becomes nonblocking while minimizing the total number of crosspoints in the network.

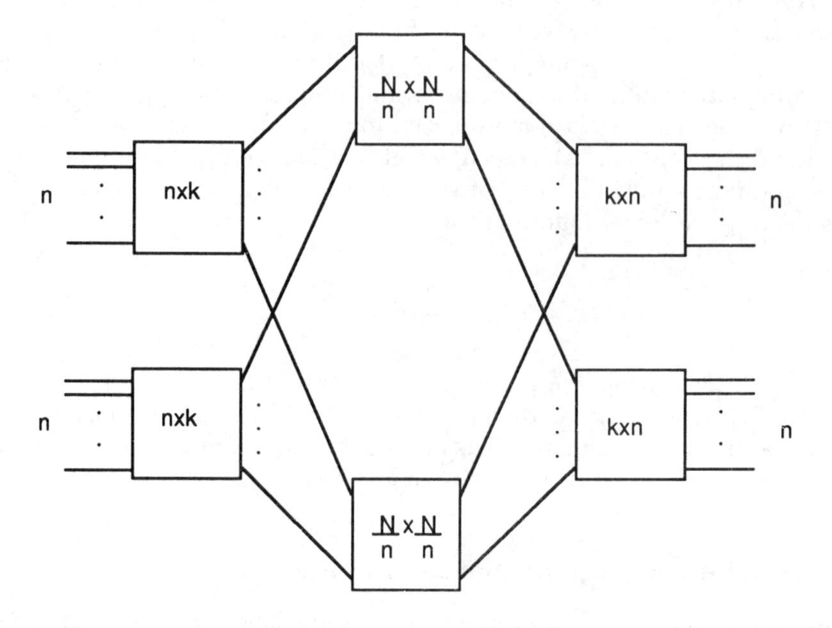

Figure 5.15 A three-stage nonblocking switch.

At the first stage, at most $n-1$ connections may already be established from the switching element where the connection attempt is being made. Similarly, at the third stage, $n-1$ connections may have already been established to the output ports of the switching element where the connection is to be made. In the worst case, all $n-1$ originating and $n-1$ terminating connections may have already been established using all distinct second-stage switching elements. Hence, a total of $2n-2$ switching elements at the second stage may not be available to make the connection. However, if $k \geq 2n-1$, then the switch becomes internally nonblocking, since, in this case, there is always a second-stage switching element available to establish the connection.

Next we determine the value of n to minimize the total number of crosspoints in the switch, given that $k = 2n - 1$.

The total number of switching elements and the number of crosspoints in each is given in Table 5.1, with the total number of crosspoints C given by

$$C = 2N(2n-1) + (2n-1)(N/n)(N/n) \tag{5.1}$$

The value of n that minimizes C can be found by taking the first derivative of C and equating it to zero. The value of n cannot be given in closed form, but it can be shown that $n = \sqrt{N/2}$ is an upper bound on the optimum value of n and becomes a tight bound as N increases. Substituting n into (5.1), we have $C = 4N(\sqrt{2N} - 1)$. This is a considerable savings on the total number of crosspoints used in the switch compared with the crossbar, particularly as the number of ports in the switch increases. However, the nonblocking switch is no longer self-routing, requiring end-to-end connections to be established for each cell (and therefore address translation tables).

5.4.3 Examples of Space-Division Switch Architectures

Various variants of Banyan network-based switch architectures have been developed for ATM networks. In this section, we briefly introduce two examples of such switch architectures: Batcher-Banyan and load-sharing Banyan.

Table 5.1
Crosspoint Complexity of the Three-Stage Internally Nonblocking Switch

Stage	*1*	*2*	*3*
No. of switching elements	N/n	2n – 1	N/n
No. of crosspoints per switching element	n × (2n – 1)	(N/n) × (N/n)	(2n – 1) × n

5.4.3.1 Batcher-Banyan Switch

The Batcher-Banyan switch, proposed by Bellcore [32], consists of a Batcher sorter connected to a router network with input buffers preceding the sorter network. Its operation is based on the following path reservation principle. At the start of a cycle, a small packet containing only the source and the destination addresses is passed through a Batcher sorter. At the output ports of the sorter, small packets with the same destination address as the packet above them are deleted to eliminate output contention. The remaining packets are circulated back to the input queues. Only the input queues that receive back the small packets are allowed to transport their cells through the switch. These cells are guaranteed to arrive at their destination ports without any conflict within the switch. Cells that are not sent in a cycle due to output contention remain at the HOL of the input buffer and are retried during consecutive cycles.

In this switch architecture, the path followed through the network is not deterministic and is a function not only of the cell under scrutiny, but also of the destinations of other HOL cells present at other input queues.

5.4.3.2 Load-Sharing Banyan

This switch architecture is based on a Banyan variant known as the *load-sharing* (or alternate routing) Banyan network [45], which eliminates internal buffering with minimal loss of throughput capacity, but at the expense of increased wiring between the switching elements. An example of a four-stage load-sharing network made of 2 × 2 switching elements is illustrated in Figure 5.16.

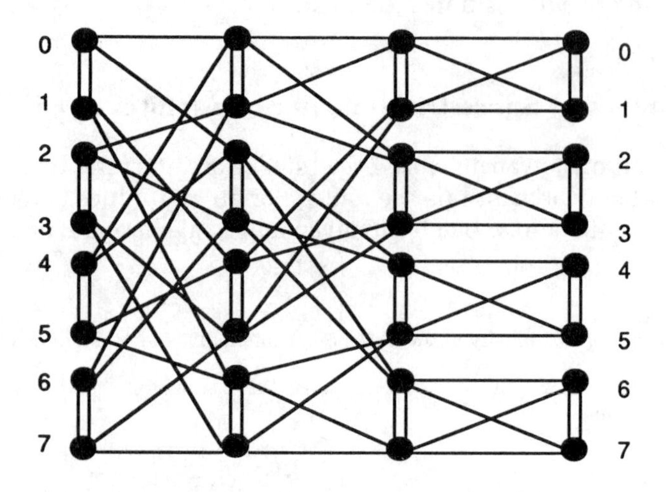

Figure 5.16 A load-sharing Banyan network.

The network operates in a similar way to that of Banyan networks. The major difference is the pairing of switch elements in a stage. Links connecting paired switch elements allow traffic to the outlets of other switching elements at the same stage, thereby increasing the probability of finding an idle output link to the next stage.

A cell must reserve a path to traverse the switch, a task that is accomplished in a distributed fashion as follows. At each stage, the output interface unit of a switching element that receives a cell sends a handshake to the interface unit of an input port of a switching element at the next stage along a separate channel. The operation of switching elements is synchronized and a cell is transmitted at each time unit from one stage to the next from each port of every switching element in that stage. Cells proceed in this manner from stage to stage as long as they can be moved toward their output ports. At each stage, a fraction of cells will not be able to proceed due to the lack of a path toward the destination. In this case, cells can either be stored at the switching elements or be dropped. The load-sharing Banyan structure does not grow by adding stages like regular Banyan networks.

Nevertheless, reasonable growth provisions can be made by properly partitioning the network so that it can be built up in large increments (cf. [45]).

5.5 PERFORMANCE ANALYSIS OF ATM SWITCH ARCHITECTURES

Performance analysis of the switch architecture used in the network is a crucial step in the design and analysis of an ATM network. For example, various bandwidth allocation procedures, discussed in Chapter 4, are based on the analysis of a single queuing station, corresponding to output buffers of the switch. Similarly, in determining whether the delay requirements of a connection can be met on a particular path or not, it is generally assumed that the delay through each switching node is constant along the path. These two assumptions are realistic only if the switch is internally nonblocking and the output contention is resolved through output buffering. However, the number of ports in a switch with output buffering is limited by the speed of the buffer and the complexity of the real estate connecting the output ports to the output buffers. Furthermore, various switch fabrics are internally blocking, and, in the case of Banyan networks, not all $N!$ permutations of concurrent connections are realizable, so input buffering is required. Finally, when cells are buffered, the delay through the switch is no longer constant; it is a function of the traffic characteristics and output destinations of connections at each incoming link.

Internal buffers are integrated into switching elements and built into a chip. Accordingly, they are relatively small buffers with a capacity to store only a few cells at a time. On the other hand, either FIFO or memory is used for input and output buffering, and their sizes need to be determined according to the cell loss and delay requirements of the applications. For example, the maximum delay at each node for a real-time application such as voice and video is restricted to a fraction of a millisecond. In this case, the size of a buffer is restricted to few tens of cells. However, as discussed in Chapter 4, small buffers restrict the amount of statistical multiplexing that can be achieved in the network, particularly for delay-insensitive applications with long bursts.

Hence, the analysis of switch fabrics is essential in the efficient design and implementation of network management functions. There are four types of techniques used to analyze a switch architecture: simulation, numerical analysis, analytical techniques, and heuristics. Simulation models are generally complex to build and require very long run times to achieve the desired accuracy in estimating the performance metrics of a switch when the required cell loss probabilities of some applications are on the order of 10^{-8} or less. Numerical analysis is restricted to Markovian models. Since the transmission and cell switching times in ATM networks are constant, this requirement mainly implies that the arrival process at the switch is modeled by a Markovian process, such as IBP or MMBP. The numerical analysis process is a three step procedure: (1) generation of the states of the underlying Markov process, (2) generation of the transition rate matrix Q, which describes the stochastic behavior of the process in steady state, and (3) numerical solution of the linear system of equations $\pi Q = 0$ and $\pi e = 1$, where π is the steady-state probability vector of states and $e = (1, 1, \ldots, 1)^T$. Numerical analysis of a switch architecture is then restricted to networks with a small number of ports, since the state space of the process (i.e., the number of unknowns) increases exponentially as the number of ports in the switch increases. Analytical results of closed-form solutions for the performance metrics of switch architectures, on the other hand, can be obtained only under a set of assumptions that do not realistically model the traffic in ATM networks. Hence, most of the techniques used to analyze ATM switch fabrics are based on heuristics, which produce fast but approximate results. Although there are various heuristics proposed in the literature that have fairly good accuracy, in general it is not possible to establish an error bound on their performance.

Early models of switch architectures were mostly developed in the context of multiprocessor systems in which a number of processors and memory modules are interconnected to each other to reduce the time it takes to solve complex problems. The two assumptions commonly used in these models are:

- Incoming traffic is uniformly distributed over output links (i.e., there is a probability that a cell at incoming link i will have output link j as its destination, p_{ij}, which is equal to $1/N$ for all $i, j = 1, \ldots, N$).
- The rate at which cells arrive at the incoming link i, p_i, is equal to p for all i, where $0 \leq p < 1$.

These two assumptions may be considered realistic in multiprocessor systems if the total system load is nearly equally partitioned between the processors. However, they do not hold in ATM networks, where the traffic is bursty and the cell arrival stream is correlated. Nevertheless, models developed based on these assumptions allow us to obtain analytical expressions to compare different schemes on a common ground. The analytical results obtained under these two assumptions can be extended to include the case in which $p_i \neq p$ for all i. However, the analysis breaks down when the uniform destination assumption is relaxed.

With nonuniform destination traffic, two cases have received the most attention. The first case corresponds to an incoming link in which the destination ports of cells are restricted to a subset of all output ports. In the second case, a relatively large portion of incoming cells from different links are destined for a particular output port, referred to as a hot spot. Depending on the destinations of cells, several hot spots can exist simultaneously in a switch.

We now present various analytical techniques and heuristics developed to analyze the different switch architectures introduced in this chapter. Instead of presenting the details of various heuristics recently proposed for the analysis of ATM switches, we introduce only the basic concepts used to construct them.

5.5.1 Shared-Medium Switch Architectures

In a shared-medium switch, cells arriving from incoming links are multiplexed into a common medium, such as a bus or a ring, and demultiplexed into the FIFOs of outgoing links. The model of a shared-medium switch depends on whether it is blocking or not. If the bandwidth of the switch is greater than the sum of the individual incoming link bandwidths, referred to as a nonblocking shared-medium switch. In this case, cell losses at the incoming ports can practically be eliminated by employing small input buffers with a capacity to hold only a few cells. If the switch bandwidth is less than the sum of incoming link bandwidths, then this is no longer the case and relatively large input buffers are required. In either case, output buffers are dimensioned such that the packet loss probability due to a momentary overload situation is less than the desired QOS requirement.

In this framework, cells are lost if they arrive at a time when there is no space available at the output buffer. Conceptually, it is possible to develop a handshake mechanism between the input and output buffers so that cells are stored at the input buffers until a space becomes available at the output buffer, thereby eliminating the cell loss at output buffers. However, doing so complicates the switch design, so it is not justified when compared to its value.

5.5.1.1 Nonblocking Shared-Medium Switch Architectures

First we will consider the case in which the switch bandwidth is greater than the total bandwidth of incoming links. In this case, for modeling purposes, the small input buffers and the switch itself do not need to be explicitly modeled, since their contribution to the performance metrics of interest (i.e., cell loss probability and delay) is negligible. We have N streams of incoming cells demultiplexed into N output queues, as illustrated in Figure 5.17. The traffic on an incoming link consists of multiplexed cell streams generated by different sources in the network. Each cell may potentially be destined for any output. Hence, the arrival stream at a particular output port is the superposition of individual thinned arrival streams from different incoming links. The term *thinned* refers to the demultiplexed traffic from an incoming link to a particular output port. A snapshot of the arriving cell stream is shown in Figure 5.17, where the first two cells are destined for output port 1, the third for port 2, and so on. One approach commonly used to model the thinned processes from an arrival stream is to assign the destination ports of cells probabilistically. That is, a cell at incoming link i is destined for output port j with probability p_{ij}.

For example, let m and m_j respectively denote the total average cell arrival rate at a link and the average cell rate of connections on that link destined for output port j. Then p_{ij} may be set to be equal to m_j/m. This assumption implies that the destination ports of consecutive cells are independent. However, the interarrival times of cells are correlated, since the consecutive cells generated by a source may still be close to each other on a link, despite being multiplexed with cell streams from other sources. If this is the case, then the probability of a cell at incoming link i routed to output port j, given that the previous cell is routed to output port j, is greater than p_{ij}. This assumption may be considered realistic if the peak rates of connections multiplexed on a link are only a fraction of the link bandwidth. Even with this assumption, characterizing the individual

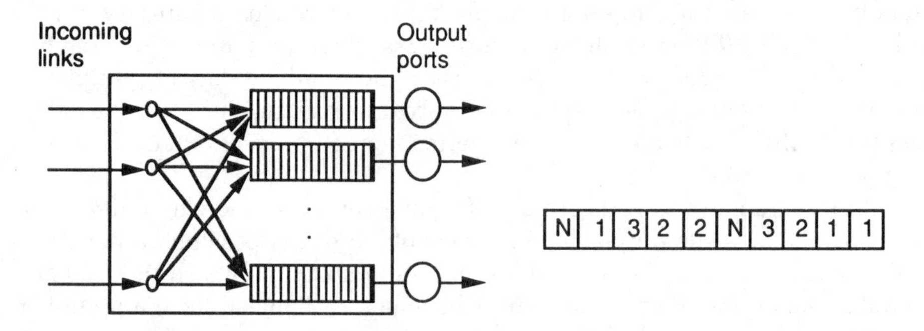

Figure 5.17 Conceptual view of a nonblocking shared-medium switch model.

bit streams from a particular input link to a particular output port is a complex task and is often approximated.

As an alternative to attempting to characterize the thinned process, the source characterization at the source may be used. Since this assumes all cells belonging to a burst arrive consecutively at an output port (i.e., much stronger correlation than the actual incoming cell stream), doing so would result in overestimation of cell loss and average delay values. Therefore, the actual performance observed from the system is better than what the model predicts. The drawback of this approach is that the resources will be underutilized. This approach, being safe, is preferred over an approach that produces a more accurate estimation of performance metrics most of the time, but underestimates at other times, because the consequences of the latter may be catastrophic under some conditions. Assuming now that the thinned processes are characterized somehow, each output queue can be analyzed numerically independent of other queues with a deterministic service time, which corresponds to the cell transmission time and N arrival streams. The solution of the queue becomes impractical as the number of ports in the switch increases. As discussed in Chapter 4, this problem can be overcome by superposing N arrival streams onto one stream. Then the queue can be analyzed numerically, for example, as an MMBP/D/1/K queue.

5.5.1.2 Blocking Shared-Medium Switch Architectures

Let us now assume that the bandwidth of the shared media used to provide connectivity between the input and output ports is less than the sum of the peak bandwidths of incoming links. In this case, the switch becomes a bottleneck and the probability that cells are dropped at the input ports is no longer negligible when there is a capacity to hold only a few cells. When a shared-media switch is blocking, cells are lost due to lack of space in buffers at both input and output ports.

Cells at the input buffers are typically served in a round-robin type of service policy, in which each input buffer is served in a cyclic manner. Multiqueue systems served by a single server, referred to as polling systems, have been studied extensively in the literature (cf. [65,71] and references therein). Mainly, three types of service disciplines are considered: exhaustive, gated, and limited services. Each time a queue is visited (see Figure 5.18), all customers waiting in the queue (including the ones that arrive during the service) are served in the exhaustive service. In the gated service, only the customers that are waiting when the server visits the queue are served. In the limited-service discipline, each queue is served until the queue is empty or a prespecified number of customers are served, whichever occurs first. Any one of the three types of service disciplines may be used in ATM networks.

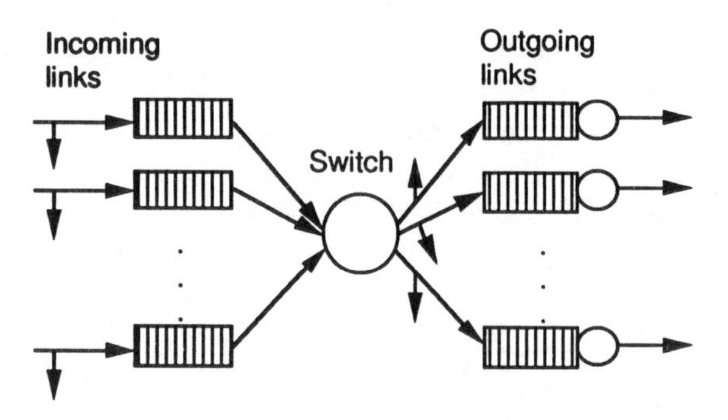

Figure 5.18 A queuing model of a shared-medium buffer blocking switch.

Techniques developed to analyze polling systems in the literature cannot be directly applied to the study of the performance metrics of ATM switches, since the arrival processes used in these approaches do not take the bursty nature of ATM traffic into consideration. The approach presented next to approximately analyze the input buffers is based on the framework developed in [36]. Consider N input queues served by a single server. Time is slotted, with the slot length being equal to the switching time of a cell at the shared media. The arrival process at each link is assumed to be the same (i.e., symmetric arrival process), with the interarrival time of cells being distributed IBP with parameters (p, q, α). Each input queue has a finite capacity B. A cell arriving at an incoming link when there are B cells waiting in the corresponding input buffer are lost and cleared from the system. The service discipline is limited-1; that is, each time the server visits a queue, only one cell is switched, if there is at least one waiting in the queue. Furthermore, the time it takes for the server to switch from one queue to the next is assumed to be zero.

Let N denote the number of incoming links; π_A the probability that a source is in an active state, $\pi_A = (1 - q)/(2 - p - q)$; π_S the probability that a source is in a silent state, $\pi_S = (1 - p)/(2 - p - q)$; and, λ_{in} the arrival rate at an incoming link, $\lambda_{in} = \alpha(1 - q)/(2 - p - q)$. Then the probability that k sources are in active states $P(k)$ is given by

$$P(k) = \frac{N!}{k!\,(N-k)!}\pi_A^k \pi_S^{N-k} \tag{5.2}$$

Similarly, the state transition probability of having k' incoming links in an active state in a slot, given that there were k active incoming links in an active state during the previous slot $P(k'|k)$, is

$$\sum_{j=0}^{k} \frac{k!}{j!(k-j)} p!^{j}\,(1 - p)^{k-j} \frac{(N-k)!}{(N-k-k'+j)!(k'-j)!} q^{j}(1 - q)^{k-j} \tag{5.3}$$

Given that k sources are active, m cells arrive during one slot with probability $P(m|k)$, where

$$P(m|k) = \frac{k!}{m!(k-m)!}\alpha^m(1-\alpha)^{k-m} \tag{5.4}$$

Let us now define a two-dimensional state variable (k, q) such that at the beginning of a slot, there are k active sources and the queue length becomes q as a result of m' cells arriving during the previous slot, of which m'' cells enter the queue while $m' - m''$ are lost. Then the probability $P(k', q')$ of being in a state (k', q') is the solution of the following system of linear equations:

$$\begin{aligned} P(k', q') &= \sum_{k=0}^{N}\sum_{m'=0}^{k'}\sum_{q=0}^{Q} P(k, q)\, P(k'|k)\, P(m'\ k')\, P(m''|m', q''), \\ &\text{for all } (k, q) \sum_{k=0}^{N}\sum_{q=0}^{Q} P(k, q) = 1 \end{aligned} \tag{5.5}$$

where $q'' = \text{maximum}(q-1, 0)$, $q' = m'' + q''$, $Q = NB$ is the sum of the buffer capacities of N input buffers, and $P(m''|m', q)$ is the probability that m'' cells will enter the queue, given that m' cells arrived during the slot and there were q cells in the system just before the arrivals. The queue length distribution $P(q)$ can then be obtained by summing the $P(k, q)$'s over k. Since the time for the server to switch from one queue to the next is assumed to be equal to zero, the rate at which cells depart from the server, λ_{out}, is equal to

$$\lambda_{out} = 1 - P(0) = \lambda_{in}(1 - P_{loss}) \tag{5.6}$$

where P_{loss} is the cell loss probability given by

$$P_{loss} = 1 - \frac{\lambda_{out}}{\lambda_{in}} \tag{5.7}$$

Using Little's relation and the mean queue length L being calculated from the queue occupancy distribution, that is,

$$L = \sum_{q=1}^{B} P(q)$$

the average delay W is equal to

$$W = L/\lambda_{out} \tag{5.8}$$

The only unknown set of variables $P(m''|m', q)$ in (5.5) is approximated using a multiple-urn model, which finds the total number of ways to place r indistinguishable balls into n distinguishable boxes, given that the capacity of each box is equal to B. Assume for a moment that there is no capacity limitation (i.e., $r \leq B$), and let $A_{n;r}$ denote the number of indistinguishable distributions and $C_{n;r;k}$ the number of ways of having at least k_i balls in the ith box, where

$$k = \sum_{i=1}^{n} k_i$$

Then we have

$$A_{n;r} = \binom{n + r - 1}{r} = \binom{n + r - 1}{n - 1} \tag{5.9}$$

$$C_{n;r;k} = \binom{n + r - k - 1}{r - k} = \binom{n + r - k - 1}{n - 1} \tag{5.10}$$

The total number of different configurations in the multiple-urn model with capacity constraints $T_{n;r;B}$ is

$$\begin{aligned} T_{n;r;B} &= \sum_{i=0}^{n} (-1)^i \binom{n}{i} \binom{n + r - i(B + 1) - 1}{n - 1} \\ &= \sum_{i=0}^{n} (-1)^i \binom{n}{i} C_{n;r-i(B+1);0} \end{aligned} \tag{5.11}$$

where $C_{n;r;0}$ denotes the total number of solutions, given that there is no limit imposed on the capacity of boxes, and $C_{n;r-i(B+1);0}$ represents the number of configurations in which at least one of the n boxes contains $B + 1$ or more balls.

We now proceed to find $X_{n;r;B}$, the number of ways of having at least one full box. If r is less than B, then $X_{n;r;B} = 0$. For $B \leq r \leq 2B$,

$$X_{n;r;B} = n\, C_{n-1;r-B;0} \tag{5.12}$$

In general, for $kB \leq r < (k + 1)B$, we have

$$X_{n;r;B} = \sum_{i=1}^{k} \binom{n}{i} (T_{n-i;r-i^*B;B} - X_{n-i;r-i^*B;B}) \tag{5.13}$$

from which $X_{n;r;B}$ can be obtained recursively. The probabilities $P(m''|m', r)$ can now be approximated as follows:

$$P(m''|m', r) = \frac{\displaystyle\sum_{i=0}^{(r/B)^+} \frac{\binom{N-i}{m''}\binom{i}{m'-m''}}{\binom{N}{m'}} \binom{N}{i} \left(T_{N-i;r-i^*B;B} - X_{N-i;r-i^*B;B}\right)}{T_{N;r;B}} \quad (5.14)$$

where $(r/B)^+$ is the largest integer less than or equal to r/B.

5.5.2 Shared-Memory Architectures

In a shared-memory architecture, there is an output queue associated with each output port, and the total number of cells waiting in the memory is bounded by the memory size M. Cells arriving from incoming links are placed in the memory, their destination addresses are decoded, and the memory addresses they are stored at are linked to the corresponding output port queues. A shared-memory switch can then be modeled as a multiserver queue as illustrated in Figure 5.17.

The queuing model of an $N \times N$ shared-buffer switch consists of N single-server queues, representing the logical queues associated with output ports. In general, each queue has a finite capacity B_i. The individual queue sizes are less than or equal to the memory size (i.e., $B_i \leq M$). If $B_i = M$, then we have a fully shared buffer. In a fully shared buffer, a large portion of the memory can potentially be used for a single output port when a burst of cells from different incoming links destined for one particular output link arrives within a short period of time. Accordingly, this allocation scheme is not fair, since the cells routed to output ports may be dropped due to the use of large buffers for a single port. To solve this problem, the number of cells that can be queued for a particular queue is often restricted such that $B_i < B$ with $NB > M$. However, at any given time, the total number of cells waiting in the different queues is always bounded above by M.

The server at each queue corresponds to an output port. Time is slotted and the service time at each server is equal to one cycle, where the duration of a cycle is constant and is equal to one slot length. The service times at output ports are synchronized; that is, one cell from each output port may depart from the switch during one cycle if there is at least one waiting in each queue. Similarly, the cell arrival times at incoming links are also slotted. If the incoming link speed is such that potentially one cell can arrive at every slot, then each slot may contain one cell or may be an empty slot, defined probabilistically by the arrival process. For example, if the arrival process is an IBP with parameters (p, q, α), the probability that a slot contains a cell is equal to α. The parameter α can be used to differentiate between incoming links with different speeds.

For example, assuming that the switch runs at OC-3 speed per port, then $\alpha = 1$ for OC-3 links, where as $\alpha = 1/3$ for OC-1 links.

Upon arrival from an incoming link, a cell joins one of the output queues according to its destination port. In reality, as discussed in the context of shared-medium switches, the destination addresses of arriving cells from an incoming link are correlated. If for the simplicity of the analysis such correlations are not modeled, then cells are routed to destination ports with deterministic probabilities: upon arrival at the *i*th input port, a cell joins the *j*th output queue with probability r_{ij}, such that

$$\sum_{i=1}^{N} r_{ij} = 1$$

Under the realistic assumptions of modeling ATM switches, this queuing system does not have a closed-form queue length distribution. As discussed previously, numerical analysis is often restricted to small switches, with a few ports and simulations requiring long running times to estimate the system performance with desired cell loss probabilities of 10^{-8} or lower.

An approximation algorithm to analyze the queuing system under consideration is developed in [27,76]. In this framework, the shared buffer is decomposed into individual queues and each queue is analyzed in isolation. Instead of going into the details of the algorithm, we only present the underlying concepts used in the heuristic, with rather small deviations. Without loss of generality, let us consider the system with the same slot size for both arrival and service processes. The slot boundaries of the arrival streams lie between the two boundaries of service slots. Hence, even if it arrives at a time the server is idle, the cell waits until the beginning of the next service slot to start receiving service.

The queuing structure of subsystem *i* in isolation is shown in Figure 5.19. The subsystem consists of queue *i* and subsystem $N - i$, which approximately models the dynamics of all N queues other than queue *i*. The latter is necessary to capture the effects of the upper bound (i.e., M, imposed on total number of cells that can wait in the shared memory).

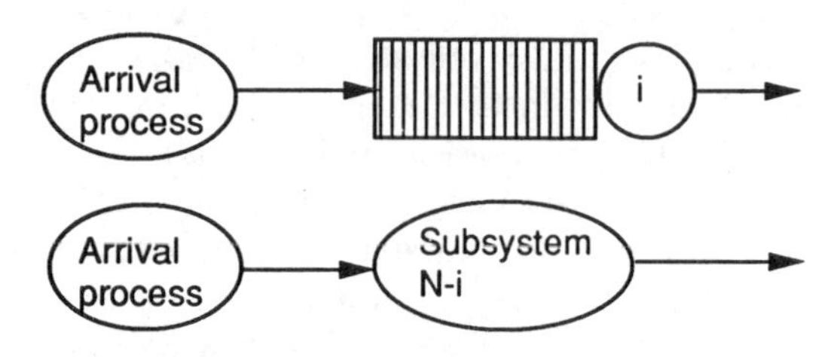

Figure 5.19 The queuing structure of subsystem *i*.

Let us consider the arrival process for incoming link i. The cell stream is decomposed and routed to corresponding output queues according to probability distribution b_{ji}. The Laplace (or z) transform of the arrival stream at link j to queue i, $\phi_{ji}(z)$, can be approximated as illustrated in Figure 5.20, where $\phi_j(z)$ corresponds to the transform function of the interarrival time of cells at incoming link j.

Then $\phi_{ji}(z)$ is obtained in terms of $\phi_j(z)$:

$$\phi_{ji}(z) = \frac{\phi_j(z)b_{ji}}{1 - (1 - b_{ji})\phi_j(z)} \tag{5.15}$$

The next step is to obtain the moments of $\phi_{ji}(z)$ and fit a distribution as discussed in Chapter 4 to obtain the arrival process of cells from incoming link j to output port i. Once this procedure is repeated for each j with $b_{ji} > 0$, we have up to N arrival streams of cells to queue i. Similarly, the characteristics of the cell arrival streams to queues other than queue i can be obtained after b_{ji} is replaced by $(1 - b_{ji})$ for each j. At the end of this step, we have a queuing system with $2N$ arrival streams. The dimensionality of the solution of the subsystem can be reduced if each N arrival stream is superposed on a single stream approximately.

As long as there is a cell waiting in the queue at the beginning of a cycle, one cell departs from the ith queue. The subsystem $N - i$ corresponds to all queues other than queue i. In this decomposition, we only know the total number of customers in these $N - 1$ queues but not how these cells are distributed to individual queues. At one extreme, all these cells may be in the same queue in which case only one cell can depart in one cycle. At the other extreme, cells may be uniformly distributed over these $N - 1$ queues and a cell departs from each busy queue, up to $N - 1$ of them depending on the total number of customers in subsystem $N - i$. Hence, it is necessary to estimate the number of busy queues in $N - i$ when there are n cells in the system and n_i cells at queue i. This probability distribution can be approximated iteratively from the solution of N subsystems in isolation with each subsystem consisting of queue i and subsystem $N - i$, $i = 1, \ldots, N$. In [76], this probability distribution is assumed to depend only on $n - n_i$ (i.e., $p_i(n - n_i, k)$), as opposed to both n and n_i, and approximated iteratively. Finally, we note that the arrival processes at each

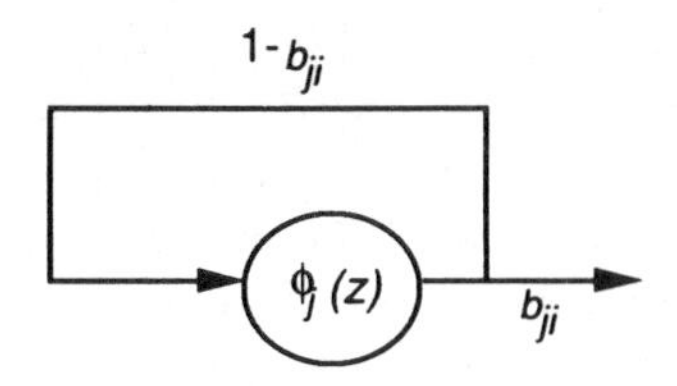

Figure 5.20 Interarrival process of cells from link j to queue i.

queue i and subsystem $N-i$ are estimated once and do not change in subsequent iterations. The steps of the decomposition algorithm can now be summarized.

In initialization, calculate the two arrival processes at each subsystem: one for queue i and another for the subsystem $N-i$. Initialize the probabilities $p_i(n-n_i, k)$ for each i and k, where $p_i(n-n_i, k)$ is the probability that there are k busy queues in $N-i$ when there are $n-n_i$ cells in it. While convergence is not established, solve subsystem i numerically, $i = 1, \ldots, N$ and calculate the new values of $p_i(n-n_i, k)$ using the solutions of each subsystem in isolation.

Since the values of $p_i(n-n_i, k)$ are estimated iteratively, the convergence criteria used depend on how these probabilities are calculated from the solution of each subsystem in isolation. One alternative is to use the mean queue lengths of each queue (cf. [76]). Then the convergence criterion is the difference between the values of the mean queue lengths of queues calculated at the current iteration and the ones calculated at the previous iteration. If the difference is less than or equal to some predetermined value, then the convergence is established; otherwise another iteration is required.

5.5.3 Space-Division Switch Architecture

For modeling purposes, a space-division switch is a collection of a large number of single queuing stations which interact with each other in a fairly complex manner. In this framework, the output port of each switching element is a server.

The queuing structure of servers depends on the type of buffering used in switching elements: input buffering, output buffering, shared-memory, and complex buffering schemes. Queuing network models of a 4×4 Banyan network consisting of 2×2 switching elements are illustrated in Figure 5.21 with different types of queuing. For presentation purposes, the switch is shown with both input and output buffers. Queuing models of larger networks are constructed similarly.

The service time at each server is slotted and is equal to one slot, referred to as *service slots*. Service begins and ends at the slot boundaries. Arrivals occur at the queuing network at the first-stage queues. Each arrival stream is also slotted, referred to as *arrival slots*. The slot boundaries of the arrival streams are in between the two boundaries. Hence, even if a cell arrives when the server is idle, the cell waits until the beginning of the next service slot and then begins service. If the internal links of the switch run at the same speed (plus the speed to accommodate the switch overhead for self-routing readers) as an incoming link, then one cell may potentially arrive from each such link at every slot. For incoming links with slower speeds, the service time at the input buffer is proportional to the speed of the switch speed. For example, it takes three slots for a cell to arrive on an OC-1 link if the internal switch links run at OC-3

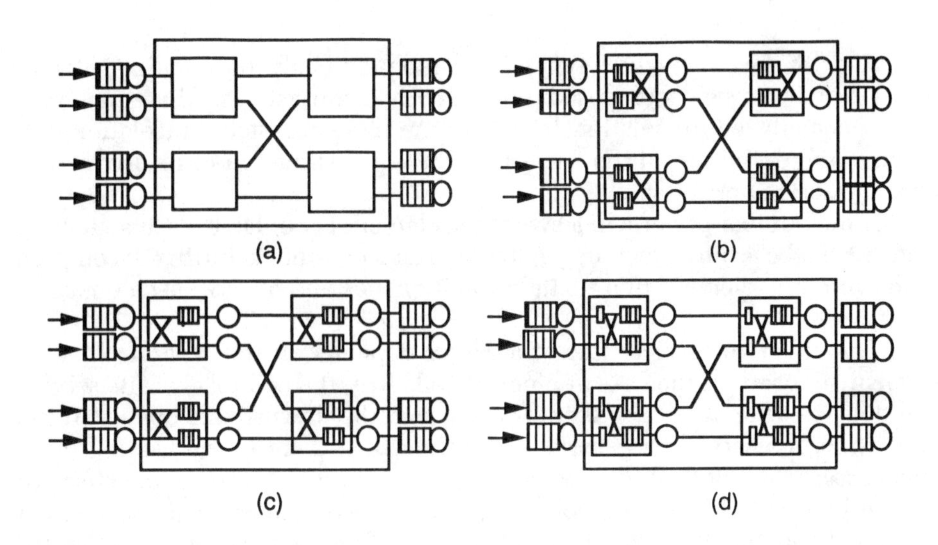

Figure 5.21 Queuing network models of a 4 × 4 Banyan network with I/O buffers: (a) 4 × 4 Banyan network; (b) input buffering; (c) output buffering; (d) input/output buffering.

speeds. Similarly, the service time at the output buffers would require more than one slot for links that run at a slower rate than the internal switch links.

Cells that arrive at the input ports go to their output ports by passing through a number of switching elements, one at each stage. Let us consider a k-stage Banyan switch. The input and output queues external to the switch are indexed 0 and $k + 1$, respectively. Consider a cell that completes its service at switching element s_j at stage j and attempts to enter the next switching element s_{j+1} along its path at stage $j + 1$. If there is a space available at s_{j+1} to accommodate the cell, the arriving cell joins the queue and the corresponding server at stage j starts serving a new cell, if there is one waiting in its queue. What happens if there is no space to accommodate the cell depends on the internal operation of the switch. A cell at switching element s_j cannot enter its destination switching element s_{j+1}, if there is no space available at the latter. When this happens, a cell is called blocked. A blocked cell is forced to wait in the switching element s_j for another cycle before it attempts to enter at s_{j+1} again. If at that moment there is a space at s_{j+1}, the cell is forwarded to the next stage. Otherwise, it will wait for another cycle, and so on. This type of blocking is referred to as repetitive service. Alternatively, the blocked cell may be dropped, eliminating the HOL blocking, referred to as blocked-cells lost.

Queuing models of Banyan networks are, in general, more complex than the other two types of switches discussed previously. Exact numerical analysis is restricted to switches with a small number of ports, whereas simulation models are often more complex to build and require very long run times to obtain accurate estimation of the performance metrics of interest.

As with the other switch types discussed previously, the queuing network of the switch can be decomposed into individual subsystems, where each subsystem is analyzed in isolation. The isolation of a subsystem in isolation first requires the definition of the subsystem and the estimation of its arrival and effective service processes.

If the number of ports in a switching element is not large, each switching element can be a subsystem. In general, it is also possible to further decompose each switching element into its individual queues and analyze each queue in isolation.

The effective service time at a server depends on the type of blocking mechanism used. In the case of blocked-cells lost, the cell receives its service and departs, whether or not there is a space at the downstream queue. At the next cycle, the server starts serving another cell if there is one waiting in its queue, regardless of the buffer occupancy at other queues. Hence, the effective service process at a node in isolation is the same as the original process. On the other hand, there is a complex relationship between a server and its upstream queues in the case of repetitive service that needs to be taken into account when a subsystem is analyzed in isolation. In particular, the effective service process at a server is equal to one slot if the cell in service finds a space at the next stage. If the cell is blocked, it is necessary to approximate the probability distribution function of the number of cycles it takes for the cell to depart from the server. This requires the buffer occupancy distribution at the destination queue to be known.

The arrival process at each subsystem is obtained from the departure processes of its upstream subsystems. For presentation purposes, let us consider a switch with no input buffering (i.e. stage 0) and the arrival process at stage 1 is the external arrival process at the switch. Furthermore, let us assume for a moment that the switching elements at stage 1 are solved and that their departure processes are determined. The arrival process at the input port of a particular switching element at stage 2 can be obtained approximately as the superposition of the departure processes of the corresponding subsystems at stage 1. If the effective service processes are known, then the subsystems at stage 2 can be solved numerically and their departure processes can be approximated, giving the necessary information to determine the arrival processes of subsystems at the third stage, and so on. Without going into the details of how each function is implemented, the steps of solving a space-division switch with blocked-cells lost can be summarized as follows. First, for $i = 0$ to k, solve numerically each subsystem in stage i in isolation, obtain the parameters of the departure process from each server, and determine the parameters of the arrival process at each input port at stage $i + 1$. Second, solve each subsystem at stage $k + 1$ in isolation. Hence, with a blocked-cells lost, the decomposition algorithm requires one pass from the first stage to the last. This is due to the fact that the effective service process at a server does not depend on what is happening at the downstream nodes and only the parameters of the arrival processes at each stage need to be estimated.

This is not the case in repetitive service blocking, where the probability distribution of the number of cycles a cell receives service, and the specification of the arrival processes are required for the solution of the corresponding subsystem in the preceding and the proceeding stages in isolation. The steps of solving a switch with repetitive service blocking can now be summarized.

Until convergence is established:

1. Initialize the probability distribution function of the number of cycles a cell receives service at each subsystem in the switch and determine the effective service process at each server.
2. Forward iteration (update the arrival processes at each stage):
   a. For $i = 0$ to k:
      (1) Solve numerically each subsystem in stage i in isolation.
      (2) Obtain the parameters of the departure process from each server.
      (3) Determine the parameters of the arrival process at each input port in stage $i + 1$.
   b. Solve each subsystem at stage $k + 1$ in isolation.
3. Backward iteration (update the blocking probabilities at each stage):
   a. For $i = k + 1$ down to 1:
      (1) Use the solution of the subsystems at stage i to obtain the effective service times at stage $i - 1$.
      (2) Use the arrival processes obtained at the previous iteration and solve numerically the subsystems at stage $i - 1$.

Different convergence criteria may be used to include convergence on the parameters of the effective service processes, arrival processes, or probability distribution of subsystems. Although in practice these algorithms are observed to be convergent, this class of algorithms generally cannot be proven to be convergent.

5.5.3.1 Internally Nonblocking Switches

An $N \times N$ internally nonblocking space-division switch permits switching between N pairs of I/O ports concurrently as long as these pairs are disjoint. However, as discussed previously, if two or more cells are destined for the same output port, then an output conflict occurs and only one cell can pass through the output port, while other cells are blocked and either dropped or temporarily stored in buffers.

Let us for a moment assume that cells arrive at the input links in a Bernoulli fashion. Each slot at input link i contains a cell with probability p, and it is an empty slot with probability $(1 - p)$. Furthermore, the traffic is distributed uniformly over the output ports and the probability that a cell is routed to an output port is $1/N$. Hence, the probability of a cell arriving at an incoming link destined for a particular outgoing link is equal to p/N. Let us first consider the case in which blocked cells are dropped. In a crossbar switch, the probability that all N incoming links will not select a particular output line is equal to $(1 - p/N)^N$. Then the probability that a particular output port is requested by any of the incoming link is equal to $1 - (1 - p/N)^N$, which is also equal to the throughput of a particular outgoing link. For $p = 1$ (i.e., the input links are 100% utilized), and as N approaches infinity, the throughput is equal to $1 - 1/e = 0.632$. Furthermore, the expected number of busy outgoing links is equal to $N\{1 - (1 - p/N)^N\}$. With the expected number of cells arriving during a slot (i.e., expected number of busy lines) being equal to Np, the probability that an incoming cell will be successfully transmitted to its destination output port P_s is equal to

$$\begin{aligned} P_s &= \frac{\text{Expected number of busy output lines}}{\text{Expected number of busy input lines}} \\ &= \frac{N\{1 - (1 - p/N)^N\}}{Np} = \frac{\{1 - (1 - p/N)^N\}}{p} \end{aligned} \tag{5.16}$$

The analysis can be easily extended to the case in which there is the probability that each slot at input link i will contain a cell with probability p_i. Incoming cells, as before, are assumed to be equally likely destined for N output ports. The probability that all incoming links will not select a particular output line is equal to

$$\prod_{j=1}^{N}(1 - p_i/N)$$

Then the probability that a particular output port will be requested by any of the incoming link is equal to

$$1 - \prod_{j=1}^{N}(1 - p_i/N)$$

Since the total offered traffic rate to the switch is

$$\sum_{i=1}^{N} p_i$$

P_s is now given as follows:

$$P_s = \frac{1 - \prod_{j=1}^{N}(1 - p_i/N)}{\sum_{i=1}^{N} p_i} \quad (5.17)$$

Another approach is to use input buffering. We recall that input buffering suffers from HOL blocking. Let us again consider the system with symmetric traffic and Bernoulli arrivals at each incoming link. Arriving cells are stored in a FIFO. When, during a cycle, there are k HOL cells at k different FIFOs contending for the same output port, one of the k cells is chosen randomly and delivered to the output port. All other cells occupy the HOL positions in their respective FIFOs and wait for the next cycle. Following the framework developed in [31], let us assume that all input queues are saturated (i.e., the probability of an input buffer being empty is equal to zero). Let B_n^i denote the number of cells destined for output port i during the nth cycle, B_i the average number of cells destined for output port i in steady state, and A_n^i the number of new cells destined for output port i at the HOL during the nth cycle. During the nth cycle, one of the B_n^i cells is switched to output port i. A_n^i, at the beginning of the nth cycle, represents the queues that have cells destined for output port i at the second position in the queue such that the HOL cells are successfully switched to their respective output ports. Then

$$B_n^i = \text{maximum } \{B_{n-1}^i - 1 + A_n^i, 0\} \quad (5.18)$$

Let L_n denote the total number of cells to be served by the switch during the $(n + 1)$st cycle and L be the average number of cells served in steady state. If no blocking occurs, since all queues are saturated, N cells will be switched. However, since B_{n-1}^i cells are blocked, we have

$$L_n = N - \sum_{i=1}^{N} B_{n-1}^i \quad \text{and} \quad L = N - \sum_{i=1}^{N} B_i \quad (5.19)$$

Furthermore, if ρ denotes the throughput of output link i, L is also equal to $L = N\rho$. Hence,

$$\sum_{i=1}^{N} B_i = N(1 - \rho) \quad (5.20a)$$

and since N goes to infinity,

$$B_i = (1 - \rho) \quad (5.20b)$$

Finally, since N goes to infinity, the number of cells moving to the head of the line in steady state destined for output port i is Poisson distributed with rate ρ. Then, from the analysis of M/D/1 queues [42],

$$B_i = p^2/\{2(1 - p)\} \tag{5.21}$$

Equating (5.20a,b) with (5.21), since N approaches infinity, the maximum throughput at an output port with input queuing is equal to $\rho = 2 - \sqrt{2} = 0.586$. It is interesting to note that dropping cells may produce higher throughput than input buffering in cases where the rate at which cells arrive is high enough to keep the output ports busy. Table 5.2 illustrates the range of values when the throughput is higher with FIFO, both stable and saturated, and when packets are dropped (cf. [37]).

It is possible to increase the throughput with input queuing beyond 0.586 by reducing the effect of HOL blocking. Let us consider the case in which at the beginning of each cycle each input line submits up to w cells to the switch until one succeeds. If $w = 1$, then we have FIFO queuing. This minimizes the effect of HOL blocking, and as w increases the throughput increases, as illustrated in Table 5.3 (cf. [25]).

We first note that the increase in throughput becomes less significant as w increases beyond 4 and that the throughput as N increases remains bounded by 0.88, even for w as large as 8. Despite the increase in throughput achieved by considering not only HOL, but, say, the first four cells in input buffers in a cycle, this scheme has a limited practical use. In particular, it not only complicates the switch design considerably, but also destroys the distributed control property, since the controller is required to have a global knowledge of the cells at all incoming links (at which output ports there are cells present).

We now consider the case of output buffering, as analyzed in [37]. Let us assume that the output buffer size is infinite (i.e., arriving cells are always allowed to enter the queue) and that zi denotes the probability that i cells will

Table 5.2
Maximum Throughput as a Function of Load with FIFO and Dropping Packets

N	*FIFO (stable)*	*FIFO (saturated)*	*Dropping*
3	$0 \le \rho \le 0.682$	$0.683 \le \rho \le 0.953$	$0.954 \le \rho \le 1$
4	$0 \le \rho \le 0.655$	$0.656 \le \rho \le 0.935$	$0.936 \le \rho \le 1$
5	$0 \le \rho \le 0.639$	$0.640 \le \rho \le 0.923$	$0.924 \le \rho \le 1$
6	$0 \le \rho \le 0.630$	$0.631 \le \rho \le 0.916$	$0.917 \le \rho \le 1$
7	$0 \le \rho \le 0.623$	$0.624 \le \rho \le 0.911$	$0.912 \le \rho \le 1$
8	$0 \le \rho \le 0.618$	$0.619 \le \rho \le 0.907$	$0.908 \le \rho \le 1$
∞	$0 \le \rho \le 0.585$	$0.586 \le \rho \le 0.881$	$0.882 \le \rho \le 1$

Table 5.3
The Throughput as a Function of w

	Window Size							
N	1	2	3	4	5	6	7	8
2	0.75	0.84	0.89	0.92	0.93	0.94	0.95	0.96
4	0.66	0.76	0.81	0.85	0.87	0.89	0.91	0.92
8	0.62	0.72	0.78	0.82	0.85	0.87	0.88	0.89
16	0.60	0.71	0.77	0.81	0.84	0.86	0.87	0.88
32	0.59	0.70	0.76	0.80	0.83	0.85	0.87	0.88
64	0.59	0.70	0.76	0.80	0.83	0.85	0.86	0.88
128	0.59	0.70	0.76	0.80	0.83	0.85	0.86	0.88

arrive at a particular output buffer simultaneously. Then, for $i = 0, \ldots, N$, we have

$$z_i = \binom{N}{i} (p/N)^i (1 - p/N)^{N-i} \tag{5.22}$$

The probability generating function $f(s)$ of the random variable is defined by

$$\sum_{i=0}^{N} s^i z_i$$

and it is equal to

$$f(s) = (1 - p/N + sp/N)^N \tag{5.23}$$

This function can be used to obtain the probability generating function Q(s) of the number of cells in the queue (cf. [42]):

$$Q(s) = \frac{(1 - p)(1 - s)}{f(s) - s} \tag{5.24}$$

Then the average queue size L is given as

$$L = \frac{N - 1}{N} \frac{p^2}{2(1 - p)} \tag{5.25}$$

Since the switch is internally nonblocking and the output buffer is assumed to have an infinite buffer, the switch is work conserving. Furthermore, due to the assumption of symmetric traffic, the utilization of each output queue is equal to p. Then, using Little's relation, the average delay in the queue W is equal to L/p; that is,

$$W = \frac{N-1}{N} \frac{p}{2(1-p)} \tag{5.26}$$

It is interesting to note that $p^2/\{2(1-p)\}$ is the mean queue length of an M/D/1 queue, where interarrival times are exponential, the service time is constant, and the queue has an infinite buffer size. Not only the average queue length, but also the probability distribution of the number of cells in the queue can be shown to converge to that of an M/D/1 queue (cf. [37]), since N approaches infinity.

This analysis is based on the assumption that each output buffer has an infinite capacity. In reality, the buffer is finite, which imposes an upper bound on the number of cells that can be stored in it. Interpolating the results of the infinite buffer model (though it is not correct to do so), it may be concluded that the throughput with output buffering can approach 0.8 to 0.85, after which the average queue length (and therefore the delay) increases exponentially (i.e. (5.25) or (5.26)). Considering the bursty nature of the traffic in ATM networks, these utilization levels may not be achievable, since the cell loss probabilities, in this case, would be intolerable. Nevertheless, this analysis illustrates that output buffering improves the throughput over the other two approaches of dropping cells and input buffering.

5.5.3.2 Internally Blocking Switches

The analysis presented in the context of crossbar switches can be easily extended to Banyan networks with no internal buffering. As before, let us assume that each slot at input link i contains a cell with probability p and the traffic is distributed uniformly over the output ports. Furthermore, consider the case in which blocked cells are dropped. In a crossbar switch, the throughput of a particular outgoing link is then equal to $1 - (1 - p/N)^N$. A Banyan network is multiple stages of crossbar switches connected according to a certain topology. Let us consider a Banyan network consisting of $n \times n$ crossbar switches with p_i denoting the average output link utilization of a switching element at the ith stage. Then, seeing that the output links of a switching element at one stage are the incoming links of a switching element at the next stage, with $p_0 =$ p for the simplicity of the notation, we have

$$p_i = 1 - (1 - p_{i-1}/n)^n, \quad i = 1, 2, 3 \tag{5.27}$$

The probability that an incoming cell will be successfully transmitted to its destination output port in a Banyan network, P_s, is equal to

$$P_s = \frac{\text{Expected number of busy output lines}}{\text{Expected number of busy input lines}} = \frac{Np_3}{Np} = \frac{p_3}{p} \tag{5.28}$$

This analysis, together with the results obtained for crossbar switches, can be easily extended to the case of asymmetric traffic, where the probability that a slot in incoming link i will contain a cell is π_i. Let $p_i^m(k)$ denote the probability that the mth output line of switching element k (indexed from top to bottom) at stage i will contain a cell. As before, assuming that incoming cells are uniformly distributed over N output ports, we have $p_i^m(k) = p_i(k)$, $m = 1, \ldots, n$ (i.e., the output links have the same throughput). Let $S_i(k)$ denote the set of indexes of the switching elements connected to the kth switching element at stage i. For simplicity of notation, $S_1(k)$ denotes the set of incoming links to the kth switching element at the first stage and $p_0(j) = \pi_j$. Then,

$$p_i(k) = 1 - \prod_{j \hat{I} S_i(k)} (1 - p_{i-1}(j)/n), \quad i = 1, \ldots, \log_n N;\ k = 1, \ldots, N/n \tag{5.29}$$

As with the two other cases, the probability that an incoming cell will be successfully transmitted to its destination output port in a Banyan network, P_s, is equal to

$$P_s = \frac{\text{Expected number of busy output lines}}{\text{Expected number of busy input lines}} = \frac{\sum_{k=1}^{N} p_{\log_n} N^{(k)}}{\sum_{k=1}^{N} p_0(k)} \tag{5.30}$$

Let us now consider a Banyan network with input buffering and no internal or output buffering. In order to solve this system approximately using the decomposition approach, the system can be decomposed into N input queues and the Banyan network. Under the assumptions presented above, the Banyan network can be solved efficiently if the utilization of each incoming link is known. In order to analyze an input buffer in isolation, it is necessary to revise the service process to accommodate the blocking delay a cell may go through due to HOL blocking in Banyan networks. In particular, the service time at an input port is equal to one slot if the cell goes through the switch with no contention. If, on the other hand, the cell cannot be delivered to its destination port during the current cycle, it waits for the next cycle and tries again. This is repeated until the cell is switched to its destination port. Hence, it is possible that a cell may be forced to wait in front of the buffer a number of slots until it is switched. Assuming for a moment the service time is geometrically distributed, each input buffer can be solved in isolation as, for example, with an MMBP arrival process, MMBP/Geo/1/K queue, where K is the buffer size in cells. The parameter of the geometric distribution is the blocking probability in the switch p_b. That is, the probability that the service will be x slots, $Pr(X = x)$, is equal to $p_b^{x-1}(1 - p_b)$. Once the queue is solved, the throughput of the queue gives the effective link utilization. Hence, with this framework the solution of an input queue requires the blocking probability in the Banyan network to be known, whereas the solution of the Banyan network requires effective link utilization. Then the following iterative scheme can be used to solve a Banyan network with input buffering.

1. Initialization: set up the initial values of link utilizations for each incoming link and compute the probability of blocking in the Banyan network.
2. Using the value of the blocking probability, solve each input queue numerically and calculate the effective utilization of each incoming link, which is equal to the product of the probability that the queue is full and the average utilization of the incoming link (i.e., offered load).
3. Given the new values of effective link utilizations, solve the Banyan network to obtain the blocking probability.
4. If the absolute value of the difference of the blocking probability between the current and the previous iterations is less than some predefined value, then stop; otherwise go to step 2.

5.5.4 Other References

The state of the art in ATM switching and various performance models of ATM switch architectures are given in survey papers [69,1,8,56]. In [69], the various switch architectures proposed for ATM switching are surveyed, and performance and implementation issues underlying such architectures are discussed. An overview of the major high-performance telecommunications switching fabrics that have been proposed and experimentally developed during the past few years are presented in [1]. An extensive survey of nonblocking ATM switch architectures focusing on their performance issues is given in [56]. Finally, the state of the art in the performance analysis of ATM switching architectures is presented in [8].

In addition to the ATM switch fabrics and the prototypes discussed above, several switch architectures have been proposed in the literature. The Roxanne switch architecture was proposed by Alcatel researchers [24]. The basic switching element is a shared-memory switch with scalability achieved by interconnecting a number of switching elements to each other in a multistage configuration. The Athena switching element is also proposed by Alcatel [10,11]. The switch architecture is a Banyan network in which each switching element is a shared-medium switch with output buffering. A multistage routing network, developed by Fujitsu [23], is constructed by connecting switch-routing modules in a three-stage link configuration. There are multiple paths between any I/O pair. Each switch-routing module is an 8×8 crossbar switch with a limited buffer at each crosspoint. An internally nonblocking space-division switch architecture is reported in [72,74]. The switch consists of a copy network, a distribution network, a routing network, and a number of broadcast and group translator modules to provide multicast capabilities. It uses internal buffering and employs a back-pressure mechanism in which an arriving cell is not allowed to enter the switch if there is no buffer available along the path from the input port to the output port. The first switch based on the Batcher-Banyan concept was proposed in [29]. To overcome the output contention, the switch adds a trap network between the Batcher and the Banyan networks which feeds back all cells (but one) contending for the same output port at the output of the Batcher network to the input ports. Another Batcher-Banyan-based switch fabric was proposed in [34]. Input buffering with a three-phase handshake-type algorithm between the I/O ports is used to resolve the output contention.

Approximations reported in the literature for the analysis of shared-memory switch architectures are in general based on the analysis of a single queue with geometric interarrival times, deterministic service times, and infinite queue sizes, referred to as the Geo/D/1 queue. The results obtained from this single queue are then used to estimate the performance of the shared-memory switch, and the switch itself is modeled with N identical Geo/D/1 queues. In [25], the queue length distribution of the shared memory is obtained as the N-fold convolution of the probability distribution of individual queues. The first two moments of the queue length distribution of the individual queues are used to estimate the queue length distribution of the switch that matches these two moments (cf. [14,51,57]). Other references in this area include [44,47], in which the inequality properties of large numbers are used to provide an upper bound to the probability that queue occupancy at the shared memory will exceed a given number. More recently, [76,43] have proposed a more realistic model of shared-memory architecture. The former is discussed in Section 5.5.2, and the latter presents an efficient and fairly accurate model for providing an upper bound on the cell loss probabilities in this architecture.

Initial studies of Banyan networks were restricted to unbuffered switches with identical Bernoulli input sources and uniform destinations [58]. These models were first extended to include single buffered switch architectures (cf. [35]) and multiple buffers (cf. [61,67,39]). The latter models include FIFO buffering either at the input or the output ports, but not both. Models that include more complex buffering and nonuniform traffic distributions are reported in [39,15,40]. These models were developed assuming Bernoulli arrival processes at the switch. This assumption is not realistic in ATM networks. More recently, approximation algorithms for Banyan networks with bursty arrival processes, nonuniform traffic patterns, and/or complex buffering have been developed (cf. [54,19,52].

References and Bibliography

[1] Ahmadi, H., and W. E. Denzel, "A Survey of Modern High-Performance Switching Techniques," *IEEE JSAC*, Vol. 7, No. 7, September 1989, pp. 1091–1103.

[2] Banwell, T. C., et al., "Physical Design Issues for Very Large ATM Switching Systems," *IEEE JSAC*, Vol. 9, No. 8, 1991, pp. 1227–1237.

[3] Bubenik, R., and J. Turner, "Performance of a Broadcast Switch," *IEEE Trans. Comm.*, Vol. 37, No. 1, 1989, pp. 60–69.

[4] Chen, J. S.-C., and T. E. Stern, "Throughput Analysis, Optimal Buffer Allocation, and Traffic Imbalance Study of a Generic Nonblocking Packet Switch," *IEEE JSAC*, Vol. 9, No. 3, April 1991, pp. 439–449.

[5] Chen, J. S.-C., and T. E. Stern, "Throughput Reduction Due to Non-Uniform Traffic in a Packet Switch with Input and Output Queueing," *ICC '91*, 1991, pp. 413–417.

[6] Clos, C., "A Study of Non-Blocking Switching Networks," *BSTJ*, Vol. 32, No. 3, pp. 406–424.

[7] Coudreuse, J. P., and M. Servel, "Prelude: an Asynchronous Time Division Packet Switched Network," *ICC '87*, Seattle, 1987, pp. 769–773.

[8] D'Ambrosio, M., and R. Melen. "Performance Analysis of ATM Switching: a Review," *CSELT Tech. Rep.*, Vol. 20, No. 3, pp. 265–284.

[9] Decina, M., P. Giacomazzi, and A. Pattavina, "Shuffle Interconnection Networks with Deflection Routing for ATM Switching: the Open-Loop Shuffleout," *ITC-13*, 1991, pp. 27–34.

[10] De Prycker, M., and J. Bauwens, "A Switching Exchange for an Asynchronous Time Division Based Network," *ICC '87*, Seattle, 1987.

[11] De Prycker, M., et al., "An ATM Switching Architecture with Instrict Multicast Capabilities for the Belgian Broadband Experiment," *Proc. ISS '90*, Vol. A8.2, 1990.

[12] Descloux, A., "Stochastic Models for ATM Switching Networks," *IEEE JSAC*, Vol. 9, No. 3, April 1991, pp. 450–457.

[13] Devault, M., J. Y. Cochennec, and M. Servel, "The Prelude ATD Experiment: Assesments and Future Projects," *IEEE JSAC*, Vol. 6, No. 9, 1988, pp. 1528–1537.

[14] Eckberg, A. E., and T. C. Hou, "Effects of Output Buffer Sharing on Buffer Requirements in an ATDM Packet Switch," *INFOCOM '88*, 1988, pp. 459–466.

[15] Eliazov, T., et al., Performance of an ATM Switch: a Simulation Study," *INFOCOM '90*, 1990, pp. 140–147.

[16] Eng, K., M. Hluchyj, Y. Yeh, "Multicast and Broadcast Services in a Knockout Packet Switch," *INFOCOM '88*, Vol. 1A.4, 1988.

[17] Eng, K., M. Hluchyj, Y. Yeh, "A Knockout Switch for Variable Length Packets," *IEEE JSAC*, Vol. 5, No. 9, 1987.

[18] Eng, K., M. Hluchyj, Y. Yeh, "A Modular Broadband (ATM) Switch with Optimum Performance," *Proc. ISS '90*, Stockholm, 1990.

[19] Fitzpatrick, G. J., et al., "Analysis of Large Scale Three Stage Networks Serving Multirate Traffic," *ITC-13*, 1991, pp. 905–910.

[20] Fujiyama, Y., et al., "ATM Switching System Evolution and Implementation for B-ISDN," *ICC '90*, 1990, pp. 1577–1580.

[21] Giacopelli, J., et al., "Sunshine: a High Performance Self-Routing Broadband Packet Switch Architecture," *Proc. ISS '90*, 1990.

[22] Gopal, I. S., I. Cidon, and H. Meleis, "PARIS: an Approach to Integrated Private Networks," *ICC '87*, Seattle, 1987, pp. 764–773.

[23] Hajikano, K., et al., "Asynchronous Transfer Mode Switching Architecture for Broadband ISDN—Multistage Self Routing Switching," *ICC '88*, 1988.

[24] Henrion, M., et al., "Switching Network Architecture for ATM Based Broadband Communications," *Proc. ISS '90*, Stockholm, 1990.
[25] Hluchyj, M. G., and M. Karol, "Queueing in High Performance Packet Switching," *IEEE JSAC*, Vol. 6, No. 9, 1988, pp. 1587–1597.
[26] Hluchyj, M.G., et al., "The Knockout Switch: a Simple Modular Architecture for High Performance Packet Switching," *Proc. ISS, 1987.*
[27] Hong, S.-W., H. G. Perros, and H. Yamashita, "An Approximate Analysis of an ATM Multiplexer with Multiple Heterogeneous Bursty Arrivals," *CCSP Tech. Rept.* TR-91-13, North Carolina State University, 1991.
[28] Hong, S.-W., Modeling and Analysis of the Shared Buffer ATM Switch Architecture for B-ISDNs, Ph.D. Thesis, Computer Science Dept., North Carolina State University, 1992.
[29] Huang, A., and S. Knauer, "Starlite: a Wideband Digital Switch," *GLOBECOM '84*, Atlanta, 1984.
[30] Hui, J., and E. Arthurs, "Starlite: a Wideband Digital Switch," *GLOBECOM '84*, Atlanta, Vol. 5.3, 1984.
[31] Hui, J., and E. Arthurs, "A Broadband Packet Switch for Integrated Transport," *IEEE JSAC*, Vol. 5, 1991, pp. 1264–1273.
[32] Hui, J. Y., "A Broadband Packet Switch for Multi-rate Services," *ICC '87*, Seattle, 1987, pp. 783–788.
[33] Hui, J. Y., *Switching and Traffic Theory for Integrated Broadband Networks*, Kluwer Academic Publishers, 1990.
[34] Hui, J., "A Broadband Packet Switch for Multirate Services," *ICC '87*, Seattle, 1987.
[35] Jenq, Y., "Performance Analysis of a Packet Switch Based on Single Buffered Banyan Network," *IEEE JSAC*, Vol. 1, 1983, pp. 1014–1021.
[36] Jiang, X., and J. S. Meditch, "Integrated Services Fast Packet Switching," *GLOBECOM '89, 1989, pp. 1478–1482.*
[37] Jou, Y. F., A. A. Nilsson, and F. Lai, "A Refined Approximation of a Finite Capacity Polling System under ATM Bursty Arrivals," *CCSP Tech. Rept. TR 92/16, North Carolina State University, 1992.*
[38] Karol, M. J., et al., "Input Versus Output Queueing on a Space Division Packet Switch," *IEEE Trans. Comm., Vol. 35, No. 12, 1987, pp. 1347–1357.*
[39] Kim, H., and A. Leon-Garcia, "Performance of Buffered Banyan Networks under Non-uniform Traffic Patterns," *IEEE Trans. Comm., Vol. 38, No. 5, 1990, pp. 648–658.*
[40] Kim, H., and A. Leon-Garcia, "Performance of Self Routing ATM Switch under Non-Uniform Traffic Pattern," *INFOCOM '90, 1990, pp. 140–147.*
[41] Kim, Y. M., and K. Y. Lee, "PR-Banyan: A Packet Switch with a Pseudo Randomizer for Nonuniform Traffic," *ICC '91, 1991, pp. 408–412.*
[42] Kleinrock, L., *Queueing Systems, Vol. 1: Theory, New York: John Wiley and Sons, 1975.*
[43] Kouvatsos.
[44] Kuwahara, H., et al., "Shared Buffer Memory Switch for an ATM Exchange," *ICC '89, Boston, 1989.*
[45] Lea, C. T., "The Load Sharing Banyan Network," *IEEE Trans. Comm., Vol. 35, No. 12, 1986, pp. 1025–1034.*
[46] Lee, T. T., "Non-blocking Copy Networks for Multicast Packet Switching," *IEEE JSAC, Vol. 6, No. 9, 1988.*
[47] Lee, H., et al., "A Limited Shared Output Buffer Switch for ATM," 4th Int. Conf. Data Comm. Sys. and Their Performance, Barcelona, 1990, pp. 163–179.
[48] Li, S.-Q., "Performance of a Non-Blocking Space-Division Packet Switch With Correlated Input Traffic," *GLOBECOM '89, 1989, pp. 1754–1763.*
[49] Liew, S. C., and K. W. Lu, "Comparison of Buffering Strategies for Asymmetric Packet Switch Modules," *IEEE JSAC, Vol. 9, No. 3, April 1991, pp. 428–438.*

[50] Lin, T., and L. Kleinrock, "Performance Analysis of Finite-Buffered Multistage Interconnection Networks with a General Traffic Pattern," manuscript.

[51] Louvion, J. R., P. Boyer, and A. Gravey, "A Discrete Time Single Server Queue with Bernoulli Arrivals and Constant Service Times," *ITC-12, Torino, Italy, 1988.*

[52] Morris, T.,

[53] Nilsson, A. A., F. Lai, and H. G. Perros, "An Approximate Analysis of a Bufferless N × N Synchronous Clos ATM Switch," *ITC-12, 1990.*

[54] Nilsson, A. A., F. Lai, and H. G. Perros, "Performance Evaluation of a Bufferless Synchronous Clos ATM Switch with Priorities and Space Preemption," *ICC '91, 1991, pp. 379–384.*

[55] A. A. Nilsson, H. G. Perros and F. Lai, "A Queueing Model of a Bufferless Synchronous Clos ATM Switch with Head of Line Priority and Push Out," J. High Speed Networking, Vol. 1, 1992, pp. 255–279.

[56] Oie, Y., T. Suda, et al., "Survey of Switching Techniques in High Speed Networks and Their Performance," 1990, pp. 1242–1251.

[57] Petit, G. H., and E. M. Desmet, "Performance Evaluation of Shared Buffer Multiserver Output Queue Switches Used in ATM," *7th ITC Specialist Seminar, New Jersey, 1990.*

[58] Patel, J. H., "Performance of Processor Memory Interconnections for Multiprocessors," *IEEE Trans. Comp., Vol. 30, No. 10, 1981, pp. 771–780.*

[59] Petersen, J., "Throughput Limitation by Head-of-Line Blocking," *ITC-13, 1991, pp. 659–663.*

[60] Rooth, J., and I. Gard, "An ATM Switching Implementation: Technique and Technology," *Proc. ISS '90, Stockholm, 1990.*

[61] Saha, A., and M. Wagh, "Performance Analysis of Banyan Networks Based on Buffers of Various Sizes," *INFOCOM '90, 1990, pp. 157–163.*

[62] Sakurai, Y., et al., Large Scale ATM Multi Stage Switching Networks with Shared Buffer Memory Switches," *Proc. ISS '90, Stockholm, Vol. A6.3, 1990.*

[63] Shaikh, S. Z., M. Schwartz, and T. H. Szymanski, "Analysis, Control, and Design of Crossbar and Banyan Based Broadband Packet Switches for Integrated Traffic," *ICC '90, 1990, pp. 761–765.*

[64] Suzuki, H., et al., "Output Buffer Switch Architecture for Asynchronous Transfer Mode," *ICC '89, Boston, 1989.*

[65] Takagi, H., "Queueing Analysis of Polling Models: an Update," in *Stochastic Analysis of Computer and Communication Systems, H. Takagi, ed., Amsterdam: North Holland, 1990, pp. 267–318.*

[66] Takeuchi, T., et al., "Synchronous Composite Packet Switching for B-ISDN," *IEEE JSAC, Vol. 5, 1987, pp. 1365–1376.*

[67] Theimer, T. H., E. Rathgeb, and M. Huber, "Performance Analysis of Buffered Banyan Networks," Symp. on Performance of Distributed and Parallel Systems, Kyota, Japan, 1988, pp. 55–72.

[68] Theimer, T. H., "Performance Comparison of Routing Strategies in ATM Switch Fabrics," *ITC-13, 1991, pp. 923–928.*

[69] Tobagi, F. A., "Fast Packet Switch Architectures for Broadband Integrated Services Digital Networks," *Proc. IEEE, Vol. 78, No. 1, January 1990, pp. 133–167.*

[70] Tobagi, F. A., and T. Kwok, "Fast Packet Switch Architectures and the Tandem Banyan Switch Fabric," manuscript.

[71] Tran-Gia, P., and T. Raith, "Performance Analysis of Finite Capacity Polling Systems with Nonexhaustive Service," Performance Evaluation, Vol. 9, 1988, pp. 1–16.

[72] Turner, J. S., "Design of a Broadcast Packet Network," *INFOCOM '86, Miami, 1986.*

[73] Turner, J. S., "Design of a Broadcast Packet Switching Network," *IEEE Trans. Comm., Vol. 36, No. 6, 1988.*

[74] Woodworth, C. B., M. J. Karol, and R. D. Gillin, "A Flexible Broadband Packet Switch for a Multimedia Integrated Network," *ICC '91, 1991, pp. 78–85.*

[75] Wulleman, R., and T. van Landegem, "Comparison of ATM Switching Architectures," Int. J. Dig. and Anal. Cabled Systems, Vol. 2, No. 4, 1989, pp. 211–226.
[76] Yamashita, H., H. G. Perros, and S.-W. Hong, "Performance Modeling of a Shared Buffer ATM Switch Architecture," *ITC-13, 1991.*
[77] Yeh, Y.-S., M. G. Hluchyj, and A. S. Acampora, "The Knockout Switch: a Simple, Modular Architecture for High Performance Packet Switching," *IEEE JSAC, Vol. 5, No. 8, 1987.*
[78] Yoon, H., K. Lee, and M. Liu, "Performance Analysis of Multibuffered Packet Switching Networks in Multiprocessor Systems," *IEEE Trans. Comp., Vol. 39, No. 3, 1990, pp. 319–327.*
[79] Denzel, W. E., A. P. J. Engbersen, and G. Karlsson, "A Highly Modular Packet Switch for GB/S Rates," *14th Int. Switching Symp., October 1992.*

6 ATM Interfaces

ATM standards provide the basic framework for the interoperability among ATM systems built by different vendors. The term *system* in this context is used generically to refer to each component that constitutes an end-to-end ATM solution, including the ATM architecture (user, control, and management planes); physical, ATM, and ATM adaptation layers used by these planes; network and service interoperability among non-ATM networks and ATM networks; and emerging and existing network applications (users) such as frame relay, ISDN, IP, and circuit emulation.

Standardization provides users, manufacturers, and service providers with freedom of choice. Manufacturers benefit from standards, which give them more opportunity to compete for the business of all the potential purchasers of telecommunications equipment. Service providers or organizations that build private networks take advantage of selecting their networking equipment from multiple vendors in a cost-effective manner. Users benefit the most, since standard services give users freedom in both selecting service providers and purchasing equipment that will best fit their networking requirements in the most cost-effective way.

ATM standards are defined based on different types of interfaces that address the connectivity and interoperability issues between the different components of ATM networks: for example, an ATM end station and an ATM switch, two switches, and service interfaces. Various interfaces that have been defined by ITU-T and the ATM Forum are illustrated in Figure 6.1. The ATM Forum produced its first implementation agreement in June 1992. This document, entitled "User-to-Network Interface Specification Version 2.0," included definitions of various physical layer interfaces and functions to support permanent ATM connections. The second round of ATM Forum specifications were approved in the summer of 1993 and included the following implementation agreements:

- "Data Exchange Interface (DXI) Specification Version 1.0";
- "Broadband Intercarrier Interface (B-ICI) Specification Version 1.0" (B-ICI 1.0);
- "User-to-Network Interface (UNI) Specification Version 3.0" (UNI 3.0).

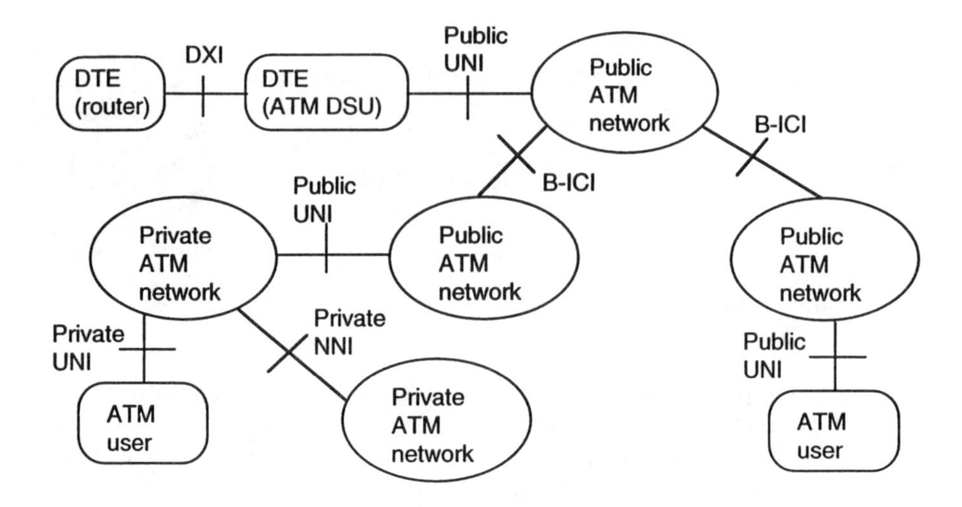

Figure 6.1 The ATM Forum interfaces.

UNI 3.0 and B-ICI 1.0 were updated to UNI 3.1 and B-ICI 1.1 in mid-1994.

ITU-T is the international authority for ATM standards and has been working on the definition of public UNI and NNI. Various aspects of the NNI have been either finalized or are in the process of being completed. First versions of UNI and NNI signaling are expected to be approved by the end of 1995.

The UNI is the demarcation point between an ATM end station and the network. An ATM end station in this context refers to any device that transmits ATM cells to the network. Accordingly, an ATM end station may be an IWU that encapsulates data into ATM cells, an ATM switch, or an ATM workstation. Depending on whether the attached ATM network is private or public, the interface is referred to as a *private UNI* or *public UNI,* respectively. If two switches are connected to each other across a UNI and one switch belongs to a public network and the other to a private network, the corresponding demarcation point is a public UNI. The UNI specifications include the definitions of various physical interfaces, ATM layer, management interface, and signaling.

NNI is used in different contexts. It may be the demarcation point between either two private networks or two public networks, respectively referred to as *private NNI* (P-NNI) and *public NNI.* The same interface may also be used as a switch-to-switch interface. In this case, NNI is a switch-to-switch interface in private networks and a network node interface in public networks. Like UNI specifications, NNI standards include the definitions of various physical interfaces, ATM layer, management interface, and signaling. The NNI framework in private networks also includes the specification of the P-NNI routing framework.

The ATM DXI was developed to allow current routers to interwork with ATM networks without requiring special hardware. In this specification, data terminal equipment (DTE) (a router) and a data communications equipment (DCE) (an ATM data service unit (DSU)) cooperate toward providing a UNI. The DXI specification includes the definitions of the data link protocol and physical layers that handle the data transfer between DTE and DCE, as well as a local management interface and management information base for the ATM DXI.

The B-ICI is a carrier-to-carrier interface (i.e., the ATM Forum version of public NNI). The B-ICI specification includes the various physical layer interfaces, ATM layer management, and higher layer functions required at the B-ICI for the interworking between ATM and various services that include switched multimegabit data service (SMDS), frame relay, circuit emulation, and cell relay. The specification also has the general principles for network provisioning, management, and accounting.

6.1 UNI SPECIFICATIONS

An ATM bearer service provides transportation of user data across an ATM network. Implementation of an ATM bearer service may be based on virtual paths (VP service) or virtual circuits (VC service), or combined VP/VC service.

Connection requirements of various types of applications an ATM bearer service should support include point-to-point and point-to-multipoint connections. Furthermore, different applications have different QOS requirements, and an ATM bearer service should be capable of supporting a variety of service levels. In a VP service, QOS associated with the VPC is selected to accommodate the most demanding QOS requirement of any VCC expected to be multiplexed on that VP. In the case of VC service, there is a 1:1 correspondence between the service provided and the connection QOS requirements (i.e., the QOS requirement of the application must be the QOS provided by the VC service).

Connections in ATM networks are established either dynamically (SVCs) or preconfigured (PVCs). The SVC support requires the definition of signaling protocols, procedures, and parameters for the dynamic management of ATM connections across a UNI.

In summary, end station connection characteristics can be summarized as follows:

- VP/VC service;
- Point-to-point and point-to-multipoint connections;
- Different traffic characteristics and service requirements;
- Permanent or switched connections.

Various features included in the UNI specification to address these requirements are categorized into four sections:

- Physical layer interfaces;
- ATM layer;
- Interim local management interface (ILMI);
- UNI signaling.

The details of the physical layer interfaces are discussed in Chapter 2. UNI signaling is quite involved and is discussed in detail in Chapter 7. The ILMI and the details of the ATM layer that are not covered in Chapter 2 are discussed next.

6.1.1 ATM Layer

When a network user wishes to communicate with another user, a signaling message is generated to request a connection from the network. In the case of permanent connections, this could be a telephone call to an operations center or a network management flow. In the case of switched connections, these signaling messages are generated dynamically, on demand as connection requests arrive. In either case, it is necessary to provide the network with enough information about the connection request, such as characterization of the connection traffic behavior and its service requirements. The network, upon accepting the connection request, provides the user with the VPI/VCI the end station uses at the UNI for the requested connection. If the network cannot accept the connection, then it informs the user that its request is rejected. When the communication is over, the user terminating the connection should inform the network to release resources associated with its connection. The dynamics of this process are presented in Chapter 7 in the context of UNI signaling. Next we review various types of information used to characterize connections and their service requirements.

6.1.1.1 Information Used to Characterize Connections

Each ATM end station requires a unique ATM address for ATM end stations to uniquely identify each other. Unique ATM addresses are also required by ATM networks to locate the destination end nodes of connection requests.

Private and public networks use different ATM address formats. Public ATM networks use E.164 addresses (i.e., telephone numbers), whereas ATM private network addresses are based on the Open Systems Interconnection (OSI) network service access point (NSAP) format [4]. NSAP addresses are based on the concept of hierarchical addressing domains, as illustrated in Figure 6.2. At

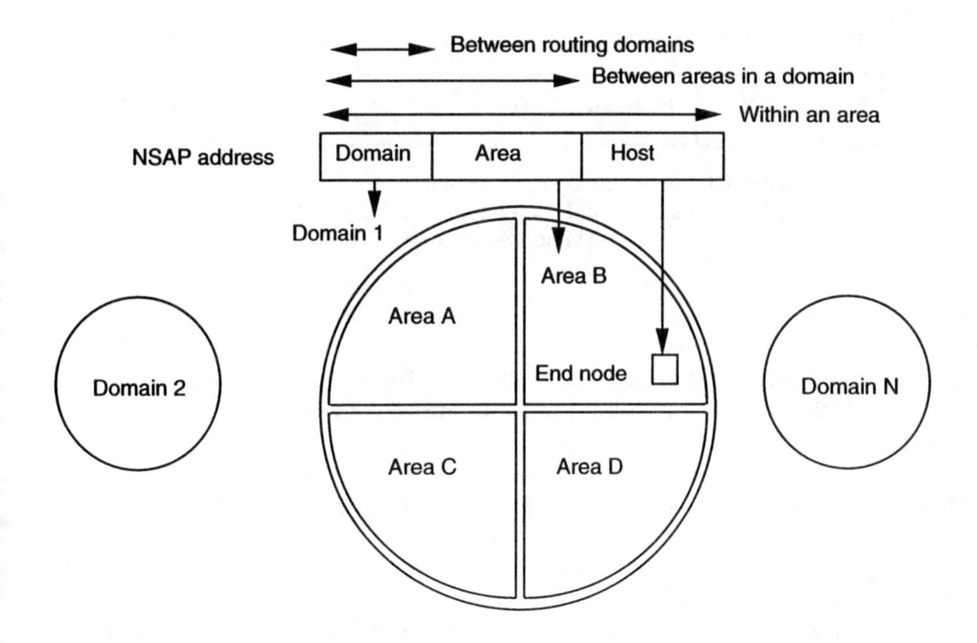

Figure 6.2 NSAP address structure.

any hierarchical level, an initial part of the address unambiguously identifies a subdomain and the rest is allocated by the authority associated with the subdomain to unambiguously identify either a lower level subdomain or an access point within the subdomain. Details on NSAP addresses are given in RFC 1237 [4].

Figure 6.3 illustrates the high-level format of an ATM address. A private ATM address is 20 bytes long and consists of a network addressing domain referred to as *initial domain part* (IDP) and *domain-specific part* (DSP). The IDP specifies a subdomain of the global address space and identifies the network addressing authority responsible for assigning ATM addresses in the specified subdomain. Accordingly, the IDP is further subdivided into two fields: the authority and format identifier (AFI) and the initial domain identifier (IDI). The AFI specifies the format of the IDI, the network addressing authority responsible for allocating values of the IDI and the abstract syntax of the DSP. The

IDP		DSP
Authority and format identifier (AFI)	Initial domain identifier (IDI)	Domain specific part

Figure 6.3 Private ATM address general format.

IDP		DSP		
AFI	IDI	High order DSP	ESI	Selector

IDI=DCC: Data country code format
IDI=ICD: International code designator format
IDI=E.164: E164 ATM address format

Figure 6.4 ATM address format.

IDI specifies the network addressing domain from which values of the DSPs are allocated and the network addressing authority responsible for allocating values of the DSP from that domain.

Three AFIs are defined in [1], namely, the data country code (DCC), international code designator (ICD), and E.164. The corresponding private ATM address format is illustrated in Figure 6.4.

A private ATM address includes a 13-byte network part and a 7-byte end station part. The end station part is divided into two fields: a 6-byte end station identifier and a 1-byte selector for use solely by the end station. For example, an ATM end station may support a number of terminal equipment to access an ATM network, and the selector field may be used to differentiate between different devices. The specifics of the network part is defined by the network administrator. In particular, the network part of the address can be subdivided into a number of subfields (on the bit boundary if so desired) to create multiple levels of hierarchy. The network part of an ATM address is configured by the network administrator to each switch port the end stations are connected to. An end station has its own separate identifier information (i.e., its MAC address).

The client registration mechanism defined in UNI 3.1 allows end stations to exchange their station identifiers for the ATM address information configured at the switch port. As a result of this exchange, the end station automatically acquires the ATM network address of the switch port it is attached to without any requirement for that address to be manually provisioned into the end station. The end station then appends its own identifier and forms its full ATM address. Similarly, the end station part of the address is registered in the network and associated with its respective network part. With this scheme, several ATM addresses with the same network-defined part (with distinct ESIs) can be registered at the network side of the UNI. It is also possible to assign more than one prefix (network portion of the address) to a UNI. Similarly, an ATM end station may require more than one network address part, if needed.

6.1.1.2 AAL Information

An ATM network provides only ATM layer connectivity among ATM layer users in two or more ATM end stations. There is no awareness of the specifics

Table 6.1
B-ISDN Service Classes

Class	*Timing Relationship*	*Bit Rate*	*Connection-Oriented/Connectionless*
A	Yes	Constant	Connection-oriented
B	Yes	Variable	Connection-oriented
C	No	Variable	Connection-oriented
D	No	Variable	Connectionless

of applications or traffic types at the ATM layer. The ATM layer does not look inside the cell payload, nor does it know or care what is carried. Hence, AAL in the user plane is used only at ATM end stations, as illustrated in Figure 6.5.

Accordingly, it is irrelevant to the network which AAL is used by the application. However, the end stations need to know which AAL is used by the user application. Hence, the AAL type and related information are required to be exchanged between the end stations.

As discussed in Chapter 2, B-ISDN services are categorized into eight possible service classes, four of which are explicitly defined in Table 6.1.

In addition, class X service is defined to allow either a vendor-specific AAL or an option to bypass the adaptation layer and go directly to the ATM layer. In the former case, class X service allows a proprietary AAL used by "nonstandard" applications. In the latter case, applications sit directly on top of the ATM layer and pass the 48 byte payloads to the ATM layer by themselves.

UNI 3.1 supports classes A, C, and X with AAL types 1, 3/4, and 5. Different AAL-specific data that may be included in the information exchanged across the UNI differ for each AAL, as illustrated in Table 6.2.

6.1.1.3 Connection Traffic Parameters

When a connection request is accepted by the network, the user and the network agree on a traffic contract for the duration of the connection. With this contract,

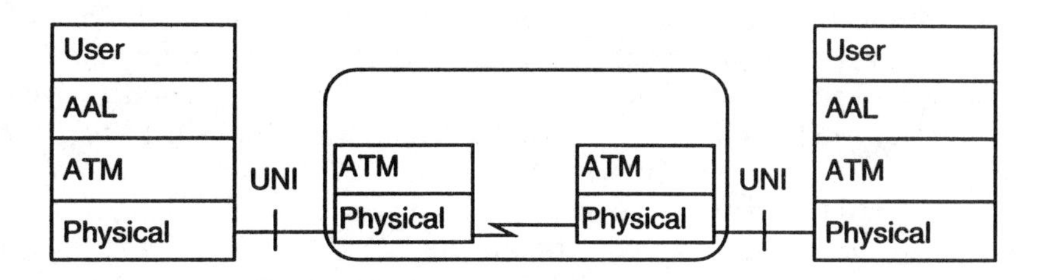

Figure 6.5 An end-to-end connection in an ATM network.

Table 6.2
AAL Parameters Exchanged Between End Stations

AAL	*Content of AAL Parameter Information Element*
AAL 1	Subtype (i.e., video)
	CBR rate (i.e., bit rate per second or is a multiplier)
	Source clock recovery method
	Error correction method
	Structured data transfer block size
	Partially filled cells method
AAL 3/4	Forward maximum CPCS SDU size
	Backward maximum CPCS SDU size
	MID range lowest MID value
	MID range highest MID value
	SSCS type
AAL 5	Forward maximum CPCS SDU size
	Backward maximum CPCS SDU size
	Mode
	SSCS type
User-defined	4 bytes for use by the terminal equipment

the network guarantees the requested service demand of the connection as long as the source traffic stays within the specified limits. Accordingly, a traffic contract includes traffic descriptors, service requirements, and the conformance definition. All of these are required to be well defined and understood by both the network and the user.

B-ISDNs are expected to support a variety of services with different traffic characteristics. Design and operation of network control functions such as call admission, bandwidth reservation, and congestion control require accurate source characterization to achieve high resource utilization. However, a source, in general, cannot provide a detailed description of its traffic behavior. Hence, there is a tradeoff between how much information should and can be defined to characterize a source.

A traffic parameter is a specification of a particular traffic aspect of the requested connection. The four parameters currently defined are the peak cell rate (PCR), CDV tolerance, sustainable cell rate (SCR), and burst tolerance (BT).

The PCR of a connection is the inverse of the minimum time between two cell submissions to the network. The transmission in the physical link is often slotted (in all framed-based interfaces). Accordingly, although cells may be generated by applications at a particular peak rate, the observed PCR of a connection at the switch may be different than the PCR at the ATM layer of the end station. For example, if the PCR of a connection is 75 Mbps over a 155-Mbps link, then cells may arrive at the switch at the rate of one cell in every two cell times. However, if the PCR of the connection is (2/3) × 155 Mbps, then two cells will arrive in every three cell slots. This would mean that some cells will arrive in consecutive slots. Based on the above definition, the PCR of the connection observed at the switch in this case would be 155 Mbps.

The interarrival times of cells of a particular connection as monitored at the switch may also be affected when cells from two or more connections are multiplexed at the ATM layer or when physical layer overhead or idle cells are inserted into the cell stream at the physical layer. In particular, when cells from different sources are multiplexed into the physical media, the cells of a particular connection may be delayed at the end station physical or ATM layer while cells from other connections are being transmitted. This phenomena is referred to as *clumping.*

A metric referred to as *cell delay variation tolerance* (CDVT) is defined to provide a solution to these two problems. Conceptually, CDVT defines how much deviation from the specified minimum cell interarrival time is allowed at the UNI (i.e., as observed by the network). CDVT is specified by the network provider and it is not negotiated between the user and the network during the connection setup time.

Given the PCR and CDVT, the maximum number of cells that can arrive back to back at the transmission link rate N is given as follows:

$$N = \lfloor 1 + \text{CDVT}/[1/\text{PCR} - \delta] \rfloor \quad \text{for } 1/\text{PCR} > \delta \tag{6.1}$$

where $\lfloor x \rfloor$ denotes the integer part of x and δ is the cell transmission time at the physical link.

The average rate of a connection is equal to the total number of cells transmitted divided by the duration of the connection. Based on this definition, the network can know the average rate of a connection only after the connection terminates. Accordingly, the average rate of a connection cannot be used by the network. The SCR is an upper bound on the average rate of an ATM connection. The SCR is used with another metric, BT, and the PCR. BT is the duration of the period in which the source is allowed to submit traffic at its peak rate. Given these parameters, the maximum number of cells that can arrive at the switch back to back at the PCR, the maximum burst size (MBS), is given as follows:

$$\text{MBS} = \lfloor 1 + \text{BT}/[(1/\text{SCR}) - (1/\text{PCR})] \rfloor \tag{6.2}$$

The CLP bit at the ATM cell header defines two loss priorities: high priority (CLP = 0) and low priority (CLP = 1). This bit may be used mainly for two purposes. As discussed in Chapter 3, an application may want to tell the network which cells of its traffic are more important than the others. Another use of high and low loss priority at the cell header is referred to as *tagging.* The network polices arrival cell streams based on the traffic contracts of connections to determine whether or not connections stay within the source traffic parameters agreed on at the call establishment phase. When a cell is detected to be nonconforming, there are two choices: either drop the cell at the interface or allow the cell to enter the network, hoping that there might be enough resources to deliver the cell to its destination. In the latter case, it is necessary to make sure that nonconforming cells do not cause degradation to the service provided to connections that stay within their negotiated parameters. Tagging is allowing nonconforming cells with CLP = 0 to enter the network with CLP = 1 (i.e., the CLP bit is changed from 0 to 1). Cells with CLP = 1 are discarded in the network whenever necessary so that the service provided to the conforming traffic (i.e., in this context, CLP = 0 traffic) is not affected. Under what conditions cell discarding is activated at an intermediate switch is not a part of the ATM standards.

In addition to the service requests in which a traffic contract is agreed on during the connection setup, ATM networks support *best effort service,* whereby explicit guarantees are neither required nor negotiated between the user and the network. For this class of service, the only traffic parameter used is a PCR with CLP = 0 + 1. CLP = 0 + 1 refers to the aggregated traffic consisting of cells with CLP = 0 and CLP = 1.

Based on this framework, the allowable combinations of traffic parameters in UNI 3.1 are defined in Table 6.3.

Table 6.3
Allowable Combinations of Traffic Parameters in Signaling Messages

Combination	*Traffic Parameters*
1	PCR for CLP = 0 PCR for CLP = 0 + 1
2	PCR for CLP = 0 PCR for CLP = 0 + 1 with tagging requested
3	PCR for CLP = 0 + 1 SCR for CLP = 0 BT for CLP = 0
4	PCR for CLP = 0 + 1 SCR for CLP = 0 BT for CLP = 0 with tagging requested
5	PCR with CLP = 0 + 1
6	PCR for CLP = 0 + 1 SCR for CLP = 0 + 1 BT for CLP = 0 + 1
Best effort service	PCR for CLP = 0 + 1

Users at the setup phase request an ATM layer QOS selected from the QOS classes the network provides for ATM layer connections. A QOS class has either specified performance parameters (referred to as *specified QOS class*) or no specified performance parameter (referred to as *unspecified QOS class*). The former provides a QOS to an ATM connection in terms of a subset of performance parameters to be defined next. In the latter case, there is no explicitly specified QOS commitment to the cell flow. This class is intended to be used for best effort service.

Currently supported classes at the UNI are:

- Unspecified QOS class 0 supports best effort service in which no explicit service requirement is specified.
- Specified QOS class 1 supports a QOS that will meet class A service performance requirements (i.e., circuit emulation).
- Specified QOS class 2 supports a QOS that will meet class B service performance requirements (i.e., VBR video).
- Specified QOS class 3 supports a QOS that will meet class C service performance requirements (i.e., frame relay interworking).
- Specified QOS class 4 supports a QOS that will meet class D service performance requirements (i.e., IP over ATM).

Different B-ISDN applications have different performance requirements. Accordingly, B-ISDNs are expected to support a variety of performance levels within a particular QOS class. The application requirements cannot be signaled to the network explicitly in UNI 3.1. The next version of the ATM Forum UNI specification will address this problem (see Section 6.1.1.6). Until then, the individual performance metrics cannot be exchanged between the network and the end stations. Instead, end stations can specify one of the service classes. The values of various performance metrics of interest for a particular class of service are determined by the network provider.

In general, the QOS requirement of an application is defined with a number of parameters that may include a subset of the following:

- Cell error ratio (CER);
- Severely errored cell block ratio;
- CLR;
- Cell misinsertion rate;
- Cell transfer delay (CTD);
- Mean CTD;
- CDV.

6.1.1.4 Conformance Specification

Given the source traffic parameters, the network polices the traffic using the generic cell rate algorithm (GCRA). The GCRA defines in an operational manner a relationship between the traffic parameters and classifies each arriving cell as either conformant or nonconformant.

The following presentation of the GCRA is based on a leaky bucket implementation. Let:

X = value of the leaky bucket counter;
LCT = last cell compliance time;
$t(k)$ = arrival time of the kth cell;
I = increment;
L = limit (i.e., the capacity) of the bucket.

Upon arrival of the first cell, say at time $t(1)$, X and LCT are respectively initialized to 0 and $t(1)$. Let us consider the bucket as a counter that is bounded above by L. In the GCRA, the counter is decreased by 1 every "unit" time, whereas the counter increases by I units each time a conforming cell arrives. If upon arrival of a cell (before the counter is increased by I) the value of the counter is less than or equal to its limit L, then the cell is considered to be conforming; otherwise it is a nonconforming cell. The value of the counter

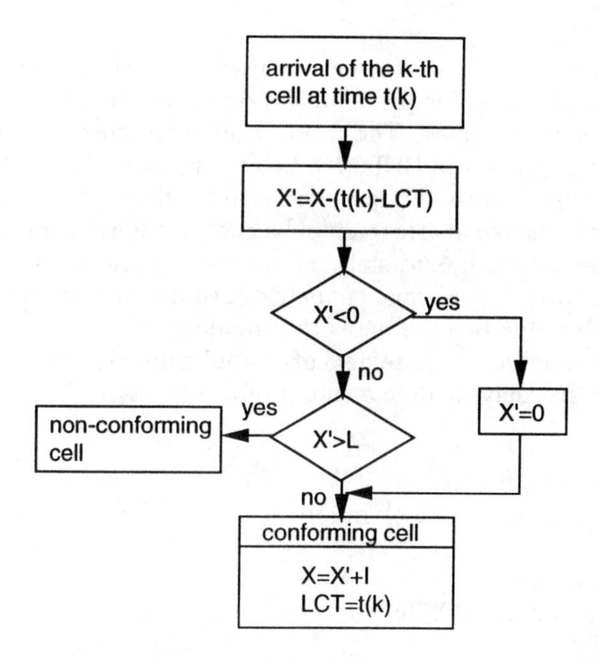

Figure 6.6 Generic cell rate algorithm.

is not updated (increased) at the arrival of a nonconforming cell, whereas a conforming cell causes the counter to be increased by I.

The dynamics of the GCRA are illustrated in Figure 6.6. At the arrival time of the kth cell, the content buffer is temporarily updated to X'. X' is equal to the value of the counter after the arrival of the last conforming cell (X) minus the amount the bucket has drained since then. The content bucket is always nonnegative. If X' is less than or equal to L, then the cell is conforming and X is set to X' plus I and LCT is set to the current time $t(k)$. If X' is greater than L, then the cell is nonconforming and the two parameters X and LCT remain unchanged. Hence, the GCRA depends only on the two parameters I and L (i.e., GCRA(I, L)).

When the GCRA is used to check PCR conformance, the two parameters used are 1/PCR and CDVT (i.e., GCRA(1/PCR, CDVT)). When used for the SCR and BT, the two parameters used are 1/SCR and BT (i.e., GCRA(1/SCR, BT)).

6.1.1.5 Traffic Contract

Given this framework, a traffic contract that is agreed on between the ATM user and the network across the UNI at the call setup phase includes the following:

- QOS class of connection;
- Connection traffic descriptor;
- Conformance definition;
- Compliant connection definition.

As discussed in Section 6.1.1.3, the requested QOS class can be one of the entries in Table 6.4.

Connection traffic descriptors include application-specified PCR, SCR, BT, and network-specified CDVT for different traffic combinations with CLP = 0 and CLP = 1. In addition, a tagging option may be requested.

Table 6.4
Available QOS Classes

QOS Class	*Description*
0	Unspecified QOS service class
1	Specified QOS class 1
2	Specified QOS class 2
3	Specified QOS class 3
4	Specified QOS class 4

The conformance definition is based on one or more applications of the GCRA. Figure 6.7 illustrates the application of GCRAs to the six combinations of traffic parameters allowed in UNI signaling.

Combination 1 corresponds to traffic in which PCR(0) and PCR(0 + 1) are specified and tagging is not requested. Conformance checking of this traffic requires application of two GCRAs, as illustrated in Figure 6.7(a). Conformance of CLP = 0 traffic is checked at the first application of the GCRA with GCRA(1/PCR(0), CDVT). Both conformant CLP = 0 and CLP = 1 traffic go through the second application of the GCRA with GCRA(1/PCR(0 + 1), CDVT). Conformant cells are those that are classified as conformant at the end of the second application of the GCRA. Nonconformant cells at each application of the GCRA can be dropped, or some of them may be allowed to enter at the network's option. The reason for specifying PCR(0 + 1) as opposed to PCR(1) is to provide the source more flexibility in allocating bandwidth to its CLP = 0 and CLP = 1 traffic. For example, with PCR(0) = 5 and PCR(0 + 1) = 8, traffic submitted to

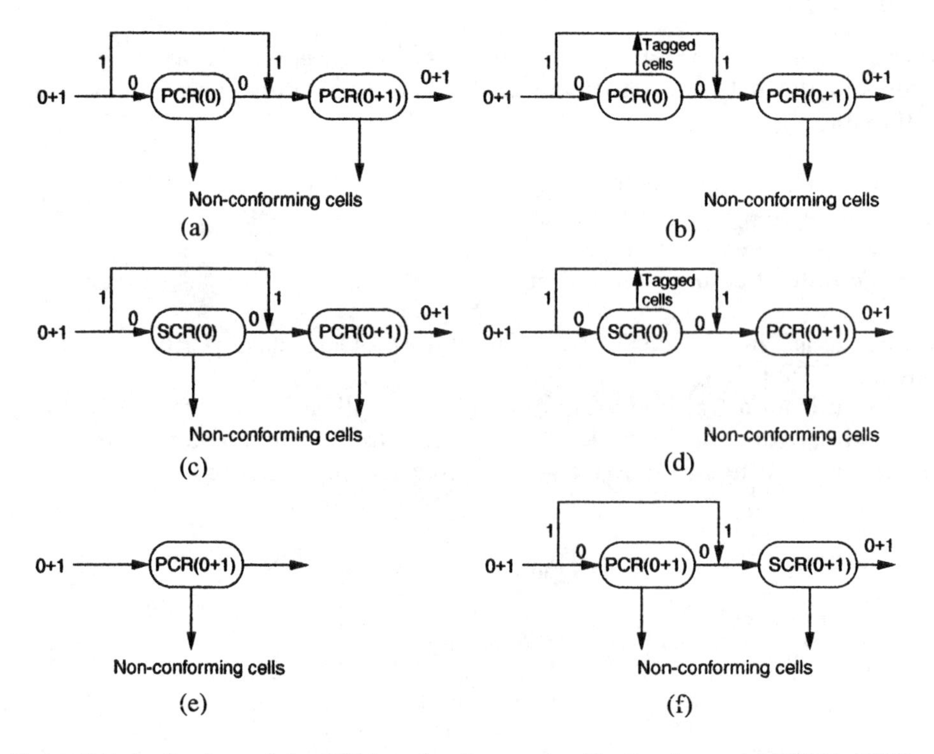

Figure 6.7 Applications of the GCRA to six classes of traffic descriptors: (a) PCR(0), PCR(0 + 1), tagging not requested; (b) PCR(0), PCR(0 + 1), tagging requested; (c) SCR(0), PCR(0 + 1); (d) SCR(0), PCR(0 + 1), tagging requested; (e) PCR(0 + 1); (f) SCR(0 + 1), PCR(0 + 1).

the network with PCR(0) = x and PCR(1) = 8 - x is conformant for any value of $0 \leq x \leq 5$. If the two types of traffic are characterized with PCR(0) = 5 and PCR(1) = 3, then the only allowed value of x is 5.

Combination 2 corresponds to traffic in which PCR(0) and PCR(0 + 1) are specified and tagging is requested. Conformance checking of this traffic requires application of two GCRAs, as illustrated in Figure 6.7(b), which is essentially the same as combination 1. The main difference is that cells that are observed to be nonconformant at the first application of the GCRA (for CLP = 0 traffic) are not classified as nonconformant. Instead, these cells are tagged and passed to the second application of the GCRA. Hence, the second application of the GCRA is applied to the combined traffic with conformant CLP = 0 and nonconformant CLP = 0 traffic at the end of the first application of the GCRA, together with CLP = 1 traffic submitted by the source.

Combination 3 corresponds to traffic in which SCR(0), BT, and PCR(0 + 1) are specified and tagging is not requested. As with the previous two combinations, conformance checking of this traffic requires application of two GCRAs as illustrated in Figure 6.7(c). Conformance of SCR = 0 traffic is checked at the first application of the GCRA with GCRA(1/SCR(0), BT). Both conformant CLP = 0 and CLP = 1 traffic go through the second application of the GCRA with GCRA(1/PCRA(0 + 1), CDVT). Conformant cells are those that are classified as conformant at the end of the second application of the GCRA. Nonconformant cells at each application of the GCRA can be dropped, or some of them may be allowed to enter at the network's option.

Combination 4 is the same as combination 3 with tagging requested. Hence, cells that are observed to be nonconformant at the first application of the GCRA (for CLP = 0 traffic) are not classified as nonconformant. Instead, these cells are tagged and passed to the second application of the GCRA. Conformance checking for this combination is illustrated in Figure 6.7(d).

Combination 5 corresponds to traffic in which only PCR(0 + 1) is specified. Conformance checking of this traffic requires a single application of the GCRA, as illustrated in Figure 6.7(e). Cells that are not conformant to GCRA(1/PCR(0 + 1), CDVT) can be dropped, or some of them may be allowed to enter at the network's option.

Combination 6 corresponds to traffic in which SCR(0 + 1), BT, and PCR(0 + 1) are specified and tagging is not requested. Conformance checking of this traffic requires application of two GCRAs, as illustrated in Figure 6.7(f). Conformance of PCR = 0 + 1 traffic is checked at the first application of the GCRA with GCRA(1/PCR(0 + 1), CDVT). Conformant traffic goes through the second application of the GCRA with GCRA(1/SCR(0 + 1), BT). Conformant cells are those that are classified as conformant at the end of the second application of the GCRA. Nonconformant cells at each application of the GCRA can be dropped, or some of them may be allowed to enter at the network's option.

In all likelihood, a source may not be able to submit all of its cells in a fully conformant manner. The term *compliant* is used generically for connections in which some of the cells may be nonconforming. The exact definition of a compliant connection is left to the network operator. The traffic contract is agreed on based on a compliant connection.

A traffic contract is an agreement between the network and the end user in which the network guarantees the QOS requirements of a compliant connection.

6.1.1.6 A Review of Traffic Management Specification Version 4.0

This section is based on the current status of ATM Forum traffic management specification version 4.0. Although the details may be modified when the specification is finalized, the underlying framework presented next is not expected to change.

Based on this traffic management specification, currently defined service classes in UNI 4.0 will include:

1. CBR;
2. Real-time variable bit rate (RT-VBR);
3. Non-real-time variable bit rate (NRT-VBR);
4. Available bit rate (ABR);
5. Unspecified bit rate (UBR).

The CBR service class is used for real-time applications that require tightly constrained delay and delay variation, such as voice and video. Cells that are delayed in the network more than the specified CTD are considered to be of less or no value to the application. The traffic characteristics of CBR sources are characterized by their PCR and CDVT and CBR traffic is policed by using GCRA(1/PCR, CDVT). Accordingly, a consistent availability of fixed quantity of bandwidth equal to PCR (or more, depending on the CDVT specification, internal switch architecture, and so forth) may be required in the network for CBR applications. QOS parameters specified for this service class include CTD, CDV, and CLR.

The RT-VBR service class, similar to the CBR service class, is intended for real-time applications that require tightly constrained delay and delay variation, such as VBR voice and video. Cells that are delayed in the network more than the specified CTD are considered to be of less or no value to the application. RT-VBR sources generate traffic at a rate that varies over time (i.e., bursty sources). As discussed in Chapter 3, VBR service may allow statistical multiplexing and consistent service quality in applications. The traffic characteristics of applications that use RT-VBR service are characterized by their PCR, CDVT, SCR, and BT, and the traffic generated at each such source is policed by using

GCRA(1/PCR, CDVT) and GCRA(1/SCR, BT). The amount of bandwidth required to support this type of traffic in the network is determined by various factors such as buffer size, scheduling discipline, and CLR requirements. It is a value between the SCR and the PCR. The specific bandwidth required to support this type of traffic is vendor-specific and it is not a part of any standards activity. QOS parameters specified for this service class include CTD, CDV, and CLR.

The NRT-VBR service class is defined for non-real-time applications with bursty traffic characteristics. The traffic characteristics of applications that use NRT-VBR service are characterized by their PCR, CDVT, SCR, and BT and the traffic generated at each such source is policed by using GCRA(1/PCR, CDVT) and GCRA(1/SCR, BT). As with RT-VBR service, the amount of bandwidth required to support this type of traffic in the network is determined by various factors such as buffer size, scheduling discipline, and CLR requirements. It is a value between the SCR and the PCR. The specific bandwidth required to support this type of traffic is vendor-specific and it is not a part of any standards activity. QOS parameters specified for this service class include the CTD and CLR. Applications that request the NRT-VBR service class expect an upper bound on the CTD and CLR, and they are not sensitive to CDV. In general, however, these bounds are expected to be less stringent than that of the RT-VBR service class.

The UBR service class supports delay-tolerant applications that are not sensitive to delay variation such as traditional computer communications applications. Unlike the other three service classes discussed so far, there is no bandwidth reservation required to support UBR service in the network (i.e., it provides best effort service). Accordingly, there is no service guarantee provided at the ATM layer. The end-to-end data integrity may be achieved by the higher layer protocols. The traffic characteristics of applications that use the UBR service are characterized by their PCR and CDVT and the traffic generated at each such source is policed by using GCRA(1/PCR, CDVT). The main reason for the specification of these parameters is to help in identifying the physical bandwidth limitation along the path of the connection. No QOS parameters are specified for this type of service.

The UBR service is intended to support connectionless data traffic that does not require QOS guarantees from the network. This type of traffic essentially increases the utilization of network resources beyond what can be achieved by reserved traffic alone (i.e., CBR, RT-VBR, and NRT-VBR services). Conceptually, cells belonging to the UBR service are transmitted only when there is no cell waiting to be transmitted in the transmission link buffer. If service guarantees are required, the NRT-VBR service may be used instead. However, this service requires the specification of the SCR and BT, which may not be known a priori, or the cost associated with the VBR service may not be justified (or preferred) by some applications (i.e., e-mail). The UBR service, however, may not be the right way to support these applications, particularly if they have the capability to adapt to network congestion status. When cells of UBR applications are lost in the network, they will be aware of this only through the use of higher layer protocols at the end stations. The consequence of this might be that most of the UBR traffic carried might be retransmissions, causing significant throughput degradation eventually reaching a catastrophic state in which all cells carried by the UBR service are retransmissions.

Various applications (i.e., TCP) have the ability to react to congestion in the network by adjusting their information transfer rates based on feedback from the network. Such sources may increase their rate if there is extra bandwidth available and scale it down if there is a lack of it. If mechanisms in the network to inform such sources are available, service provided to these applications can be significantly improved.

The data communications industry historically had a different perspective on the use of bandwidth than the telecommunications industry. Telecommunication applications use a fixed, dedicated bandwidth, whereas data communication applications can adapt to a time-varying available bandwidth. In particular, most traditional data applications can tolerate delays, while they are usually sensitive to cell losses (i.e., TCP slow start). Under these conditions, there is a need for the network to inform the application of impending congestion via a feedback mechanism to the sender to reduce loss so that the source can vary its information transmission rate up and down between a specified lower (i.e., zero) and an upper (i.e., PCR) bound based on this feedback. The ABR service is being defined to support this type of traffic (cf. [9]).

In the ABR service, the information transfer rate is adjusted at the end stations based on the information provided by the network. The transfer parameters specified are the PCR and CDVT, which are used to define the upper bound on the information transfer rate, and the minimum cell rate (MCR), which specifies the minimum usable bandwidth. The MCR could be any value between zero and the PCR, depending on the application requirements. The network (implicitly) guarantees to provide a low CLR to sources that adjust their cell flows according to the feedback information. Although the ABR traffic is bursty, the transfer parameters SCR and BT are not used in ABR service. Similarly, no QOS parameters are (implicitly) specified. In particular, this service is intended for non-real-time applications that can tolerate delay and delay variation.

The feedback is provided using the RM cell (i.e., PTI = 110). Related information in the cell payload may include the network-specified explicit rate (current maximum information transfer rate) that the source is allowed to transmit, the congestion indicator, the current cell rate specified by the source, the MCR, and so forth. The details of the ABR service are still being worked on. When completed, the specification will include the source behavior, switch behavior, and destination behavior. The basic procedure is based on sources to generate RM cells and the switches and destination stations to specify the values of various types of control information and return the RM cells to the source. Source behavior describes the specifics of the RM cell generation process and the regulation of the information transfer rate based on the feedback from the network. The main function at the destination end station is to turn all RM cells around to return to the source end station. Switch behavior defines the generation of network-specific feedback information which will eventually be processed by the source and the destination stations. It may be based on the use of the EFCI bit (mainly for backward compatibility with the currently deployed ATM switches) or the specification of the explicit rate. Given the current status of its resources, the process of determining the explicit rate to be allowed for a particular ABR connection is not a part of the specification and is left to vendors to solve. This by itself is a very complicated problem. Let us consider a switch that supports CBR, VBR (both real-time and non-real-time), and ABR services on a link provisioned to, say, 85% of the link bit rate. The total bit rate used by CBR traffic will be subtracted from this provisioned capacity. The remaining bandwidth, however, is used by the bursty VBR and ABR traffic, and determining the explicit rate available at this link for ABR connections requires the solution of a two-class priority queue with unknown traffic characteristics of ABR applications. When an equivalent capacity or Gaussian approximation (cf. Chapter 3) is used for call admission, it would be possible to estimate the reserved bandwidth for VBR traffic, which may be subtracted from the available link bit rate (after the reserved CBR bandwidth is deducted). The outcome of this is the bandwidth (a conservative bound) available for ABR connections. It is not clear how this information may be used to determine the explicit rate allowed to ABR connections, since their traffic behavior (within given bounds) is unpredictable. This bandwidth may be treated as the current peak rate of aggregate ABR connections. However, this would result in significant underutilization of resources.

Various criteria adopted at the ATM Forum for the ABR service mechanisms to be tested against include [9]:

- The feedback mechanism should scale with speed, distance, number of users, and number of nodes.
- No set of connections should be arbitrarily discriminated against or favored. The available bandwidth for ABR service should be shared fairly among all the service users.
- The network should not depend on correct implementation of the service (i.e., source and destination behavior) at the end stations.
- The calculations of the control loop should actually converge to steady state.
- The service model should accommodate multiple subnetworks, including LANs and private and public networks. It should be easy to address and implement, should offer a flexible choice of operating points, should not

dictate any particular switch architecture and resource allocation strategies, and should minimize buffer sizes used for ABR traffic.

UNI 4.0 will allow different values of QOS parameters to be offered on a per-connection basis. A network may support any subset of the possible values for each of the QOS parameters. The QOS parameters defined are:

- Peak-to-peak CDV;
- Maximum CTD;
- CLR;
- CER.

These QOS parameters can be specified individually (unlike the case in UNI 3.1, where only the classes are specified). The originating end station selects the QOS by specifying the requested minimum and maximum acceptable values of each of the individual parameters. As discussed in Chapter 3, the end-to-end delay is the sum of various deterministic (fixed) delay components (i.e., propagation delay, fixed component of switching delay) and random delay components (i.e., queuing delay).

The maximum CTD specified for a connection is the $(1 - \alpha)$ quantile of the CTD. That is, $100(1 - \alpha)\%$ of cell delays are required to be less than or equal to the specified value. CTD can vary from 0.1 μs to 504 sec (or can be unknown or unspecified).

The peak-to-peak CDV is the $(1 - \alpha)$ quantile of CTD minus fixed CTD that could be experienced by any delivered cell on a connection during the entire connection holding time. CDV can vary from 0.1 μs to 504 sec, with a granularity of 10 μs, or can be unknown or unspecified. The value of α has not yet been specified.

The CLR is the value of the ratio of the number of cells lost to the total number of cells transmitted by the source end station that the network agrees to offer as an objective over the life time of the connection. It is expressed as an order of magnitude having a range of 10^{-1} to 10^{-15} and can be unknown or unspecified.

The CER is the ratio of the number or errored cells to the total number of cells transmitted by the source end station. An errored cell in this context is a cell delivered with payload errors. Like the CLR, the CER is the value the network agrees to offer as an objective over the life time of the connection. It is expressed as an order of magnitude having a range of 10^{-1} to 10^{-15} and can be unknown or unspecified. Since it is not possible to know what types of error conditions will occur during the life time of the connection, the network can only estimate the CER value.

Not all types of QOS parameters apply to all traffic types. For example, the CDV is an important parameter for CBR and real-time service, whereas it

Table 6.5
QOS Parameters and Service Classes

	CDV	*CTD*	*CLR*	*CER*
CBR	Specified	Specified	Specified	Specified
RT-VBR	Specified	Specified	Specified	Specified
NRT-VBR	Not specified	Specified	Specified	Specified
ABR	Not specified	Not specified	Further study	For further study
UBR	Not specified	Not specified	Not specified	Not specified

is not needed for NRT-VBR, UBR, and ABR services. Table 6.5 summarizes the use of QOS parameters in different service classes.

6.1.2 Interim Local Management Interface

The UNI 3.1 ILMI provides an ATM user with the status and configuration information concerning connections (both VP and VC) available at its UNI. The ILMI communication protocol is based on the simple network management protocol (SNMP) network management standard. The term *interim* refers to the usage of this interface until related standards are completed by the standards organizations.

The main functions provided by the ILMI include:

- Status, configuration, and control information about the link and physical layer parameters at the UNI;
- Address registration across the UNI.

Various types of information available in the ATM UNI management information base (MIB) include:

- Physical layer;
- ATM layer and its statistics;
- VPCs;
- VCCs;
- Address registration information.

The ILMI supports all physical layer interfaces defined by the ATM Forum. It provides a set of attributes and information associated with a particular physical layer interface, as well as status information on the state of the physical link connecting two adjacent UNI management entities (UME) at each side of the UNI.

Configuration information at the ATM layer provides information on the size of the VPI and VCI address fields that can be used by an ATM user, the number of configured VPCs and VCCs, and the maximum number of connections allowed at the UNI.

VPC ILMI MIB (VCC ILMI MIB) status information indicates a UME's knowledge of the VPC (VCC) status (i.e., end-to-end, local, or unknown). Configuration information, on the other hand, relates to the QOS parameters for the VPC (VCC) local end point.

Address registration identifies the mechanism for the exchange of identifier and address information between an ATM end station and an ATM switch port across a UNI. ATM network addresses are manually configured by the network operator into a switch port. As a result of this exchange, the end station automatically acquires the ATM network address as configured by the network operator without any requirement for the same address to have been manually provisioned. Similarly, the ATM end station identifier is registered in the network and it is associated with the respective network part of the address. Chapter 11 includes additional information on ILMI.

6.1.3 UNI Signaling

Signaling in a communication network is the collection of procedures used to dynamically establish, maintain, and terminate connections. For each function performed, the corresponding signaling procedures define the sequence and the format of messages exchanged that are specific to the network interface across which the exchange takes place.

UNI 3.0 is built upon ITU-T Q.2931 broadband signaling standard currently under development. It includes extensions to Q.2931 to support capabilities important for the early deployment and interoperability of ATM products. Three major areas of extensions are point-to-multipoint signaling, private ATM address formats, and additional traffic management capabilities.

UNI signaling is discussed in detail in Chapter 7. Various capabilities defined in UNI 3.1 signaling include the following:

- Establishment of point-to-point VCCs;
- Establishment of point-to-multipoint VCCs;
- Three different ATM private address formats;
- One ATM public address format;
- Symmetric and asymmetric QOS connections with a declarable QOS class;
- Symmetric and asymmetric bandwidth connections with a declarable bandwidth;
- Transport of network-transparent parameters;
- Support of error handling.

6.2 DATA EXCHANGE INTERFACE SPECIFICATION

The physical layer interfaces for ATM over WANs (at the time the DXI was developed) included DS-3, OC-3, and OC-12. Although it is envisioned that in the near future a large portion of the physical infrastructure will be fiber, these interfaces are still expensive and it is necessary for the success of ATM to enable ATM services within the current infrastructure. The main objective of the DXI is to provide installed equipment with access to ATM networks without costly (hardware) upgrades. Toward this goal, DXI allows DTE (i.e., a router) and DCE, usually called an ATM data service unit (ATM-DSU), to cooperate in providing a UNI for ATM networks. The DXI framework defines the protocols for a DTE to transport a DTE-SDU to a corresponding peer entity via an ATM network, as illustrated in Figure 6.8.

The DXI defines a data link control protocol and physical layers which handle data transfer between DTE and DCE. The DXI local management interface (LMI) and the MIB are also specified as part of the DXI. The DXI supports V.35, RS449, and HSSI physical layer interfaces at speeds ranging from several kbps to 50 Mbps. The physical layer interface used between the DSU and the ATM network across the UNI can be any one of the physical layers specified in UNI 3.1.

The data link layer defines the method by which the DXI frames and their associated addressing are formatted for transport over the physical layer between the DTE and the DCE. The protocol is dependent on the mode selected. In particular, three operational modes are defined across a DXI: 1a, 1b, and 2.

6.2.1 Mode 1a

In this mode, transport of a DTE-SDU is based on AAL 5 common part convergence and SAR sublayers, as illustrated in Figure 6.9(a).

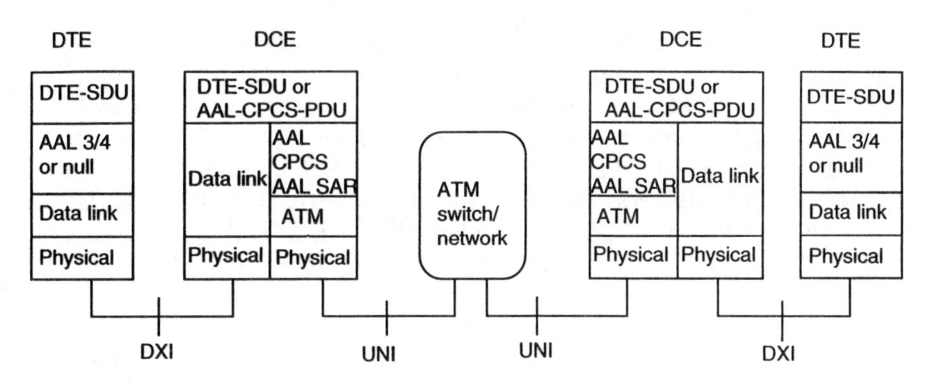

Figure 6.8 DXI framework. DXI-FCS, DXI frame check sequence; rsvd, reserved.

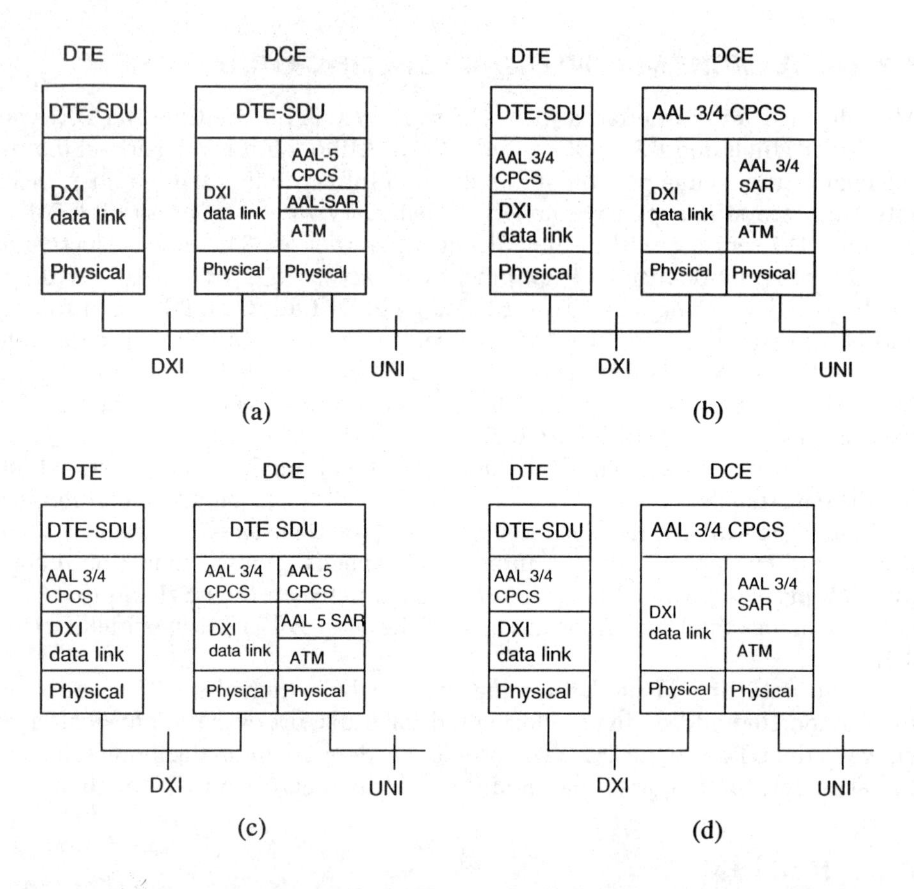

Figure 6.9 Three operational modes of a DXI: (a) modes 1a and 1b protocol architecture for AAL 5; (b) mode 1b protocol architecture for AAL 3/4; (c) mode 2 protocol architecture for AAL 5; (d) mode 2 protocol architecture for AAL 3/4.

At the origination node, the data link control layer receives DTE-SDU and encapsulates into DXI data link control frame (DXI-PDU), as illustrated in Figure 6.10. The resulting PDU is transmitted to DCE across the DXI. The DCE strips off the DXI encapsulation and obtains the values of the DXI frame address (DFA) and CLP. The DCE then encapsulates the DTE-SDU into an AAL 5 CPCS-PDU and segments the resulting PDU into 48-byte AAL 5 SAR-SDUs. The DCE also maps the DFA to the appropriate VPI/VCI of each cell. The CLP bit value at the DXI header is also copied to the CLP bits of the transmitted cells.

For data transmission from the ATM network to the destination DTE, the reverse process is followed. The only exception is the use of the congestion notification (CN) bit at the DXI header, which is set to 1 if one or more cells of the DTE-PDU experienced congestion in the network as specified in their cell header, EFCI bit.

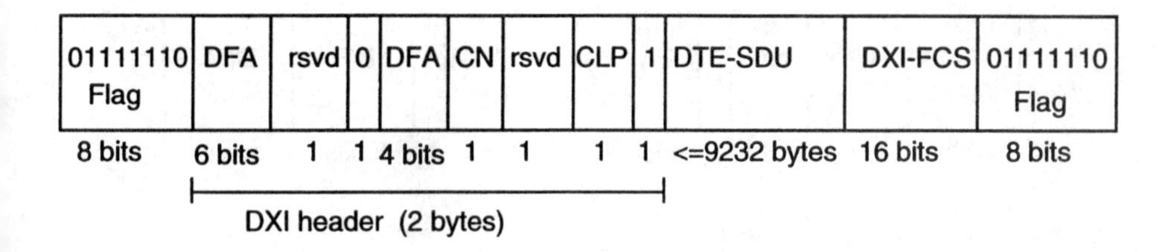

Figure 6.10 DXI data link PDU format in mode 1a.

6.2.2 Mode 1b

This mode consists of mode 1a plus transport of DTE-SDU service based on AAL 3/4, as illustrated in Figure 6.9(b). When AAL 3/4 is used, DTE first encapsulates the DTE-SDU into an AAL 3/4 CPCS-PDU by appending the corresponding CPCS header and trailers to the DTE-SDU. The CPCS-PDU is then encapsulated into a DXI data link control frame, as with mode 1a. The resulting frame format is illustrated in Figure 6.11.

The resulting frame is transmitted to the DCE, which strips off the DXI header and DXI FCS, thereby obtaining the DFA and CLP bit. The DCE segments the received frame into AAL 3/4 SAR-SDUs and translates the DFA into the appropriate VPI/VCI value. The CLP bit value is also copied from the DXI header to the cell header. For data transmission from the ATM network to the destination DTE, the reverse process is followed, as with mode 1a.

6.2.3 Mode 2

In this mode, DTE operations are the same as in mode 1b with AAL 3/4. That is, DTE encapsulates the DTE-SDU into an AAL 3/4 CPCS-PDU. The resulting PDU is then encapsulated into a DXI data link control frame and transmitted to the DCE.

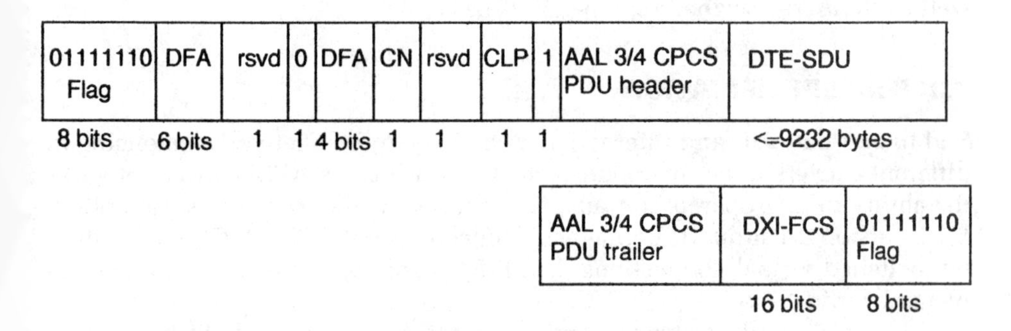

Figure 6.11 DXI data link PDU format in mode 1b with AAL 3/4.

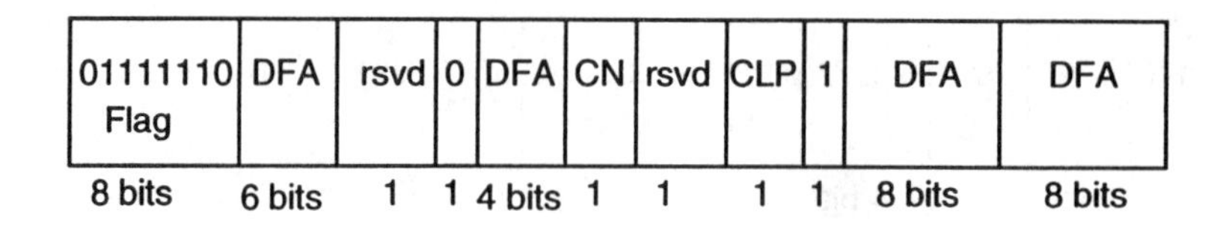

Figure 6.12 DXI header in mode 2.

The main difference between mode 2 and modes 1a and 1b are the use of CRC-32 (as opposed to CRC-16) and the fact that the DXI header in mode 2 is 4 bytes long, as illustrated in Figure 6.12.

At the DCE, both AAL 5 and AAL 3/4 connections are allowed. If AAL 5 is used, then the DCE strips off both the DXI and AAL 3/4 CPCS encapsulations. The remaining PDU (i.e., DXI-DSU) is then encapsulated into an AAL 5 CPCS-PDU, segmented into AAL 5 SAR-SDUs, and transmitted to the ATM network using the services of ATM and physical layers, as in mode 1a (and 1b with AAL 5).

If the AAL 3/4 connection is used, then the DCE strips off only the DXI encapsulation and segments the AAL 3/4 CPCS-PDU into AAL 3/4 SAR-SDUs, which are transmitted to the ATM network through the ATM layer.

6.2.4 DXI LMI

The ATM DXI is managed by the DTE through an LMI. Some of the ATM UNI ILMI messages are also exchanged between the DTE and the ATM switch the DCE is attached to. DCE does not pass the UNI status back to the DTE. The LMI defines the protocol for exchanging management information across a DXI. It is designed to support a management station running an SNMP or switch running the ILMI protocol. The LMI supports the exchange of DXI-specific, AAL-specific, and ATM-UNI-specific management information.

The LMI allows the DTE to set or query the operation mode of the DXI as well as the AAL assigned on a per VCC basis.

6.3 B-ICI SPECIFICATION

End-to-end national and international service requires networks belonging to different carriers to be interconnected. The B-ICI gives ATM carrier networks the ability to interoperate in transporting different services across each other. B-ICI version 1.1 supports permanent connections only. The B-ICI specification for switched virtual connections (i.e., B-ICI signaling) is expected to be completed by mid-1995.

Figure 6.13 illustrates a generic example of carrier networks having service-specific UNIs connected to other carrier ATM networks. The B-ICI specifi-

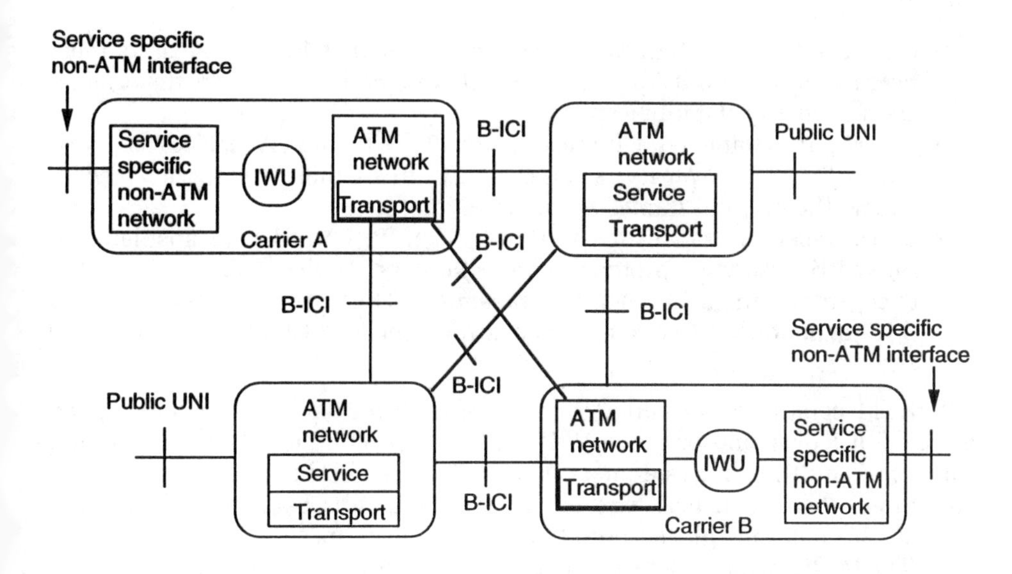

Figure 6.13 A generic view of service-specific networks interconnected via ATM networks.

cation includes physical layer, ATM layer, and service-specific functions above the ATM layer required to transport, operate, and manage a variety of intercarrier services across a B-ICI. The document also includes traffic management, network performance, and operations and maintenance specifications.

The physical layer of the interface is based on an ITU-T-defined network node interface, which includes SONET/SDH physical and ATM layers, with the addition of the DS-3 physical layer.

The initial B-ICI is a multiservice interface that supports the following intercarrier services:

- Cell relay service (CRS);
- Circuit emulation service (CES);
- Frame relay service (FRS);
- SMDS.

A service-specific non-ATM network is connected to an ATM network via an IWU that allows a non-ATM service to be mapped into an ATM service and vice versa. The main function performed at an IWU for each service supported across a B-ICI is as follows:

- CRS receives ATM cells from one network and transmits them to another over a permanent connection.

- CES receives DS-n frames encapsulated in AAL 1 PDUs, transmits them over PVCs across a B-ICI to another network, and reconstructs the original DS-n frames at the other end.
- FRS receives frame relay frames, encapsulates them in AAL 5 PDUs, transmits ATM SDUs over PVCs across a B-ICI to another network, and reconstructs the original frames at the other end.
- SMDS receives SMDS interface protocol (SIP) L3 PDUs encapsulated in intercarrier service protocol connectionless service (ICIP-CLS) PDUs, encapsulates them in AAL 3/4, transmits ATM SDUs over PVCs across a B-ICI to another network, and reconstructs the ICIP-CLS PDUs at the other end.

The ATM cells of a particular service are multiplexed together and passed across B-ICI over one or more VPCs or VCCs preconfigured at subscription time. This multiplexing is done at the interface level. That is, for each service supported, there is at least one connection, and cells belonging to different services are not multiplexed onto the same connection.

The traffic management and congestion control framework across a B-ICI follows very closely the framework developed in the UNI specification. The traffic parameters and connection traffic descriptor definitions in the UNI specification also apply to the B-ICI, with some simplifications. For example, the source traffic descriptor across the B-ICI is required to include service type, conformance definition, PCR, and CDVT, while the inclusion of SCR and BT in the traffic contract is optional. The network parameter control (NPC) function monitors and controls offered traffic and the validity of the ATM connection. Its main purpose is to protect network resources and the QOS of connections already established in the network during periods of congestion, which might be caused by equipment malfunction or misoperation. Each carrier may use any implementation of the NPC function as long as it does not violate the QOS objectives of a valid and compliant connection (or may not use any NPC function at all).

The interconnectivity across a B-ICI is obtained by establishing a VCC or VPC at the B-ICI with a QOS suitable to meet the service requirements for end-to-end connections. Current B-ICI specification version 1.1 supports PVCs only. That is, both the source and destination points of each PVC are predefined and are fixed for the duration of the connection. A consequence of this is that there is a fixed route between two IWUs communicating with each other across two or more public networks.

PVC-based intercarrier support for each service supported across a B-ICI is reviewed next.

6.3.1 Cell Relay Service

CRS is a cell-based information transfer service that offers its users direct access to the ATM layer at rates up to the access link rate. Both VPC and VCC connec-

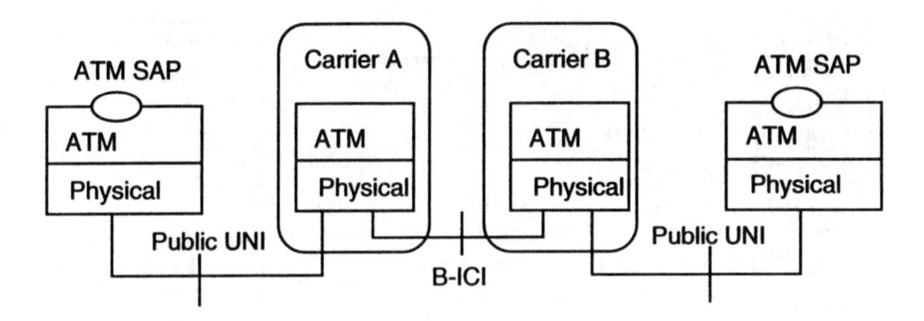

Figure 6.14 CRS across a B-ICI.

tions are supported. In CRS, a point-to-point PVC denotes an ATM layer VPC or VCC from a source ATM-SAP to a destination ATM-SAP, as illustrated in Figure 6.14.

6.3.2 Circuit Emulation Service

CES supports the transport of CBR signals using ATM technology. Figure 6.15 shows examples of CES and its role in supporting CBR services. In Figure 6.15(a), two end stations support voice service over ATM using AAL 1 and generate DS-*n* frames. The B-ICI supports the transport of DS-n signals across two public networks, thereby connecting two ATM end stations with UNIs at each end. In Figure 6.15(b), one user is connected to an ATM network via a UNI, whereas a DS-n interface is used at the other end. In this case, the end station that supports voice over ATM communicates with another end station possibly attached to a voice network through an IWU. The IWU provides interoperability between a native DS-*n* network and an ATM network. Finally, in Figure 6.15(c), both users are connected with DS-*n* interfaces, requiring interworking functions at both ends.

6.3.3 Frame Relay Service

FRS is a connection-oriented data transport service. Two frame relay interworking scenarios are illustrated in Figure 6.16. These are two network interworking scenarios originally defined by ITU-T and now adopted by the Forum. In the first scenario, two frame relay networks/CPE are connected via an ATM network, and in the second scenario a frame relay network/CPE is connected with broadband CPE emulating frame relay.

6.3.4 Switched Multimegabit Data Service

SMDS is a public packet-switched service that provides for the transport of data packets without the need for call establishment procedures. Customer

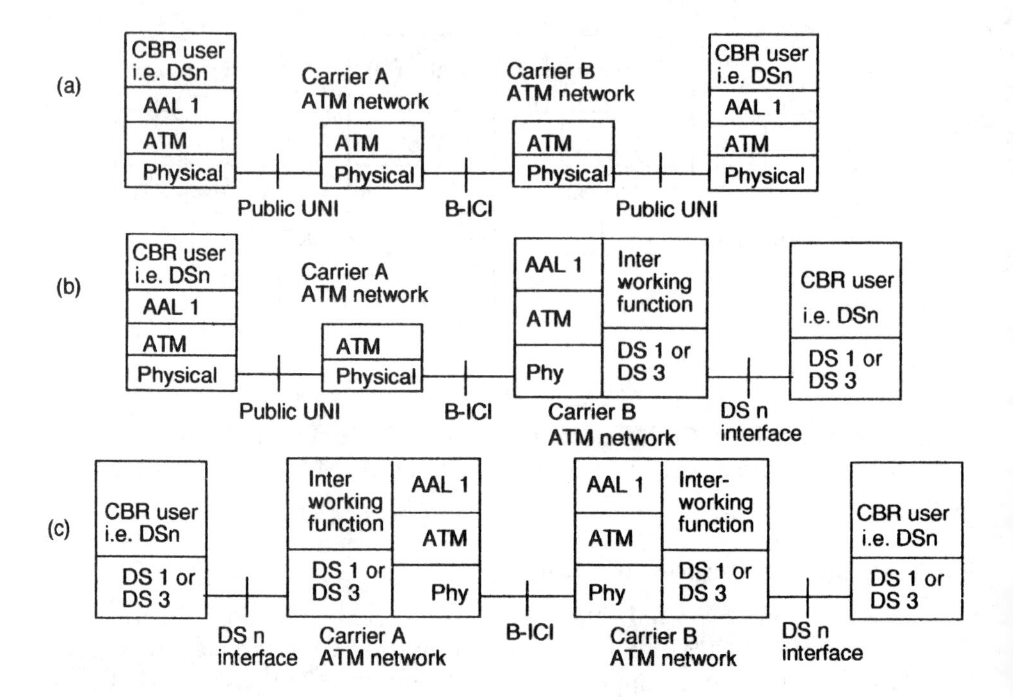

Figure 6.15 Circuit emulation service.

access to SMDS will be over the SMDS subscriber network interface (SNI). The carrier network provides an internetworking function for the encapsulation of the SMDS L3-PDU within an ICIP-CLS PDU. AAL 3/4 is used to support the transfer of these PDUs within ATM cells. Figure 6.17 shows an example of SMDS/ATM interworking functions.

6.4 NETWORK-TO-NETWORK INTERFACE

The ATM NNI is the point of demarcation between two ATM networks. The same interface may also be used as a network node interface (i.e., as a switch-to-switch interface). If the two networks interconnected through an NNI are private, the interface is referred to as a P-NNI. The interface is a public NNI if the two networks are public. The P-NNI specification is currently being worked on at the ATM Forum. The P-NNI routing framework is introduced in detail in Chapter 8, whereas P-NNI signaling framework is discussed in Chapter 7.

ITU-T standardization on NNI includes the specifications of physical and ATM layers; control, user, and management planes; and NNI signaling in the context of public networks. The details of NNI signaling are presented in Chapter 7.

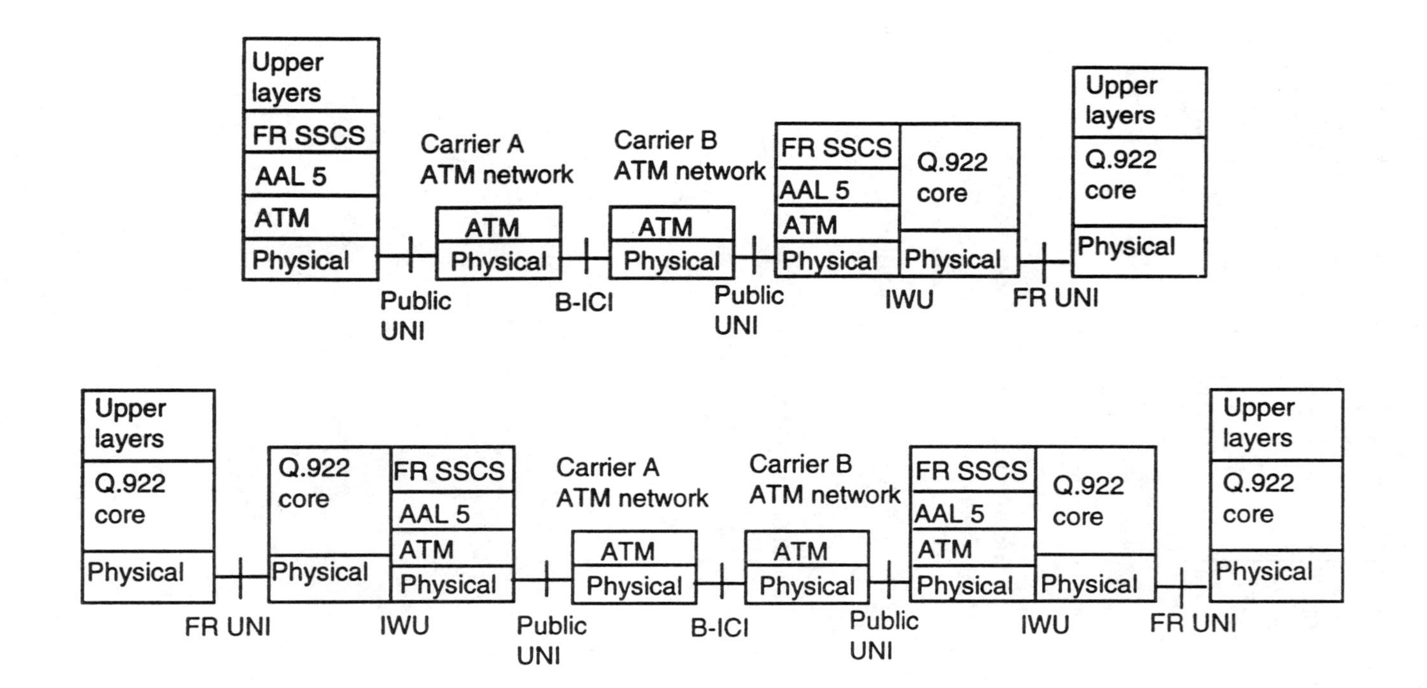

Figure 6.16 Frame relay service.

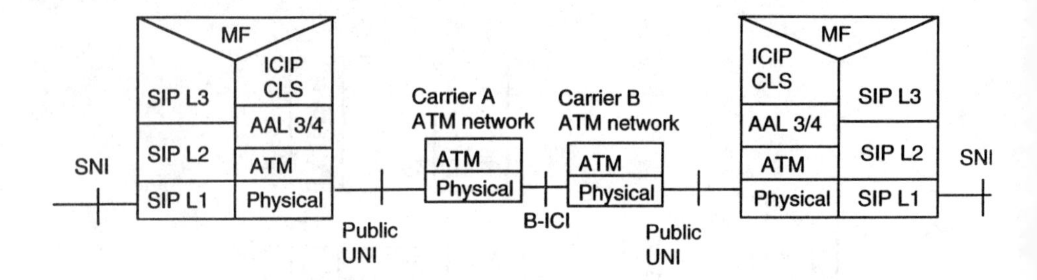

Figure 6.17 SMDS/ATM interworking. MF, mapping function.

References and Bibliography

[1] ATM User-Network Interface Specification, Version 3, ATM Forum, August 1993.

[2] ITU-T Recommendation Q.2931: B-ISDN User-Network Interface Layer 3 Specification for Basic Call/Bearer Control, December 1993.

[3] Bellcore TA-NWT-001111, Broadband ISDN Access Signaling Generic Requirements, Issue 1, Livingston, NJ, August 1993.

[4] Colella, R., E. Gardner, and R. Collan, "RFC 1237: Guidelines for OSI NSAP Allocation in Internet," Network Working Group, July 1991.

[5] ITU-T Recommendation: SSCF and SSCOP, 1993.

[6] ATM DXI, Version 1.1, ATM Forum, July 1994.

[7] ATM B-ICI, Version 1.1, ATM Forum, July 1994.

[8] ATM Forum newsletters: 53 Bytes, 1993–1994.

[9] ATM Forum Traffic Management Specification, Version 4.0, 1995.

7 Signaling in ATM Networks

Signaling in a communication network is the collection of procedures used to dynamically establish, maintain, and terminate connections, requiring information exchange between the network users and the switching nodes, and between switching nodes. For each function performed, the corresponding signaling procedures define the sequence and the format of messages exchanged that are specific to the network interface across which the exchange takes place.

The B-ISDN architecture model consists of separate signaling (control plane) and user-to-user connections (user plane). For signaling purposes, the two B-ISDN interfaces defined are the UNI and the NNI. The UNI is the interface between an ATM end station and the network. The ATM end station in this context is used to refer to any device capable of generating ATM cells and can be a workstation, an IWU, or an ATM switch. Depending on whether the network is private or public, the interface is respectively referred to as a private UNI or a public UNI. When an ATM end station is attached to a public network, the interface is always referred to as a public UNI.

The NNI is used for different types of interfaces in different contexts. An NNI can be a switch-to-switch interface or an NNI. In ITU-T terminology, it refers to a network node interface in public networks. Signaling in public networks is defined between two networks. It has, however, the capabilities and the features required in switch-to-switch signaling and is often used internally in a public network as well. Both types of interfaces are referred to as public NNIs. A P-NNI is the corresponding interface between two private networks. It is being defined by the ATM Forum as the interface between two switching systems in which a switching system may be a single switch or a collection of switches under one administration. This framework permits vendors to design and develop their internal proprietary signaling for their switches while guaranteeing interoperability among networks belonging do different organizations across standard interfaces. In this context, a network consists of one or more switching nodes managed by the same organization, and the interoperability is achieved across this demarcation point between two networks.

The four types of interfaces are illustrated in Figure 7.1. The standardization and specification activities on the UNI and NNI have been undertaken by

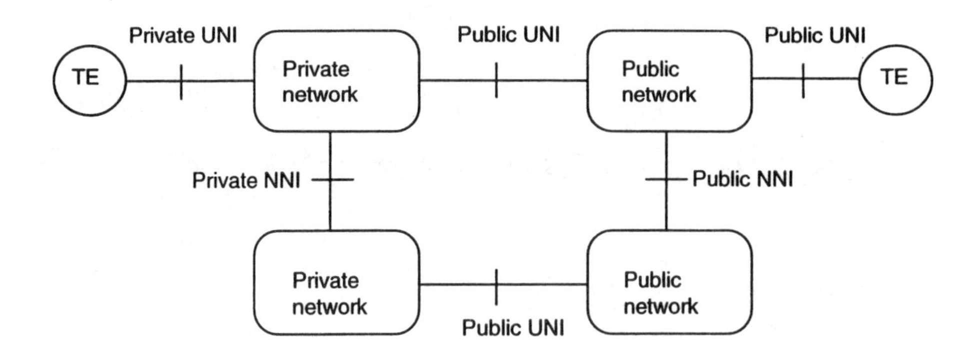

Figure 7.1 B-ISDN signaling interfaces.

both ITU-T and the ATM Forum. ITU-T's focus is on public networks. The ATM Forum addresses both private and public networks. Current UNI signaling specifications are referred to as ATM Forum UNI signaling version 3.1 (UNI 3.1) and ITU-T Q.2931. Similarly, signaling specifications at the NNI are referred to as ATM Forum private NNI signaling and ITU-T B-ISDN user part (B-ISUP).

In this chapter, we discuss the structure and use of signaling across the UNI, public NNI, and P-NNI.

7.1 USER-TO-NETWORK SIGNALING

UNI signaling is used to dynamically establish, maintain, and terminate connections at the edges of the network (i.e., between an ATM end station and a public or private network). The various capabilities specified in UNI signaling include:

- Establishment of point-to-point VCCs;
- Establishment of point-to-multipoint VCCs;
- Three different ATM private address formats;
- One ATM public address format;
- Symmetric and asymmetric QOS connections with a declarable QOS class;
- Symmetric and asymmetric bandwidth connections with a declarable bandwidth;
- Transport of user-to-user information;
- Support of error handling.

Next we discuss these capabilities in detail.

7.1.1 Signaling Architecture at the UNI

The UNI signaling architecture is shown in Figure 7.2. UNI signaling is a layer 3 protocol that runs on top of signaling AAL (SAAL). SAAL consists of AAL

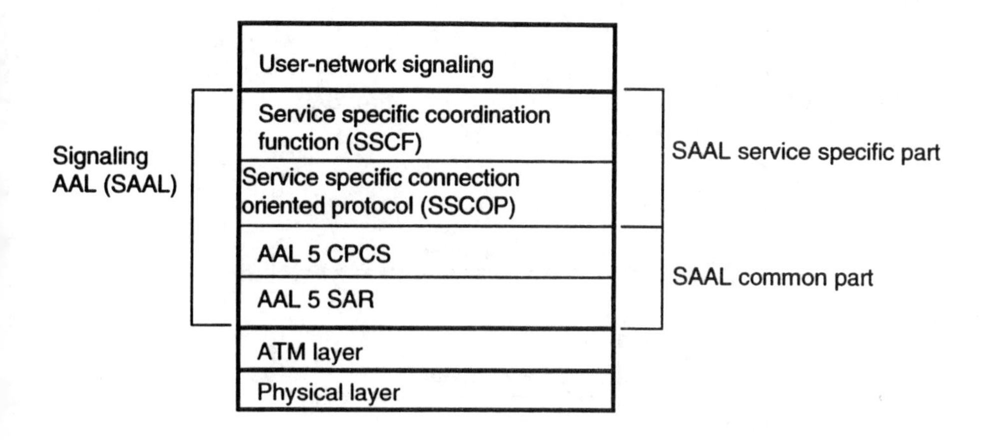

Figure 7.2 ISDN signaling structure.

5 common part and service specific convergence sublayers (SSCSs). The main function of the SSCS is to ensure reliable transfer of the signaling messages for call/connection control using the services of the ATM layer on signaling VCs.

Since ATM requires connections to be established before any information can be sent, it is necessary to establish signaling channels before signaling messages can be exchanged between the ATM end station and the network. UNI signaling uses a dedicated point-to-point signaling VC with VCI = 5 and VPI = 0.

The service-specific part consists of a service-specific coordination function (SSCF) and a service-specific connection-oriented protocol (SSCOP). All SAAL functions are accessed by UNI signaling through a service access point (SAAL-SAP). Signaling SSCF maps the particular requirements of UNI signaling to the requirements of the ATM layer. SSCOP, on the other hand, provides the mechanisms for the establishment, release, and monitoring of signaling information exchange between peer signaling entities. The ATM-SAP provides a bidirectional flow of information and allows ATM functions to be accessed by the AAL. There is always a 1:1 correspondence between a connection end point within the SAAL-SAP and a connection end point within the ATM-SAP. The interaction between the two sides of the UNI is illustrated in Figure 7.3.

The UNI signaling layer requests service from the SAAL via a set of four service primitives illustrated in Table 7.1: request, indication, response, and confirm. The same set of primitives is also used between the SAAL and ATM layers.

Based on these primitives, the information exchanged through the SAAL-SAP is defined as in Table 7.2.

SSCOP receives variable-length SDUs from the signaling layer, forms the PDUs, and transfers them to the peer SSCOP. At the receiving end, the SSCOP

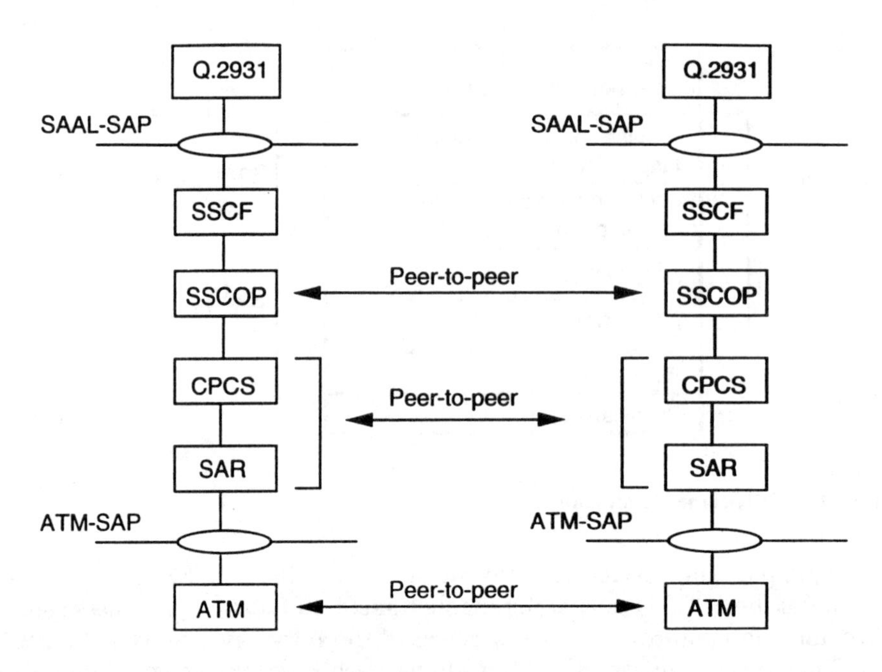

Figure 7.3 Peer-to-peer communication across the UNI.

Table 7.1
UNI Signaling/SAAL Primitives

Primitive	*Description*
Request	Used when a higher layer is requesting a service from the next lower layer
Indication	Used by a layer providing a service to notify the next higher layer of any specific activity that is service-related
Response	Used by a layer to acknowledge receipt from a lower layer of the primitive type indication
Confirm	Used by the layer providing the requested service to confirm that the activity has been completed

delivers the SDU to the signaling layer. The SSCOP uses the services of the CPCS, which provides an unassured information transfer and a mechanism for detecting corruption in SSCOP PDUs. The set of functions provided by the SSCOP are summarized as follows:

- *Sequence integrity* preserves the order of SSCOP-PDUs submitted by the signaling layer.

Table 7.2
SAAL-SAP Primitives

Primitive Name	*Type*	*Description*
AAL-ESTABLISH	Request Indication Confirm	Used for establishing assured information transfer between AAL entities at the UNI
AAL-RELEASE	Request Indication Confirm	Used for terminating assured information transfer between AAL entities at the UNI
AAL-DATA	Request Indication	Used in conjunction with assured data transfer at UNI
AAL-UNIT-DATA	Request Indication	Used in conjunction with unassured data transfer at UNI

- *Error correction by retransmission* allows the receiving SSCOP to detect the missing PDUs and the sending SSCOP to correct sequence errors through retransmission.
- *Flow control* allows the receiver to control the rate at which the peer SSCOP transmitter entity may send information.
- *Keep alive* ensures that two peer SSCOP entities participating in a connection remain in a link-connection-established state even in the absence of data transfer for long periods of time.
- *Local data retrieval* allows the local SSCOP user to retrieve in sequence SDUs that have not yet been transmitted by the SSCOP entity or those that are not yet acknowledged by the peer SSCOP entity.
- *Connection control* permits the establishment, release, and synchronization of an SSCOP connection.
- *Transfer of user data* is used for the conveyance of user data between the users of the SSCOP.
- *Protocol control information (PCI) error detection* detects errors in the PCIs.
- *Status reporting* allows the transmitter and the receiver entities to exchange status information.

In order to perform these functions, the different PDUs listed in Table 7.3 are used for peer-to-peer communications between two SSCOP entities.

For completeness, the signals defined between the SSCF and SSCOP are given in Table 7.4.

Given this framework, an example of a sequence of events that takes place while sending a signaling message across the UNI is illustrated in Figure 7.4.

Table 7.3
SSCOP PDUs

PDU Name	*Description*
Begin	Used to initially establish an SSCOP connection or to reestablish an existing SSCOP connection
Begin acknowledge	Used to acknowledge the acceptance of an SSCOP connection request by the peer SSCOP entity
End	Used to release an SSCOP connection between two peer entities
End acknowledge	Used to confirm the release of an SSCOP connection that was requested by the peer SSCOP entity
Reject	Used to reject the connection establishment of peer entity
Resynchronize	Used to resynchronize the buffers and the data-transfer state variables in the transmit direction of a connection
Resynchronize acknowledge	Used to acknowledge the resynchronization of the local receiver in response to a resynchronize PDU
Sequenced data	Used to transfer sequentially numbered PDUs containing user information
Status request	Used to request status information about the peer SSCOP entity
Sequenced data with poll	This PDU is the functional concatenation of sequenced data and status PDUs
Solicited status response	Used to respond to a status request PDU; contains information on reception status of sequenced data, credit information for the peer transmitter, and sequence number of status request being responded to
Unsolicited status response	Used to respond to detection of a new missing sequenced data and credit information for the peer transmitter

Signaling messages are used to characterize the connection and its service requirements to the network and for the network to inform the user whether or not the connection request is accepted (i.e., the network has the resources to support the connection). This process requires a sequence of signaling messages to be exchanged between the user and the network. Each signaling message is a request for a specific function or it is a response to a specific request. In either case, a signaling message is composed of a number of information elements (IE) with each IE identifying a specific aspect of the function requested by the message (or the response to a request).

7.1.2 Signaling Message Format

The ATM bearer service provides a sequence-preserving, connection-oriented cell transfer capability between source and destination end stations with an

Table 7.4
SSCF/SSCOP Primitives

Signal Name	*Type*	*Description*
AA-ESTABLISH	Request Indication Confirm	Used for establishing assured point-to-point information transfer between peer user entities
AA-RELEASE	Request Indication Confirm	Used for terminating assured point-to-point information transfer between peer user entities
AA-DATA	Request Indication	Used for the assured point-to-point transfer of SSCOP-SDUs
AA-ERROR	Indication	Indicates SSCOP detected an error during the assured transfer of an SDU
MAA-ERROR	Indication	Used to report errors to layer management

agreed-on QOS. The offered service can be either point-to-point bidirectional or point-to-multipoint unidirectional. Point-to-point call control messages consists of call establishment, call clearing, and maintenance signaling messages. Point-to-multipoint connections are established first by establishing a point-to-point connection. It then uses special messages to add new end points and to clear end parties of the connection.

Regardless of its type, the format of a signaling message is defined as in Figure 7.5. The *protocol discriminator* is used to distinguish B-ISDN UNI signaling messages from the other types of messages of other protocols. The *length of call reference* field indicates the length of the *call reference* in octets, identifying the call at the UNI to which the message applies. Its value is assigned by the originating side of the interface for a call and remains fixed and unique for the duration of the call. The call reference *flag* indicates which end of the signaling VC (i.e., the network or the end point) assigned the call reference value. This provides the capability to distinguish between incoming and outgoing messages in case the same call reference value happens to be used at the two ends of the connection. In addition, a global call reference value of zero indicates that the message received pertains to all call references associated with the corresponding signaling VC.

The *message type* identifies the function of the message being sent. Various messages associated with point-to-point and point-to-multipoint connections are defined in the next two sections. Finally, the *message length* is the number of octets of the contents of the message. A signaling message may contain one or more IEs as needed. The format of an IE is given in Figure 7.6. Its fields are defined in a way similar to that of the corresponding signaling message fields.

We now proceed with a brief description of point-to-point and point-to-multipoint call processing in UNI signaling [2].

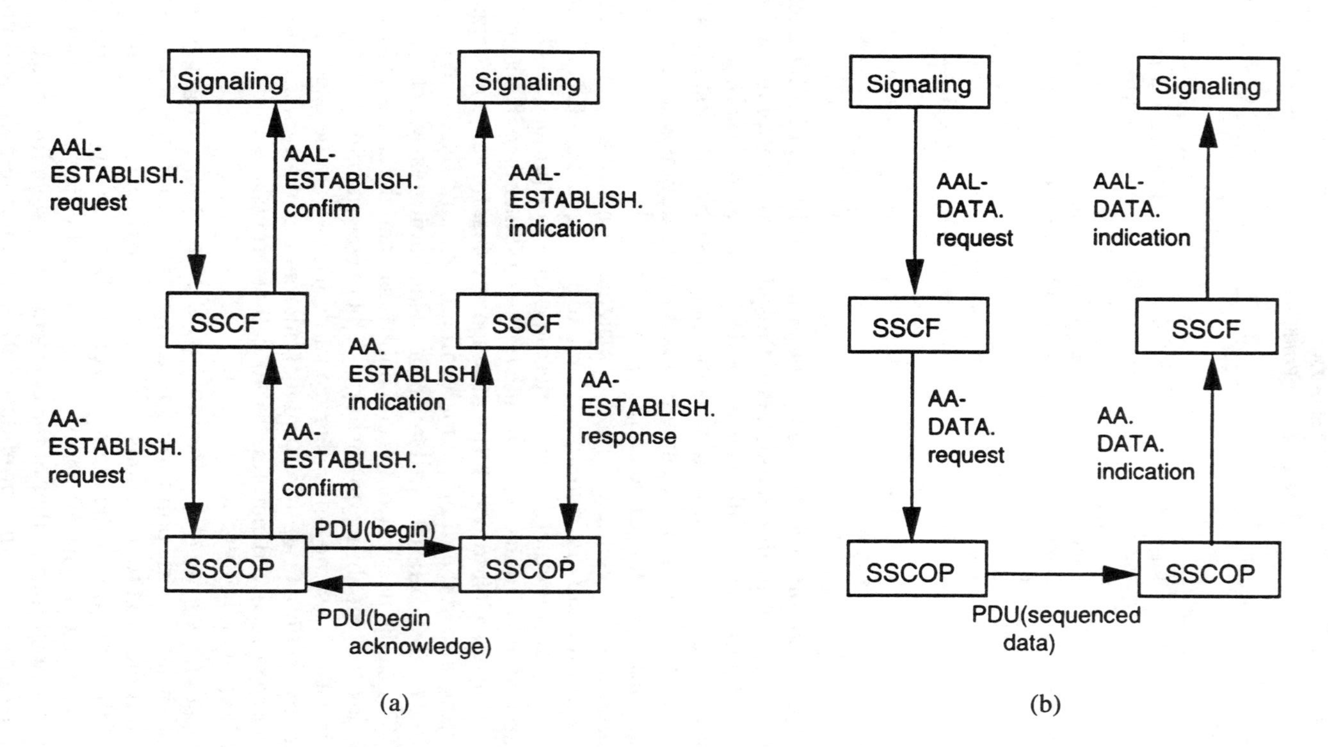

Figure 7.4 SSCF/SSCOP primitives exchanged across a UNI: (a) SAAL connection establishment, and (b) message transfer.

Octets	8	7	6	5	4	3	2	1
1	Protocol discriminator							
2	0	0	0	0	Length of call reference			
3	Flag	Call reference value						
4	Call reference value (continued)							
5	Call reference value (continued)							
6	Message type							
7	Message type (continued)							
8	Message length							
9	Message length (continued)							
etc.	Variable length information elements as required							

Figure 7.5 Signaling message format.

Octets	8	7	6	5	4	3	2	1
1	Information element identifier							
2	1	coding standard		IE information field				
3	Length of information field							
4	Length of information field (continued)							
5, etc.	Contents of the information field							

Figure 7.6 General IE format.

7.1.3 Point-to-Point Call Processing

A point-to-point connection is a collection of VC or VP links that connect two end points of a connection. As discussed previously, UNI signaling takes place between end stations and the network at the edges of the network. Signaling messages provide the network with enough information to characterize the source and to locate the destination UNI. Network services required to establish, manage, and terminate connections internal to the network, such as finding a path between the call originator and the destination node, belong to internal network control mechanisms.

The three groups of point-to-point call processing messages are given in Table 7.5.

7.1.3.1 Origination of an ATM Connection from the User

In this section, we review the sequence of events and the messages exchanged to establish a point-to-point connection at the calling-user UNI.

Table 7.5
Point-to-Point Signaling Messages

Message Type	*Name*	*Definition*
	SETUP	Initiates call establishment *Calling user to network and network to called user*
Call establishment messages	CALL PROCEEDING	The requested call establishment has been initiated and no more call establishment information will be accepted *Called user to network and network to calling user*
	CONNECT	Call acceptance *Called user to network and network to calling user*
	CONNECT ACKNOWLEDGE	User has been awarded the call *Network to called user and calling user to network*
Call clearing messages	RELEASE	Equipment sending the message has disconnected the virtual connection and intends to release the VC and call reference *User and network*
	RELEASE COMPLETE	Indicates the VC and call reference are released and the VC is available for reuse *Recipient of RELEASE message . . . user and network*
	RESTART	Request to release all resources associated with the indicated VCs controlled by the signaling channel *User and network*
Miscellaneous messages	RESTART ACKNOWLEDGE	Response to a RESTART message to indicate requested restart is complete *User and network*
	STATUS	Response to STATUS ENQUIRY message or sent at any time to report certain error conditions *Recipient of STATUS ENQUIRY message . . . user and network*
	STATUS ENQUIRY	Solicits a STATUS message *User and network*

Table 7.6
Point-to-Multipoint Signaling Messages

Message Name	*Definition*
ADD PARTY	Add a party to an existing connection
ADD PARTY ACKNOWLEDGE	Response to ADD PARTY message to acknowledge that the add-party request was successful
ADD PARTY REJECT	Response to ADD PARTY message to acknowledge that the add-party request was not successful
DROP PARTY	Clear a party from an existing point-to-multipoint connection
DROP PARTY ACKNOWLEDGE	Response to DROP PARTY message to indicate that the party was dropped from the connection

Before proceeding with the details, we will present a high-level description of the signaling message flows. For presentation purposes, consider two end systems and a generic network node that represents a subnetwork providing a connection between the two systems, as illustrated in Figure 7.7.

End systems 1 and 2 (ES1 and ES2) are connected to the ATM network across two separate UNIs. The sequence of message exchange that takes place can be summarized as follows:

1. ES1 wishes to set up a connection with ES2. It sends a SETUP message across its UNI to the network, which contains pertinent information that identifies the two end nodes and connection characteristics.
2. The network node replies to ES1 with a CALL PROCEEDING message indicating that the SETUP message is received and a connection setup is being processed. This message also includes the VPI/VCI value to be used for user traffic. The network sends much of the same information contained in the SETUP message sent by ES1 to the destination network node. The network node that the end station is attached to formulates a SETUP message and delivers to ES2 across the UNI the ES2 is attached to the network. This SETUP message includes the VPI/VCI value used for data transmission.
3. If necessary, ES2 sends to the network a CALL PROCEEDING message indicating that it has received the SETUP message but it will take longer to process.
4. ES2 decides to accept the SETUP request and forwards a CONNECT message to the network node.

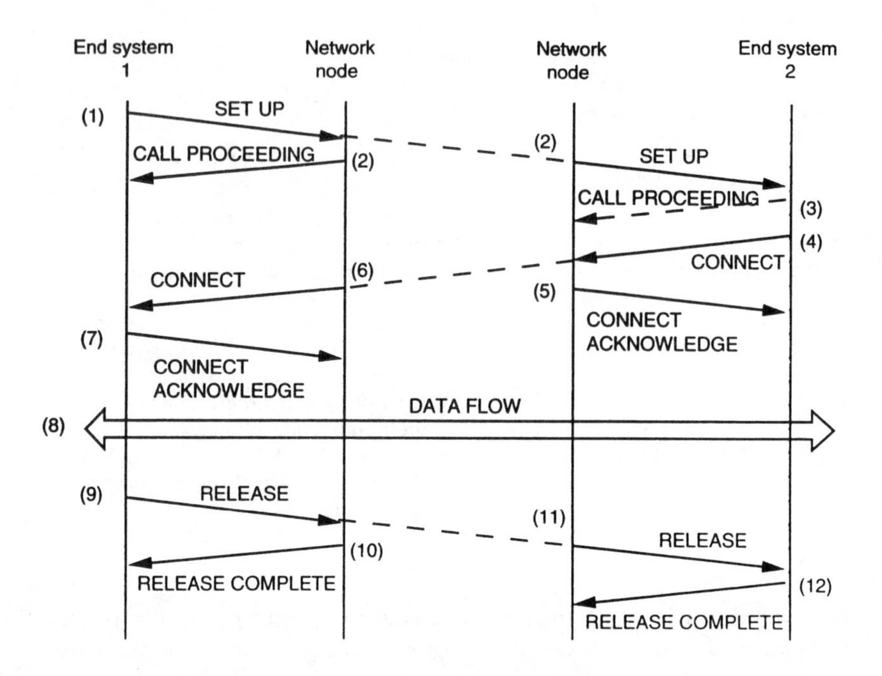

Figure 7.7 Principal signaling message flows for point-to-point ATM connections.

5. The network node sends a CONNECT ACKNOWLEDGE message to ES2. ES2 can at this time start sending data on the ATM connection using the VPI/VCI label received in the SETUP message.
6. The network conveys the acceptance information to the network node the call originates from. Then a CONNECT message is sent to ES1. This message contains information related to the connection setup.
7. ES1 processes the CONNECT message and accepts the connection setup. It sends a CONNECT ACKNOWLEDGE message to the network node. ES1 can now start sending data cells on the ATM connection using the VPI/VCI label received in the CONNECT message.
8. Data cells flow across the UNIs in both directions.
9. ES1 wishes to terminate the connection by sending a RELEASE message across its UNI.
10. The network acknowledges the receipt of the RELEASE message by sending a RELEASE COMPLETE message. The portion of the connection between the network and ES1 is cleared.
11. The network sends a RELEASE message to ES2 across its UNI.
12. ES2 sends a RELEASE COMPLETE message to the network to acknowledge the receipt of the RELEASE message. The portion of the connection between the network and ES2 is cleared.

We next discuss further details on the dynamics associated with the exchange of signaling messages at the UNI that the call originates.

Each message contains a number of IEs that are the parameters describing various aspects of the interaction. The inclusion of IEs may be mandatory (M) or optional (O). When an element is optional, notes explaining the circumstances under which such elements are included are specified in the standards documents.

The sequence of events taking place at the user and the network sides of the UNI the call originates is illustrated in Figures 7.8 and 7.9, respectively. In order to initiate a call at the UNI, the calling user puts together a SETUP message and sends it to the network. This message identifies the destination node and characterizes the connection. A SETUP message may contain the following IEs:

- *AAL parameters* (O) identify the requested AAL and its parameter values. As discussed previously, the different AALs are supported in signaling are AAL 1, AAL 3/4, AAL 5, and user-defined AAL (which can be null).

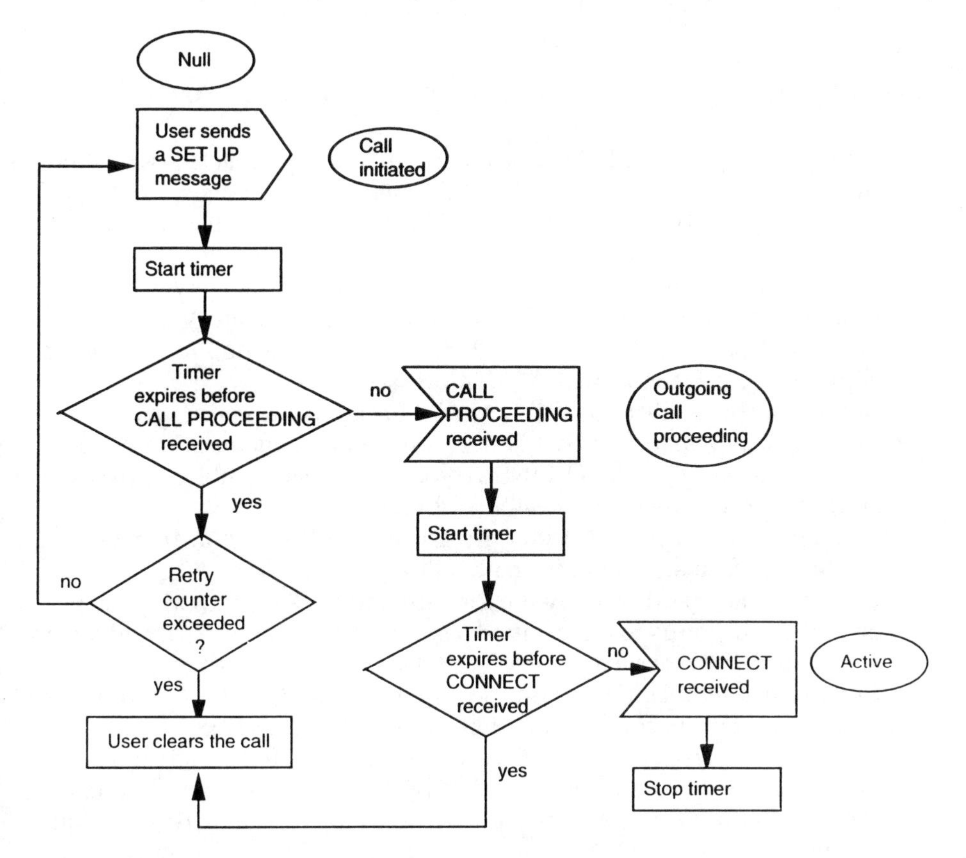

Figure 7.8 Sequence of call origination events at the user side of the UNI.

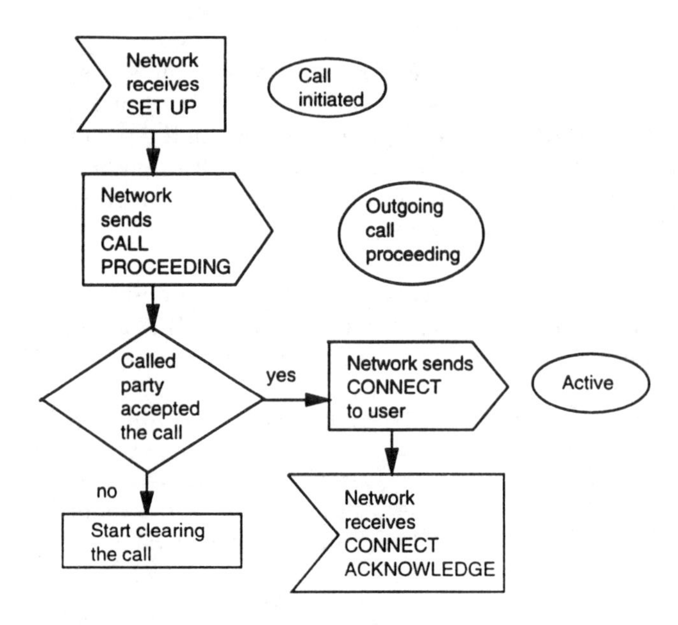

Figure 7.9 Sequence of call origination events at the network side of the UNI.

- *ATM user cell rate* (M) specifies a set of traffic parameters that include the forward and backward PCRs, SCRs, maximum burst size, and a traffic control capability such as the best effort service indicator and the use of tagging.
- *Called-party number* (M) identifies the called party (i.e., its ATM address).
- *Called-party subaddress* (O) is used only to convey an ATM address of the called party in the OSI NSAP format across a public network that supports only E.164-format ATM addresses.
- *Calling-party number* (O) is the ATM address of the calling party.
- *Calling-party subaddress* (O) is used only to convey an ATM address of the calling party in the OSI NSAP format across a public network that supports only E.164-format ATM addresses.
- *Connection identifiers* (M) identify the local ATM connection resources on the interface and includes both VPI and VCI.
- *QOS parameters* (M) specify the forward and backward QOS class, where class 0 corresponds to unspecified service and classes 1, 2, 3, and 4 correspond to B-ISDN service classes A, B, C, and D, respectively.
- *End-point reference* (O) is used to uniquely identify the individual end point of a point-to-multipoint connection at the interface.
- *Broadband bearer capability* (M) indicates a requested broadband connection-oriented service to be provided by the network, including class of service, traffic type (i.e., constant or variable rate), and timing requirements.

- *Broadband high-layer information* (O) provides a means for compatibility checking for the higher layer information type (i.e., International Standards Organization (ISO), user-specific, vendor-specific).
- *Broadband low-layer information* (O) provides a means of compatibility checking for the layer 2 protocol used (i.e., Q.921, Q.922, X.25 link layer, HDLC, LAN logical layer, and window size). It is also used to specify the layer 3 protocol used (i.e., X.25 packet layer, OSI connectionless mode protocol, user-specified, and packet size).
- *Transit network selection* (O) identifies the requested transit network service by including the national network and carrier identification codes.

Following the transmission of the SETUP message, the call at the calling party enters the call-initiated state. Upon receiving the SETUP message, the network knows the called-party address and the characteristics of the connection request. The network enters the outgoing-call-initiated state and selects an available VPI and VCI, which identify the connection at the interface. The VPI used in the signaling protocol (VPI = 0) and the actual VPI used for information flow have different values. If no identifier is available or the network cannot establish the connection for whatever reason, it sends a RELEASE COMPLETE message, and the user, upon receiving it, clears the call.

In response to a SETUP message, the network sends a CALL PROCEEDING message and enters the outgoing-call-proceeding state. Similarly, upon receiving the CALL PROCEEDING message, the call at the user side of the interface enters the outgoing-call-proceeding state. If the resources required to provide the QOS requested by the call are available and the calling party accepts the connection, the network side of the interface the call originated from sends a CONNECT message to the calling user and enters the active state. Upon receiving the CONNECT message from the network, the user sends a CONNECT ACKNOWLEDGE message and enters the active state. The CONNECT message includes IEs such as AAL parameters, broadband low-layer information, connection identifier, and end-point reference, all optional, whereas CONNECT ACKNOWLEDGE does not include any IEs.

7.1.3.2 Call/Connection Establishment at the Destination Interface

After the network determines a path in the network that can provide the requested service to the call, a SETUP message is transmitted to the network side of the destination interface. Internal signaling used in the network currently is proprietary (P-NNI/B-ISUP may be used once specified). The sequence of events taking place at this interface between the called user and the network is illustrated in Figures 7.10 and 7.11.

The network indicates the arrival of a call at the UNI by transferring a SETUP message to the called user with a call reference value unique to its

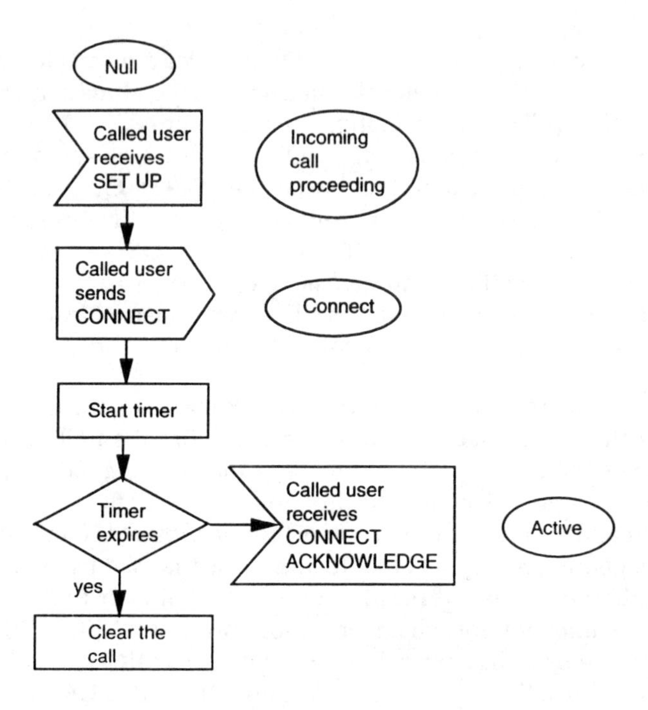

Figure 7.10 Sequence of incoming call events at the network side of the UNI.

interface and enters the incoming-call-present state. Upon receiving this SETUP message, the called user enters the incoming-call-present state.

If the user cannot accept the call due to lack of resources or incompatibility with the specified service, it sends a RELEASE COMPLETE message and the network clears the call. If the call is accepted, then the called user either sends a CALL PROCEEDING message and enters the incoming-call-proceeding state or sends a CONNECT message and enters the connect-request state. CALL PROCEEDING is sent only if a response to a SETUP message cannot be sent immediately. Upon receiving the CONNECT message, the network sends a CONNECT ACKNOWLEDGE message and enters the active state. Similarly, upon receiving the CONNECT ACKNOWLEDGE message from the network, the called user enters the active state.

7.1.3.3 Network Side Call-Clearing Procedures

In order to release a connection, the user sends a RELEASE message, disconnects the VC, and enters the release-request state. Upon receiving a RELEASE message, the network enters the release-request state, disconnects the associated VC, and initiates procedures for clearing the connection to the remote user. After

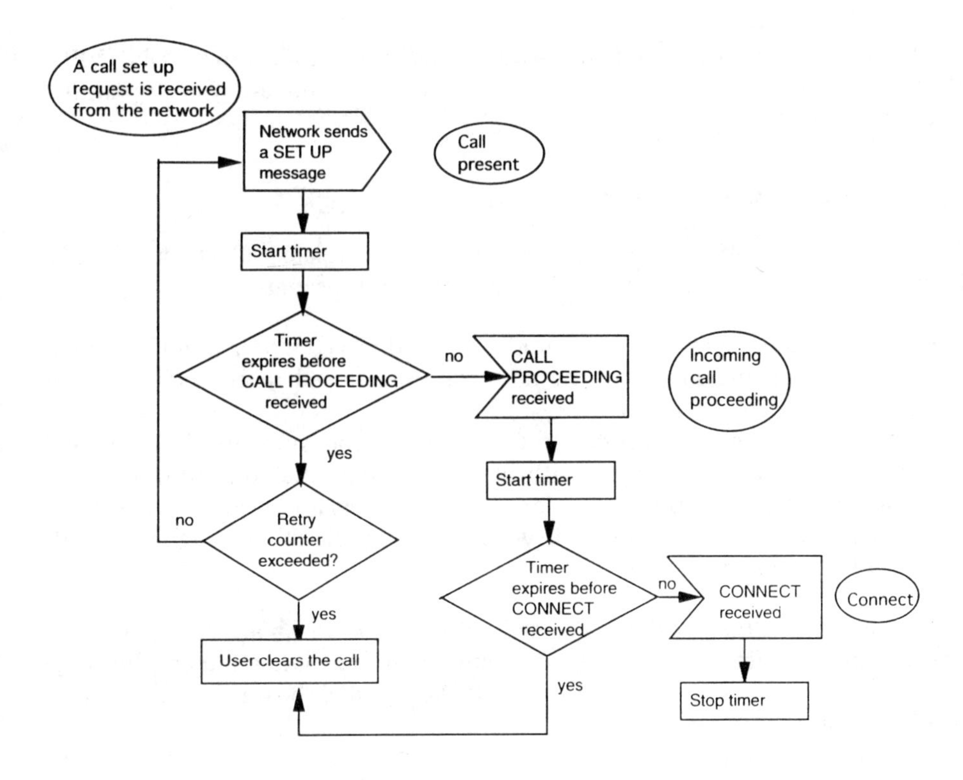

Figure 7.11 Sequence of incoming call events at the user side of the UNI.

disconnecting the VC, the network sends a RELEASE COMPLETE message, releases both the call reference and the connection identifier, and enters the null state. Upon receiving the RELEASE COMPLETE message, the user release both the connection identifier and the call reference and returns to the null state. If no response to the RELEASE message is received, then it is retransmitted. If no response to the second transmission is received, then the connection identifier and call reference are cleared and the user enters the null state. When a connection release is initiated by the network, events that take place are similar to the user-initiated release. The difference is that in this case, the network enters the release-indication state, as opposed to the release-request state. Similarly, the user, upon receiving the RELEASE message, enters the release-request state. After clearing the connection identifier and the call reference, both sides of the interface eventually return to the null state.

7.1.3.4 Restart Procedures

The restart procedures return one or more VCs controlled by a signaling channel to the null state. All resources associated with these VCs are freed and are available for reuse. There is currently one signaling channel across a UNI (VPI = 0, VCI = 5). Hence, a RESET message causes all communication to stop at that interface.

7.1.4 Point-to-Multipoint Call Processing

A point-to-multipoint connection is a collection of associated VC or VP links with associated end nodes, and it has the following properties:

- Traffic is generated by a single node, referred to as the *root,* and is received by all other end nodes of the connection, referred to as *leaf nodes* (or *leaves*).
- There is no bandwidth reserved from the leaf nodes toward the root. Leaf nodes cannot communicate with the root through the point-to-multipoint connection (i.e., point-to-multipoint connections are unidirectional).
- Leaf nodes cannot communicate with each other directly through the point-to-multipoint connection.

The same VPI/VCI values are used at the UNI to reach all leaf nodes. The traffic characteristics and service requirements of the connection are the same for all leaf nodes (i.e., the same PCR, SCR, MBS, QOS, bearer capability, and ATM cell rate).

A point-to-multipoint connection is set up by first establishing a point-to-point connection from the root to a leaf node. The first SETUP message to a leaf node has an end-point reference value of zero specifying that this is a point-to-multipoint connection. The SETUP message also contains a so-called broadband bearer capability IE, indicating in this case a point-to-multipoint connection in its user plane connection configuration field. Accordingly, in order to set it up, end stations are required to know ahead of time that this is a point-to-multipoint connection. A regular point-to-point connection cannot be converted to a point-to-multipoint connection dynamically. Additional leaves are added to the existing connection after the establishment of a point-to-point connection one at a time or simultaneously using ADD PARTY messages.

To differentiate the point-to-point states defined in Sections 7.1.3.2 and 7.1.3.3 from the point-to-multipoint connection states, the states of the former are, in this context, referred to as *link states,* whereas the states of point-to-multipoint connections are referred to as *party states.* Current party states include null, active, add party initiated, add party received, drop party initiated, and drop party received. The party states for each party are maintained at the root along with the link states. A leaf node, however, maintains only the link states.

After the SETUP message is sent by the root, the party state at the user side of the interface changes from null to add party initiated. Similarly, the party state at the network side of the root interface changes to add party initiated upon reception of the SETUP message. The respective party states change to active at the network side when the network sends a CONNECT message, and at the user side when the CONNECT message is received. No party state changes occur upon receiving or sending other messages (i.e., CALL PROCEEDING, CONNECT ACKNOWLEDGE, or CONNECT message sent by the leaf node).

The root adds leaves to an existing connection when the link is in an active state by transferring ADD PARTY messages across its interface. Each ADD PARTY message has the same call reference value as in the original SETUP message and an end-point reference value. Following the transmission of an ADD PARTY message, the party state is the add-party-initiated state. Furthermore, the connection identifier, QOS, bearer capability, and ATM user cell rate used for the new party are the same as they are in the original call. Accordingly, the ADD PARTY message includes the following IEs, defined previously in the context of point-to-point signaling:

- AAL parameters (O);
- Called-party number (M);
- Called-party subaddress (O);
- Calling-party number (O);
- Calling-party subaddress (O);
- End-point reference (M);
- Broadband high-layer information (O);
- Broadband low-layer information (O);
- Broadband sending complete (O);
- Transit network selection (O).

If the network determines that the requested service can be provided, then it sets up a path to a new leaf (i.e., extends the current tree to the new node). After receiving an indication that the add-party request is accepted, the network side of the interface the root is connected to sends an ADD PARTY ACKNOWLEDGE message and enters the active party state for that party. Each party is identified by its end reference value. Upon receiving this acknowledgment, the calling user also enters the active party state.

For the incoming call (i.e., at a leaf node), the network indicates the arrival of an add-party request by transferring a SETUP or ADD PARTY message across the interface. The SETUP message is sent only if the link is in the null state and point-to-point signaling procedures are used. When the network sends an ADD PARTY message, it enters the add-party-initiated party state. Similarly, the user side enters the add-party-initiated party state upon receiving the ADD PARTY message. If the user accepts the connection, then it sends an ADD PARTY ACKNOWLEDGE message and enters the active party state. An add-party reject is sent and the null party state is entered if the user rejects the connection request.

In order to release the connection to a particular leaf, the root sends a DROP PARTY message to the network, which changes the party state to drop party initiated. This message prompts the network to release the end-point reference and to initiate procedures for dropping the party to the corresponding remote user. After releasing the end-point reference, the network sends the DROP PARTY ACKNOWLEDGE and enters the null party state. Upon receiving the DROP PARTY ACKNOWLEDGE, the user releases the end-point reference and enters the null party state. If all parties on the call at the interface are in the null party state, then the user releases the call by sending a RELEASE message (used for point-to-point signaling to release the call). When the network receives a RELEASE message, all parties that are in the drop-party-initiated or drop-party-received state enters the null state. Parties in add-party-received or add-party active party states are also cleared. A similar sequence of events takes place if the network initiates the DROP PARTY or RELEASE message.

7.2 Private NNI Signaling

Let us consider that an ATM end station attached to one private network is requesting a connection to another ATM end station attached to another (or the same if P-NNI is used internal to the network) private network as illustrated in Figure 7.12.

The end station that originated the connection request is referred to as *calling user,* and the destination end stations are referred to as *called users.* The direction from the calling user to the called user is referred to as the *forward direction.* The backward direction is from the called user to the calling user.

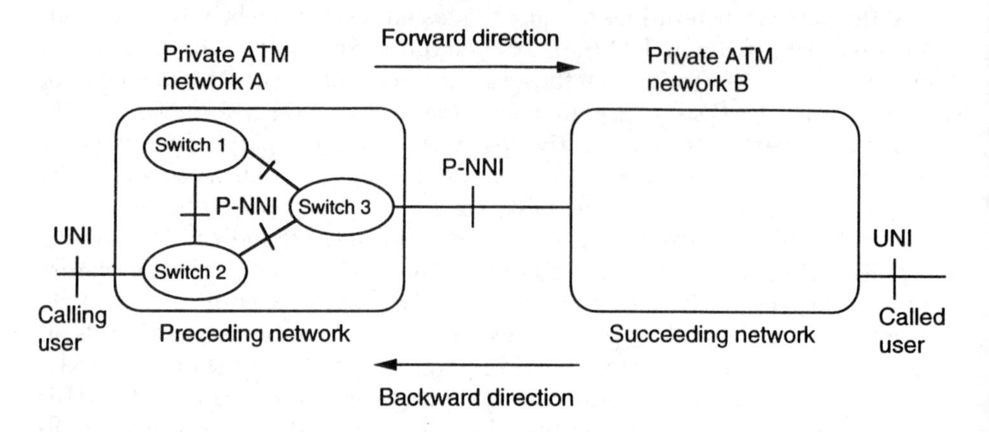

Figure 7.12 Private network-to-network interface.

A network that precedes another network from the calling user toward the called user is said to be the *preceding network*. A *succeeding network* comes after a preceding network along the same direction.

Both stations are attached to their respective networks across UNIs. Hence, the calling user uses UNI signaling to request a connection from its network. The network the called user is attached to uses UNI signaling to request a connection to the called user. Each network may use its own proprietary signaling to establish connections internally among its ATM switches. The interoperability between two private networks requires the standardization of this interface. P-NNI signaling is used to establish connections between each preceding and succeeding networks along the end-to-end path.

The P-NNI in private networks is the demarcation point between two switching systems. A switching system can be a single switch in an ATM network or it can be a collection of switches under one administration. In the former case, the P-NNI is a switch-to-switch interface, whereas it is an NNI in the latter case. Current network services emerging within the framework are not yet sufficient to fully manage and control a private network. This is further elaborated on in Section 7.4. Accordingly, for presentation purposes here, we use P-NNI to refer to an NNI. However, switch-to-switch interface use of P-NNI signaling is the same. A switch may be considered for signaling purposes as a network without any loss of generality. The ATM Forum P-NNI signaling implementation agreement specifies the procedures and messages used to dynamically establish, maintain, and release ATM connections at the P-NNI between two private networks. Figure 7.13 shows the P-NNI signaling structure.

The ATM Forum P-NNI specification includes both routing and signaling frameworks. P-NNI routing includes various "generic" network control functions, such as call admission, path selection, and topology distribution. P-NNI signaling is used to carry P-NNI routing-related protocol information as well as perform call control functions for connection establishment, management, and termination.

P-NNI routing is reviewed in Chapter 8. Upon receiving a P-NNI signaling message, the switch passes the signaling information to its protocol control entity for processing. It is the responsibility of this entity to process the control aspects of the request and inform the signaling entity how to proceed (i.e., accept or reject the connection request, the path to the next P-NNI node along the path). P-NNI signaling is being built on UNI signaling. In fact, a large portion of the P-NNI signaling messages listed in Table 7.7 are the same as UNI signaling messages. The main difference is that while the signaling procedures are asymmetric at the UNI, P-NNI signaling procedures are symmetric. In addition, P-NNI signaling messages are used to carry topology and routing information between P-NNI nodes.

UNI signaling procedures at the user side (end station) differ from the network side (switch). For example, connection identifiers are assigned by the

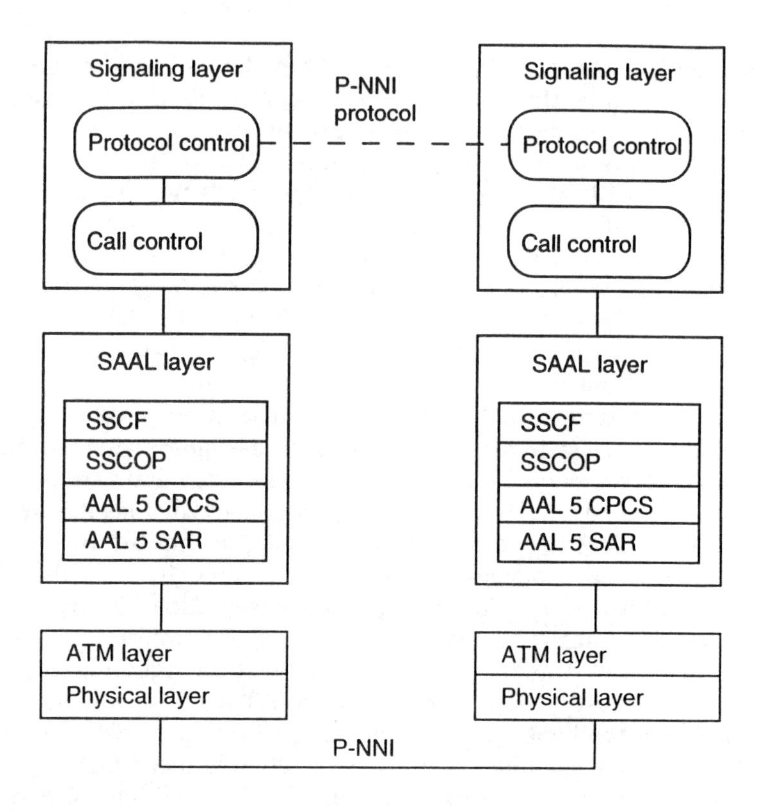

Figure 7.13 P-NNI signaling structure.

Table 7.7
P-NNI Signaling Messages

Point-to-Point Signaling Messages	*Point-to-Multipoint Signaling Messages*
CALL PROCEEDING	ADD PARTY
CONNECT	ADD PARTY ACKNOWLEDGE
CONNECT ACKNOWLEDGE	ADD PARTY REJECT
SETUP	DROP PARTY
RELEASE	DROP PARTY ACKNOWLEDGE
RELEASE COMPLETE	
STATUS	
STATUS ENQUIRY	

network only. When a SETUP message for a connection request is received by the network, the network needs to perform a set of functions including path selection, target location, and bandwidth allocation. When a destination node receives a SETUP message from the network, it essentially checks the compatibility information between the end stations (e.g., AAL and low-layer bearer capability). Hence, the procedures performed upon receiving a SETUP message differ depending on the recipient, user, and network.

In the case of a P-NNI, signaling procedures are required to be symmetric. That is, the same set of functions are performed when a SETUP from the preceding network is received whereby either network can be the preceding network depending on where the connection originated. P-NNI signaling is based on the network side procedures of UNI signaling.

The call state of the P-NNI is specific to each call, since there can be multiple calls simultaneously at the P-NNI at any given time. These states are listed in Table 7.8.

The relationship between the call states and connection establishment/release messages is illustrated in Figure 7.14.

Let us consider a successful call setup at a P-NNI. Call establishment is initiated when the preceding network sends a SETUP message. Upon sending this message, the preceding network enters the call-present state. When the succeeding network receives the SETUP message, it enters the call-initiated state and sends a CALL PROCEEDING message indicating that the connection setup request is received. Upon sending the CALL PROCEEDING message, the call state is changed to the call-proceeding-sent state. The succeeding network

Table 7.8
ATM Point-to-Point Call States

Call States	*Description*
Null	No calls exist
Call initiated	State of succeeding network after it has received the SETUP message from the preceding network
Call proceeding sent	After a network has acknowledged the receipt of the information necessary to establish a call
Call present	After a network has sent the SETUP message to the succeeding network but has not yet received a response
Call proceeding received	After the preceding network has received acknowledgment that the succeeding network has received the SETUP request
Active	When the ATM connection is established
Release request	When a network has received a RELEASE message from the other network to release the connection
Release indication	After a network has received a RELEASE message while it is waiting for a response from call control

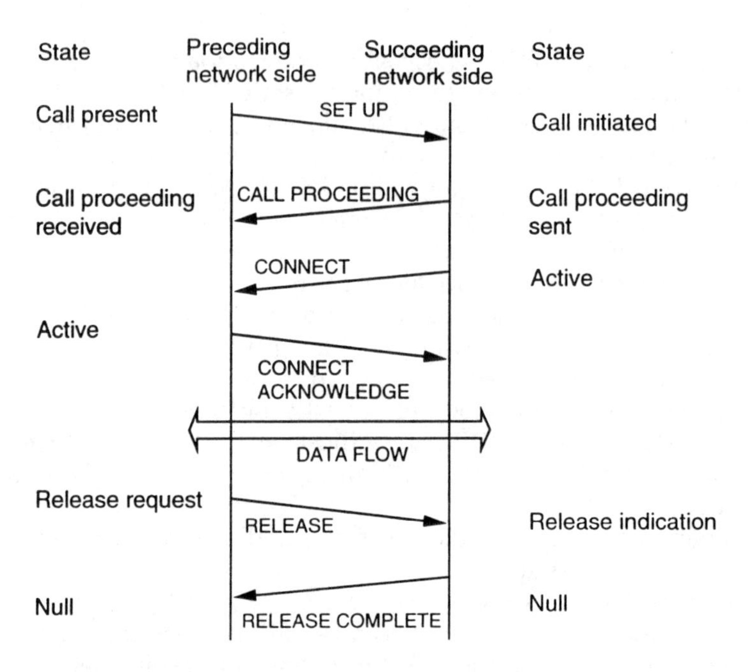

Figure 7.14 P-NNI finite state machine.

eventually sends a CONNECT message and enters the active state. The CONNECT message indicates that the connection request is accepted by the called user and the end-to-end path from the P-NNI to the called user is established. Upon receiving the CONNECT message, the preceding network (which requested the connection establishment) sends back a CONNECT ACKNOWLEDGE message and enters the active state.

The established connection in the network is used for some period of time until one of the two networks (due to a request from one of the end users) issues a RELEASE message. Upon receiving this message, the other network enters the release-request state, sends a RELEASE COMPLETE message, and enters the null state. Upon receiving the RELEASE COMPLETE message, a network changes its call state to null state. At this stage, network resources reserved for the corresponding connection are released.

Figure 7.15 illustrates the sequence of P-NNI signaling messages exchanged between the two sides of P-NNIs for a successful end-to-end connection across three networks.

Another feature that is different in P-NNI signaling than in UNI signaling is the incorporation of the crankback function (refer to Chapter 8 for details). In P-NNI routing, the end-to-end path is determined by the originating switch. There is no guarantee that an intermediate network along the path will have

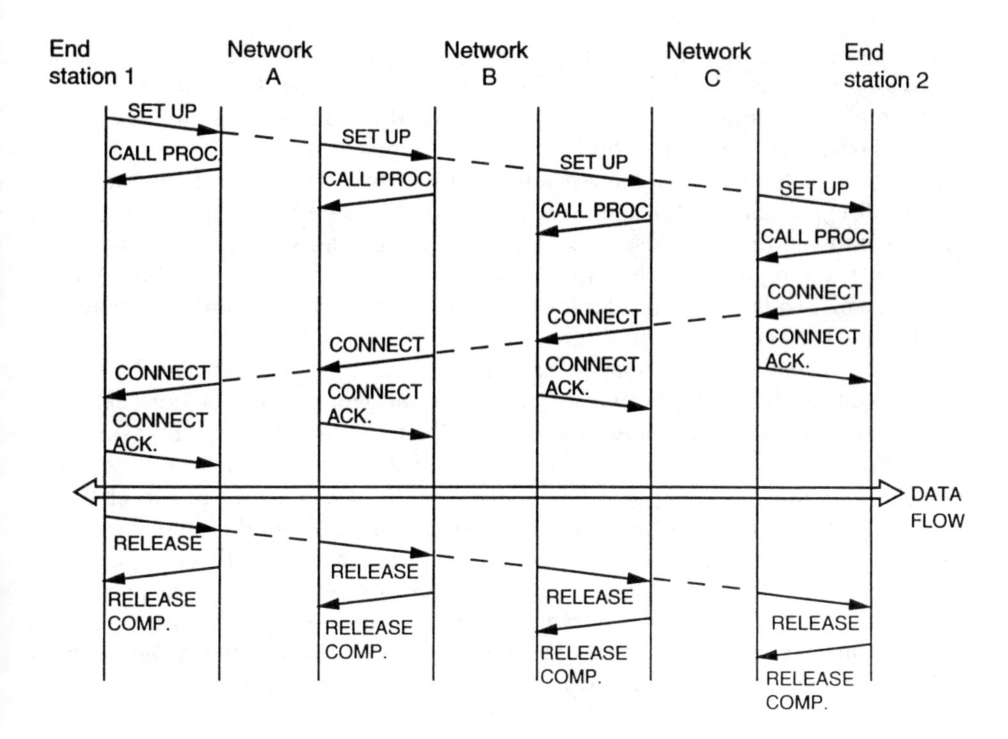

Figure 7.15 Call establishment and release signaling at the P-NNI.

the resources to support the requested connection. When a call connection request is rejected (i.e., insufficient network resources), the reason for the rejection is sent toward the originating network for alternate routing. A crankback IE is used in RELEASE, RELEASE COMPLETE, and ADD PARTY REJECT messages to indicate the node or link where the call (or the party in point-to-multipoint connection setup) was not be accepted.

P-NNI signaling work at the ATM Forum has not yet been completed [1].

7.3 PUBLIC NNI SIGNALING

Service provider networks are quite different from private networks. Traditionally, service provider networks have been circuit-switching networks and provided mainly voice services (as well as fax and modem traffic). Essentially, a 64-kbps voice channel is dedicated to each connection, and voice channels are time-division multiplexed on digital transmission systems. With ISDN, multirate channels supporting higher rates became possible based on multiple 64 kbps given to an application (i.e., low-bit-rate video).

Although things have changed during the last decade or so, typical service provider networks (public networks) have had a hierarchical structure consisting of five classes of switches, as illustrated in Figure 7.16.

Customers are connected to level 5 switches (end offices). Switches that are higher in the hierarchy aggregate the traffic from the lower level switches. Hence, the switches at the higher levels in the hierarchy are larger and support higher bandwidth transmission links. At each level, however, switches operate only at the single-voice-channel level. The current trend is to minimize the number of levels in the hierarchy and move toward a distributed network of fully interconnected switches.

End-to-end connections are established, managed, and released by signaling. Signaling in public networks used to be inchannel. That is, channels that were used to carry voice frames were are also used to carry signaling information. With inchannel signaling, the signaling information transfer rate is rather limited; signaling takes place when there is no user activity in the channel. Another disadvantage is the delay in establishing end-to-end connections. The signaling information to set up a channel can be transmitted only after a channel is set to the next switch along the path of the connection (i.e., the process of setting up the channel and forwarding signaling messages to the next hop cannot be overlapped in time). One consequence of these two disadvantages of inchannel signaling is that different types of services that can be deployed in the network are limited by the signaling capability.

To address these problems, common channel signaling (CCS) was developed. Control signals in the CCS framework are transmitted over signaling links

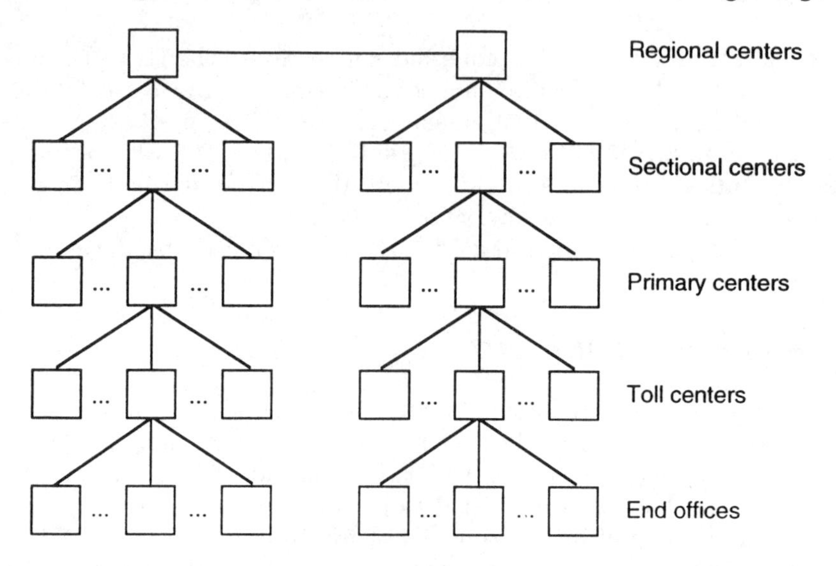

Figure 7.16 Hierarchical structure of public networks.

that are dedicated to carrying signaling information only. These signaling links are shared by a large number of channels, making it cost-efficient to use.

7.3.1 Common Channel Signaling

CCS is a specialized data communication architecture designed for the transfer of signaling information between processors in telecommunications networks. A telecommunication network is composed of a number of switching and processing nodes interconnected by transmission links. To communicate using CCS, each one of these nodes requires the implementation of the necessary *within-the-node* features of ITU-T signaling system 7 (SS7).

An SS7 network is composed of a set of signaling points with signaling links interconnecting the signaling points. A signaling point is a switching and processing node that implements the required within-the-node features of SS7. Signaling links are used to convey the signaling messages between two signaling points.

Various types of signaling points defined in SS7 include switching centers (SC), OAM centers, service control points (SCP), and signaling transfer points (STP). A switching center is a network node that switches traffic from incoming links to outgoing links. In circuit switching, this function is performed at the TDM-slot level; that is, the switching function performs the mapping from an (incoming link, slot number) to (outgoing link, slot number). The STP exclusively transfers signaling messages from one signaling link to another (i.e., a node that is not the origin or the destination of the signaling message). The SCP is a real-time processing system hosting one or more applications that provide enhanced network services. Figure 7.17 illustrates a typical public network topology with CCS.

In SS7 terminology, when two nodes are capable of exchanging signaling messages through the signaling network, a *signaling relation* is said to exist between them. A signaling route in SS7 terminology is a *predetermined* path (subject to failures) the signaling messages take in the signaling network between each pair of origin-destination signaling points with a signaling relation. The term *signaling mode* refers to the association between the path taken by a signaling message and the signaling relation to which the message refers. Three types of signaling modes are defined in SS7: associated mode, nonassociated mode, and quasiassociated mode.

In the associated mode of signaling, the messages exchanged between two adjacent signaling points are conveyed over a signaling link that directly interconnects them to each other. In the case of nonassociated mode of signaling, the messages relating to a particular signaling relation are conveyed over two or more signaling links in tandem passing through one or more signaling points other than the origin and the destination of the messages. The quasiassociation mode is a limited case of the nonassociated mode in which the path taken by

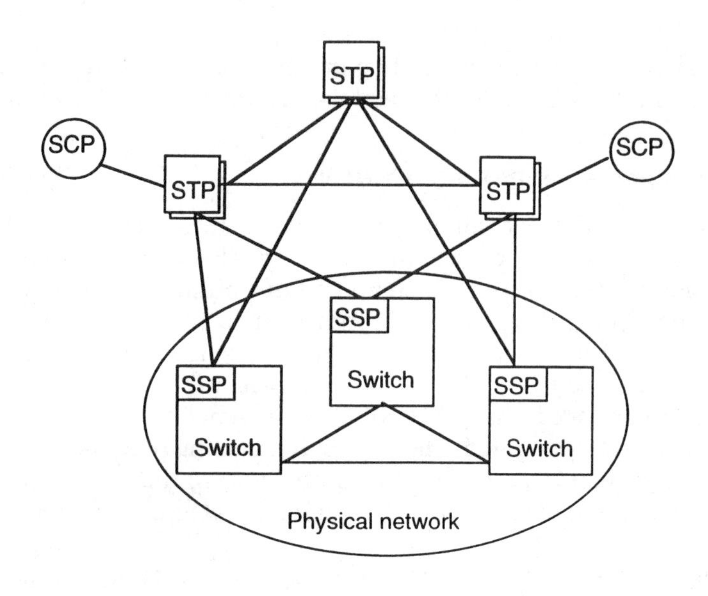

Figure 7.17 A public network with CCS.

signaling messages relating to a particular signaling relation is fixed, except for rerouting caused by failure and recovery events.

The accessibility of user functions through a signaling network depends on the signaling modes. In the case of the associated signaling mode, only user functions located at adjacent signaling points may be accessed. When quasiassociated signaling is employed, user functions located at any signaling point may be accessed provided that the corresponding message routing data are present.

We now define the primary components of a typical public network: *service switching points* (SSP), SCPs, *switched management systems* (SMS), STPs, *intelligent peripherals* (IP), and *vendor feature nodes* (VFN).

SSP is a network switch that contains a table and additional software-based functions. It takes information provided by the caller and searches its table. A match between the caller's information and an entry in the table triggers a specific action in the SSP. For enhanced network services, the action would be to obtain instructions from an SCP. This is achieved by sending and receiving query messages to the SCP containing the service application.

An example is the 800 service (also known as free telephone) support. When a calling user dials an 800 number, the dialed number does not correspond to a particular telephone number. For example, an airline reservation system may have centers throughout the United States with a single 800 number. Depending on where the call originated (geographically), the call may be routed to the closest center. Hence, upon receiving an 800 number, the switch may send a query to an SCP to find out where the call will be routed to.

SCP is a real-time processing system hosting one or more applications that provide enhanced services. It is an addressable entity or a node in the CCS network. Its primary task is to provide highly reliable access to and processing of database inquiries. Examples of enhanced services include service intelligence for 800 calls, 900 calls, and credit card calls.

The IP is a stand-alone computer that provides specific network capabilities such as announcement, digit collection, voice recognition, and speech synthesis. The IP performs its functions as a result of commands from the SCP. An IP has at least one direct connection to an SSP, but it may be accessed by any SSP in the network.

The SMS provides a framework by which new services can be added to the network. It detects traffic overloads and failure conditions in the network and provides surveillance activities, alerting the SCP of node alarms, network element failures, and trouble spots. The SMS is mainly used to update various databases and tables in the network, primarily those residing in the SSP, IP, and SCP.

The STP is an essential part of the CCS network. It is a packet switch that routes signaling traffic among the various network elements by using the address associated with each message. When a call triggers a transaction to an SCP, the SSP has to suspend processing of the caller's connection. The CCS network enhances the routing time so that the SCP obtains the information quickly and is able to route the appropriate results more quickly.

The VFN is an off-network node connected to a service provider network. VFNs may be third-party service providers offering services that the service providers do not offer. The VFN allows a service provider to participate in introducing services required by end users, stimulates network traffic, and provides new revenue from third-party service providers.

The SS7 structure is composed of two major functions: MTP and user part. The overall function of the MTP is to serve as a transport system providing reliable transfer of signaling messages between the communicating user functions. The term user in SS7 is used to refer to any functional entity that uses the transport capability provided by the message transport part. A user part is composed of a set of functions of a particular type of a user that is part of the CCS system. Various functions defined in SS7 and the corresponding series of Q.700 to Q.795 Recommendations are summarized in Table 7.9.

7.3.2 Overview of the SS7 Architecture

Figure 7.18 illustrates the SS7 architecture and the relationship between the various functional blocks.

Table 7.9
SS7 Functions and Related ITU-T Recommendations

SS7 Function	*Recommendation*
Message transfer part (MTP)	Q.701–Q.704, Q.706, Q.707
Signaling connection control part (SCCP)	Q.711–Q.714, Q.716
Transaction capabilities (TC)	Q.771–Q.775
Operations maintenance and administration part (OMAP)	Q.795
Telephone user part (TUP)	Q.721–Q.725
Data user part (DUP)	Q.741
ISDN user part (ISUP)	Q.761–Q.764, Q.766
B-ISDN user part (B-ISUP)	Q.2761
Supplementary services	Q.730

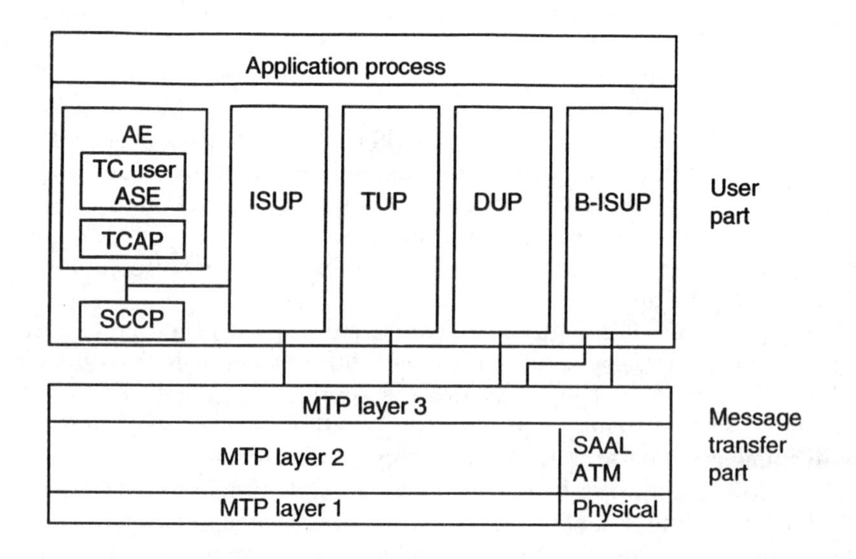

Figure 7.18 SS7 architecture: SCCP, application element, transaction capabilities application part, and SAAL.

The application process is a set of functions and features that supports a particular network requirement. It may be a coordinator of specific aspects of network operation, an individual service, or a supplementary service control function. Application elements (AE) are the elements representing the communication functions of the application processes that are pertinent to internodal communication using SS7 application protocols.

The user part currently includes the following functions (excluding B-ISUP, which is discussed later):

- TUP;
- DUP;
- ISUP;
- TC;
- SCCP.

TUP recommendations define telephone signaling functions used in SS7 for international telephone call control signaling. TUP standards include the specifications of telephone signaling messages, their encoding, signaling procedures, and cross-office performance. DUP recommendations define the protocol to control interexchange circuits used on data calls, data call facility registration, and cancellation. ISUP encompasses signaling functions required to provide switched services and user facilities for voice and nonvoice applications in ISDN. The ISDN recommendations series defines ISDN network signaling messages, their encoding and signaling procedures, and cross-office performance.

Similarly, TC provides the means to establish noncircuit related communication between two nodes in the signaling network. The TC recommendation series defines the TC signaling messages and their encoding and signaling procedures. The TC user part consists of two elements: *transaction capabilities application part* (TCAP) and *intermediate service part* (ISP). The ISP provides connection-oriented TC services. Other services using connectionless network service do not use ISP. The TCAP provides the mechanisms for transaction-oriented (as opposed to connection-oriented) applications and functions.

The SCCP provides the means to control logical signaling connections in an SS7 network and transfer signaling data units across the SS7 network with or without the use of logical signaling connections. The SCCP provides a routing function that allows signaling messages to be routed to and from a signaling point based on, for example, dialed digits. The SCCP also provides a management function that controls the availability of the subsystems and broadcasts this information to other nodes in the network requiring this knowledge.

The original (non-ATM) MTP illustrated in Figure 7.18 is composed of three layers: MTP layers 1, 2, and 3. MTP layer 1 defines the physical, electrical, and functional characteristics of a signaling data link and the means to access it. The signaling data link is a full-duplex physical link used solely by SS7 traffic. The maximum bit rate currently standardized for such links is 64 kbps. MTP layer 2 defines the functions and procedures for and relating to the reliable transfer of signaling messages over a single signaling data link. This is the data link layer protocol used in SS7 and designed for links with large bit error rates.

The major functions provided by MTP layer 2 are alignment and delimitation using start and end of frame delimiters, error detection, error correction, error monitoring, and flow control. Error detection is performed on all frames using CRC. Error correction is accomplished either through a go-back-N scheme using positive/negative acknowledgment transmission or positive acknowledgment cyclic retransmission. Error monitoring is used to monitor the quality of transmission on signaling links. MTP layer 2 flow control is used by the receiving end of a signaling link to inform the transmitting end of congestion or a failure condition. In the case of congestion, the receiving end sends a special message and withholds acknowledgment of all incoming messages. When level 2 detects a failure 3 failure condition, it informs the STPs it can reach that it cannot deliver messages to level 3.

A signaling message carries information pertaining to a call, such as management transaction, which is transferred as an entity by the message transfer function. Figure 7.19 illustrates the message signal unit (MSU) format in SS7.

The flags at the two ends of an MSU delimit the signal unit at both ends with a unique pattern. The check bits (CK) field contains an error detecting code (based on CRC-16) used by the receiver to detect transmission errors in the MSU.

The service information field (SIO) includes a 4-bit service indicator (SI) and a 2-bit network indicator (NI). The SI is used to determine the user of the message, whereas NI specifies whether it is an international or a national network service.

The signaling information field (SIF) consists of a user part and the routing label. The user part carries data from some SS7 application (i.e., B-ISUP) or network management data. The routing label includes the origin point code (OPC) and the destination point code (DPC) (addresses). Both point codes are 14 bits long. In MTP management messages, the routing label also includes the signaling link code (SLC), whereas this field is used for signaling link selection (SLS) in other user part messages. The SLC indicates the particular signaling link which the management message refers to among those interconnecting two involved signaling points. SLS is used to perform load sharing among different signaling links. This field is 4 bits long.

Every signaling point (SP) is allocated its own unique point code. The combination of the OPC and DPC determines the signaling relation. The DPC is used by the receiving SP to determine whether the message is addressed to that SP or requires further routing by means of the STP signal transfer capability. The DPC, defined by the MTP user, is also used by the MTP routing function to direct outgoing messages toward their destination. The OPC may either be determined by the MTP user or it may be fixed and inserted at MTP layer 3.

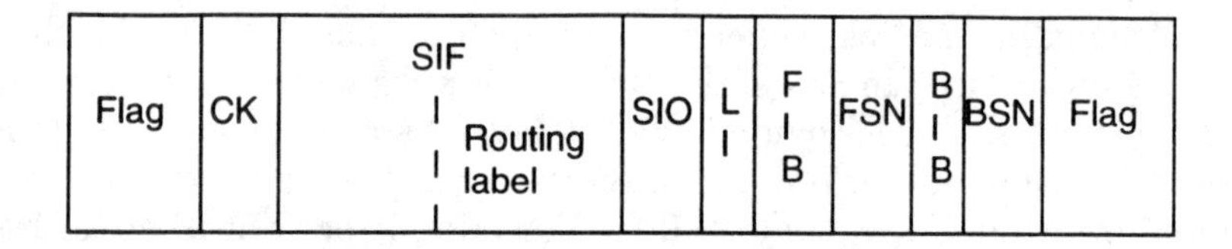

Figure 7.19 SS7 signaling message structure.

The forward sequence number (FSN) is used to uniquely number MSUs. It is a 1-byte field providing modula 127 sequencing. The forward indicator bit (FIB) is used to indicate whether the MSU contains an original user part or a retransmission. The backward sequence number (BSN) contains the number of the last MSU received successfully from the other side providing an acknowledgment. The backward indicator bit (BIB) is used to indicate negative acknowledgment by inverting the BIB. That is, the BIB value is maintained in a number of MSUs to indicate positive acknowledgment until an error is detected. When this occurs, the BIB is inverted in the next outgoing MSU.

Finally, the LI field is used to specify the length of the MSU following the LI field. It is also used to indicate the type of signal unit format.

MTP layers 1 and 2 are being replaced by ATM physical, ATM, and SAAL layers in the emerging SS7 architecture for B-ISDN. The design and use of various functions for links that are restricted to 64 kbps with large bit error characteristics are restrictive in ATM networks and B-ISDN services.

B-ISDN provides a VC-based transport mechanism based on ATM that can be used for transporting signaling messages as well as all other types of user traffic (i.e., voice, video, and data) and management data. In this environment, ATM VCs may be used as logical signaling links to route signaling messages in the network. Hence, there may not be any need to have a separate physical signaling network in B-ISDNs.

In the transition from ISDN to B-ISDN, two types of network signaling links (MTP signaling links and ATM virtual signaling channels) are likely to coexist. This would require interoperability between the two signaling networks, which is envisioned to be achieved at MTP layer 3. Keeping the MTP layer 3 common in the emerging signaling framework allows the use of current network processors (i.e., SCPs). However, various functions defined in the current MTP layer 3 are not needed in ATM networks and can be eliminated, while new functions such as multicast need to be defined to support emerging B-ISDN applications.

7.3.2.1 MTP Layer 3

MTP layer 3 defines a set of transport functions and procedures common to and independent of the operation of signaling links. These functions are summarized in two categories:

- *Signaling message handling* functions that direct signaling messages to the proper signaling link or user part;
- *Signaling network management* functions that, on the basis of predetermined data and information about the status of the signaling network, control the current message routing and configuration of the signaling

network facilities. These functions also include the ability to control the reconfigurations and other actions to preserve or restore the normal message transfer capability.

MTP recommendations specify methods by which different forms of signaling networks can be established. All information to be transferred by the MTP are assembled into messages. From a transportation point of view, each message is self-contained and handled individually. Figure 7.20 illustrates various functions defined in MTP layer 3.

The purpose of the signaling message handling functions is to ensure that the signaling messages originated by a particular user part at an originating point are delivered to the same user part at the destination point. Depending on the message route, this delivery may be made through a signaling link directly interconnecting the origin and destination signaling points or via one or more STPs. The signaling message handling subsystem includes message routing, message discrimination, and message distribution functions.

Each message contains a routing label, which assumes that each signaling point in a signaling network is allocated a unique code (within a domain) according to a code plan established for the purpose of labeling. *Message routing* is the process of selecting the signaling link used for each signaling message. It is based on the analysis of the routing label in combination with the predetermined routing data at the signaling point concerned. Message routing is destination-code-dependent.

Load sharing may be used when there is more than one signaling link connecting the two signaling points with a signaling relation. Two basic cases of load sharing are defined:

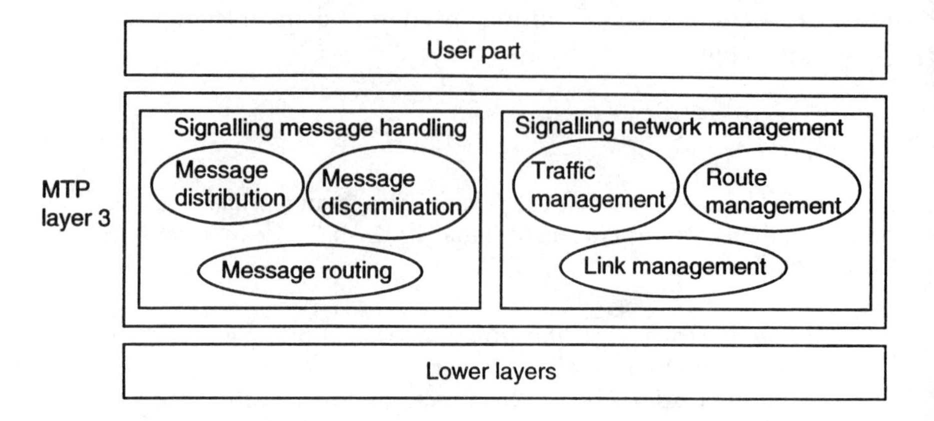

Figure 7.20 Signaling message handling functions.

- Load sharing between links belonging to the same link set;
- Load sharing between links not belonging to the same link set.

Each succession of signaling links that may be used to convey a message from the originating point to the destination point constitutes a *message route*. In SS7, the message route taken by a message with a particular routing label is predetermined and, at a given point of time, fixed. In the event of failures in the signaling network, the routing of messages that were previously using the failed message route is modified in a predetermined manner under the control of the signaling traffic management function.

Message discrimination is the process of determining whether or not the signaling point is the destination point of arriving messages. This decision is based on the analysis of the destination code in the routing label of the message. If the signaling point is the destination point, then the arriving message is delivered to the message distribution function. If it is not the destination point, then the message is routed to the next hop along the message route to its destination signaling point.

Message distribution is the process of determining at the destination point of the message which user part or level 3 function the message is destined for.

The purpose of the signaling network management functions is to provide reconfiguration of the signaling network in the case of failures and to control traffic in case of congestion. This requires communications between signaling points (and the STPs) concerning the occurrence of the failures. In some circumstances it may be necessary to activate and align new signaling links in order to restore the signaling traffic capacity between two signaling points. When the faulty link or signaling point is restored, the opposite actions and procedures take place in order to reestablish the normal reconfiguration of the signaling network. Signaling network management functions include signaling traffic management, signaling route management, and signaling link management.

The tasks of the signaling traffic management function include:

- Control of message routing, which includes the modification of message routes when required in the event of failures to preserve accessibility of all destination points and the restoration of normal routing when failure conditions are removed;
- Control of the resulting transfer of rerouted signaling traffic in a manner that avoids irregularities in the message flow;
- Flow control.

Message routing control is based on the analysis of predetermined information about all potential routes between any two origin-destination signaling point pairs with signaling relation and the status of the resources in the signaling network. Once the current message route becomes unavailable, the transfer of the signaling traffic is performed in accordance with specific procedures: *changeover, changeback, forced rerouting,* and *controlled rerouting.* These procedures are designed to avoid missequencing or multiple delivery of messages, to the extent it is possible based on current circumstances.

The changeover procedure transfers signaling traffic from one signaling link to one or more different signaling links when the link in use fails or is required to be cleared of traffic. When a failed regular signaling link becomes available, the changeback procedures are used to transfer signaling traffic from the alternate signaling link to the original signaling link. A forced rerouting procedure transfers signaling traffic from one signaling route to another when the route in use fails or is required to be cleared of traffic. When a failed regular signaling route becomes available, the controlled rerouting procedures are used to transfer signaling traffic from the alternate signaling route to the original one.

A signaling network, in general, has a signaling traffic capacity much higher than necessary to support normal traffic load, mainly to satisfy stringent reliability requirements. However, in overload conditions that may arise due to resource failures or extremely high traffic periods, the signaling traffic management function takes flow control actions to minimize the problem. For example, when a particular signaling link is overloaded, the flow control function informs the route management function, which in turn may use alternate paths. Congestion at a signaling link is determined by monitoring the link buffer occupancy. A number of buffer thresholds are defined to detect the onset of congestion.

Signaling route management is used only in the quasiassociated mode of signaling. The purpose of this function is to ensure a reliable exchange of information between the signaling points about the availability of signaling links. An STP, for example, may send messages indicating unavailability of a signaling point via that STP, thus enabling other signaling points to stop routing messages to an incomplete route. The different procedures pertaining to signaling route management include *transfer-prohibited, transfer-allowed, transfer-restricted, transfer-controlled, signaling-route-set-test,* and *signaling-route-set-congestion-test* procedures.

The task of the signaling link management function is to control the locally connected set of signaling links. It supplies to the traffic management functions information on the availability of signaling links and initiates and controls actions aimed at restoring the normal availability of failed signaling links. The different procedures pertaining to signaling link management are *restoration, activation, deactivation,* and *inactivation* of signaling links.

7.3.2.2 MTP Layer 3 in ISDN versus B-ISDN

In the near future, SS7 networks using both ATM-based and MTP layer 2–based technologies may coexist depending on network requirements such as access to remote nodes. SCPs in ATM networks are more likely to offer sophisticated services. The TCAP and SSCP are likely to be used in public ATM networks to support TC between switches and SCPs and between remote switches. No change of these protocols is required if they are used on top of MTP layer 3.

Considering MTP layers 2 and 1 being replaced by SAAL, ATM, and physical layers of B-ISDN architecture, various functions provided in MTP layer 3 on MTP layer 2–based technologies versus what might happen in ATM-based technologies are listed in Table 7.10.

7.3.2.3 MTP Layers 1 and 2 in ISDN versus B-ISDN

In the current SS7 protocols, the maximum message length is 272 bytes, including MTP layer 3 headers. The maximum standardized signaling bit rate is 64 kbps. In B-ISDN, emerging applications are likely to generate much longer

Table 7.10
MTP Layer 3 Subset in ATM Networks

	MTP Layer 2–Based	*Need in B-ISDN*
Message discrimination		Yes
Message routing	SLS	Yes
	Load sharing	Yes
	Link set	No
Signaling link management	Link activation	Yes
	Link restoration	Yes
	Link deactivation	Yes
	Linkset activation	No
	Link block/unblock	Yes
	Automatic allocation of signaling terminals	No
Signaling traffic management	Changeover	Yes
	Changeback	Yes
	Forced rerouting	No
	Controlled rerouting	No
	MTP restart	No
	Management inhibit	Yes
	Signaling traffic flow	No
Signaling route management	Transfer prohibited	No
	Transfer allowed	No
	Transfer restricted	No
	Signaling route test	No
	Signaling route congestion test	No

messages and require higher bandwidth in the network mainly due to response time requirements.

Error detection and correction for B-ISDN signaling messages require specific functions at the AAL SSCS (SAAL). The SAAL used at the UNI has the same structure as that at the NNI. It is composed of the AAL 5 CPCS, the SSCOP, and the SSCF. The SAAL provides reliable transmission of signaling messages and flow control. The SAAL together with the help of the ATM layer corresponds to MTP layer 2 while taking advantage of having more reliable transmission links in the network.

Error monitoring is one of the unique features of the ISDN signaling framework. A similar function is required in B-ISDN due to the critical nature of signaling transport and its reliability requirements.

7.3.3 B-ISDN User Part

The B-ISUP is the SS7 user part protocol that provides the signaling functions required to support basic bearer services and supplementary services for B-ISDN applications. The B-ISUP is being developed for international applications as an NNI. However, it is also suitable for national applications as a network node interface. Most signaling procedures, IEs, and message types specified for an NNI are also required in typical national applications.

Furthermore, coding space is reserved in order to allow national administrations and private operating agencies to introduce network-specific signaling messages and elements of information within the standardized protocol structure.

Table 7.11 illustrates the set of ITU-T recommendations for the B-ISUP.

The B-ISUP can be viewed as a set of functional blocks, each representing a particular function, as illustrated in Figure 7.21.

The B-ISUP makes use of the services provided by MTP layer 3. Information is transferred to and from the MTP in the form of parameters carried by primitives. The general form of a primitive is illustrated in Figure 7.22, where *X* specifies the service provided, *generic name* describes an action by *X*, *specific name* indicates the purpose of the primitive, and the *parameter* contains the elements of supporting information transferred by the primitive.

The four primitives used across the B-ISUP-MTP interface are the MTP_TRANSFER, MTP_PAUSE, MTP_RESUME, and MTP_STATUS. MTP_TRANSFER is used either for the B-ISUP to access message handling functions of the MTP or by MTP to deliver signaling message information to the B-ISUP. MTP_PAUSE is used by the MTP to indicate its inability to transfer messages to the destination. MTP_RESUME is used by the MTP to indicate its ability to resume unrestricted transfer of messages to the destination. Finally, MTP_STATUS is used by the MTP to indicate that the signaling route to a

Table 7.11
ITU-T Recommendations for B-ISDN User Part

Recommendation	*Description*
Q.2761	*Functional Description of B-ISUP* Specifies an overview of the signaling capabilities and functions required to support basic bearer services and supplementary services for capability set 1 B-ISDN applications
Q.2762	*General Functions of Messages and Signals of B-ISUP of SS7* Describes the elements of signaling information and their function used by the B-ISUP to support basic bearer services and supplementary services for capability set 1 B-ISDN applications
Q.2763	*B-ISUP Formats and Codes* Specifies the formats and codes of B-ISUP messages and parameters required to support basic bearer services and supplementary services for capability set 1 B-ISDN applications
Q.2764	*SS7 B-ISUP Basic Call Procedures* Describes the basic B-ISUP signaling procedures for the setup and cleardown of national and international B-ISDN release 1 network connections
Q.2730	*B-ISUP Supplementary Services* Describes the supplementary services supported by B-ISUP capability set 1

specific destination is congested or that the B-ISUP at the destination is unavailable.

7.3.3.1 B-ISUP Signaling Messages

Recommendation Q.2762 describes the elements of signaling information and their functions used by the B-ISUP to support basic bearer services and supplementary services for capability set 1 B-ISDN applications. Capability set 1 includes the following features:

- Demand (switched virtual) channel connections;
- Point-to-point switched channel connections;
- Connections with symmetric or asymmetric bandwidth requirement;
- Single connection (point-to-point) call;
- Basic signaling functions via protocol messages, IEs, and procedures;
- Class X, class A, and class C ATM transport services;
- Request indication of signaling parameters;
- VCI negotiation;
- Out-of-band signaling for all signaling messages;

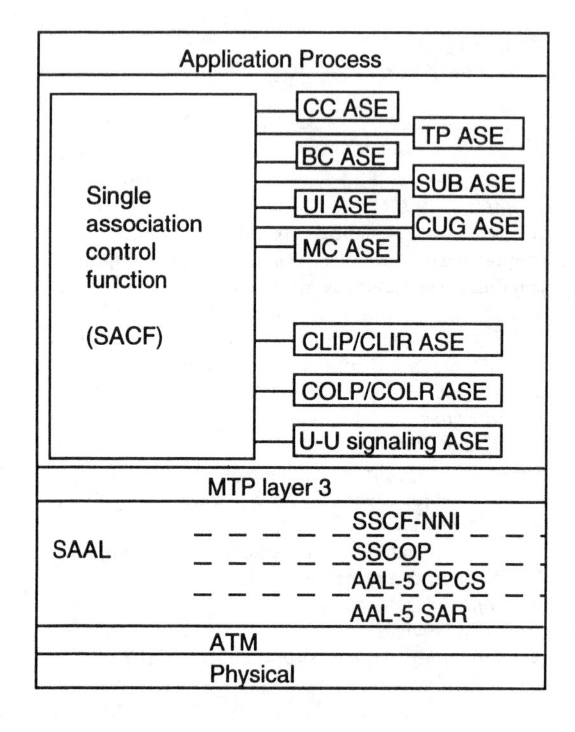

ASE: application service element
CC: call control
BC: bearer control
SUB: subaddressing
UI: unrecognized information
CUG: closed user group
MC: maintenance
CLIP: calling line identification presentation
CLIR: calling line identification restriction
COLP: connected line identification presentation
COLR: connected line identification presentation
U-U: user to user

Figure 7.21 B-ISUP specification model.

X	Generic name	Specific name	Parameter

Figure 7.22 Format of the primitives used between B-ISUP and MTP layer 3.

- Error recovery;
- Public UNI addressing formats for unique identification of ATM end points;
- End-to-end compatibility parameter identification;
- Signaling interworking with ISDN and provision of ISDN services.

Table 7.12 lists the B-ISUP messages and gives a brief description of each.

Recommendation Q.2763 specifies the formats and codes of B-ISUP messages and their parameters required to support basic bearer services and supplementary services for capability set 1 B-ISDN applications. Figure 7.23 illustrates the B-ISUP message format.

The *routing label* includes the origin and destination codes as well as the signaling link selection field. For each virtual connection, the same routing

Table 7.12
B-ISUP Messages

B-ISUP Message	*Description*
Address complete (ACM)	Sent in the backward direction indicating that all the address signals required for routing the call to the called party have been received
Answer (ANM)	Sent in the backward direction indicating that all the address signals required for routing the call to the called party have been received and the call has been answered
Blocking acknowledgment (BLA)	Sent in response to a BLO message indicating that the VP is blocked
Blocking (BLO)	Sent for maintenance purposes to the exchange at the other end of a VPC to cause an engaged condition of that resource for subsequent calls outgoing from that exchange
Confusion (CFN)	Sent in response to any message if the exchange does not recognize the message or parts of it
Call progress (CPG)	Sent during the setup or active phase of the call indicating that an event of significance to the originating or terminating access has occurred
Forward transfer (FOT)	Sent when the outgoing international exchange operator wants to help an operator at the incoming international exchange (i.e., operator assistance)
IAM acknowledge (IAA)	Sent in response to an IAM message indicating that IAM has been accepted and the requested bandwidth is available in both directions
Initial address (IAM)	Sent to initiate seizure of an outgoing VC and to transmit number and other information relating to the routing and handling of the call
IAM reject (IAR)	Sent in response to an IAM message indicating call refusal due to resource unavailability
Network resource management (NRM)	Sent to modify network resources associated with a call
Reset acknowledge (RAM)	Sent in response to an RSM message indicating that the resources have been released
Release (REL)	Sent to indicate that the call/connection is released due to the reason supplied and that the resources are available for new traffic upon receipt of the RLC message
Resume (RES)	Sent to indicate that the calling or called party, after being suspended, is reconnected
Release complete (RLC)	Sent in response to a REL message when the resources of the call/connection concerned have been made available for new traffic
Reset (RSM)	Sent to release a virtual connection when a REL or RLC message is inappropriate
Subsequent address (SAM)	Sent following an IAM message to convey additional called-party information

Table 7.12 (Continued)
B-ISUP Messages

B-ISUP Message	*Description*
Segmentation (SGM)	Sent to convey an additional segment of an overlong message
Suspend (SUS)	Sent to indicate that the calling or called party has been temporarily disconnected
Unblocking acknowledgment (UBA)	Sent in response to an UBL message indicating that the resource has been unblocked
Unblocking (UBL)	Sent to the exchange at the other end of a VPC to cancel the engaged condition of the resource caused by a previously sent BLO message
User part available (UPA)	Sent as a response to a UPT message to indicate that the user part is available
User part test (UPT)	Sent to test the status of a user part marked as unavailable for a signaling point
User-to-user information (USR)	Used for the transport of user-to-user signaling independent of call control messages

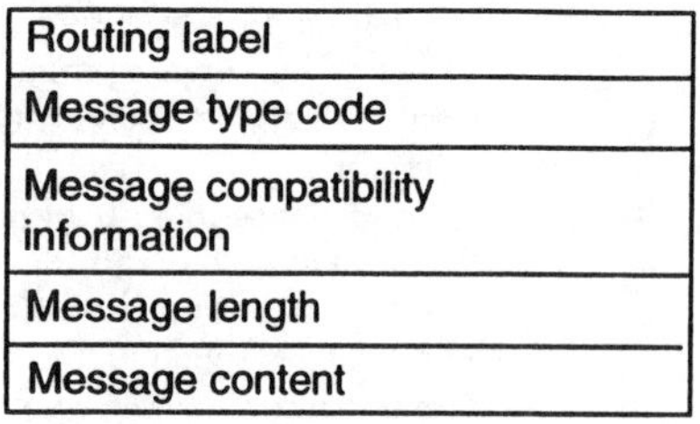

Routing label
Message type code
Message compatibility information
Message length
Message content

Figure 7.23 B-ISUP message format.

field is used for each message transmitted for that connection. The 1-byte *message type code* uniquely defines the function and the format of each B-ISUP message. The *message compatibility information* field is a 1-byte field that specifies the behavior of a switch if the message is not understood. The *message length* field indicates the number of bytes included in the B-ISUP message content. The message content of each message may contain a number of parameters. The generic format of a parameter is illustrated in Figure 7.24.

The content of each parameter contains a number of subfields specific to each parameter. Each field is defined similarly to the corresponding fields in the B-ISUP message format.

7.3.3.2 B-ISUP Procedures

The generalized model for the B-ISUP basic call application process is illustrated in Figure 7.25. The B-ISUP application entity (B-ISUP AE) provides all

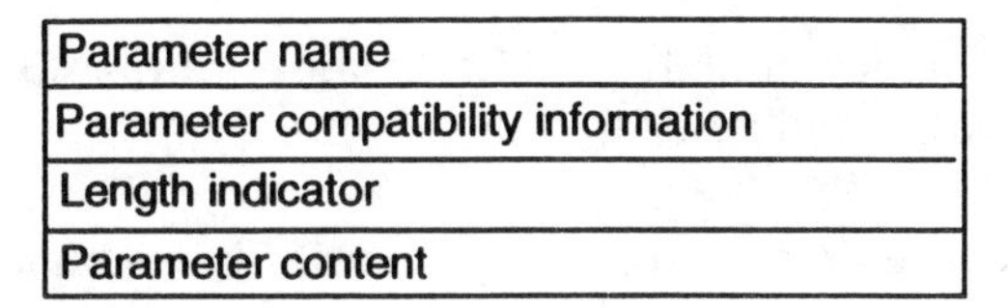

Parameter name
Parameter compatibility information
Length indicator
Parameter content

Figure 7.24 B-ISUP message parameter format.

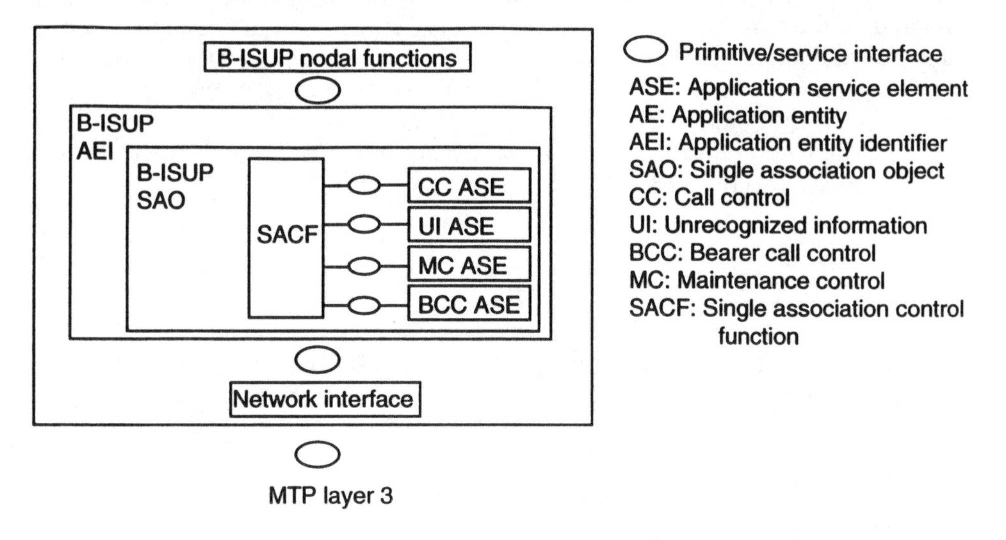

Figure 7.25 B-ISUP specification model.

the communication capabilities required by the B-ISUP nodal functions. An instance of the B-ISUP AE is created for each signaling association required, identified by a unique signaling identifier (SID) value. The SID is used to label signaling messages relating to a particular B-ISUP AE instance. Based on the SID value, the network interface distributes messages to correct the B-ISUP AE. The coordination among B-ISUP AEs is provided by the B-ISUP nodal functions.

When it is determined at an exchange that the requested signaling function requires the B-ISUP, B-ISUP nodal functions create an instance of the B-ISUP (i.e., a B-ISUP AE instance that contains a single association object (SAO) of the appropriate type). Three types of SAOs are defined:

- *Incoming call and connection control,* which contains incoming call control, incoming bearer connection control (BCC), maintenance control, and the unrecognized information application service elements (ASE) and single association control function (SACF);
- *Outgoing call and connection control,* which contains outgoing call control, outgoing BCC, maintenance control, and the unrecognized information ASEs and SACF;

- *Maintenance,* which contains maintenance control and the unrecognized information ASEs and SACF.

In order to communicate with a peer entity in another exchange, the B-ISUP nodal function uses the services of the application entity identifier (AEI) using its primitive and service interfaces. In the originating exchange, nodal functions pass a set of related primitives across primitive interfaces to ASEs, related actions are taken at the corresponding ASE, and the outcome is put in a packet and passed to MTP layer 3 through a service interface. At the destination exchange, the message is received by MTP layer 3, and with the SID value, it is delivered to the corresponding B-ISUP SAO.

Various call control primitives between the B-ISUP nodal functions and the SACF are listed in Table 7.13.

7.3.3.3 Call Control

Recommendation Q.2764 describes the basic B-ISUP signaling procedures for the setup and cleardown of network connections in the context of six types of exchanges:

- Originating exchange;
- Intermediate national exchange;
- Outgoing international exchange;

Table 7.13
Call Control Primitives Between AP and SACF

Primitive Name	*Corresponding B-ISUP Messages*
Set-Up	Initial address (IAM)
Address-Complete	Address complete
Incoming-Resources-Accepted	IAM acknowledge
Incoming-Resources-Rejected	IAM reject
Subsequent-Address	Subsequent address
Release	Release, Release complete
Answer	Answer
Progress	Call progress
Suspend	Suspend
Resume	Resume
Forward-Transfer	Forward transfer
Network-Resource-Management	Network Resource Management
Segment	Segmentation
Error	—

- Intermediate international exchange;
- Destination exchange;
- Incoming international exchange.

A subset of different B-ISUP signaling procedures at these six interchanges are reviewed next.

Successful Call/Connection Setup

Before a route between two exchanges can be put into service, the VPIs are required to be assigned unambiguously and identically at both ends of each VPC. It is also necessary to define which exchange controls this VPI. This is achieved by assigning each exchange one-half of the VPI values: the exchange with the higher signaling point code is the assigning exchange for all even-numbered VPI values and the other exchange for all odd-numbered VPI values. The exchange that assigns the VPI also controls the VPC.

When an exchange attempts to set up a call/connection, it first uses a VPI which it controls. If there is no bandwidth and/or VCI available, then the exchange asks the other exchange to assign the connection identifier.

After the originating exchange receives the complete information from the calling party and has determined that the call/connection is to be routed to another exchange, it executes the route and VC selection procedures. Routing information is stored at the originating exchange or at a remote database. The selection of a route depends on the called-party number, broadband bearer capability, ATM cell rate, and maximum end-to-end delay (together with propagation delay counter). The originating exchange creates an instant of the B-ISUP AE and issues a Set-Up request primitive to the AE. This primitive implicitly confirms that resources for the call/connection are available.

If the exchange is a nonassigning exchange, the Set-Up primitive includes all the parameters of the set-up message assigned in an assigning exchange except the Connection Identifier Element parameter.

The information used to determine the route is included in the Set-Up request primitive for the correct routing at intermediate exchanges. If this exchange is the assigning exchange, then the Connection Element Identifier parameter is also included in the Set-Up request primitive, specifying the VPI/VCI value used for the call/connection.

After receiving a Set-Up indication primitive, an intermediate national assigning exchange performs the VPI/VCI and bandwidth assignment procedures. If this is successful, then the Incoming-Resources-Accepted primitive is issued. In the case of a nonassigning exchange, the Incoming-Resources-Accepted primitive is issued immediately after receiving the Set-Up indication primitive.

After issuing the Incoming-Resources-Accepted primitive, an intermediate national exchange analyzes the called-party number and other routing information to determine the routing of the call/connection. If the exchange can route the call/connection, it creates an instance of the B-ISUP AE and issues a Set-Up request primitive to the AE.

Actions required at the other types of exchanges are similar to that of originating and intermediate national exchanges except the destination exchange. After issuing the Incoming-Resource-Accepted request primitive, the destination exchange analyzes the called-party number to determine which party the call/connection should be connected to. It also checks the called party's access condition and performs various checks to verify whether the connection is allowed. If the connection is allowed, the destination exchange will proceed to offer the call/connection to the called party.

Unsuccessful Call/Connection Setup

If at any time a call/connection leg cannot be completed due to lack of resources (incoming or outgoing side) or if the maximum end-to-end transit delay is exceeded, then the exchange immediately starts the release of the call/connection and issues a Release request primitive toward the preceding exchange with the appropriate cause value specifying the reason for rejection of the call/connection (i.e., lack of available VPI/VCI, SID, bandwidth or maximum end-to-end transit delay exceeded).

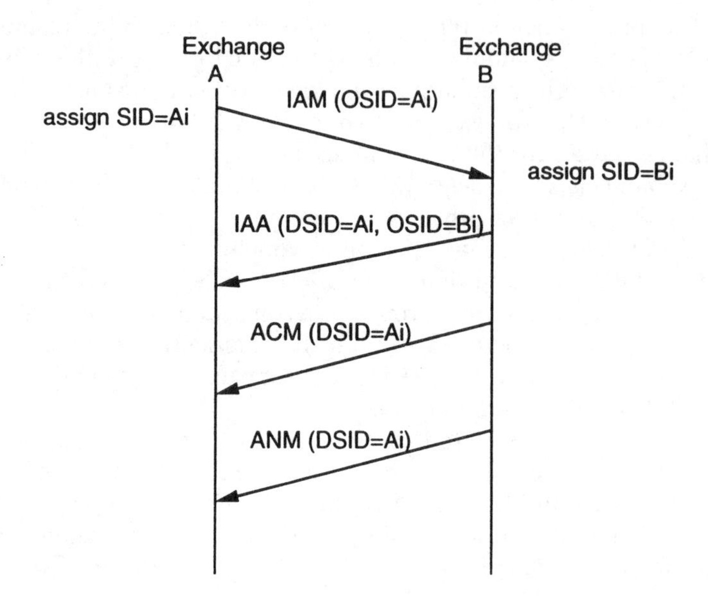

Figure 7.26 B-ISUP signaling message flows for a successful connection setup.

Upon receiving the primitive from the succeeding exchange, which indicates that the call/connection leg cannot be completed (Incoming-Resources-Rejected indication primitive, Release indication primitive), an exchange releases the VPI/VCI and the bandwidth and terminates the outgoing signaling association; that is, the associated AEI is deleted. The exchange may attempt to reroute the call/connection. If these attempts fail, then the exchange releases the call/connection and issues a Release request primitive with the received cause value toward the preceding exchange. If this is the originating exchange, then an indication is sent to the calling user.

Normal Call/Connection Release

When a connection is released by either the calling or the called party, Release request/indication received by an exchange initiates the release of the call

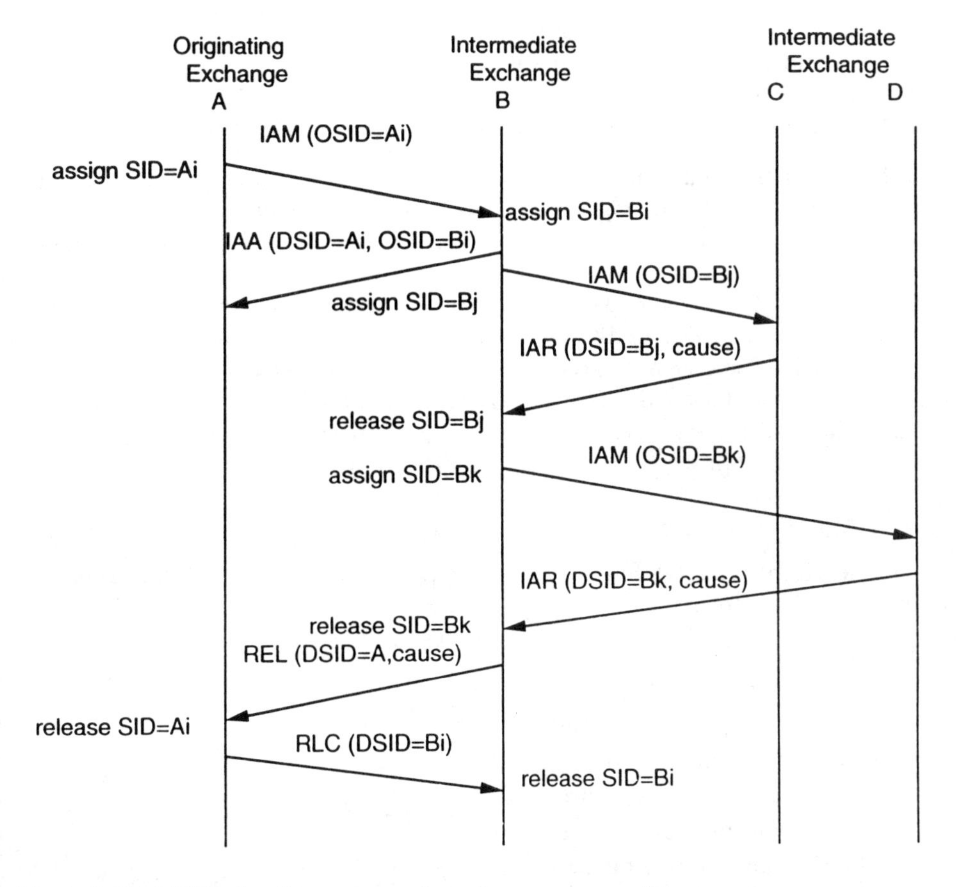

Figure 7.27 B-ISUP signaling message flows for an unsuccessful connection setup.

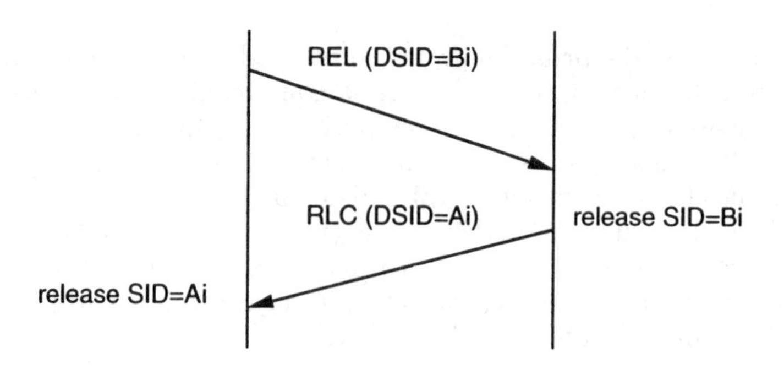

Figure 7.28 B-ISUP signaling message flows for a successful connection release.

and the VCC. Response/confirmation to this primitive completes the release procedure. The following actions are performed by any exchange receiving a Release indication primitive:

- The associated VPI/VCI is made available for new traffic;
- The bandwidth is made available for new traffic;
- Release response primitive is sent;
- The signaling association is terminated.

Upon receiving a request to release the call/connection from the calling party, the originating exchange immediately starts the release of the ATM connection and sends a Release request primitive toward the succeeding exchange. On receipt of the Release indication primitive, an intermediate exchange issues a Release request primitive toward the succeeding exchange. Finally, the destination exchange releases all the resources toward the called party upon receiving the Release request primitive. If the release is initiated by a called party, then the same set of procedures applies, except that the functions at the originating and destination exchange are transposed. If the release is initiated by the network, then the procedures can be initiated at any exchange.

7.3.3.4 Maintenance Control

Recommendation Q.2764 defines various application process functions relating to maintenance control:

1. Reset of resources;
2. Blocking of VPs;
3. Remote user part availability;
4. Transmission alarm handling;
5. Signaling congestion control;
6. Destination unavailability control;
7. VPI/VCI consistency check.

Reset of Resources

The reset procedure is used to return signaling identifiers and connection elements to the idle condition. The procedure is invoked under abnormal conditions: when SIDs or CEIs are unknown or ambiguous. The reset procedure is initiated for:

- Signaling anomalies detected by the B-ISUP signaling system;
- Maintenance action due to memory mutilation (e.g., losing the association information between a SID and a CEI);
- Maintenance action involving startup and restart of an exchange and/or signaling system.

A Reset-Resource request primitive, which contains a resource identifier, is issued to initiate reset. After issuing the primitive, the exchange should stop sending any traffic on the connection. On receiving the parameter, the exchange places the referenced resource it controls in an idle state and returns all resources associated with the VP the exchange controls.

Blocking of Virtual Paths

The VP blocking procedure is provided to prevent a VP from being selected for carrying new nontest call/connections. This procedure can be initiated by the exchange at either end of a VP automatically under a fault condition or manually to permit other exchange management procedures. Unblocking makes the VP available for nontest traffic.

When an exchange initiates the blocking procedure, it issues the Block-Request primitive with the resource identifier identifying the blocked VPI. The VP is put into the locally blocked state so that no new nontest calls/connections can be completed over this path in either direction.

When an exchange receives this primitive, the VP is put into the blocked state and the bandwidth becomes unavailable so that no new nontest calls/connections can be completed over this VP.

Similarly, when an exchange issues an unblocking procedure, it issues the Unblock-Resource request primitive with the resource identifier identifying the unblocked VPI.

Remote User Part Availability

The user part availability procedure is used to check the availability of a particular B-ISUP. On receipt of a Remote-Status indication primitive with the cause "user part unavailability-inaccessible remote user," the B-ISUP marks the concerned user part unavailable, informs the management/overload function, and starts the availability test procedure by issuing a User-Part-Available request primitive.

On receipt of a User-Part-Available confirmation primitive, the B-ISUP marks the concerned user part available, informs the management/overload function, and deletes the associated maintenance AEI.

Transmission Alarm Handling

Since fully digital transmission systems are provided between two exchanges, the switching system will inhibit selection of the concerned VPs for the period the fault conditions persist. Transmission alarm handling requires no special action for active calls/connections.

Signaling Congestion Control (SCC)

Automatic congestion control is used when an exchange is in an overload condition. Two levels of congestion are (1) a less severe congestion threshold (congestion level 1), and (2) a more severe congestion threshold (congestion level 2). If either of the congestion levels is reached, an Automatic Congestion Level parameter is included in all Release request primitives, which indicates the level of congestion to the adjacent exchanges. Upon receiving the Automatic Congestion Level parameter in the Release request primitive, the B-ISUP passes the appropriate information to the signaling system independent network management/overload control function within the exchange and discards the parameter.

Upon receiving this parameter, an exchange should reduce the traffic to the overload exchange. When the first congestion indication is received by the B-ISUP, the traffic load into the affected destination point code is reduced by one step. At the same time, two timers, short SCC and long SCC, are started. During the short SCC, all received congestion indications are ignored so that traffic is not reduced too rapidly. Receiving the indication after short SCC but before long SCC causes the traffic load to be reduced by one more step and both timers are restarted. If no indication is received until the long SCC expires, then the traffic load is increased by one step. The number of steps, the amount of increase/decrease of traffic load, and the definition of thresholds are considered to be implementation-dependent.

Destination Availability Control

Destination-Unavailable and Destination-Available primitives are used to respectively block and unblock VPs/channels to the affected destination (signaling point) for new connections. Existing connections need not be released upon receiving the Destination-Unavailable primitive, even though signaling messages cannot be sent to the affected exchange.

VPI/VCI Consistency Check

The VPI/VCI consistency check is provided to verify the consistent and correct allocation of a logical VPI to a VP on an interface in both connected exchanges. The check is performed to guarantee that a user plane information flow is possible between two adjacent exchanges using the bilaterally agreed-on logical VPI. This is done using the loopback capability of Recommendation I.610 that operates on the VP level.

7.3.3.5 Unrecognized Information

Unrecognized messages are passed to and from the B-ISUP AEI using the Unrecognized-Message-Type request/indication primitive. This primitive carries all the parameters received in the unrecognized message.

As a general rule, every message contains a message compatibility information field and every parameter contains a parameter compatibility information field. An unrecognized message may be received because the message or the parameter might be intended for use by another exchange in the network. The compatibility information fields provide the exchange instructions specific to the handling of the complete message. Various actions defined for unrecognized messages are defined as follows:

- Pass on parameter transparently;
- Pass on message transparently;
- Discard parameter;
- Discard message;
- Discard parameter and send notification;
- Discard message and send notification;
- Release call/connection.

7.3.3.6 Bearer Connection Control

BCC ASE protocol procedures provide the capability to set up and clear down bearer connections between adjacent exchanges. For presentation purposes, these procedures are described in two parts: outgoing BCC ASE and incoming BCC ASE. The BCC ASE uses SACF indication/request primitives to communicate with SACF.

Outgoing BCC ASE procedures commence when a Link-Set-Up request primitive is received, after which the parameters received in the primitive are sent to the SACF for passing to the succeeding exchange in an IAM message, and a timer is started to await the IAM acknowledge or IAM reject message. If an IAM acknowledge message is received, the setup of the bearer connection to the succeeding exchange is considered to be successfully completed. In this case, the timer is stopped and the contents of the IAM acknowledge message are passed on in a Link-Accepted indication primitive. On the other hand, if an IAM reject message is received, the connection attempt is considered to be a failure. The timer is stopped and the contents of the IAM reject message are passed on in a Link-Rejected indication primitive.

At any time after issuing the Link-Accepted indication, various messages and/or primitives may be received, causing respective actions to be taken as follows:

- If a Transfer indication primitive containing a network resource management message is received, then it is passed on as a Link-Resource-Management indication primitive.
- If a Link-Resource-Management request primitive is received, then it is passed on as a network resource management message in a Transfer request indication.
- If a Transfer indication primitive is received containing parameters from an address complete, call progress, or answer message, it is passed on as a Link-Information indication primitive.
- If a Link-Information request primitive is received, then the contents of this message are passed to the SACF in a Transfer request primitive.

The outgoing BCC ASE initiates forward release of a bearer connection when a Link-Release request primitive is received from the calling user or when a release message is received from the SACF as a result of a request from the called user.

The outgoing BCC ASE initiates forward release of a bearer connection upon receiving a Link-Release request primitive. In this case, a release message is sent to the SACF for sending to the succeeding exchange. When a release message is received from the SACF, the message is passed on as a Link-Release indication primitive for backward release. When the connection is released, a release complete message is sent to the SACF for sending to the succeeding exchange.

When an IAM message is received at the incoming BCC ASE in a Transfer indication primitive, it is passed on in a Link-Set-Up indication primitive. The response to this primitive may be Link-Accepted or Link-Rejected primitive, indicating respectively the success or the failure to accept this incoming bearer connection request. For a successful setup, the IAM acknowledge message is sent, whereas the incoming bearer setup is terminated and the IAM reject message is sent if the bearer setup request is rejected.

At any time after the reception of a Link-Accepted request, various messages and/or primitives may be received, causing actions to be taken as follows:

- If a Transfer indication primitive containing a network resource management message is received, then it is passed on as a Link-Resource-Management indication primitive.
- If a Link-Resource-Management request primitive is received, then it is passed on as a network resource management message in a Transfer request indication.
- If a Link-Information request primitive is received, then the contents of this message are passed to the SACF in a Transfer request primitive.

7.3.3.7 Network Interface Function

The network interface function provides a transport interface for instances of the B-ISUP AE. Messages between MTP layer 3 and the network interface are exchanged using MTP-3 service primitives across the MTP-3 service primitive interface. This interface serves multiple instances of signaling associations within one exchange.

Four primitives are defined at the MTP-3 service primitive interface. MTP-Transfer indication primitives received from MTP layer 3 are distributed to AEIs based on the destination SID parameter in the message. If the destination SID corresponds to an existing B-ISUP AEI, the message is passed to that AEI. If the SID does not correspond to an existing AEI, or it is an original exchange SID and there is no destination SID in the message, then an instance of B-ISUP is created. The other three primitives, MTP-Status, MTP-Pause, and MTP-Resume, are passed to the maintenance ASE of an existing AEI, which is determined by examining the indicated affected remote signaling point code. Each primitive corresponds to remote-status indication, destination-unavailable indication, and destination-available indication, respectively. When a Transfer request primitive is received from a B-ISUP AEI, it is mapped into an MTP-Transfer request primitive and it is passed to MTP layer 3 for routing.

7.3.3.8 Single Association Control Function

The service primitive technique used to define the B-ISUP ASE and the SACF is a way of describing how the services offered by an ASE can be accessed by the user of the service (i.e., the SACF or the B-ISUP nodal functions). For outgoing messages, the SACF receives primitives from the B-ISUP nodal functions. The SACF then issues appropriate primitives to the corresponding ASEs.

The request primitives are listed in Table 7.14 and the response primitives are listed in Table 7.15. The B-ISUP messages constructed at the ASE corresponding to these primitives are given in italics.

For incoming messages, the SACF constructs the corresponding primitives and passes them to appropriate ASEs, as illustrated in Table 7.16.

7.4 HOW DOES IT GET TOGETHER?

Let us start with an application at an ATM end station wanting to establish an SVC to a partner application in another ATM end station. To initiate the process, the originating end station sends a SETUP message to the network across its UNI using UNI signaling. This message is received by the switch at the other side of the interface. The SETUP message carries various IEs used by the network to establish the VCC. In addition, the SETUP message includes IEs used by the destination ATM end station. From the network point of view, the most important IEs are:

- Called-party IE: the destination ATM end station.
- ATM traffic descriptors: a collection of traffic parameters that describe the traffic flow for the forward and reverse paths; the values for the forward and reverse path may be different.
- Broadband bearer capability: describes the type of service that the user application requests for this connection. For example, a user might indicate bearer service class A for a voice circuit or a CBR connection with end-to-end timing requirements.
- QOS parameters: the QOS class of the connection.

The first IE in the list is used by the network to locate the destination node, whereas the latter three are used to define the application requirements for a VCC.

Upon receiving the SETUP message, the first thing the ATM switch must do is locate the destination end station. To accomplish this, the switch employs some type of a directory service. After the destination end station is located, the switch uses the source traffic characteristics and application requirements in the SETUP message IEs to determine the amount of resources required to support the application in the network. Then the path selection algorithm is run to determine a path from the origin node to the destination nodes. In determining this path, the end-to-end delay requirement of the application might be taken into consideration explicitly.

This type of path computation in which complete paths are determined by the node where the connection originates from is referred to as *source routing*. This is the type of routing chosen by the ATM Forum P-NNI working group. Source routing requires that the switch the connection request originates

Table 7.14
Messaging Between B-ISUP Nodal Functions and the SACF for Request Primitives

Nodal Functions to SACF	*SACF to CC ASE*	*SACF to BCC ASE*	*SACF to MC ASE*
Set-Up	Call-Set-Up *IAM*	Link-Set-Up *IAM*	—
Address-Complete	Call-Address-Complete *Address Complete*	Link-Information *Address Complete*	—
Incoming-Resources-Accepted	—	Link-Accepted *IAM acknowledge*	—
Incoming-Resources-Rejected	—	Link-Rejected *IAM Reject*	Congestion-Level *IAM Reject*
Subsequent-Address	Call-Subsequent-Address *Subsequent Address*	—	—
Release	Call-Release *Release*	Link-Release *Release*	Congestion-Level *Release*
Answer	Call-Answer *Answer*	Link-Information *Answer*	—
Progress	Call-Progress *Call Progress*	Link-Information *Call Progress*	—
Suspend	Call-Suspend *Suspend*	—	—
Resume	Call-Resume *Resume*	—	—
Forward-Transfer	Call-Forward-Transfer *Forward Transfer*	—	—
Network-Resource-Management	—	Link-Resource-Management *Network Resource Management*	—
Segment	Call-Segment *Segmentation*	Link-Information *Segmentation*	—
Block-Resource	—	—	Block *Blocking*
Unblock-Resource	—	—	Unblock *Unblocking*
User-Part-Available	—	—	User-Part-Test *User Part Test*
Reset-Resource	—	—	Reset *Reset*
Check-Resource-Begin	—	—	Check-Begin *Consistency Check Request*

Table 7.14 (Continued)
Messaging Between B-ISUP Nodal Functions and the SACF for Request Primitives

Nodal Functions to SACF	*SACF to CC ASE*	*SACF to BCC ASE*	*SACF to MC ASE*
Check-Resource-End	—	—	Check-End *Consistency Check End*
	SACF to UI ASE		
Unrecognized-Message-Type	Unrecognized-Message *as contained in the primitive*		
Confusion	Confusion *Confusion*		

CC = Call control.
MC = Maintenance control.
UI = Unrecognized information.

Table 7.15
Messaging Between B-ISUP Nodal Functions and the SACF for Response Primitives

Nodal Functions to SACF	*SACF to BCC ASE*	*SACF to MC ASE*
Release	Link-Release *Release Complete*	—
Block-Resource	—	Block *Blocking Acknowledgment*
Unblock-Resource	—	Unblock *Unblocking Acknowledgment*
User-Part-Available	—	User-Part-Test *User Part Test Acknowledgment*
Reset-Resource	—	Reset *Reset Acknowledgment*
Check-Resource-Begin	—	Check-Begin *Consistency Check Request Acknowledgment*
Check-Resource-End	—	Check-End *Consistency Check End Acknowledgment*

from determine the end-to-end path. This may require that various nodes in the network know the topology of the network and the utilization and current reservation levels of network resources. The topology of the network may change at times due to link and node failures. Although the topology of the network does not usually change that often, utilization levels change frequently. This

Table 7.16
Distribution of Received B-ISUP Messages to ASEs

Received Message	*To BCC ASE*	*To CC ASE*	*To MC ASE*
Address complete	Yes	Yes	No
Answer	Yes	Yes	No
IAM Acknowledge	Yes	No	No
IAM	Yes	Yes	No
IAM Reject	Yes	No	Yes
Call Progress	Yes	Yes	No
Release	Yes	Yes	Yes
Resume	No	Yes	No
Release Complete	Yes	No	No
Subsequent Address	No	Yes	No
Suspend	No	Yes	No
Forward Transfer	No	Yes	No
Network Resource Management	Yes	No	No
Segmentation	Yes	Yes	No

necessitates an efficient means of distributing network control information that maximizes the amount of information available to each node while minimizing the amount of network control traffic overhead. These two objectives are contradictory and the design and development of this feature is a complex task.

Once a path is found, the origin switch proceeds with the end-to-end connection establishment using some type of internal signaling. P-NNI signaling may also be used after some proprietary extensions. Internal signaling needs to provide quick connection setup. Each node along the path determines if it can support the required service (bandwidth, delay, loss) and then responds somehow (i.e., using switch/vendor-specific set of algorithms) to the origin ATM switch. Similarly, if the destination ATM switch can support the connection, it takes the information received in the internal signaling message, builds a SETUP message similar to the one built by the source user, and sends it to the destination end user across the UNI that connects the network to the destination end station.

The destination end station receives and processes the SETUP message. If it accepts the call, then it responds with a CONNECT message that includes any end-to-end information that had to be negotiated. The destination ATM switch forwards this information back to the origin ATM switch. The destination ATM switch also sends a CONNECT ACKNOWLEDGMENT message to the destination end user to indicate that the CONNECT was received and processed. If every switch along the path can support the connection, the origin ATM switch sends a CONNECT message back to the source user. The source responds with a CONNECT ACKNOWLEDGE message. At this time, the connection is fully enabled and user ATM cells can start flowing.

In summary, UNI signaling essentially provides the basic instructions for an ATM end station and an ATM switch to talk and understand each other. Based on this standards framework, it is the responsibility of the network to provide a comprehensive set of control services to provide an ATM service that would include:

- Location of the destination ATM end station;
- Determination of the amount of resources required in the network to support the service requirements of the application using the source traffic characteristics and other related information;
- Determination of a path from the source end station to the destination end stations; Internal signaling used to establish a connection on a path.

After the connection is set up, how does the network monitor the connection to ensure that the end stations do not send more cells than what they agreed upon at the connection establishment time? Let us now consider the network operating with traffic flowing through a large number of connections. At the edge of the network, it is necessary to monitor the traffic generated by each source to ensure that each source stays within the parameters negotiated at the call establishment phase. This may be done using the GCRA specified in the UNI specification.

Nonconforming traffic, if allowed to enter the network, might cause the QOS provided to conforming sources to degrade if additional care is not taken in the network. Let us now focus on the nonconforming cells. There are various reasons that a source might not stay within the parameters negotiated at the call setup time: the source might not be able to characterize its traffic behavior accurately, there might be an equipment malfunction, or the source might simply be cheating. It is a value-added service, therefore, to have network services that provide some amount of forgiveness to nonconforming sources.

Tagging is applied only to high-loss priority cells. Marking CLP = 0 bits of nonconforming cells with CLP = 1 allows (some) nonconforming traffic to enter the network to achieve higher utilization of network resources. Doing so, however, requires the development of mechanisms in the network so that the services provided to conforming traffic are not degraded by the nonconforming traffic. Tagging cells provide the basic framework for making the network more flexible with nonconforming traffic. Mechanisms used to guarantee that this excess traffic does not cause degradation to the services provided to conforming traffic are not a part of the standards (a major problem in the operations of switches/networks).

Another feature required in the network is to minimize the negative effects of resource failures to the services provided to network applications. A closely associated function with that is providing support for different priority connections. A path-switching mechanism may be used to reroute connections established on a link (node) around failures in such a way that the impact of the failure to the service provided to affected connections is minimal. Similarly, different connections may have different priorities. Depending on the availability of resources in the network, it may be necessary to take down some low-priority connections in order to accommodate a high-priority connection in the network. For example, it may be necessary to take down a number of batch applications in order to establish a new connection for an interactive multimedia application such as video conferencing. The main challenge here is to minimize the cascading effect that may occur when a connection that was taken down tries to reestablish a new connection.

Currently predominant networking technologies are designed to emphasize service for one or two of the QOS classes; for example, Telex (delay- and loss-tolerant), telephony (delay- and jitter-intolerant), TV and CATV (delay- and jitter-intolerant), and packet-switched networks like X.25 and frame relay (delay-tolerant). The promise of ATM is that networks can be built on a single technology. This imposes additional requirements on the network:

- Transmission scheduling and hardware-based switching with multicast capability, which provides for the integration of different services with different QOS requirements.
- Group management, which provides the mechanism to group user resources for the purpose of defining multiple logical (i.e., virtual private) networks. This is a critical requirement in networks, where resources belonging to one customer may have to be isolated from those of other customers. The same mechanism can also be used to define multiple virtual LANs on a network.
- Control services such as bandwidth management, congestion control, and path selection enable a network to provide QOS guarantees for measures such as end-to-end delay, delay jitter, and packet (or cell) loss ratio, while significantly reducing recurring bandwidth cost. Path switching makes it possible for a network to reroute network connections automatically in the face of node and/or link failures, while minimizing user disruption.
- Call preemption, which enables a network to provide different levels of network availability to users with different relative priorities.

Public networks have operated under the rule of "keep it simple" due to the very large amount of connections traditionally supported in such networks. In particular, in the case of telephone networks, each connection requires a fixed 64-kbps channel. Routes in the network are predetermined at provisioning time. That is, based on an estimate of the traffic between two switches, a route in the network is determined and the number of channels are allocated (logically) to meet a desired connection blocking probability (i.e., the probability that an end-to-end channel is not available when a connection is requested).

Although this provisioning work is extremely complicated, once it is completed and the network topology is designed, routing in the network becomes static until the traffic assumptions used in provisioning change significantly. Routing in this context is referred to end-to-end path selection in the network between two particular end points (not to include call routing to different locations based on time-of-day routing), excluding routing around failures. Based on this framework, the congestion control problem is simply reduced to a call admission procedure. That is, if there is a channel available in the network to support the connection, then the connection request is accepted; otherwise it is rejected. There is no possibility of congestion in the network, since resources are dedicated to each connection for their duration and are not shared by any other connection.

With B-ISDN, this picture will change dramatically. Because applications in B-ISDNs have different traffic characteristics and different service requirements, it becomes a practically unfeasible task to provision the network while achieving high resource utilization in it.

Efficient resource utilization is becoming more important than it ever was due to the changing nature of the industry. In particular, service providers are facing competition from other service providers as well as other industries (e.g., cable, private networks) causing the service providers to squeeze every bit out of their networks. This is very different from their old mind set when they monopolized the telecommunications services. One argument against this would be that with the availability of fiber in the backbone, hundreds of megabits (up to several gigabits) per second transmission rates are much more than what applications can generate at the edges of the network. Although this is true now and is expected to be true for the next few years, one argument against bandwidth becoming unlimited is just to remember the similar arguments made a couple of decades ago and where we are now.

One possible scenario in service provider networks is to use the SCPs to monitor the resource usage and update routing tables in the network in a semidynamic manner. This would still provide the basic framework to support the high service request rates in the network while allowing routes to be changed more dynamically as resources available along particular paths in the network become more limited from the increasing connections multiplexed onto them. Private networks are different from service provider networks. First, they are usually designed for closed user groups (i.e., departments of an organization). Accordingly, the traffic demand in private networks is much lower than in public networks. This changes the nature of provisioning. The P-NNI framework includes routing and topology distribution protocols that may also be used in a switch-to-switch environment. These functions, however, have not necessarily been optimized. Rather, as with any other standard, they represent a compromise among different vendors who are interested in providing their solutions as value-added services in a standards-compliant manner.

Most of the network services discussed above, however, are not part of any standards and are expected to be available as vendor-specific solutions. Signaling, together with other ATM standards, however, are required to guarantee interoperability among switches built by different vendors.

7.5 CONCLUSIONS

This chapter has discussed the basics of signaling in ATM networks. How can an end station define its specific service requirements? How can an end station define its traffic characteristics? How can two networks interoperate when connected to each other? How does signaling take place in private networks; in public networks? How do they differ?

Signaling and related standards provide the basic framework for connection establishment, management, and termination. Additional network services are needed to complement ATM standards in order to support different applications with different characteristics and requirements in ATM networks. Some of these services are reviewed using a connection setup as an example.

References and Bibliography

[1] P-NNI Draft Signaling Specification, September 1994.
[2] P-NNI Draft Specification, September 1994.
[3] ATM User-Network Interface Specification, Version 3, ATM Forum, August 1993.
[4] ITU-T Recommendation Q.2931, "B-ISDN User-Network Interface Layer 3 Specification for Basic Call/Bearer Control," December 1993.
[5] ITU-T Recommendation Q.2100, "Q.SAAL.0 B-ISDN Signaling ATM Adaptation Layer Overview Description," 1994.
[6] ITU-T Recommendation Q.2110, "Q.SAAL.1 SSCOP Specification," 1994.
[7] ITU-T Recommendation Q.2130, "Q.SAAL.2 B-ISDN Signaling ATM Adaptation Layer—SSCF for the Support of Signaling at the UNI."
[8] ITU-T Recommendation Q.2140, "Signaling at the NNI," 1994.
[9] Bellcore TA-NWT-001111, *Broadband ISDN Access Signaling Generic Requirements,* Issue 1, Livingston, NJ, August 1993. [10] Colella, R., E. Gardner, and R. Collan, "RFC 1237: Guidelines for OSI NSAP Allocation in Internet, Network Working Group," July 1991.
[10] Onvural, R. O., H. Sandick, and R. Cherukuri, "Structure and Use of B-ISDN Signaling," to appear in *Computer Networks and ISDN.*

Routing 8

A communication network is a collection of devices wishing to communicate and a set of nodes providing a connection between such devices. Since, in general, there is no direct connection between devices, the network must route the traffic from node to node until the information is delivered to the destination node. A path in the network is then defined as a collection of sequential communications links ultimately connecting two devices to each other. In general, there is more than one path connecting any two devices in the network. The process of selecting a path in the network is referred to as the *routing function.* There are a number of desirable attributes of routing functions: correctness, simplicity, robustness, stability, fairness, and reliability [42].

The first two attributes are self-explanatory. Robustness is the ability of the technique to cope with changes in the topology of the network due to failures of network resources and congestion. Ideally, the network should react to such contingencies without affecting the service that is provided to current user traffic. Moreover, reaction to changing conditions should take place fast. Otherwise the network may experience an unstable swing between extremes. For example, some traffic from one congested area may be shifted to another area. This may cause congestion in the second area, while the first area may now be underutilized. Then some traffic may be rerouted back to the first area trying to balance the network load in the network. Swinging between different areas, the network may never reach steady state.

Performance criteria to be optimized may depend on the type of traffic carried in the network. Usually the criteria may be the minimization of the average end-to-end delay or the maximization of the network throughput. Since these two objectives are conflicting, the goal may be to maximize the network throughput while meeting end-to-end delay requirements of the traffic carried in the network. Furthermore, it may not be possible to achieve the optimum and be fair to all users. Hence, some compromise between the global efficiency and fairness to individual connections are needed.

8.1 ROUTING IN CURRENT NETWORKS

Various routing techniques have been implemented in current networks such as TYMNET, TRANSPAC, ARPANET, SNA, and DNA. These networks are all examples of store-and-forward packet-switched networks in which data packets are buffered and processed at each node along their paths in the network. The link speeds of these networks are on the order of several kilobits per second. With these slow links, the main design criterion becomes the efficient use of network bandwidth.

In general, routing algorithms implemented in these networks are variants of shortest path algorithms that route packets from source to destination over a least-cost path, and they mainly differ in the cost criteria used. Some networks use fixed cost for each link in the network, while some others use some measured metrics such as congestion, mean delay, and link utilization.

Given the performance criteria, the routing technique can be classified according to the time and place of the routing decision [41]. In particular, the decision time is either at the packet level or virtual-circuit level. Routing decisions in datagram networks are made individually at each node for all arriving packets. A virtual circuit is defined as a logical connection between two users that wish to communicate. It is set up and released on a per-connection-request basis. In virtual-circuit networks, the routing decision is made at the circuit establishment phase and all packets belonging to a virtual circuit follow the same path in the network (as long as there are no failures along the path causing rerouting of connections). The decision place also varies among different networks. In some networks, each node has the responsibility of selecting an output link for routing packets as they arrive, referred to as *distributed routing.* On the other hand, in *centralized routing,* a central node is responsible for making all routing decisions. Another alternative is *source routing,* in which the originating node determines the complete path.

The amount of information required on the topology of the network, such as link utilization, queue lengths, and average delay figures, depends on the performance criteria used and the time and the place of the decision. Some techniques may use no information at all (i.e., static routing), some may use local information (i.e., queue length for each outgoing link), and some others may require global information about the network to make a decision.

Typically, the routing information in virtual-circuit networks is given as a link-by-link logical channel number (LCN) defined uniquely for each virtual circuit in the network. In this scheme, the LCN is determined separately for each link. Accordingly, when LCN is employed as the path identifier, the node performs a table lookup to determine the output port and updates the LCN label to the one used in the output port. Alternatively, LCN may be unique from source to destination, eliminating the need to update its value at intermediate nodes. The main difference between the two methods is the size of the LCN and the processing time required by each. LCNs are usually short with locally defined values resulting in transmission efficiency. This method requires routing tables at each intermediate node to be set up at every connection establishment phase. A global LCN eliminates this problem. However, this scheme necessitates large routing tables at intermediate nodes, and the table lookup operation may be time consuming as the number of connections in the network increases.

8.2 ROUTING IN ATM NETWORKS

For routing purposes, ATM can be viewed as a packet-switched connection-oriented network. That is, VCs need to be allocated for users to establish end-to-end connections, and the values of LCNs need to be swapped from the ones used in the input links to the ones used in the output links at each intermediate node. Two levels of identifiers are used to route cells in ATM networks: VPI and VCI.

Let us assume that a VC consisting of a six-hop path is selected between two users A and B by a routing algorithm and consider the steps of establishing an end-to-end connection between the two users using only the VCI field of the ATM header.

After finding a path between the two nodes, the network assigns the VCI values to be used at each link along the path and sets up the routing tables at each node N1 to N7. This VCC is illustrated in Figure 8.1. The transmission starts and all cells of the connection follow the same path in the network. Upon completion of the communication, one of the two end users releases the connection, which causes the corresponding entries from the routing tables to be deleted at each intermediate node. The processing of routing tables in the call setup and release phases may be time-consuming as the number of connections at intermediate nodes increases.

The concept of VPs is introduced in ATM networks to reduce the size of routing tables and the complexity of network control functions. A VP is a logical direct link between two nodes in the network that are connected via two or more sequential physical links. The connections using the links between the two end nodes that a VP defines are bundled together and transported with an identifier common to all VCs, referred to as VPI. The cells belonging to different VCs are switched using this common VPI along the nodes of the VP, reducing the routing table size. For example, the path between nodes A and B in Figure 8.1 may consist of two VPs, as illustrated in Figure 8.2. Then the network assigns a VCI value of a2 from user A to N1 and x4 from N4 to B. This connection is multiplexed with a number of connections arriving at node N1 using the

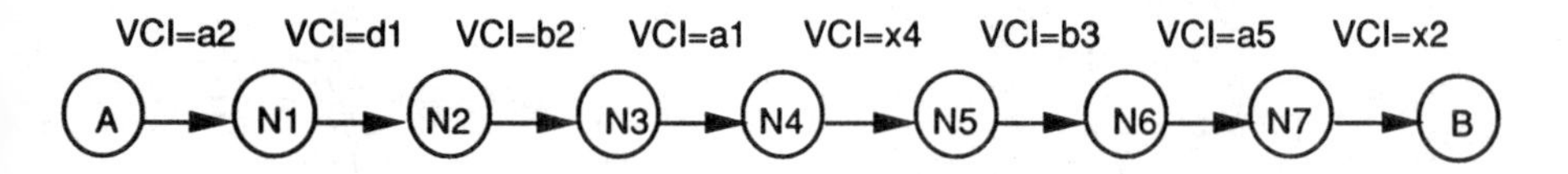

Figure 8.1 Virtual channel connection.

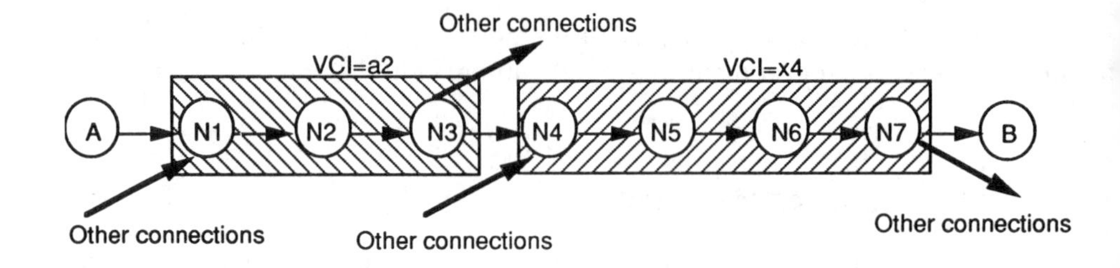

Figure 8.2 Virtual path connection.

path N1 → N2 → N3. All these connections have the same VPI value, which is swapped between the values used at the input links and the ones used at the output links in all three nodes. While the VPI values are changed at each node, the unique VCI values of connections remain unchanged. At node 4, the connection from node A to node B is switched at the VC level. That is, its VCI is swapped and a new VPI is assigned to be used at node 5 and multiplexed with other connections that use the path N4 →. . . N7. At node 7, the connection is switched at the VCI level and its cells are transmitted to node B. Hence, two levels of connections exist in ATM networks: VCCs and VPCs. Conceptually, the two types of connections differ only at the routing field used in switching cells. Accordingly, there are two types of switches used in ATM networks: VP switch and VP/VC switch, as illustrated respectively in Figures 8.3 and 8.4.

In VP switching, VCs multiplexed into VPs are switched from the input to output links using only the VPI field of cell headers. VCIs of connections pass through VP switches unchanged and have no significance in routing cells between ports. Compared with LCNs in current networks, VPIs can be viewed

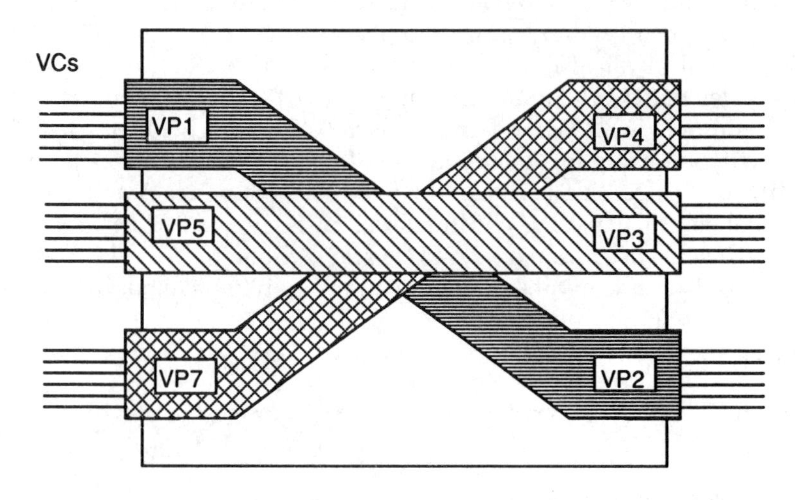

Figure 8.3 Virtual path switching.

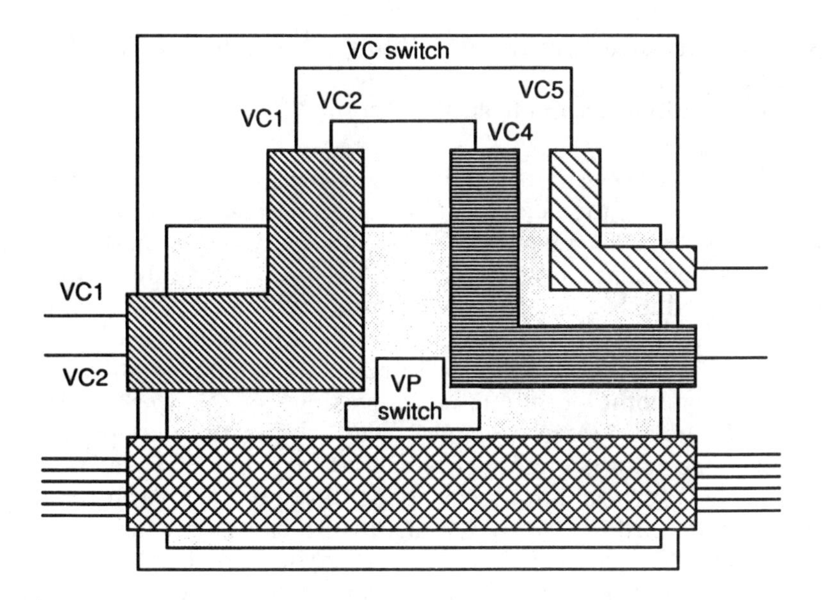

Figure 8.4 Virtual path/virtual channel switching.

as globally significant LCNs, with the term *globally* defined over a sequence of links forming a part of the end-to-end user path.

VP/VC switches have the capability of switching cells using VPIs as well as the complete routing field of VPI plus VCI. The VPs defined in the routing tables of these types of switches are classified into two classes: those that terminate at this switch and those that do not. For the latter, this type of switch is identical to a VP switch. For VPs that terminate at this switch, both the VPI and VCI fields are processed. The switch first checks the VPI to determine whether this VP terminates at this node or not. If not, then the VPI is changed to the one used at the output link and the cell is switched to the corresponding port. If the VP is to be terminated, then the VCI is used to determine the new values of VPI and VCI values used on the output link. Both fields are then modified to include the new values and the cell is switched to its output port.

VPs can be viewed as semipermanent connections in the network. The routing tables of VP switches are set up by network control processors, and changes in these tables occur in a much larger time scale compared to individual connection attempts and holding times. In this respect, VPs are static connections. However, their definitions in the network may change over time as the traffic conditions change. In summary, a VP is a logical direct link between two nodes in the network with the following characteristics:

- VPs are mapped onto a set of sequential physical links by setting the routing tables at corresponding nodes.

- A VP has its own bandwidth, limiting the number of VCs that it can accommodate.
- VPs are multiplexed on physical links.

With VPs, nodal processing per call setup is reduced compared to the pure VC scheme. However, the label overhead is now larger and the transmission efficiency is reduced. That is, the VP/VC scheme gives lower node and higher link cost. This characteristic is valuable to ATM networks because transmission links are expected to become cheaper with the use of high-bandwidth optical links. The functions needed to set up connections are reduced with the use of VPs, since no call setup processing is required at an intermediate node unless it is a termination point for the VP the connection uses. Network control functions with VPs are also more efficient, since such functions are performed on a bundle of connections as opposed to single connections. However, in addition to the additional overhead introduced, using VPs introduces various problems. First, connections within VPs are not distinguishable. Hence, an instant overflow of cells in a VC may cause degradation of service provided to other connections in the VP it is allocated. It is possible to have one VP fully utilized and another one with a large amount of available bandwidth sharing common physical links utilized relatively little. This may cause connection attempts on these links to be rejected, although there is enough bandwidth to carry them. Adaptation to changes in traffic conditions can be provided by rearrangement of the bandwidth allocated to VPs. Hence, the development of an efficient means of dynamically changing the VP bandwidth is essential to achieve high utilization of network resources. We note that this function is not as complex to implement in a pure VC scheme as in a VP/VC scheme. This is mainly due to the fact that each VC is allocated some amount of bandwidth as determined by the network necessary to meet its service requirements, not more. Finally, the problem of determining the VPs in the network and their bandwidth allocation has not been satisfactorily addressed in the literature.

8.2.1 Issues in Using Virtual Paths

A VP is often allocated some amount of bandwidth, limiting the number of VCs that it can accommodate. A number of VCs are multiplexed on a VP and a number of VPs are multiplexed on a physical link. As discussed in Chapter 4, the characteristics of bursty traffic may change as their cells go through a number of nodes. In particular, the characteristics of the traffic on a VP may change as it goes through a number of nodes due to multiplexing within a VP and multiplexing VPs on physical links. This would mean that a VP may require different amounts of bandwidth at the downstream nodes. For presentation purposes, consider a path traffic passing through a number of sequential links with the same transmission rates. In one extreme, if there is no other traffic on the path, then the path traffic does not use any buffers at the downstream nodes (the peak rate of the VP is less than or equal to the any transmission link it passes through). On the other extreme, let us assume that highly bursty traffic from various links is multiplexed onto the same physical links. Then the number of buffers required to provide, for example, the required cell loss probabilities to the path traffic increases at the downstream nodes. However, if the traffic multiplexed from other links is not bursty, then the amount of bandwidth used on a VP decreases as the path traffic goes through a number of nodes. Hence, as in the discussion in Chapter 4, the amount of bandwidth required on a VP depends on the characteristics of the traffic carried in other VPs. Due to the complex effects of multiplexing, the amount of bandwidth allocated to VPs and the efficient use of VPs without degrading the network throughput are issues that need to be satisfactorily addressed in order for ATM networks to operate efficiently.

It is desirable to define more than one VP on different paths to provide connections between pairs of (end) nodes, mainly for reliability purposes. Let us consider a case where there are *n* VPs that can accommodate a new connection request and that they are all equally desirable with respect to non-traffic-related parameters (e.g., number of hops, total propagation delay). There are three possible schemes that can be used to select a VP from a set of VPs with similar nontraffic attributes [31]:

- *Minimum Bandwidth Unused* (MinBUN): All VPs are ordered in ascending order according to the amount of bandwidth currently unused. Then a new connection is allocated to the first VP that can accommodate the connection.
- *Maximum Bandwidth Unused* (MaxBUN): All VPs are ordered in descending order according to the amount of bandwidth currently unused. Then a new connection is allocated to the first VP that can accommodate the connection.
- *First in, First Allocated* (FIFA): VPs are ordered initially in some arbitrary manner. A new connection is allocated to the first VP on the list that can accommodate the connection, independent of the current status of VPs.

Choosing the VP with the minimum allocated bandwidth for existing connections (MaxBUN) attempts to balance the traffic carried at each link in the network. This scheme, used in traditional networks, minimizes the delay at intermediate nodes. However, the average delay values in two different queues with high and low utilization may become negligible in ATM networks as link speeds increase. Furthermore, due to the small buffer capacities at switching nodes, the difference in delay may not be significantly different in the two queues. Accordingly, it may be possible to allocate the new connection to a highly utilized VP and still deliver its requested grade of service without degrading the performance of existing connections (MinBUN). Doing so would utilize a VP as much as possible while leaving big chunks of bandwidth available for future connection requests. This may prove to be a desirable feature in ATM networks where various applications with significantly differing bandwidth requirements are integrated in the same network. Finally, it may be possible to choose the first VP from the list of all VPs defined between the two devices that can accommodate the connection, minimizing the processing times in establishing connection requests. The performance characteristics of these strategies are investigated in [40] in the context of computer memory management proposed to minimize the amount of external storage fragmentation. The problem studied here differs slightly from these studies. One scheme is better than the other if it maximizes the total throughput between two nodes while treating all classes fairly. The performance of three allocation schemes is summarized in [31] as follows:

- MinBUN favors connections with larger bandwidth requirements over connections with smaller bandwidth requirements.
- MaxBUN favors connections with smaller bandwidth requirements over connections with larger bandwidth requirements.
- Although a more thorough investigation is necessary, it is concluded that MaxBUN discriminates against connections with larger bandwidth requirements more than MinBUN discriminates against connections with smaller bandwidth requirements.
- FIFA does not discriminate against any type of connection. This scheme causes fewer connection requests to be lost for low-bandwidth connections than MinBUN, and fewer losses for high-bandwidth connections than MaxBUN. However, FIFA causes more connection requests to be rejected for low-bandwidth connections than MaxBUN and MinBUN for high-bandwidth connections.

It is well known that traffic characteristics in communications networks vary over different time frames (hour to hour, day to day, month to month). Similar behavior is expected in ATM networks. This requires a flexible means of adding and deleting VPs to adjust to the variability of traffic over the different time frames that need to be provided in this environment. Although some of the concepts developed for telephony networks can be used in ATM networks, there are major differences between the two, limiting the applicability of the techniques developed for low-speed networks. For example, in telephony networks, the total bandwidth is equally divided into a number of VCs, since each VC requires a constant bandwidth. This fact alone relatively simplifies the problem considerably. In ISDN, multirate connections (multiples of 64 kbps) may be used for connections that require more bandwidth than a 64-kbps channel (i.e., 384 kbps). However, ATM networks are capable of supporting VCs with a fine granularity and different (and not necessarily dedicated) bit rates, changing the face of and complicating the problem.

8.2.2 Other Issues in Routing

The two major routing functions in connection-oriented packet networks are the establishment of a VC upon connection request and rerouting of VCs to account for changes in the network operating conditions.

In order to establish a connection, a call setup message is sent to the user's local network node (i.e., across its UNI). The setup message includes the source and destination ATM addresses, its traffic characteristics, service requirements, and other related information for the network to be able to find a path from source to destination. The network attempts to find a path in the network that meets the service requirement of the connection, and if there is one, it establishes the connection. If there is more than one path that can accommodate the new connection, one of them is chosen based on some type of optimization criteria specific to the network.

The path chosen may consist of one or more VPs. Some VPs may not have enough bandwidth to accommodate the connection, whereas there might be some others that are lightly used. Hence, it may be necessary (either at the connection request time or as a background task that runs based on the related data collected on resource utilization) to change the amount of bandwidth allocated to VPs semidynamically.

The establishment of a VC will be based on performance criteria. However, unlike traditional networks, ATM networks will support various types of traffic with different performance requirements, such as average delay, instantaneous variation in delay, cell loss, and number of consecutive cells lost, as discussed in Chapter 3. These service requirements are rather conflicting, and achieving all these performance objectives in an integrated network is a very complex problem, if not impossible. For example, it is possible to treat all classes with the strictest performance objectives. This would satisfy the requirements of all types of services, but will cause the network to be underutilized. While the network operator provides resources to satisfy user requirements, its performance objective is to maximize network throughput. Minimizing the number of hops between two communicating nodes minimizes the amount of network resources, thereby maximizing the network throughput. The criteria used in the path selection process have not been addressed satisfactorily in the literature. The problem differs significantly from the ones solved in current networks due to integration of various types of traffic with different service requirements and due to the concept of VPs introduced in ATM networks.

Another major function of routing techniques is to reroute VCs/VPs to account for changes in the network operating conditions, such as congestion and node/link failures. Three desired attributes in the maintenance of routes in ATM networks are:

- Fast changes to the allocated bandwidth of a VC/VP;
- Quick detection of failures and fast rerouting of VCs and VPs that were using failed components;
- Flexible means of adding and deleting VPs to adjust to the variability of traffic over different time frames.

In addition to the need to be able to change the amount of bandwidth allocated to VPs in a semidynamic manner, it may be necessary to change the bandwidth allocated to a connection (i.e., VC). This may be done at the user's request or the network may estimate the connection traffic characteristics by taking measurements of actual traffic and determine traffic parameters (or the amount of resources required to support the connection) more accurately. If this change is a decrease in the reserved bandwidth, then it may not introduce a significant problem, since the connection is already allocated to a path that can support this bandwidth. However, if it is necessary to increase the connection bandwidth, it may no longer be possible to use the same path (i.e., the path cannot accommodate the additional bandwidth). This may require rerouting one or more connections to different paths, an operation that needs to be performed in a nondisruptive manner. Algorithms to achieve this have not received much attention in the literature. Similarly, network failures are unavoidable. When they occur, fast recovery mechanisms are needed to minimize the effects of failures on existing traffic. Rerouting decisions can be made at a central node, source nodes, or intermediate nodes in a distributed manner.

The resources consumed by the routing algorithms are the amount of storage for the topology database, processing capacity to compute and evaluate routing tables, and the bandwidth to propagate the topology database updates to other nodes. These constitute a major part of network control overhead, which increases with the size of the network and the number of places such decisions are made. If the topology database includes information about all resources in the network, then optimum or close-to-optimum routing decisions can be made, assuming the optimization criteria and the constraints are well defined. However, this would require a large network control overhead. Another alternative is to divide the network into logical clusters. Nodes belonging to a cluster may have complete information about the resources in its own cluster but may have only some minimal information about other clusters. Decisions made in this environment are necessarily suboptimal when paths include a number of different clusters. The network control overhead with this scheme can be made manageable by the appropriate choice of clusters [6].

8.3 ROUTING METHODOLOGIES

In this section, we discuss various routing methodologies proposed and implemented in current public and commercial networks from an ATM network perspective.

8.3.1 Shortest Path Routing

Shortest path routing algorithms route packets from source to destination over a *least cost path*. Almost every shortest path algorithm implemented in current public and commercial networks is a variant of the shortest path algorithms proposed by Dijkstra [12] and Ford and Fulkerson [15]. The Ford and Fulkerson algorithm is a distributed algorithm that requires information on the immediate neighbors only, while the former requires a complete knowledge of the network topology at decision places. Dijkstra's algorithm converges faster when failures that do not partition the network occur. If a failure causes the network to be partitioned, then one algorithm does not necessarily behave better than the other.

In shortest path routing, decision places in the network maintain routing tables containing the length of the shortest paths to various destinations and the next node along the path (or calculate dynamically on demand). As traffic conditions in the network change, a path previously known to be the shortest may no longer be the current shortest path. Accordingly, update messages on the current status of resources are sent to decision places to keep the information current so that these decision places can make the best decisions they can possibly make with these constraints. Due to the time required to propagate these messages, the decisions, in general, are close to optimum rather than being optimal. Update messages in the network are sent out either on a periodic basis or whenever a change occurs in the routing table of a node. Then the entries in a routing table may be incorrect for a period of time until an update message sent after a failure is received by other nodes. Both shortest path algorithms are subject to the routing table looping problem, which occurs when a packet traverses a loop due to incorrect table entries. This is not an ATM-specific problem, but is pronounced where a burst of cells may be forced to loop in the network, causing significant performance degradation.

8.3.2 Fixed-Path Routing

In fixed-path routing, a set of routing tables is determined at a central control point and distributed to every node in the network. At any given time, only one path is defined in the routing table connecting any two devices. When there is a significant change in traffic conditions or in the topology of the network, the control point computes new routing tables and updates the tables of all nodes that are affected by the changes. If the traffic flow is stable, this scheme can be optimized to produce good results. Despite its desirable attributes, fixed-path routing algorithms may not be used in ATM networks due to the constantly changing traffic characteristics of applications with significantly different bandwidth requirements (except perhaps for connectionless services to set up paths from gateways to connectionless servers in ATM networks). This mainly arises out of providing guarantees to QOS requirements applications (unlike legacy packet networks).

8.3.3 Saturation Routing

Saturation routing is a call setup procedure developed to operate in circuit-switched networks. Upon receiving a new connection request, a node broadcasts the request to all its neighbors. Similarly, upon reception of the first copy of the request, an intermediate node forwards the message to all its links except the link the message came from. The process is repeated at every node until the destination node is reached [26].

The main advantages of saturation routing are its simplicity, robustness, flexibility, and reliability, since its performance is not affected by topology changes. Furthermore, it allows users to move freely in the network and guarantees to find a path from source to destination if one exists. Its main drawback is the large amount of control messages the nodes have to process and transmit. If used in ATM networks, the message must necessarily include the amount of bandwidth required by the connection so that the message is not sent out on a link that does not have enough bandwidth. This method also requires a significant load on the copy network at each node, which has to be taken into consideration in the design of switch fabrics for such networks.

8.3.4 Stochastic Learning Automata–Based Routing

In adaptive routing, the selection of a path from source to destination is done probabilistically, where the probabilities assigned to different paths depend on the network state. Since communications networks are nonstationary environments, their states vary with time. Under rapidly changing conditions, information about the state at a particular time may no longer be true at the next time instant. Choosing a path probabilistically is an attempt to minimize the effects of uncertainty in decision making.

In this scheme, a source node assigns probabilities to all paths and probabilistically chooses the minimum-cost path. The crucial step in this method is dynamically adjusting the path selection probabilities. If a selection results in a favorable outcome for the network performance, then the probability of selecting that particular path may be increased, while the probability is decreased if the outcome is not favorable.

A stochastic automata–based learning algorithm is proposed in [14], where the probabilities are adjusted as follows. In the case of a favorable outcome, the probability increase is inversely proportional to how small the cost was and how small the probability was. If the outcome is not a favorable one, then the amount of decrease in probability is directly proportional to both how large the cost and the probability were. It is noted that a change in the probability of a selected path causes the probabilities of all other paths that connect the source to a particular destination to change so that the sum of probabilities over all paths connecting two nodes is equal to one.

The key feature that makes learning automata attractive is its simplicity and decentralized operation. In practice, this scheme can lead the network to an optimum global operation. However, this is not without its shortcomings. In particular, it may take considerable time to reach the steady state (if it exists), and by the time the automation reaches the steady state, the network state may change significantly enough that the procedure may start seeking a new steady state.

8.3.5 Routing in Telephony Networks

The main objective of the routing function in telephony networks is to maximize the number of active calls in the network. One way to achieve this goal is to minimize the number of resources used for connections. Accordingly, these networks are highly connected, making it possible to establish connections over a direct link or at most two hops. Furthermore, in telephony networks, bandwidth requirements of calls are constant (i.e., the link bandwidth is equally divided), and holding times of connections are usually on the order of minutes.

Dynamic Nonhierarchial Routing (DNHR) was developed by AT&T. There are three network nodes involved in route selection: source, destination, and an intermediate node. A direct path is tried first. If a channel is not available in the direct path, two-hop paths that connect the two end nodes are tried in a predefined sequence that may depend on the time of the day. The routing tables and network configuration are reviewed and maintained by a central node, which evaluates the network data on a weekly and semiweekly basis.

Northern Telecom's Dynamically Controlled Routing (DCR) is a centralized adaptive routing system. A central node maintains the routing tables and updates its entries every 10 sec based on the information received during the last 10-sec interval. As with DNHR, a direct path is tried first, and if it cannot be used to make the connection, two-hop paths are tried in a sequence defined by the central node.

In ATM networks, connections have varying bandwidth requirements and ATM networks are not necessarily highly connected. Considering traffic characteristics of B-ISDN applications, detailed studies are necessary that will investigate how two types of networks can interact with each other; for example, routing traffic originating in an ATM network destined for a private network passing through a public network.

8.4 ROUTING MODES

Traditional networks provide point-to-point transport from any end station in the network to any other end station. Depending on the routing technique used, a point-to-point path is set up either by a control node, by the source, or in a distributed manner between two end stations. In addition to point-to-point connections, ATM networks are expected to provide point-to-multipoint connections as well as multipoint-to-multipoint connections. Point-to-point paths in current networks are determined in the network based on a single network-specific criterion (e.g., maximize network throughput, minimize average end-to-end delay). Often, these algorithms are referred to as *shortest path algorithms* and have polynomial time complexity.

The path selection problem in ATM networks is more complicated. This is mainly due to the fact that end-to-end ATM connections are required to meet the service requirements of their end users and can be formulated as follows. Maximize *network throughput* (i.e., minimize the number of hops) subject to:

- End-to-end CLR < application CLR requirement;
- End-to-end maximum delay < application maximum delay requirement;
- End-to-end maximum CDV < application CDV requirement.

Furthermore, future connections may be taken into consideration toward maximizing the long-term network throughput (i.e., MaxBUN, MinBUN, FIFA).

Path selection in ATM networks is a constrained optimization problem, whereas it is formulated as an unconstrained optimization in traditional networks. Bandwidth reservation along each path is used to guarantee end-to-end CLR requirements of end users. The two delay-related constraints can be addressed in the path selection algorithm. Doing so, however, the problem becomes, in general, intractable. That is, the constrained shortest problem, in general, cannot be solved in polynomial time (i.e., it is an NP-complete problem). This requires the development of heuristics with a polynomial time complexity so that paths in ATM networks can be determined in real time.

The problem is even more complicated for point-to-multipoint and multipoint-to-multipoint connections often required by multimedia applications. The term *multimedia* is used to refer to the concurrent presence of two or more applications such as voice, data, image, and video. Examples of multimedia applications include teleconferencing, entertainment video, medical imaging, advertising, and education.

Multimedia applications are distinguished from unimedia applications in various ways [18]:

- There are requirements for synchronization among various information types that can range from coarse synchronization, such as sequencing the transmission of various objects (i.e., image followed by voice followed by image), to a more precise, fine synchronization, such as synchronizing voice to the speaker's lip motion.
- There are performance restrictions on the average end-to-end delay values, referred to as *latency*, and instantaneous variations in latency, referred to as *jitter*.
- Multimedia applications typically take place within a group involving two or more users in two modes: point-to-multipoint connections, in which the traffic is generated from a single node and distributed to all other group members, or multipoint-to-multipoint connection, in which any group member can generate traffic to be distributed to all other group members.

There are two basic ways of establishing connections among group members. The simplest way is to connect all pairs of nodes in the group point to point. However, this would require the usage of large amounts of network resources (assuming that the network is not fully connected). A more efficient alternative is to use multicast trees. A multicast tree is a collection of transmission links forming a tree that spans all group members. Messages entering the tree from one group member are routed and copied as necessary by intermediate nodes, to be delivered to all group members. By using trees to provide connections between group members as opposed to pairwise point-to-point connections, the overhead associated with routing packets is incurred only once, when the tree is initially constructed. Routing at intermediate nodes is simpler and quicker, since there is only one unique identifier associated with a tree. Furthermore, the network control and management functions are much simpler with a single tree.

To establish a connection within a group of nodes, the network first determines the amount of bandwidth required for each member in the group and finds a set of paths connecting group members. In point-to-multipoint connections, there is a single source station generating traffic to be distributed to a number of destination nodes. Since traffic flows in one direction, forming a multicast tree is a relatively easy task. In particular, any algorithm developed in the literature to find a spanning tree may be used to construct a multicast tree after links with currently available capacities less than the traffic generated by the source node are eliminated from the network. When optimization criteria is introduced, however, the problem becomes NP-complete. In particular, let us consider a graph $G = (N, E)$, where N is the set of nodes and E is the set of edges connecting these nodes. Assuming that the links are not directional for the simplicity of the presentation, each link (i, j) is assigned is a cost $c(i, j)$. Furthermore, let V be the set of n group members that participate in a point-to-multipoint connection where V is a subset of N. Then, finding a tree T that includes all the nodes in V (and possibly nodes in $N - S$ as well to provide connection) such that

$$\sum_{(ij) \in T} c(i, j)$$

is a minimum is a Steiner tree problem known to be NP-complete. For example, with $c(i, j) = 1$, the objective function becomes the minimization of the total number of links used to construct the tree. Similarly, with $c(i, j)$ corresponding to delay at link (i, j), then the objective is to minimize the total delay on the tree.

Various heuristics have been developed in the literature to solve the Steiner tree problem, which can be used to construct a point-to-multipoint connection in ATM networks. The interested reader may refer to [35] and the references therein for a survey of these algorithms.

The minimum spanning tree heuristic presented next has the worst case time complexity of $O(N^2 n)$. Furthermore, the cost of the tree obtained by the algorithm can be at most twice the cost of the optimal tree.

Step 1: For each member node, find the minimum distance to all other group members. Construct a fully connected graph defined over the member nodes only with edge costs being equal to the corresponding minimum distance.
Step 2: Determine a minimum spanning tree of the fully connected graph.
Step 3: Construct a new graph by replacing each edge in the minimum spanning tree with the corresponding minimum-cost path in the original graph.
Step 4: Determine the minimum spanning tree on the new graph and delete successively each leaf node that is not a group member.

Although the algorithm is developed for undirectional links, it can be extended to directed graphs. The steps of the algorithm are illustrated in Figure 8.5. Consider the network given in Figure 8.5(a). The marked nodes are the group members and the problem is to determine a tree from the left-most group member to all others such that the sum of link costs is minimal.

The first step is to construct a fully connected graph defined over the member nodes only, with edge costs being equal to the corresponding minimum distance. This graph is given in Figure 8.5(b), with the shortest distance marked for each link. Then a minimum spanning tree is constructed on this graph, as illustrated in Figure 8.5(c). Replacing the edges on the spanning tree with the original paths, we have the graph given in Figure 8.5(d). Note that although in this case this graph happens to be a tree, it is not necessarily always the case.

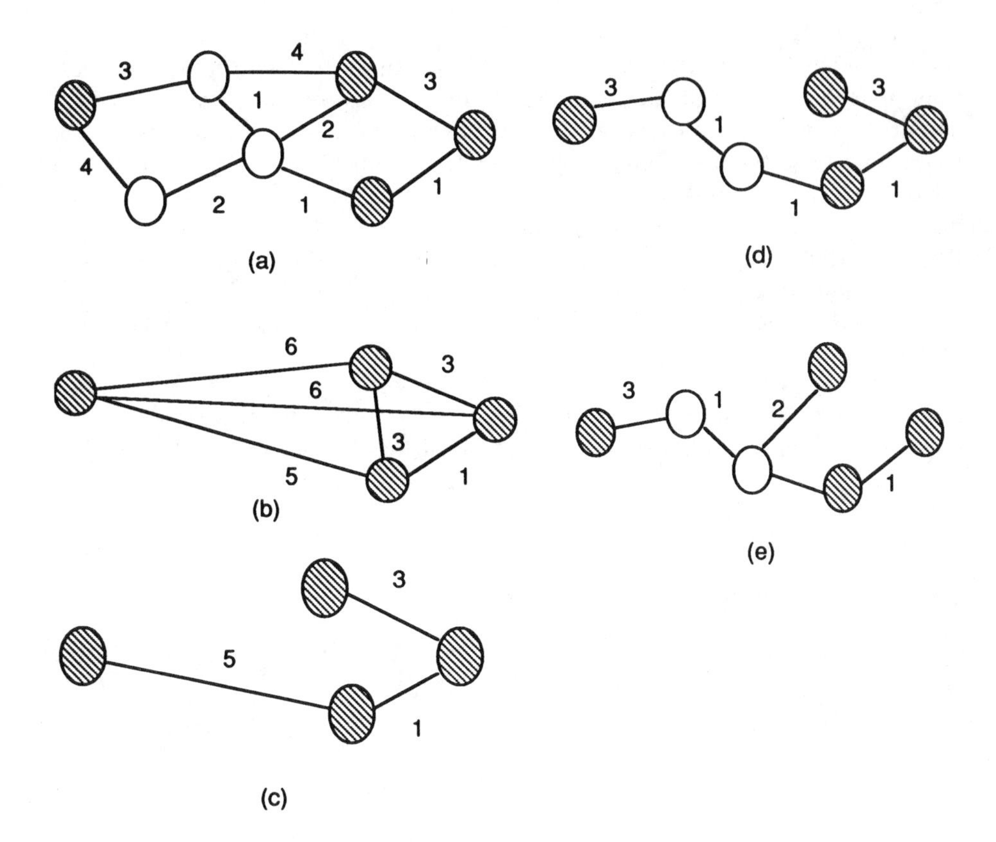

Figure 8.5 Example of a point-to-multipoint connection tree.

In particular, at this step, it is more likely to have a reduced graph with fewer nodes and links than the original graph. The multicast tree is then obtained by determining the minimum spanning tree of this reduced graph. Since in this example we had a tree at the end of step 3, step 4 is redundant and the multicast tree is given in Figure 8.5(d). To illustrate that this algorithm may not produce the optimal tree, a tree with a better cost than the one constructed is given in Figure 8.5(e).

Similarly, finding a multicast tree for multipoint-to-multipoint connections is an NP-hard problem [18]. In general, there are three possible approaches to addressing NP-complete problems. The first approach is to investigate possible special cases that may yield to a well-behaved problem. The second approach is to develop an exact algorithm with an exponential worst case time complexity that may produce a result rather quickly for most practical cases of interest. Finally, the third approach is the development of heuristics to attempt to solve the problem in a reasonable period of time. Although these types of algorithms may produce partial solutions to the original problem, they proved to be useful in attacking NP-complete problems. For example, to construct a multicast tree, a proposed heuristic may produce partial feasible trees. These partial trees can then be combined to form a complete tree by rerouting existing traffic on the edges connecting them to other areas in the network, making otherwise unfeasible links usable. Another possibility is to treat the multipoint-to-multipoint connections within n group members as n point-to-multipoint connections. In this case, various heuristics developed based on the algorithms discussed above for point-to-multipoint connections may be used.

The multicast tree problem, despite its importance, has not received much attention in the literature. There is a growing need to explore the three possible approaches discussed above to solve the multicast tree problem in real time, since multimedia networks with efficient multicasting capabilities are nearing deployment.

8.5 PRIVATE NNI

The P-NNI is the demarcation point between two switching systems. In this context, a switching system may be a single ATM switch. Alternatively, it may be a collection of two or more ATM switches managed and operated under one administration. In the former case, the P-NNI becomes a switch-to-switch interface, whereas in the later case it is referred to as an NNI. These two uses of P-NNI are illustrated in Figure 8.6.

Let us first consider a P-NNI network under one administration that uses a P-NNI as a switch-to-switch interface. In this case, a P-NNI is used between every pair of neighbor switches in the network. These switches may be built by different vendors. ATM end stations are attached to the network across UNIs.

The main advantage of this environment is that network administrators have the flexibility to work with as many switch vendors as they like and do

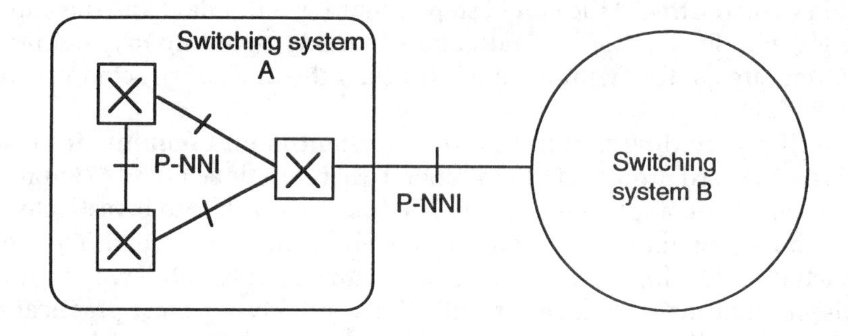

Figure 8.6 P-NNI between two switching systems.

not depend on a single vendor solution for their networking needs. For example, the network may evolve in phases. The networking requirements at each phase may be different and the best solution (based on some administrative-specific criteria) may be available from different vendors at each phase. Interoperability among different switches is achieved through the P-NNI. However, the end-to-end operation in this environment is limited by the capabilities provided at the P-NNI.

Every ATM switch will have a control point providing various networking services, connection priorities, transport services, and so forth that are not provided as part of the P-NNI specification. Instead, such services are developed and implemented as vendor-specific value-added services in a proprietary manner. As an example, let us consider the bandwidth management function. Given the source characteristics of a connection signaled with the UNI SETUP message (i.e., PCR, SCR, BT, QOS class) the amount of bandwidth reserved on a link (more accurately, the connection admission decision as discussed in the next sections) depends on the bandwidth management procedure used at the transmitting switch. Then, the amount of bandwidth reserved along a sequence of links connecting two (or more) end stations would differ from one link to another. In this case, the total number of connections that can be supported along this path would be restricted by the switch that has the most conservative bandwidth allocation method. Hence, switch-to-switch operation with different vendor switches might result in bandwidth fragmentation in the network.

Similarly, the path selection algorithms implemented in different vendor switches may have different optimization objectives. When connections originating at different switches are established using different path selection algorithms, it is more likely that the network will not operate at near optimum, which might otherwise be possible to achieve with a single vendor solution. Integration of services with different traffic characteristics and different service requirements may be supported differently in different vendor switches (e.g., scheduling algorithms, the number of "logical" buffers implemented). Supporting end-to-end services in this environment while achieving high utilization of network resources is, to say the least, a challenge.

The other alternative is to build the network with switches from a single vendor. This would eliminate the disadvantages of operating at the level provided by the P-NNI and allow it to take advantage of vendor-specific network services and solutions toward utilizing network resources efficiently. Interoperability among networks can still be provided with the P-NNI. A single vendor may not, however, provide solutions to all the requirements of network operators at the time they are needed.

In summary, the advantages of operating at the level of the switch-to-switch interface do not come free. The single-vendor solution is not necessarily the correct answer either. Networks are often built in phases as the networking requirements evolve. Most likely, networks will be built with switches from a few different vendors. Each switch vendor is expected to support standard interfaces. Most switching vendors are also expected to provide additional networking services to complement standard interfaces as value-added functions, thereby differentiating their products from the others. The interoperability among switches built by different vendors and among different networks will be achieved through the standard interfaces.

8.5.1 The P-NNI Framework

Let us consider end station 1 (ES-1) attached to network A requesting a connection to end station 2 (ES-2) attached to network C, as illustrated in Figure 8.7.

ES-1 sends across its UNI a SETUP message to switch A.1, providing the network with the information related to its connection request. In order for the two end stations to communicate with each other, it is necessary for the network to find a path in the network that has enough resources to support the connection and manage the connection. The P-NNI framework provides a standards-based framework to these two network service requirements: P-NNI signaling and P-NNI routing. P-NNI signaling, discussed in Section 7.2, is used to establish, manage, and terminate connections across a P-NNI. The main function of P-NNI routing is to find a path across a network between two end stations (point-to-point connection) or two or more end stations (point-to-multipoint connection). The selected path is required to meet the end-to-end service requirements of the connection.

8.5.2 P-NNI Routing

In P-NNI routing, the switching system that a connection request originates from its UNI is responsible for finding the end-to-end path to the destination end station. This is referred to as *source routing*. In determining the path, the

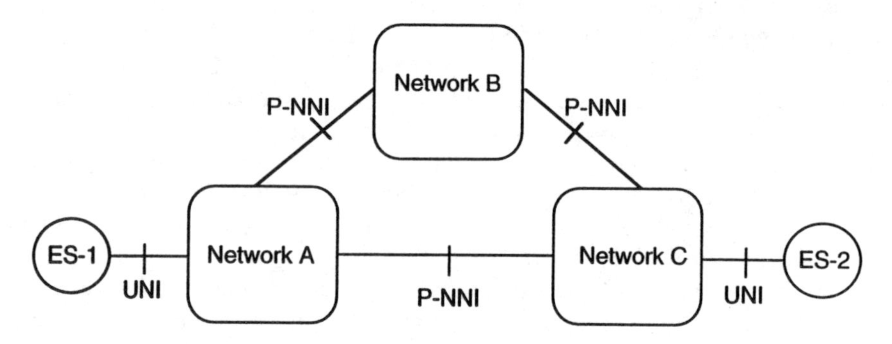

Figure 8.7 An example network.

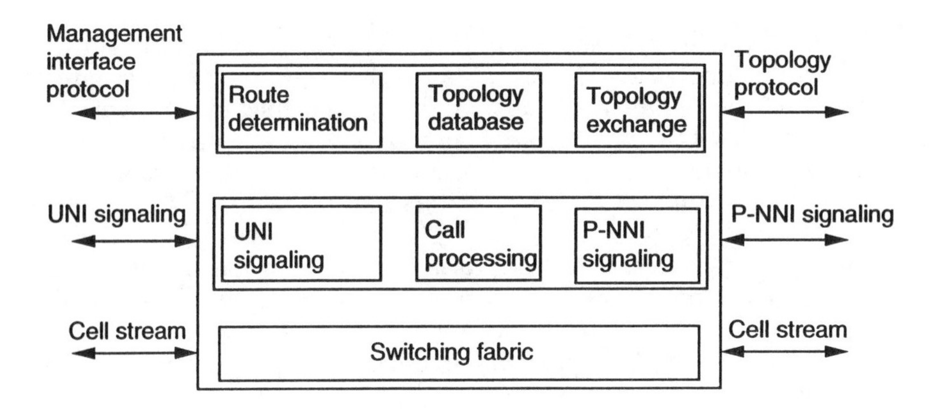

Figure 8.8 P-NNI framework.

originating switching system uses link state routing in which each switching system advertises information about the P-NNI links attached to it to other switching systems. In this context, a P-NNI link connects one switching system to another (in a given direction) across a P-NNI. Accordingly, a P-NNI link could be a logical link consisting of one or more links in sequence.

For each connection request, the source switching finds a path based on the advertised capabilities and the desirability of other switching systems to carry connections with different characteristics. After finding the path, the originating switching system uses P-NNI signaling to request connection establishment from intermediate switching systems along the path. The sequence of switching systems visited is specified in the designated transit list (DTL) stack included in the P-NNI signaling SETUP message. Each switching system processes received connection request messages, makes connection admission decisions (i.e., accept or reject), and passes the signaling message to the next switching system along the path (if accepted) or denies the connection request and sends a REJECT message to the preceding switching system.

Based on this framework, various requirements for P-NNI routing include:

- Unique identification of switching systems and the P-NNI links connecting them;
- The availability of the topological information on the P-NNI network at switching systems;
- Path selection algorithm;
- Connection admission control (CAC) procedure.

CAC is defined as the set of actions taken by a switching system during the connection setup phase in order to determine whether the connection setup request can be accepted or should be rejected. CAC procedures are vendor-specific (i.e., proprietary) and are not expected to be standardized. However, the source switching system should be able to predict the outcome of a possibly unknown CAC procedure at a switch (and/or a switching system) with some confidence so that a large portion of connection setup requests under normal operating conditions are successful (i.e., result in connection establishment). A generic CAC procedure specified for this purpose is discussed in Section 8.5.5.

Similarly, the path selection algorithm is not expected to be standardized. P-NNI routing will determine paths that will likely satisfy the service requirements of connections. In general, there would be more than one path in the network to choose from. Any optimization to achieve high utilization of network resources (e.g., maximization of the network throughput) while meeting end-to-end service requirements of connections is provided in a vendor-specific manner.

P-NNI routing is based on link state routing. Each switching system in this framework advertises a set of parameters including information about the links attached to it, QOS parameters it can guarantee, and its capability and desirability to carry particular types of connections. Source switching systems use this information during the path selection process in determining end-to-end paths.

Let us for a moment assume that each switching system sends a message advertising the status of resources it controls as soon as a change takes place (i.e., upon accepting or terminating a connection). This message will arrive at other switching systems (at the minimum) after the corresponding propagation delays. Accordingly, source switching can never have up-to-date information about the status of other switching systems. Given this constraint, the questions are how much and what internal state information each switching system should advertise, at what frequency it is sent, and how far it is distributed in what form.

Increasing the frequency and the amount of information advertised may be thought to enable "better" paths to be chosen. Doing so would also result in significant cost in bandwidth and processing complexity. At the other extreme, if no information is advertised then the chances of choosing "poor" paths would increase.

In [45], the effect of the frequency of generating topology update messages to the network throughput is investigated using a simulation model.

Define variables δ and Δ to denote the percentages of the link capacity and T_u and nT_u to denote two monitoring intervals such that:

- When a change occurs in the link reservation level, send a topology update message (TUM) if the change in the link reservation level since the last time a TUM was sent is greater than or equal to Δ.

- If no TUM is sent for T_u sec, send a TUM if the change in the link reservation level since the last time a TUM was sent is greater than or equal to δ.
- If no TUM is sent for nT_u sec, send a TUM.

This framework allows the evaluation of a variety of mechanisms in a single experimental setup. For example, if Δ is set to a value smaller than the smallest amount of reserved bandwidth (or set equal to 0%), then a TUM is sent every time the amount of reserved bandwidth at a link changes; that is, a connection setup request is accepted or a bandwidth reserved for a connection is released. If Δ is set to more than 100%, then TUMs are sent only periodically if there is more than δ% change in the amount of reserved bandwidth during the monitoring period. If δ is set to more than 100%, no TUM is sent due to periodic monitoring, T_u; instead, TUMs are sent only when a connection request causes the amount of bandwidth reserved on a link to be changed more than Δ since the last time a TUM is sent. Finally, both δ and Δ can be set to more than 100% so that TUMs are sent only periodically every nT_u sec.

The reason for sending TUMs periodically is twofold. First, if TUMs are transmitted unreliably, then periodic generation of TUMs would increase the chance for a TUM to be received by other switches in case no other TUM is sent due to other events for a long time. Second, periodic generation of TUMs would also help the value of Δ to be set to a large value while ensuring that the link state information at a node is not outdated for a long period of time.

The reported results include a variety of statistics averaged over the duration of the simulation. Three metrics of interest in this study are:

- Average link reservation level: The percentage of the reservable capacity of each link averaged over all links in the network;
- Number of TUMs sent: The total number of TUMs sent during the duration of the simulation;
- Connection completion probability: The ratio of the number of connections completed during the duration of the simulation to the total number of connections generated.

As expected, the completion probability is minimum when no TUM is sent due to changes in the link bandwidth reservation levels (in this case, TUMs are sent only periodically) and maximum when TUMs are sent every time the reserved bandwidth at a link changes. The difference between the two can be very large, up to 53% (more) or, interestingly enough, it can be 5% (or smaller). Similarly, the difference between the average link reservation levels may also be significant, about 60% or very small.

In the simulation runs reported, the performance gain in sending TUMs every time the reserved bandwidth changes compared with transmitting only periodic TUMs was achieved at the expense of transmitting and processing a little more than 112,000 TUMs (122,300 compared with 1,000).

It was also observed that there appears to be no significant correlation between the number of TUMs sent and the connection completion probability when δ and Δ are varied from low to high values. In general, the increase in the number of total TUMs transmitted does not always cause the network throughput to increase (in some cases, the network throughput was observed to decrease as the number of TUMs transmitted increased!).

Furthermore, it was concluded that the use of δ does not appear to improve the connection completion probability (although it triggers more TUMs to be sent). In general, the use of δ may even be harmful in cases where a small amount of bandwidth becomes available at a link. The generation of a TUM in this case introduces the possibility of a large number of nodes to simultaneously send connection requests to a particular link that cannot accommodate several connection requests due to the lack of bandwidth.

Hence, the topology update mechanism in ATM networks with high-speed links needs to be designed with care. Otherwise, the cost of bandwidth and the processing complexity may not be justified when compared with the amount of network efficiency achieved.

8.5.3 Domain Hierarchy

The choices of what internal state information to advertise, how often, and to where require the specification of a multilevel hierarchical routing model. In P-NNI routing, the number of hierarchical levels is not preset and may vary from one level (i.e., no use of hierarchy) up to 104 (determined by the number of bits in the address field). A very large corporate network may be expected to vary between two and four hierarchical levels of routing.

The P-NNI hierarchical model explains how each level of hierarchy works, how multiple nodes at one level can be summarized into the higher layer, and how state information among nodes within the same level and between different levels is exchanged. The model is recursive; that is, the mechanisms used to summarize the lowest level of P-NNI routing into the next higher layer are the same as the mechanisms used to summarize any layer to the next higher layer.

At each level of hierarchy, the topology is represented by logical nodes and logical links. At the lowest level of the hierarchy, each node represents a real switching system. At higher layers, each node may represent either a real system or a group of switching systems. Similarly, P-NNI links may correspond to either a real physical link or a virtual link.

Nodes are collected into peer groups. All the nodes within a peer group exchange link information and obtain an identical topology database representing the peer group. Peer groups are organized into a hierarchy in which one or more peer groups are associated with a parent peer group. Parent peer groups are grouped into higher layer peer groups. The steps for forming a peer group P-NNI hierarchy are:

1. At the lowest level, switches are arranged into peer groups. One node in each peer group becomes the peer group leader.
2. Each peer group leader is responsible for specifying the parent peer group, either through configuration of the peer group leader or by a default peer group identifier based on a portion of its address (i.e., prefix).
3. Step 2 is repeated until all nodes are under one leader.

As an example, let us consider the network in Figure 8.9 that consists of 22 nodes and 30 bidirectional links (cf. [46]).

At the lowest level, the administration of this network decides to organize it, say, into seven peer groups A.1, A.2, A.3, B.1, B.2, B.3, and C, as illustrated in Figure 8.10. Each peer group has an identifier that ranges in length from 0 to 13 bytes (i.e., 0 to 104 bits), elects a peer group leader, and is assigned a parent level (indicated in the figure by the length of the identifier).

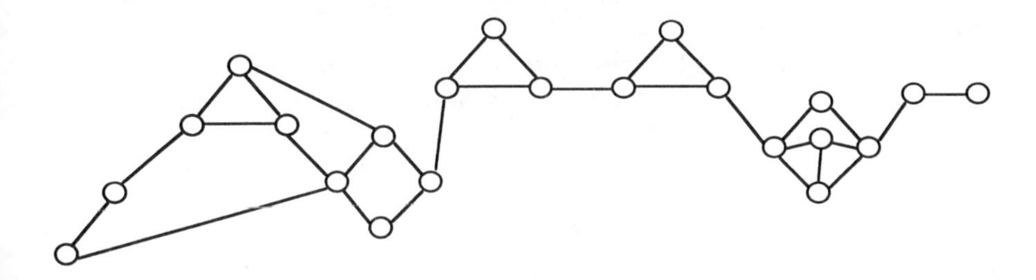

Figure 8.9 A P-NNI network.

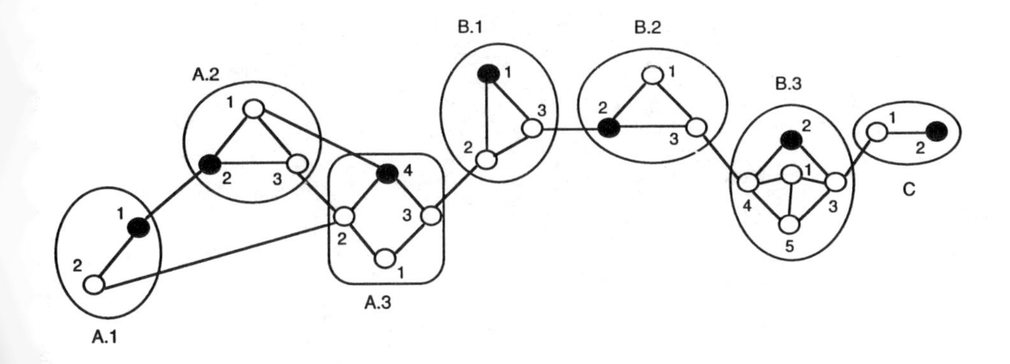

Figure 8.10 Peer groups for the network given in Figure 8.9.

Each peer group is responsible for specifying the parent peer group. This may be accomplished by configuring the peer group leader or by a default parent group identifier based on a prefix of the child peer group identifier.

Continuing with the example, the peer group leaders of A.1, A.2, and A.3 discover that they have a common parent peer group identifier and form a single peer group at level 72 (the level number of the parent peer group is less than any of the peer group identifiers), called A. Similarly, peer groups B.1, B.2, B.3 form a single peer group B at level 80. Peer groups A, B, and C discover that they all belong to the same peer group at level 64. The resulting peer groups are illustrated in Figure 8.11.

Within a peer group, each node has the description of the topology of the peer group including descriptions of all nodes, links, and destinations that can be reached from each node and the status of nodes between the nodes. Routing outside a peer group follows the same link state operation, but at the higher layers of the hierarchy. Considering a peer group as a logical node, topology information is exchanged between the logical nodes identifying each such node at the same hierarchy (i.e., level) and the logical links that connect them. An advertised link may correspond to a multihop path between real systems or may correspond to a link between logical group nodes. Each logical group node may represent an entire peer group.

The operation of P-NNI routing in a parent group attempts to collapse a child peer group into a single node. In this case, there would be different paths to cross a peer group, each with different characteristics. Hence, it is not always possible to advertise the true cost of the real physical paths. The representation

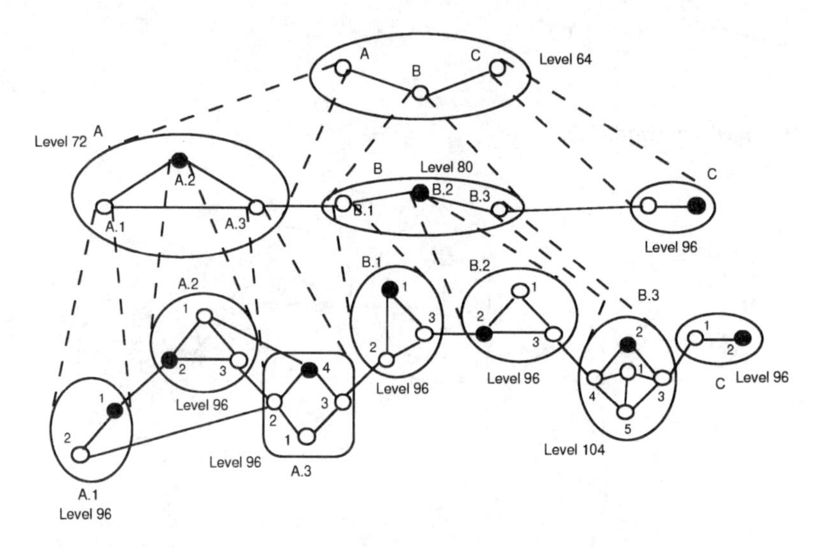

Figure 8.11 Peer groups for the network in Figure 8.9.

of peer groups and their interconnections as logical nodes and logical links implies that information distributed at higher layers is summarized. This allows the routing to scale to very large networks. Its consequence is that somewhat imperfect paths may be selected.

The list of end systems that are reachable through a logical group node is generally the complete list of systems reachable in the lower level child peer group that it represents. This information is summarized by address prefixes. The set of summary addresses defaults to a single prefix that is identical (bit for bit) with the child peer group's identifier. Summarizing end system reachability is recursive; that is, those addresses announced at one level can further be summarized into more inclusive summary addresses at higher levels (with shorter prefixes). If there are additional end systems that cannot be summarized this way, they are advertised explicitly.

The hierarchical summary discussed so far allows those switches that take part in the highest level of routing to calculate routes to any destination represented in the highest level peer group. However, it is necessary for all switches in the overall P-NNI network to be able to route calls to any destination. This requires that the results of higher level routing must be available at the lower level switches.

All switches participating in P-NNI routing maintain link state databases not only for their peer groups, but also for their parent peer group, grandparent peer group, and so forth, to the top level. This allows routes to be calculated on demand by any switch.

In order for the P-NNI hierarchical routing to work properly, it is required to be used everywhere. In practice, however, there will be routes that do not support this protocol (e.g., public networks). To address this problem and use such links in P-NNI routing, it is possible to include "external routes" and advertise them in P-NNI networks. This is achieved through nodes advertising their special external routes to a particular set of destinations. Various uses of external routes include:

1. To allow routing through public carriers that are using other routing methods.
2. In some places, particularly at boundaries between organizations, it may be desirable for security and administrative reasons to avoid the use of any automatic routing protocol and to use manually configured routes instead. This allows a "fire wall" type of capability between routing calculations used in different organizations.
3. To address policy routing requirements of organizations that are beyond the capabilities developed in P-NNI routing protocol.
4. There may be topologies that are not well suited to link state routing, such as a backbone network with connectivity to hundreds of thousands

of locations, where it is desirable to give each location a default route to the backbone rather than tell each location about all the other thousands of locations that are reachable via the backbone.

Each switch exchanges special "hello" packets with its immediate neighbors to determine its local topology. All nodes within a peer group exchange link state update (LSU) messages with each other to report their local topologies to all the others. These LSUs are exchanged reliably. Several other types of LSUs are used to allow a switch to announce its links to neighbor nodes, the metrics associated with each link, and the end systems reachable via the node. Some of these may not need to be exchanged reliably. The details of LSUs and their reliable distribution are not yet specified.

8.5.4 Link State Parameters

Link state parameter is a generic term that includes both link metrics and link attributes. A link metric is a link state parameter that requires the values of the parameter for all the links along a given path to be combined to determine whether the path is acceptable and/or desirable for carrying a particular type of connection such as maximum CTD and maximum CLR. A link attribute is a link state parameter considered individually to determine whether a given link is acceptable and/or desirable for carrying a particular type of connection such as performance-related attributes and policy-related attributes. Currently defined P-NNI link state parameters are listed in Table 8.1.

8.5.5 Generic CAC and Path Selection

P-NNI routing is used to determine paths that satisfy performance constraints. Such paths are not necessarily optimized with respect to any predetermined performance criteria. Since the path selection algorithm will not be standardized, how link parameters are used in determining paths would differ from one vendor switch to another. In general, however, link constraints (such as ACR, CRM, VF) may be used to prune the network graph during the path selection. That is, some links that are not likely to be able to support one or more connection requirements (as determined by the source switch) are not considered in the path selection algorithm. Path constraints such as MCTD(i), MCDV(i), and so forth need to be incorporated somehow in the path selection algorithm, since these are end-to-end metrics of interest to the connection.

As discussed previously in Section 8.5.1, CAC will not be standardized. Furthermore, it is not feasible for a source switching system to know and determine exactly whether a link can accommodate a new connection request or not. A generic CAC is specified in the P-NNI routing framework to determine,

Table 8.1
P-NNI Link State Parameters

Link State Parameter	*Description*
Maximum CTD for traffic class i (MCTD(i))	Represents the maximum delay a cell belonging to traffic class i will incur going through the switching system and the associated link (including switching, queuing, and transmission delays)
Maximum CDV for traffic class i (MCDV(i))	Represents the difference between the maximum and minimum delays a cell belonging to traffic class i will incur going through the switching system and the associated link
Maximum CLR for traffic class i (MCLR(i))	Represents the maximum CLR a connection of traffic class i will incur at the switching system and the associated link
Administrative weight	Set by the network operator used to indicate the level of desirability of using a link for any reason significant to operator
Available cell rate (ACR)	A measure of effective available capacity in cells per second
Cell rate margin (CRM)	A measure of the difference between the effective bandwidth allocation and the allocation of SCR in cells per second; this is an indication of the safety margin allocated above the aggregated SCR
Variance factor (VF)	A relative measure of CRM normalized by the variance of the aggregate cell rate on the link

based on its current knowledge, if a link has potentially enough resources to support a new connection. Accordingly, the generic CAC algorithm is used to determine links that will likely support the connection (i.e., exclude links that will likely reject the connection). For a connection with traffic parameters PCR and SCR, the steps for the generic CAC are given as follows:

1. If $ACR(i) \geq PCR$, then include the link.
2. Otherwise, if $ACR(i) < SCR$, then exclude the link.
3. Otherwise, if $\{ACR(i) - SCR\}\{ACR(i) - SCR + 2CRM(i)\} \geq VF(i)\ SCR(PCR - SCR)$:
   a. Then include the link;
   b. Otherwise, exclude the link.

If PCR = SCR (i.e., CBR service), steps 1 and 2 always give a conclusive answer and step 3 is not executed. If CRM and VF are zero, step 3 will always conclude with "include the link." If VF is infinity, step 3 will always result in "exclude the link."

8.5.6 P-NNI Addresses and Identifiers

The P-NNI routing framework uses ATM private addresses to uniquely identify various physical and logical resources in private networks including ATM end stations, switches, and logical groupings of switches and switching systems. In order to support interoperability between ATM end stations attached to public and private networks, the P-NNI routing framework includes the capability to work with public ATM addresses.

Private ATM addresses were introduced in Section 6.1.2.1. When used in the P-NNI context, the address prefix can be any length within the 13-byte high-order part (i.e., the network part of the address). In particular, the P-NNI routing protocol is designed to support prefixes configured on bit boundaries giving n-bit-long address prefixes with n varying from 0 to 104 (8×13).

Various components of P-NNI networks assigned addresses and identifiers are switches, end stations, peer groups, hierarchical levels, nodes, and links.

8.5.6.1 Switch Addresses

Each switching system actively taking part in P-NNI routing is assigned a private ATM address. Since this address format is not mandatory in public networks, supporting P-NNI routing in a public network (if so desired) requires E.164 addresses to be assigned to switches. In order to make the two formats compatible for the purposes of P-NNI routing, an E.164 address is embedded in the E.164 NSAP format with 4 bytes of zeros in the high-order part DSP, and a 48-bit end system identifier is assigned to the switch.

8.5.6.2 ATM End System Addresses

End system addresses in P-NNI routing could be part of a private network or a public network. Accordingly, ATM end system addresses may respectively be private ATM addresses or E.164 addresses. In addition, it is possible that more than one address can be reached from an end station. Therefore, reachable end system addresses may be a prefix of a private ATM address or a prefix of an E.164 address, where a prefix may range in size from zero bytes to the full length of the address.

8.5.6.3 Peer Group Identifiers

Peer groups are identified using the network part of the private ATM addresses. The prefix assigned to a peer group is chosen from the private ATM addresses owned by the organization that administers the peer group based on the following rules:

1. The default for the lowest level peer group identifier equals a prefix on the switch's address.
2. The default for a higher level peer group identifier equals a prefix on the lower level peer group identifier.
3. P-NNI routing allows other (nondefault) peer group identifiers to be reconfigured.
4. Parent peer group identifiers must be shorter in length than their child peer group identifiers.

The last rule requires that peer group identifiers must get smaller as the peer group levels increase and constrains the manner in which peer group identifiers are assigned. The 20-byte private ATM address (more accurately, the 13-byte high-order part) is large enough so that this may not cause a major problem in practice, while it solves the following two problems:

1. Eliminates the possibility of a peer group hierarchy loop, which occurs when a peer group is both the parent (or grandparent) and child (or grandchild) of the same peer group.
2. When comparing two peer group identifiers, it quickly and easily restricts the possible relationship that could exist between the two corresponding peer groups.

8.5.6.4 Level Indicators

Since peer group identifiers are assigned based on their levels in the hierarchy, a peer group identifier is shorter than any of its child peer group identifiers. This implies that the length (indicated in bits) of the peer group identifier indicates the relative level of P-NNI routing. That is, the level indicator specifies the exact number of bits in the prefix used for the peer group identifier. In particular, in a P-NNI network, the level indicator may be used to determine whether one level is higher than, lower than, or the same level as another level. It is not possible to make any conclusions, however, by simply comparing the level indicators defined in different P-NNI networks.

8.5.6.5 Node Identifiers

Nodes in P-NNI routing represent either a single real system or a single lower level (i.e., child) peer group. A node identifier consists of a 1-byte level indicator, 1-byte subfield length, and a 20-byte flat identifier, as illustrated in Figure 8.12.

When used to represent a single real physical system, the level indicator includes the lowest level in the P-NNI hierarchy in the network. The subfield

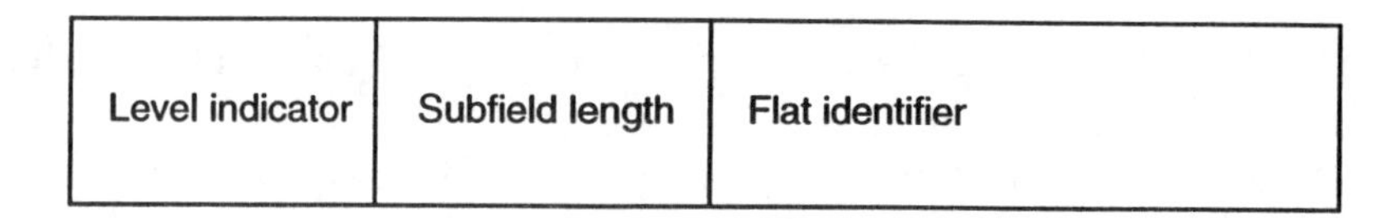

Figure 8.12 Node identifier format.

length takes the value 160 and the flat identifier consists of the private ATM address of the system represented by the node with the selector field (i.e., least significant 7 bytes) set to zero.

For nodes that represent a lower level peer group, the first byte specifies the level of the logical group node and the subfield length specifies the length of the associated lower level peer group identifier. The first 13 bytes of the flat identifier are the peer group identifier (padded with zeros in the low-order bytes, if necessary). The next 6 bytes are the partition identifier, which is currently set to zero and reserved for future use to distinguish multiple nodes at the same level of the P-NNI hierarchy representing different partitions of the same lower peer group. The last byte of the flat identifier is set to zero.

Node identifiers are used in at least two different contexts: to coordinate routing messages and to use in source routing fields (i.e., in DTL stacks) in signaling messages. When used to coordinate routing messages, the node identifier specifies the source of P-NNI routing packets. In this context, the node identifier is to coordinate multiple routing packets from a single node and indicate in P-NNI LSUs the identity of its neighbor node on the other end of the link. When used in signaling messages, it is used to determine which switching system the signaling message will be routed next.

8.5.6.6 Link Identifiers

Transmission links in ATM networks are generally duplex links. A particular link, however, may have different characteristics in each direction. Two immediate neighbor switches at the two ends of a point-to-point link may have different-size buffers, and therefore different maximum CTD and CDV. As ATM connections may have different traffic and service characteristics in the forward and backward directions, transmission links will also have different traffic loads in each direction. Hence, each duplex transmission link is treated as two simplex (unidirectional) links, each with a specified transmitting node.

Between any two nodes, there might be more than one transmission link. As discussed above, each link can have different characteristics in each direction. Hence, each link in the network is identified by its transmitting node and logical port identifier. The logical port identifier is 4 bytes long and has local significance only (i.e., meaningful only when used with the transmitting node identifier). Transmitting node identifiers are 22 bytes long, as discussed in Section 8.5.6.5.

8.5.7 P-NNI Connection Setup

After it finds a path in the network that is likely to support the connection, the source switching system must ensure that the selected path is used. This is achieved by including a DTL in the call setup request. Each DTL contains the path elements for one sequence. Hence, the end-to-end path is specified as a DTL stack consisting of one or more DTLs. The top of the stack is the lowest level (local) sequence.

Consider the network illustrated in Figure 8.10. Let us assume that it is to be set up from end station A.1.2.*x* to the end station B.3.3.*y*. A.1.2.*x* sends a UNI SETUP message to its switch A.1.2, which examines its view of the world. The destination is reachable from peer group B. Examining the topology, A.1.2 finds that there are two paths to B: (1) A.1.2, A.1.1, A.2, A.3, B, and (2) A.1.2, A.3, B. Let us assume that the first path is chosen. Then A.1.2 builds three DTLs in a stack:

DTL-1: {A.1.2, A.1.1}, destination-2
DTL-2: {A.1, A.2, A.3}, destination-1
DTL-3: {A, B}, destination-1

Each DTL lists nodes that the call setup needs to visit at a given hierarchical level. The destination pointer following each DTL specifies which element in the list is the next node to be visited at that level. When the end of the top DTL is reached, it is removed from the call setup request and the next DTL is examined.

Based on this framework, A.1.2 forwards the cell setup to its neighbor A.1.1, which looks at the top DTL, notices that the destination pointer points to itself, tries to advance it, finds it exhausted, and removes the top DTL. The current destination is A.2 (after the pointer is moved to the next entry in the DTL). Since A.1.1 is not in the peer group summarized into the logical node A.2, it starts looking to see how to get to A.2. It determines that node A.2.2 is its immediate neighbor in A.2, so the call setup is sent to that node after removing the top DTL and advancing the destination pointer to 2. The new DTL stack now looks like the following:

DTL-1: {A.1, A.2, A.3}, destination-2;
DTL-2: {A, B}, destination-1.

A.2.2 looks at the top DTL and sees that the current destination is A.2. Since A.2.2 is in A.2, it advances the DTL destination pointer and starts routing to A.3. Analyzing the topology, it finds out that the path is through A.2.3, so it pushes a new DTL onto the stack and sends the setup message to A.2.3.

DTL-1: {A.2.2, A.2.3}, destination-2;
DTL-2: {A.1, A.2, A.3}, destination-2;
DTL-3: {A, B}, destination-1.

A.2.3 determines that the top DTL target has been reached and the current DTL is exhausted. It then notices that the destination is A.3, finds A.3.2 is its immediate neighbor, and sends the setup message there with the DTL stack.

DTL-1: {A.1, A.2, A.3}, destination-3;
DTL-2: {A, B}, destination-1.

A.3.2, upon receiving the setup message, notices that it is in A.3, the current node pointed to in the top DTL, and the target is reached. The DTL is exhausted, so the top DTL is removed. That leaves the target B. A.3.2 builds a route to B through A.3.4 and pushes a new DTL on the stack.

DTL-1: {A.3.2, A.3.4, A.3.3}, destination-2;
DTL-2: {A, B}, destination-1.

A.3.2 then forwards the setup request to A.3.4, which repeats the process and forwards the request to A.3.3. Similarly, A.3.3 processes the request, notices that the DTL is exhausted, and the destination is now B. It determines that its neighbor B.1.3 is in B. A.3.3 prepares a new DTL and sends it to B.1.3.

DTL-1: {A.3.3, B.1.3}, destination-2;
DTL-2: {B.1, B.2, B.3}, destination-1.

B.1.3 then forwards the setup to B.1.2. Proceeding similarly, B.1.2 forwards the setup request to B.2.2. B.2.2 processes the request, finds a path to B.3 through B.2.1 and B.2.3 and sends the request to B.2.1.

DTL-1: {B.2.2, B.2.1, B.2.3}, destination-2;
DTL-2: {B.1, B.2, B.3}, destination-2.

B.2.2 forwards the setup request to B.2.1, which looks at the top DTL, advances the pointer, and forwards the request to B.2.3. B.2.3 removes the DTL, forwards the setup request to its neighbor in B.3. B.3.4 builds a new DTL:

DTL-1: {B.3.4, B.3.1, B.3.3}, destination-2.

Proceeding in a similar fashion, the setup request eventually reaches the last switch along the path (i.e., B.3.3), which has a reachability to the destination end station.

Crankback

Crankback is a mechanism that supports the computation of a new path after an attempt to allocate a resource along a previously selected path fails. Rerouting in the P-NNI routing framework is achieved by each node that adds a DTL into the DTL stack storing a copy of the setup request (i.e., the P-NNI SETUP message). In addition, each exit node from a peer group keeps a copy of the DTLs received in the setup request (but not the signaling message itself). These exit nodes are not involved in the rerouting decisions. However, they keep a copy of the DTL information so that related information for crankback received from the preceding node can be sent to the node responsible for rerouting.

Let us consider the setup example in Section 8.5.7 and assume that the link between A.3.3 and B.1.2 is blocked (i.e., cannot support the connection). A.1.2 in this example originally chose the path A.1.2, A.1.1, A.2, A.3, B. Before forwarding the setup request, A.1.2 stores a copy of the signaling message in case alternate routing becomes necessary. A.1.1 does not need to store a copy of the setup request, since it will not be involved in the rerouting decision. However, it stores a copy of the DTL stack for a possible crankback to A.1.2. A.2.2 pushes a new DTL and keeps a copy of the setup request. A.3.4 does not keep any copy, since it is a transit node in this peer group. At this time, the connection request is received by A.3.3, which determines that the link connecting A.3.3 and B.1.2 cannot support the connection. From the received setup signaling message, A.3.3 determines that the crankback node is A.3.2. A release message is sent to A.3.4, which determines that its address does not match the crankback node address and sends the message to A.3.2. A.3.2 receives the message, determines that it is the crankback point and starts rerouting the procedure. At this point, A.3.2 can try the path A.3.2, A.3.1, B or decide to crankback further (in this case leaving the responsibility of rerouting to A.2.2). Assuming the decision is to crankback further, A.2.2 eventually receives the release message, determines it is a crankback point, finds a new path (A.2.2, A.2.1, B) and attempts to establish the connection along the new path. If the new path can support the connection, the path is established as before.

References and Bibliography

[1] Aceveres, J. J., "A New Minimum Hop Routing Algorithm," *INFOCOM '87*, Vol. 198, pp. 170–1807.

[2] Akinpelu, J. M., "The Overload Performance of Engineered Networks with Nonhierarchial and Hierarchial Routing," *BSTJ*, Vol. 63, No. 7, 1984, pp. 1261–1281.

[3] Ash, G. R., A. H. Kafker, and K. R. Khrishnan, "Servicing and Real Time Control of Networks with Dynamic Routing," *BSTJ*, Vol. 60, No. 8, 1981, pp. 1821–1845.

[4] Ash, G. R., R. H. Cardwell, and R. P. Murray, "Design and Optimization of Networks with Dynamic Routing," *BSTJ*, Vol. 60, No. 8, 1981, pp. 1787–1820.

[5] Ash, G. R., J.-S. Chen, A. E. Frey, and B. D. Huang, "Real Time Network Routing in a Dynamic Class of Service Network," *ITC-13*, 1991, pp. 187–194.

[6] Bar-Noy, A., and M. Gopal, "Topology Distribution Cost vs. Efficient Routing in Large Networks," Proc. *SIGCOMM '90*, 1990, pp. 242–252.

[7] Burgin, J., and D. Dorman, "Broadband ISDN Resource Management: the Role of Virtual Paths, *IEEE Comm. Mag.*, September 1991, pp. 44–48.

[8] Cassandras, C. G., M. H. Kallmes, and T. Towsley, "Optimal Routing and Flow Control in Networks with Real Time Traffic," *INFOCOM '89*, 1989, pp. 784–791.

[9] Chemouil, P., M. Lebourges, and P. Gauthier, "Performance Evaluation of Adaptive Traffic Routing in a Metropolitan Network: a Case Study," *GLOBECOM '89*, 1989, pp. 314–318.

[10] Cidon, I., and I. S. Gopal, "Dynamic Tree Detection in Computer Networks," *INFOCOM '87*, 1987, pp. 181–187.

[11] Di Benedetto, A., P. La Nave, and C. Sisto, "Dynamic Routing of the Italcable Telephone Traffic: Experiences and Perspectives," *GLOBECOM '89*, 1989, pp. 309–313.

[12] Dijkstra, E. W., "A Note on Two Problems in Connection with Graphs," *Numer. Math.*, Vol. 1, 1959, 269–271.

[13] Dziong, Z., et al., "Bandwidth Management in ATM Networks," *ITC-13*, 1991, pp. 821–827.

[14] Economides, A. E., P. A. Ioannou, and J. A. Silvester, "Decentralized Adaptive Routing for Virtual Circuit Networks Using Stochastic Learning Automata," *INFOCOM '88*, 1988, pp. 613–622.

[15] Ford, L. R., Jr., and D. R. Fulkerson, *Flows in Networks*, Princeton: Princeton University Press, 1962.

[16] Gerla, M., et al., "Topology Design and Bandwidth Allocation in ATM Nets," *IEEE JSAC*, Vol. 7, No. 8, 1989, pp. 1253–1262.

[17] Granel, E., et al., "A Comparative Study of Several Dynamic Routing Algorithms with Adaptive Preselection and Selection Phases," *ITC-13*, 1991, pp. 383–3988.

[18] Gun, H., and R. O. Onvural, "On Multicast Tree Formation in Multimedia Networks," *IBM Tech. Rept.*, TR 29, Research Triangle Park, 1992.

[19] Humblet, P. A., "Another Adaptive Distributed Shortest Path Algorithm," *IEEE Trans. Comm.*, Vol. 39, No. 6, 1991, pp. 995–1003.

[20] Jueneman, R. R., and G. S. Kerr, "Explicit Path Routing in Communications Networks," *Proc. Int. Conf. Computer Communication*, 1976, pp. 340–342.

[21] Kamimura, K., and H. Nishino, "Capacity and Flow Assignment of Packet-Switched Networks with Concave Line Cost Function," *ITC-13*, 1991, pp. 285–290.

[22] Kanayama, Y., Y. Maeda, and H. Ueda, "Virtual Path Management Functions for Broadband ATM Networks," *GLOBECOM '91*, 1991.

[23] Khrishnan, K. R., "Adaptive State Dependent Traffic Routing Using On-Line Trunk Group Measurements," *ITC-13*, 1991, pp. 407–411.

[24] Lee, K. J., and B. Kadaba, "Distributed Routing Using Topology Database in Large Computer Networks," *INFOCOM '88*, 1988, pp. 593–602.

[25] Lee, K., D. Towsley, and M. Choi, "Distributed Algorithms with Constraints in Communication Networks," *INFOCOM '87*, 1987, pp. 188–199.

[26] Ludwig, G., and R. Roy, "Saturation Routing Network Limits," *Proc. IEEE*, 1977, pp. 1353–62.

[27] Maglaris, B., R. Boorstyn, S. Panwar, and T. Spirtos, "Routing in Bursty Switched Voice/Data Integrated Networks," *INFOCOM '87*, 1987, pp. 162–169.

[28] Maglaris, B., et al., "Routing of Voice and Data in Burst Switched Networks," *IEEE Trans. Comm.*, Vol. 38, No. 6, 1990, pp. 889–897.

[29] Mitra, D., R. J. Gibbens, and B. D. Huang, "Analysis and Optimal Design of Aggregated Least Busy Alternative Routing on Symmetric Loss Networks with Trunk Reservations," *ITC-13*, 1991, pp. 477–482.

[30] Ohta, S., I. Sato, and I. Tokizawa, "A Dynamically Controllable ATM Transport Network Based on Virtual Path Concept," *GLOBECOM '88*, 1988, pp. 1272–1276.

[31] Onvural, R. O., "Selecting Between Equally Desirable Virtual Paths in ATM Networks," *Proc. 1st Int. Conf. on Computer Comm. and Networks*, San Diego, CA, June 1992, pp. 221–225.

[32] Onvural, R. O., and H. G. Perros, "Performance Issues in High Speed Networks," Book of Speakers, *TELECOM '91*, Zurich, 1991.

[33] Onvural, R. O., and I. Nikolaidis, "Routing in ATM Networks," in *High Speed Networks*, H. G. Perros, ed., Plenum, 1992.

[34] Onvural, R. O., and Y. C. Liu, "On the Amount of Bandwidth Allocated to Virtual Paths in ATM Networks," *GLOBECOM '92*, 1992.
[35] Plesnik, J., "Heuristics for the Steiner Problem in Graphs," *Discrete Applied Mathematics*, Vol. 37/38, 1992, pp. 451–463.
[36] Sato, Y., and K. Sato, "Virtual Path and Link Capacity Design for ATM Networks," *IEEE JSAC*, Vol. 9, No. 1, January 1991, pp. 104–111.
[37] Sato, Y., et al., "Experimental ATM Transport System and Virtual Path Management Techniques," *Proc. GLOBECOM '91*, 1991.
[38] Schwartz, M., and T. E. Stern, "Routing Techniques Used in Computer Communications Networks," *IEEE Trans. Comm.*, Vol. 28, No. 4, 1980, pp. 265–278.
[39] Semal, P., "Performance Evaluation of the Saturation Routing Method," *ITC-13*, 1991, pp. 389–394.
[40] Shore, J. E., "On the External Storage Fragmentation Produced by First-Fit and Best-Fit Allocation Strategies," *Comm. ACM*, Vol. 18, No. 8, 1975, pp. 433–440.
[41] Stallings, W., *Data and Computer Communications*, New York: Macmillan, 1989.
[42] Tanenbaum, A., *Computer Networks*, Englewood Cliffs, NJ: Prentice Hall, 1981.
[43] Toyoshima, K., M. Sasagawa, and I. Zokizawa, "Flexible Surveillance Capabilities for ATM Based Transmission Systems," *ICC '91*, 1991, pp. 699–704.
[44] Van Hoesel, S., P. Kellenberg, and W. Schooten, "Routing and Multiplex Bundling in a Transmission Network," *ITC-13*, 1991, pp. 395–400.
[45] Peyravian, M., and R. O. Onvural, "On the Frequency of Transmitting Topology Update Messages in High Speed Networks," *IBM Tech. Rept.*, Research Triangle Park, 1994.

9 Connectionless Service in ATM Networks

A large portion of the information exchanged within an organization travels short distances, typically within an office, a building, or a campus. Despite declining processor and computer costs, the prices of peripherals such as printers and high-volume/high-speed disk drives are still high, making it unfeasible to place them on every desktop. Consequently, the need arises for a low-cost network allowing several computers and terminals to share resources and communicate with each other. A LAN is a communication network that provides interconnection between a variety of devices within a limited geographical area.

Network topology defines how communicating devices are interconnected. The two most common topologies used in LANs are the bus and ring networks, illustrated in Figure 9.1.

In both topologies, variable-length packets generated by stations are broadcast over a medium shared by all stations in the LAN. This eliminates the need for the routing function: sender accesses the medium and sends a packet with the address of the destination station, stations in the network receive the packet and filter the address, and the destination stations copy the packet after matching the address in the packet with their own. This simplicity is achieved at the cost of the complexity required to regulate the access to the shared medium.

The current data communications infrastructure is largely based on LANs. Workstations within a geographically close location (i.e., in a building) are connected to LANs. LANs are connected to each other in a campus over a campus network. Campus networks are interconnected via routers over WANs. Accordingly, there is a vast base of networking applications developed to run in the LAN environment. If ATM is to be successful in the local area, it is necessary to be able to run existing applications over ATM LANs with no changes to the applications until the networking infrastructure evolves into all ATM networks.

ATM technology, however, is quite different from the legacy LAN technology. Current LAN architectures and various architectural challenges to support legacy networking applications in ATM are discussed next. This is followed by a presentation of the three different approaches proposed to enable LAN

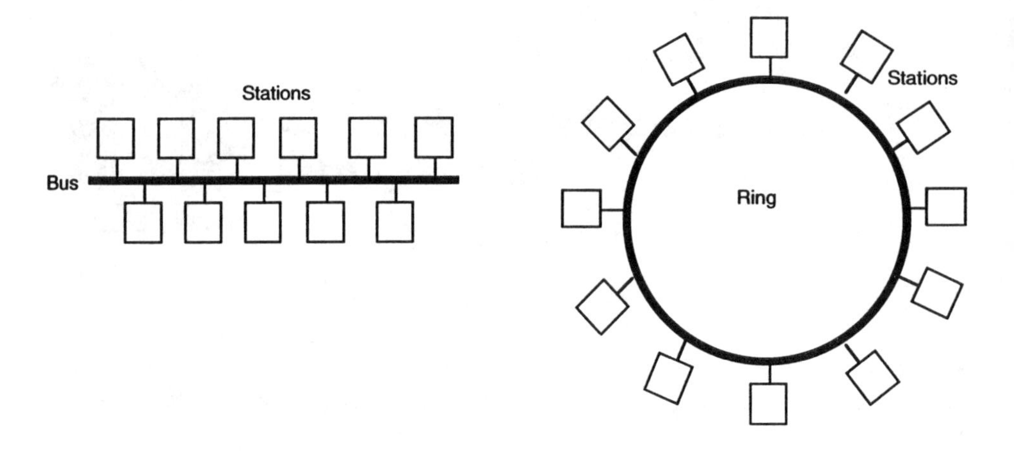

Figure 9.1 Legacy LAN topologies.

services in ATM networks: ITU-T server method, ATM Forum LAN Emulation, and IETF Classical IP over ATM.

9.1 LEGACY LAN PROTOCOLS

In 1980, the Institute of Electrical and Electronics Engineers (IEEE) Computer Society undertook the task of developing LAN standards, which eventually resulted in the 802 series defining two sublayers, the logical link layer (LLC) and medium access control (MAC), on top of the physical layer, as illustrated in Figure 9.2.

The LLC sublayer is concerned with the transmission of LLC frames between two or more stations with no intermediate switching nodes. At the sender, the LLC layer generates LLC frames from user data by appending the source and the destination addresses and a control field to facilitate error-free transmission across the LAN and end-to-end flow control. An LLC address identifies an LLC user. LLC frames are passed to the MAC layer for transmission toward their destinations. At the receiver, the LLC layer receives the frames from the MAC layer, interprets them, and performs error control.

The MAC layer manages access to the media shared by the attached multiple stations. At the originating station, the MAC layer receives LLC frames and constructs MAC frames, accesses the physical media, and passes the MAC frames to the physical layer for transmission over the physical media. A MAC frame includes, in addition to user data, framing and deframing bits and address and error detection fields. A MAC address identifies a station in the network. At the receiving station, MAC frames are disassembled to perform address recognition and error detection and are passed to the LLC layer.

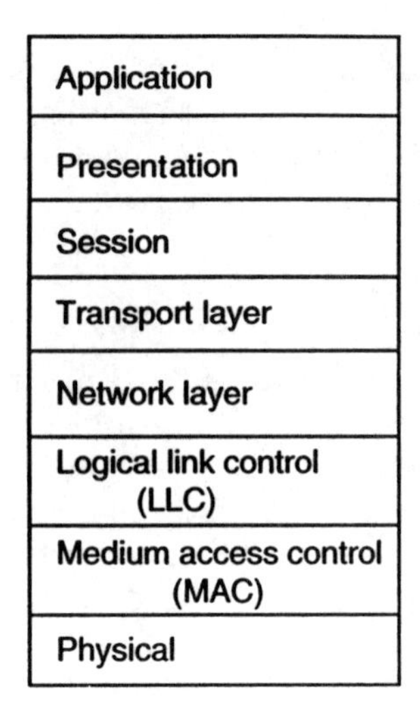

Figure 9.2 IEEE LAN architecture based on the OSI model.

Figure 9.3 illustrates the current IEEE 802 series of LAN standards. FDDI does not belong to this family of protocols, but it is included for completeness, since FDDI runs under the LLC layer. There are several products in the market that do not conform to these IEEE standards. Hence, the 802 family of LAN protocols does not imply universal compatibility.

- 802.1 deals with the relationship among 802 series of standards and their relationship to the ISO's OSI model. This specification also explains the relationship of the standards to higher layer protocols and discusses internetworking and network management issues.
- 802.2 describes the interface service specification to the network and to the MAC layers. It also provides a description of a peer-to-peer protocol defined for the transfer of information and control between any pair of data link layer SAPs on a LAN. These LLC procedures are independent of the type of medium access method used.

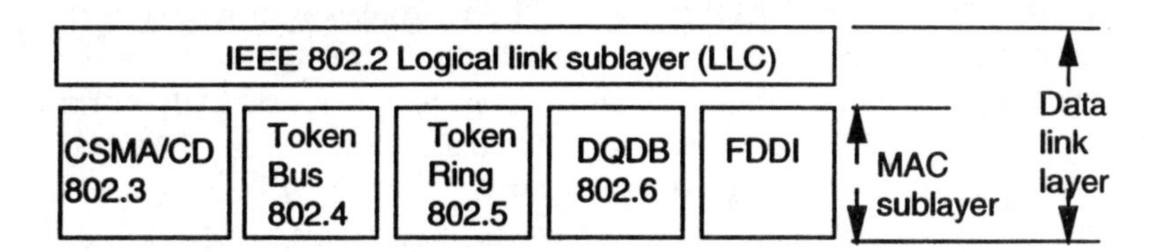

Figure 9.3 The IEEE 802 family of LAN protocols.

- 802.3 describes a physical layer bus using a Carrier Sense Multiple Access/Collision Detection (CSMA/CD) access method.
- 802.4 describes a physical layer bus using a token access method.
- 802.5 describes a physical layer ring using a token access method.
- 802.6 describes a MAN.

There are three types of services defined in 802.2. Type 1 provides an unacknowledged connectionless service, type 2 defines a connection-oriented service, and type 3 is an acknowledged connectionless service.

There is no acknowledgment, flow, or error control in type 1 service. Thus, there is no guarantee that a user frame will be correctly received. In the type 2 service, an end-to-end logical connection is established before user packets can start flowing. Each packet received at the destination is acknowledged back to the sender and flow control is exercised between the two peer LLC layers. Logical connections are terminated at the end of the session. Type 3 service is similar to type 1 service in that a connection is not established at the LLC layer before the transmission starts. However, in this case, each LLC frame is acknowledged, and using the acknowledgment scheme, flow control is exercised. Sequence is not preserved in type 3 service; hence, reordering of user frames above the LLC layer may be required.

The MAC layer regulates the access to the medium. LAN access methods are classified as either contention or noncontention. In contention-based methods, control is distributed among all stations. There is no protocol used between the stations to decide which station should transmit the next time the media becomes idle. Thus, collisions may occur if two or more stations start transmitting almost simultaneously. In the noncontention method, access to the network is prescribed to avoid collision. In particular, a unique combination of bits, called a *token,* circulates through the network following a predetermined path. When a station has data to send, it must wait for the token to arrive with a permit to transmit before it can place its packet onto the medium. This scheme eliminates the collision problem but requires token management.

9.1.1 LAN Interconnection

Since the early 1980s, networking has been changing significantly. The early computing platform was based on a centralized computing platform in which dumb terminals were used to access a central processor. Instead of a centralized processing environment, today, most computing takes place on networks of personal computers, workstations, and multiuser systems from different vendors. Accordingly, an efficient means of access to various resources at different locations has become a necessity. Together with the availability of high-speed communication links, more and more organizations are interconnecting their

LANs over WANs. Some of the benefits obtained by interconnecting LANs include enhanced system availability by providing the capability to run multiple independent LANs, coverage of large geographical area, sharing resources located in physically separated LANs, and the capability to support a large number of users.

Interconnection of LANs is achieved through a whole range of devices, including repeaters, bridges, routers, and gateways. Based on the OSI model of Figure 9.2, repeaters operate at the physical layer and are used to extend the network size by regenerating electrical signals physically.

Bridges operate at the data link layer and connect individual LAN segments together to form one large logical network, as illustrated in Figure 9.4. Unlike repeaters, bridges can regulate traffic from one network segment to another based on the destination MAC addresses of received packets. This is achieved by filtering the addresses of MAC frames and forwarding packets with nonlocal MAC addresses only. Since their operation is based on MAC addresses, all higher layer protocols are transparent to bridges. In general, the use of bridges can be summarized as follows:

- Link local and remote homogeneous LANs;
- Partition network traffic for load balancing, isolation, reliability, and network management;
- Provide physical media conversion;
- Extend maximum distances available to a single LAN segment and increase the number of nodes supported in the organization.

Two types of bridging, namely the transparent bridging and source routing bridging, are defined in the IEEE 802 framework.

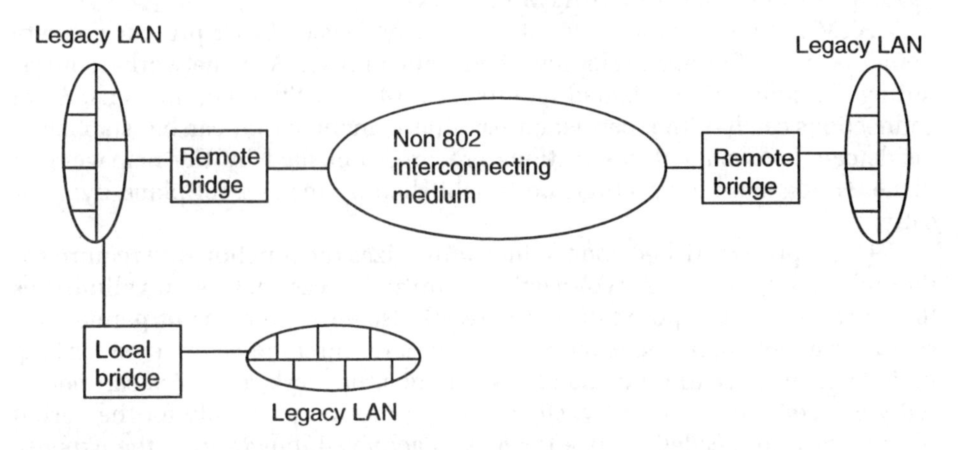

Figure 9.4 Local and remote bridging.

The transparent bridging method is mainly used by 802.3/Ethernet LANs, providing connectivity between any two stations in the network as if they were on the same LAN. The filtering and forwarding functions are handled in the bridges. Accordingly, end stations have no awareness of routes through a bridged network. In this method, only one of several routes between two stations can be active at any given time.

The source routing bridging method is used in IEEE 802.5/Token Ring LANs to provide connectivity between any two stations in the network. Unlike transparent bridging, this method allows more than one route to be active between two stations. In this case, end stations must be aware of the routes in the network and choose one prior to transmitting frames.

Routers operate at the network layer. These devices handle the integration of networks that were disjointly fabricated with different network protocols and facilitate data transfer across different media. The network layer manages data transfer between two devices through a set of intervening networks by routing and switching data through each network. Since multiple paths may exist between any two devices that wish to communicate, routers have the intelligence to determine a "best" path to the destination.

Finally, gateways operate at higher layers (i.e., transport to application layers), translating messages between dissimilar networks that employ different high-level protocols such as TCP/IP and X.25.

9.1.2 LAN Characteristics and ATM

ATM technology is quite different from legacy LAN technology. In this section, the two technologies are compared and contrasted with a view to identifying various architectural challenges required to support networking applications developed in legacy LANs in ATM networks.

ATM is a connection-oriented technology. Legacy LANs provide connectionless service. Supporting legacy LAN applications in ATM networks requires hiding the connection-oriented nature of ATM from these applications. ATM connections can be either permanent/semipermanent or they can be established on demand. Permanent connections are preestablished by the management plane. On-demand connections are established by the control plane dynamically.

Using preestablished connections minimizes the functionality required at the end stations, minimizes connection setup and release delays, and eliminates the signaling load imposed on the network. However, the use of permanent/semipermanent connections requires network resources to be reserved for long periods of time. With on-demand connections, network resources can potentially be used more efficiently, since connections are used only for the period of time they are needed. This efficiency is achieved, however, at the expense of handling signaling load in the network to set up/tear down connections.

Legacy LANs provide broadcast and multicast natively (i.e., shared media) with special MAC addresses. Various LAN protocols rely on these features, and they are needed in ATM networks if these protocols will be used in ATM networks without any changes. ATM networks are switched-based. An ATM connection can be point-to-point or point-to-multipoint. Multipoint-to-multipoint connections can be supported in different ways through the use of point-to-point and/or point-to-multipoint connections.

Connectionless service does not require explicit QOS guarantees from the network. ATM best effort service with an unspecified QOS class can be used to support legacy LAN applications. Although no explicit service guarantees are required (or provided) by legacy LAN applications, these applications implicitly require the minimization of CLR in ATM networks to reduce the possibility of retransmission and therefore effective network throughput degradation.

ATM uses fixed-size 53-byte cells to transmit data in the network. Legacy LANs, on the other hand, are frame-oriented and transmission is based on variable-length frames (with a predefined minimum and maximum frame length). The transmission of LAN frames over ATM requires the SAR functions. Two AALs may be used for connectionless traffic: AAL 3/4 and AAL 5.

Three private ATM address formats in private ATM networks and one public ATM address format are available in public ATM networks. A private ATM address includes a 48-bit end-system identifier field, the same as MAC addresses. There are special MAC addresses defined in LANs for broadcast. The same capability is required in ATM networks to support legacy LAN protocols.

There are two application programming interfaces (API) used by legacy LAN applications, as illustrated in Figure 9.5: high-level API and low-level API.

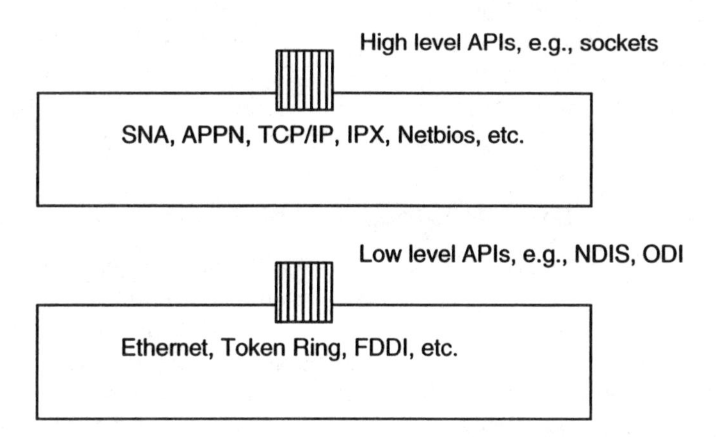

Figure 9.5 Application programming interfaces.

A high-level API provides access to transport layer services. Examples of such APIs include APPC and sockets. A low-level API, on the other hand, allows an LLC user to interact with the LLC layer. Two examples of low-level APIs are network driver interface specification (NDIS) and open data link interface (ODI).

Based on this framework, there are various means of supporting legacy LAN applications over ATM networks. ATM may be viewed as another link layer and the existing network layer protocols can be adapted to run over ATM. That is the method chosen by Internet Engineering Task Force (IETF) and defined as Classical IP over ATM. The main issue with this method is that this solution is network protocol–specific and each protocol used by legacy applications needs to be adapted independently to run over ATM. Doing so requires considerable effort.

Another approach may be to develop a new transport and network layer protocol for ATM. Higher layer protocols for B-ISDN services are expected to be designed and optimized for ATM technology with high-speed and reliable links in the network. Interworking between legacy applications and such protocols would then be provided by a gateway that performs a transformation between legacy protocols and the new protocols. However, standardization of a new protocol would take several years to complete and would not be available to address the solution in the near future.

The other option would be to use the low-level API to provide common access to higher layer protocols. In particular, legacy protocols run on top of the LLC layer. LLC, on the other hand, requires services from the MAC layer to provide services to LLC users. Hence, it might be possible to replace the MAC layer with another layer that would emulate LAN characteristics on top of ATM. If access to this layer is provided through the standard interfaces (i.e., NDIS or ODI) and all required MAC services are provided, then the LLC would not know the fact that there is an ATM network underneath as opposed to a legacy LAN. Since this framework does not require any changes in the LLC and higher layers, no changes in legacy applications and protocols are required with this framework. This is the approach taken by the ATM Forum.

9.2 ITU-T CONNECTIONLESS SERVER METHOD

SMDS is a packet-switched public data service. SMDS (referred to as *connectionless broadband data service* (CDBS) in Europe) provides a LAN-like transport across a wide area (a connectionless packet-switching service) supporting diverse legacy LAN protocols across MANs and WANs. Its features include a framework to support QOS, maximum packet size, and an addressing structure (based on E.164 addresses) that enables group addressing and address screening.

SMDS is a carrier service. As public networks start to migrate into B-ISDN, ATM will be used to provide various carrier services, including SMDS.

Providing these services in ATM networks will preserve customer investment in these services. The ITU-T connectionless service framework has been developed mainly for migrating SMDS services over ATM technology.

The ITU-T recommendations that address connectionless service in B-ISDN are listed in Table 9.1

ITU-T recommends two general approaches to providing connectionless service in B-ISDN: direct service and indirect service.

9.2.1 The Indirect Service Approach

The simplest way to provide connectionless service over ATM networks is to terminate connectionless protocol at the edges of the ATM network through the use of IWUs, as illustrated in Figure 9.6.

In this architecture, the ATM network only provides transmission paths between the IWUs and the connectionless service is provided transparently to the network. ATM connections between the IWUs can be permanent (i.e., PVC-based) or can be established on demand (i.e., SVC-based). All functions required to terminate connectionless protocols, finding target locations, and connection establishment (if connections are established on demand) are performed at the IWUs. This framework is referred to as *indirect service,* since connectionless

Table 9.1
ITU-T Recommendations on Connectionless Service in B-ISDN

ITU-T Recommendation	*Contents*
F.812	Provides a service description of a broadband connectionless data bearer service, including source address validation, addresses based on E.164 numbering, address screening, point-to-point and multicast information transfer, QOS parameters, and interworking with other connectionless and connection-oriented data services
I.211	Describes connectionless data service and identifies two configurations: type (1) connectionless service, in which a connectionless service function (CLSF) is installed outside the B-ISDN, and type (2) connectionless service, in which CLSF is installed within the B-ISDN
I.327	Describes high-layer capabilities for the support of connectionless service and gives functional architectural models for the cases defined in I.211
I.362	Specifies the use of AAL 3/4 for connectionless data services and identifies that routing and addressing are provided by the layer above AAL 3/4
I.364	Describes the support of a connectionless data service based on AAL type 3/4 on B-ISDN

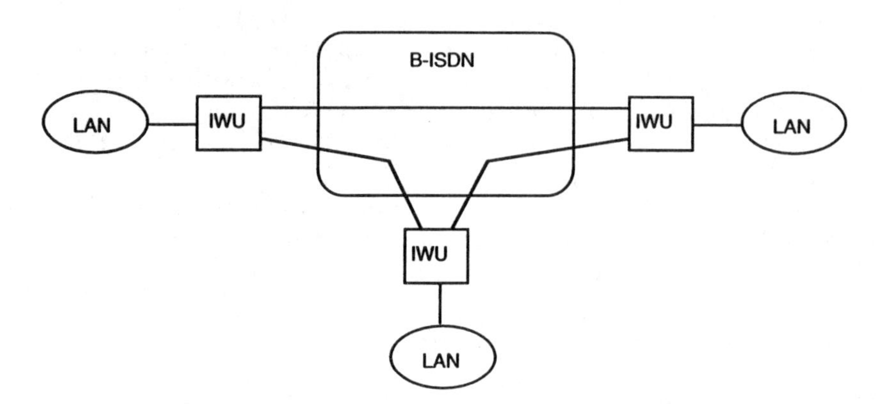

Figure 9.6 Indirect connectionless service.

protocols are not handled within the B-ISDN and the ATM network only provides transparent transfer of information among IWUs.

As discussed previously, using permanent connections reduces the intelligence required in the IWUs for signaling and reduces the signaling handling load in the network. However, this solution does not scale well as the size of the networks and/or number of IWUs used for the connectionless service increases. This is mainly due to the fact that network resources are allocated to each permanent connection for a long period of time, thereby restricting efficient network utilization. Another problem that needs to be addressed is the provisioning of these permanent connections. If too much bandwidth is allocated, then network resources are wasted. If too little bandwidth is allocated, the transmission delay increases and the possibility of losing frames at the IWU buffers increases.

When connections between IWUs are established on demand, the scalability problem is resolved as connections are established for the duration of the communication and terminated at the end. In this case, however, the network is required to handle a large signaling load. In particular, connections that are used for connectionless traffic are, in general, short connections in which few LAN frames are transmitted before they are released. As connections are established dynamically, LAN frames are buffered at the IWUs, causing additional delays. These delays increase as the signaling load in the network increases.

9.2.2 The Direct Service Approach

In the direct approach, the CLSF is provided through the use of CLSs, as illustrated in Figure 9.7.

In this framework, IWUs communicate with CLSs. An IWU simply segments the LAN frames into ATM cells and forwards them to a CLS. Frames

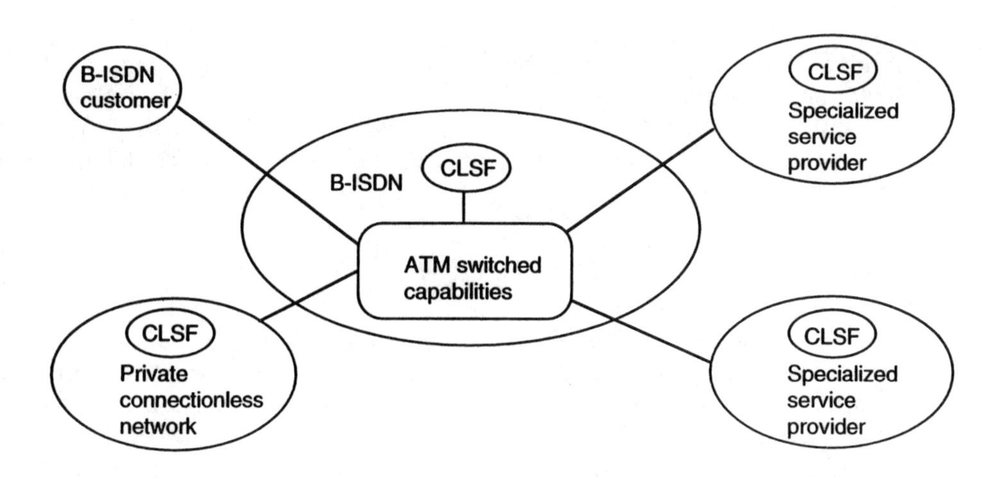

Figure 9.7 Connectionless service function configuration and the use of CLS in B-ISDN.

(carried in ATM cells) go through one or more CLSs until they are delivered to the destination IWU, which reassembles the frame and delivers them to their destination LANs.

ITU-T connectionless service in B-ISDN uses AAL 3/4. At the origin IWU, the LAN frames are encapsulated into connectionless network access protocol (CLNAP) data units before being passed to the AAL. With AAL 3/4, CLNAP-PDU is encapsulated into AAL 3/4 CPCS PDU and segmented into a number of cells. Each cell includes in its payload an indication of whether the cell is the beginning, a continuation, or the last cell of the frame. Each IWU attached to an ATM network is assigned a publicly administered E.164 address. Each connectionless frame includes the E.164 addresses of both the destination and the origin in the first cell payload. This addressing information is used at each server the frame passes through to determine the next hop (i.e., CLS or the destination IWU). The cells of LAN frames are transmitted across the ATM network to a CLS using an ATM connection.

Essentially, a CLS handles the connectionless protocol and realizes the adaptation of connectionless data units into ATM cells for transmission over a connection-oriented network. CLSs may be external or internal to the network. A CLS supports both the connectionless service function and ATM interfaces. Transmission of LAN frames (i.e., cells) uses the connectionless network interface protocol (CLNIP). The CLS consists of various functional elements that include the following:

- Termination of subscriber to CLS protocol (i.e., AAL functions such as detection of lost cells, cell payload CRC check, address screening);

- Route determination and label table setup (i.e., uses the E.164 address of the frame to determine to which CLS the cells of the frame will be routed);
- Protocol termination between CLSs (i.e., cell payload CRC check, cell sequence number of check, message length check, inter-CLS function termination, such as hop count).

Accordingly, some of the functions performed at a CLS include:

- *Connection functions,* which include all port-related functions for the termination of ATM connections. This includes access termination functions (ATF) required to receive/transmit information from/to a B-ISDN user corresponding to physical, ATM, AAL type 3/4, and CLNAP layers. Similarly, network termination functions (NTF) include the necessary functions required to receive/transmit from/to a CLS corresponding to physical, ATM, AAL type 3/4, and CLNIP.
- *Connectionless handling functions,* which include all service-specific functions required for the support of connectionless service in B-ISDN, such as address validation/screening, access class enforcement, group address handling, protocol conversion (between CLNAP and CLNIP), and routing.
- *Control functions,* which are related to connection/resource handling and service processing, including information exchange with other network elements and CLSs. Of particular interest are the access connection control, network connection control, and connection/resource allocation functions.

There are two methods proposed to process LAN frames at CLSs: message-based and cell-based. In the message-based approach, the complete LAN frame is reassembled at each CLS it passes through. After the complete message is constructed, its length and integrity are checked and the routing decision is made. The frame is then segmented into ATM cells using AAL 3/4 and sent to the next CLS (or the destination IWU) along its path.

In the message-based approach, corrupted frames are detected and discarded. Doing so increases the effective resource utilization at the downstream network nodes/links. In particular, corrupted frames are discarded at their destination nodes. Detecting such frames as early as possible along their paths from the origin to destination eliminates resource wastage (both in the network and the downstream CLSs). However, this efficiency is achieved at the expense of reassembly and segmentation delays and additional processing at each CLS.

The cell-based method is proposed to eliminate these disadvantages at the expense of potentially less effective resource utilization in the network. In this approach, the cell assembly and disassembly functions at the CLSs are not performed and all other functions are executed at the cell level. In particular, with the cell-based method, cells are transmitted to the outgoing link in a cut-through mode (as they are received after header processing).

If the message-based method is employed, then the VPI/VCI and MID values, together with the incoming link identification, are used to determine the cells of a particular frame. These cells are stored at a buffer until the whole frame is received. In either case (message- or cell-based), conceptually, two tables are used. The segment handling table correlates incoming VPI/VCI and MID values to the ones used at the outgoing link, whereas a directory table matches the E.164 addresses to VPI/VCI values used at the outgoing link. Clearly, VPI/VCI values used at outgoing link are not needed to be stored as two different entities. This operation is illustrated in Figure 9.8.

As discussed previously, CLSs terminate the connectionless protocol stack and provide functions that include addressing and routing of connectionless frames and adaptation of the connectionless protocol to the intrinsically connection-oriented ATM transport functions. These functions are defined by ITU-T in the context of the CLNAP and CLNIP. The general protocol structure for connectionless data service in B-ISDN is illustrated in Figure 9.9.

9.2.2.1 Connectionless Network Interface Protocol

CLNAP functionality provides the connectionless layer service, which includes routing, addressing, and QOS selection. Figure 9.10 illustrates the protocol architecture for supporting the connectionless layer service.

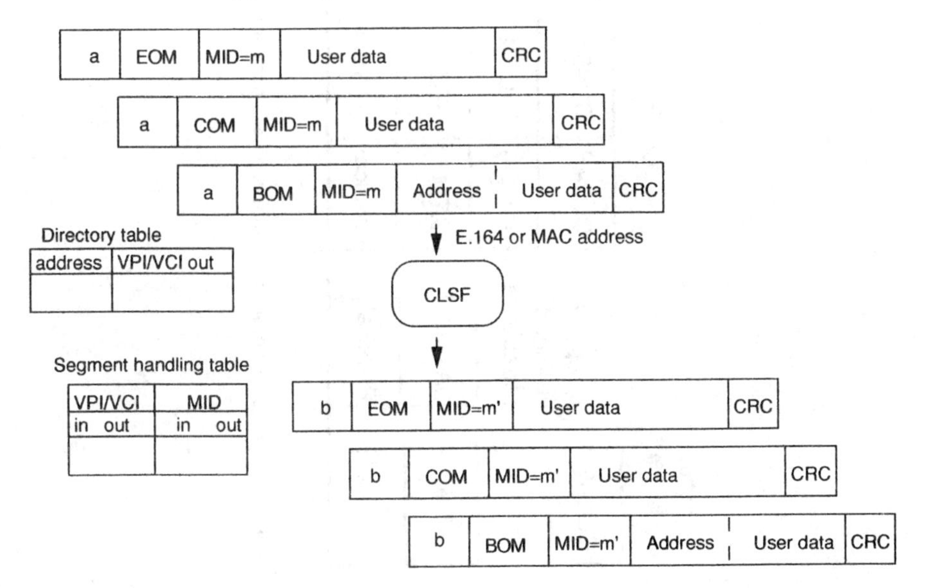

Figure 9.8 Label swapping in a CLS.

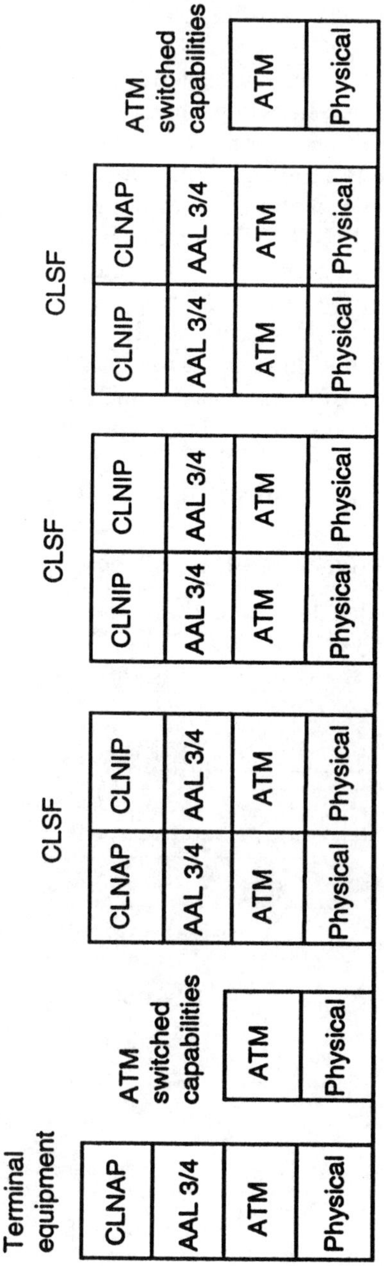

Terminal equipment
CLNAP
AAL 3/4
ATM
Physical

ATM switched capabilities
ATM
Physical

CLSF
CLNAP | CLNIP
AAL 3/4 | AAL 3/4
ATM | ATM
Physical | Physical

CLSF
CLNIP | CLNIP
AAL 3/4 | AAL 3/4
ATM | ATM
Physical | Physical

CLSF
CLNIP | CLNAP
AAL 3/4 | AAL 3/4
ATM | ATM
Physical | Physical

ATM switched capabilities
ATM
Physical

Figure 9.9 Protocol structure for connectionless data service.

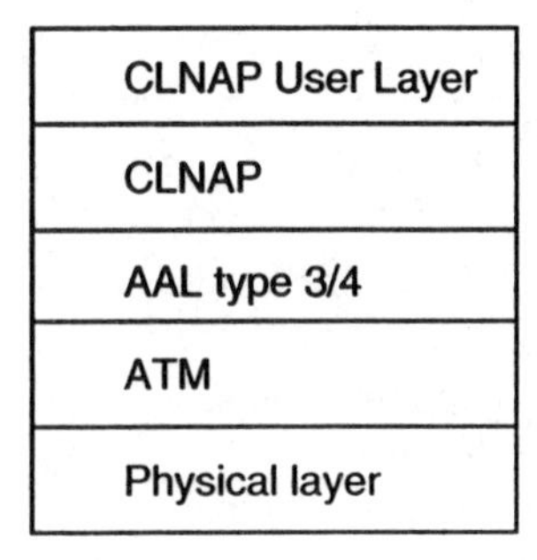

Figure 9.10 Connectionless layer service.

The CLNAP layer uses the services of AAL type 3/4 and provides its service to the CLNAP layer user. The CLNAP expects from AAL type 3/4 the transfer of variable-size LAN frames from the source to one or more destinations in a sequential manner (i.e., among CLNAP entities at different nodes). This service is performed in a connectionless manner in such a way that lost or corrupted LAN frames are not retransmitted.

Various functions performed at CLNAP include the following:

- Delineation and transfer of CLNAP data units;
- Addressing, which allows the CLNAP user layer to select to which entity (or entities) the CLNAP-SDU is to be delivered as well as to indicate to the CLNAP user at the destination node the source of the delivered CLNAP;
- QOS indication, which allows the definition of the QOS desired for the CLNAP-SDU.

Based on these functions, the structure of the CLNAP-PDU is defined as illustrated in Figure 9.11.

The user information field has a variable length of up to 9,188 bytes and is used to carry the CLNAP-SDU. Both the destination and source address fields are 8 bytes long and contain a 4-bit address-type subfield followed by a 60-bit address subfield. Address type indicates whether the address subfield contains a publicly administered individual address or a publicly administered group address. The address field, on the other hand, indicates which CLNAP entities the CLNAP-PDU is destined for.

The CLNAP-PDU also includes a 6-bit high-layer-protocol-identifier field used to identify the CLNAP entity to which the CLNAP-SDU entity is to be passed at the destination nodes. Since the PDU is defined at 4-byte boundaries, it may be necessary to pad the user information field by 0 to 3 bytes. The 2-bit PAD length field indicates the number of PAD bytes added to this field. The QOS field is 4 bits long and indicates the QOS requested for the CLNAP-PDU.

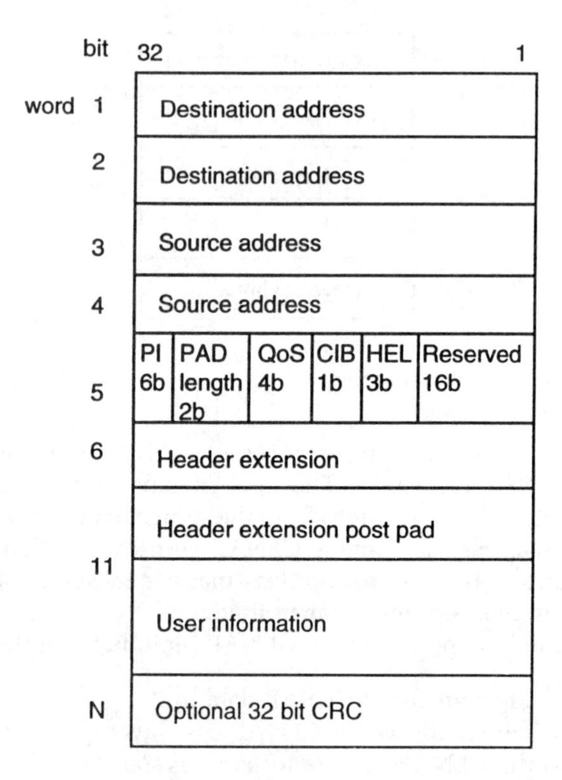

Figure 9.11 CLNAP-PDU structure.

The CLNAP-PDU also uses an optional 32-bit CRC, performed (if used) over the CLNAP-PDU. The CRC indication bit (CIB) indicates the presence or absence of the CRC at the PDU. The header extension length (HEL) is a 3-bit field that can take any value from 0 to 5 and indicates the number of 32-bit words in the header extension field.

9.2.2.2 Connectionless Network Interface Protocol

The CLNIP layer supports the transfer of connectionless data among network nodes. The protocol uses AAL type 3/4, which provides a sequential transfer of CLNIP-PDUs in the unassured mode. No retransmission capability for lost or corrupted data is provided at the CLNIP.

The CLNIP protocol data structure is the same as in Figure 9.10 and its fields are defined similar to that of CLNAP-PDU fields. In particular, the destination address may be either an individual address or a group address with the source address always being an individual address. The protocol identifier (PI) has the same value as in CLNAP-PDU if no encapsulation is performed. Otherwise, it takes a value between 44 and 47, indicating that the PDU is encapsulated. The PAD length indicates the length of the PAD field (0 to 3 bytes) at the end of the information field so that the field length is an integral multiple of 4 bytes. QOS indicate the QOS requested for the CLNIP-PDU. CIB is the CRC indication bit to specify the existence (or lack of) the optional 32-bit CRC. HEL can take any value from 0 to 5 and indicates the number of 32-bit words in the header extension field. User information can be up to 9,236 bytes in length carrying the user information. Figure 9.12 illustrates the encapsulation of a CLNAP-PDU within a CLNIP-PDU.

The CLNAP-PDU is aligned to have an integral multiple of 32-bit words in length. The CLNIP layer then appends the CLNIP-PDU header and passes the PDU to the AAL type 3/4 CPCS, which adds its own header to the PDU. Then the CPCS-PDU is passed to the SAR sublayer to be segmented into 44-byte (AAL type 3/4 payload) packets. The first 44 bytes of the PDU is marked BOM, the last cell is marked EOM, and all others are COM, which is included at the AAL type 3/4 header before the payload is passed to the ATM layer for cell header insertion for transmission.

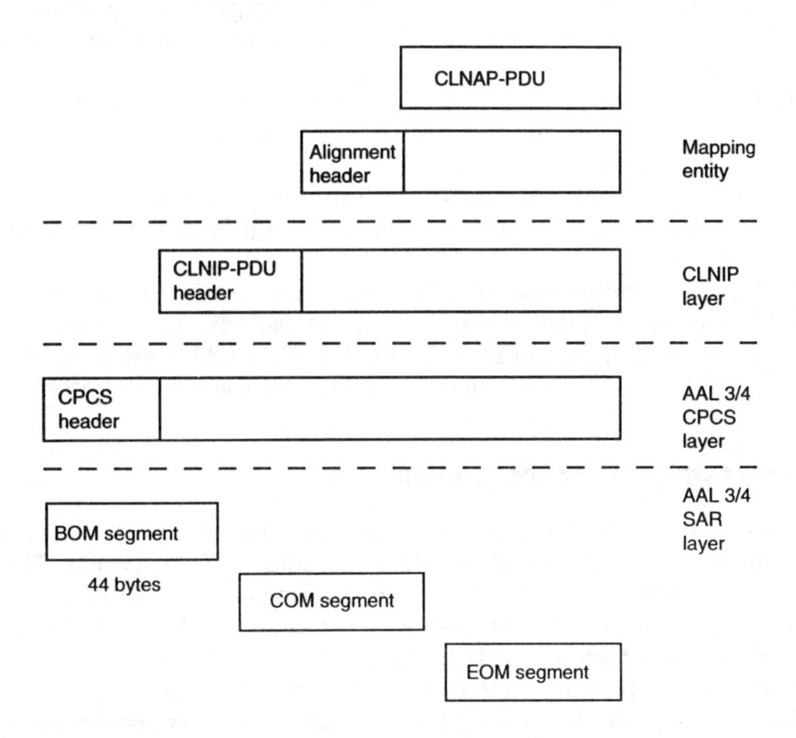

Figure 9.12 Encapsulation of a CLNAP-PDU within a CLNIP-PDU.

In the direct approach, there are two sets of connections: between an IWU and a CLS and between CLSs. These are all ATM connections and can be either permanent or switched. Any combination of the two may be used for any connection in providing the connectionless service. For example, to minimize the signaling-related processing at the IWU, IWU-CLS connections can be permanent, whereas switched-on-demand connections may be used between CLSs. The use of permanent connections at the IWUs does not introduce the scalability problem as was the case with the indirect approach. This is because only one connection to a CLS at the IWU is required for the LAN frames to be forwarded to the network.

9.2.3 Comparison Between the Direct and Indirect Approaches

The direct method has a number of advantages over the indirect method. Each IWU requires a single connection to the network (i.e., to a CLS). IWUs are not responsible for routing decisions. This responsibility is shared among CLSs. Compared with having a connection between every pair of IWUs in the indirect approach, the total number of connections required with the direct approach is reduced significantly. Since connections are aggregated into a small number of ATM connections in the network, better resource utilization is expected with the direct approach.

The main advantage of using CLSs in the direct approach is also its main disadvantage. CLSs perform a large set of functions and require considerable processing. They may cause performance bottlenecks as the connectionless traffic in the network increases. With the message mode, large buffers are required to assemble each LAN frame. This also contributes to end-to-end delay of each LAN frame.

ITU-T connectionless service framework uses AAL 3/4, mainly because it was designed with SMDS in mind. As discussed next, other connectionless service proposals (in private networks) are based on AAL 5 services. This may complicate the interoperability between ATM LANs and public ATM WANs.

9.3 ATM FORUM LAN EMULATION

The ITU-T connectionless service framework provides the capability to interconnect legacy LANs and SMDS networks across ATM networks. The ATM Forum, on the other hand, is focused on defining an ATM LAN architecture and how to support legacy LAN applications in ATM LANs. Unlike the ITU-T-defined service which uses E.164 addresses, the ATM Forum connectionless service framework is based on the use of MAC addresses and specifies how to emulate the connectionless legacy LAN services in connection-oriented ATM networks. The main objective of the LAN emulation (LE) architecture, then, is to run existing LAN applications over ATM networks without any changes and for interoperability among software applications residing on legacy LAN systems and ATM attached end systems.

9.3.1 Basic Configurations and the LAN Emulation Service

Different types of connectivity for connectionless service are desirable in ATM networks, as illustrated in Figure 9.13. In legacy LANs, connectivity between an Ethernet and a Token Ring cannot be easily achieved. Similarly, this interconnection is not a goal over ATM networks. The envisioned interconnectivity over an ATM network is among 802.3 LANs, among 802.5 LANs, among ATM end stations as well as servers, and among stations attached to a legacy LAN and ATM stations.

The main objective of the LE framework is to enable this environment so that legacy LAN applications and protocols in different types of networks can run without any changes. This is achieved based on the use of the low-level API to provide a common access to higher layer protocols. In particular, legacy applications and protocols run on top of the LLC layer. If the LLC layer is kept the same, then protocols running on top of the LLC do not require any changes with ATM. The LLC requires services from the MAC, whereas MAC uses the services of the physical layer. The ATM architecture uses the services of the ATM adaptation, ATM, and physical layers. The only set of services missing in the ATM framework is that of the MAC layer. In the LAN emulation (LE)

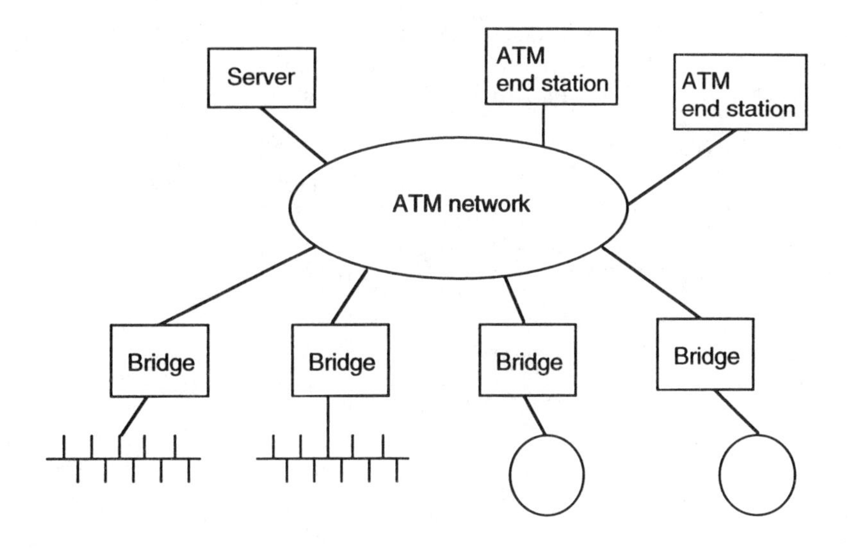

Figure 9.13 LAN emulation in ATM.

framework, the MAC layer is replaced with the LE service which provides MAC services to the LLC layer in ATM networks (i.e., it emulates LAN characteristics on top of ATM). If access to this LE layer is provided through the standard interfaces (i.e., NDIS or ODI) and all of the required MAC services are provided, then the LLC would not know the difference.

Figure 9.14 illustrates this architecture: the layers, interfaces, and the areas that need to be addressed to provide LE service based on ATM. The LE to LLC layer interface is either NDIS or ODI. The LE layer specification includes the LE layer to AAL interface, which allows access to the ATM network. The adaptation layer used in LE is AAL 5.

Various functions that need to be defined at the LE layer include initialization, address resolution, and data forwarding/receiving. Furthermore, in order for two peer LE layers to talk to each other and use the functions defined at the LE layer, an LE protocol needs to be standardized.

Extending the connectionless service environment of Figure 9.13, envisioned LE service and the related architecture are illustrated in Figure 9.15.

In the first configuration, the LE service allows protocols and applications developed for legacy LANs to run on stations directly attached to an ATM network. In the second configuration, an end station connected to an ATM network communicates with a station attached to a legacy LAN. In this case, in addition to the LE service at the ATM end station, a bridging function is needed between the legacy LAN and the ATM network. This bridge includes the ATM end station LE architecture on its interface attached to the ATM network and the MAC layer on the other side, where it interfaces to the legacy LAN. In order to provide a connection between the two, a MAC relay function is needed. This framework allows today's bridging methods such as those defined in 802.1 to be employed without any modifications if the LE service on the ATM side provides the illusion of having a traditional LAN on an ATM port.

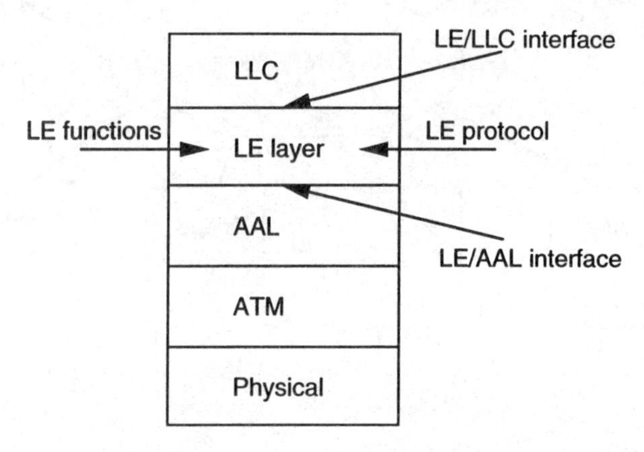

Figure 9.14 The LAN emulation architecture.

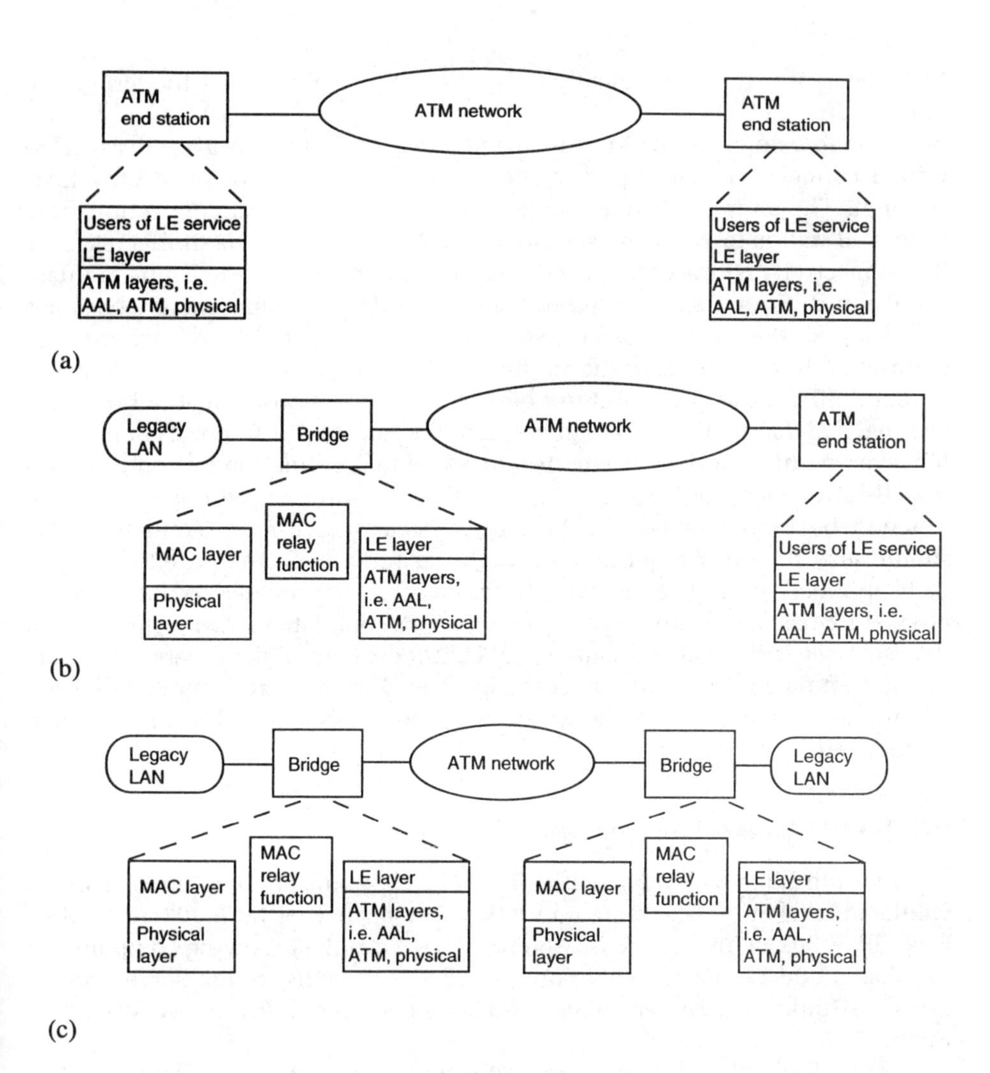

Figure 9.15 LAN Emulation among different types of stations in ATM networks: (a) configuration 1—ATM to ATM, (b) configuration 2—ATM and a legacy LAN station, and (c) configuration 3—legacy LAN to legacy LAN over ATM.

Finally, the third configuration presents the case where two legacy LANs are connected across an ATM network. Similar to the second configuration, this interconnection requires bridges to connect legacy LANs to ATM with the same bridging functionality.

9.3.2 ATM LAN Segment

In legacy LANs, the membership of an individual LAN segment is defined by a physical connection to a physical shared medium. Membership in an ATM

LAN segment is defined logically rather than physically, hence the term *emulated LAN.*

Membership in an ATM LAN segment may be defined by a multicast ATM virtual connection that emulates the broadcast channel for that ATM LAN segment. The simplest way to achieve this is to establish a connection from every end station to every end station in a LAN segment. This is similar (as far as the connectivity is concerned) to the ITU-T indirect approach to connectionless service in B-ISDN. As discussed in Section 9.3, this solution quickly becomes infeasible as the number of end stations in an emulated LAN increases. In addition, a large portion of the traffic in a LAN segment is unicast. Hence, it is more efficient to use point-to-point ATM connections than the broadcast channel that defines the LAN segment. Establishing a point-to-point connection, however, requires knowledge of the address of the destination station in a form that B-ISDN signaling can understand. This requires an address resolution function be implemented. Furthermore, the architecture of emulated LAN should also provide broadcast and multicast functions as required by legacy LAN applications and protocols. An emulated LAN provides communication of LAN frames among all its users. One or more emulated LANs could run on the same network. Each emulated LAN is independent of the others, and users in one emulated LAN cannot communicate directly across emulated LAN boundaries; communication between emulated LANs is possible only through routers and bridges.

9.3.3 LAN Emulation Framework

ATM Forum LE over ATM specification [4] defines the architecture of a single emulated LAN only. An emulated LAN is either Ethernet/IEEE 802.3 or Token Ring (IEEE 802.5). In either case, the LE service has the following components: LE clients and LE service that consists of an LE configuration server, an LE broadcast/unknown server (BUS), and an LE server (LES), as illustrated in Figure 9.16.

The LE client is the entity in end stations that performs various control functions, data forwarding, and address resolution. The LES implements the control coordination function for the emulated LAN that includes registering and resolving MAC addresses. The BUS provides services to support broadcast/multicast traffic and initial unicast frames that are sent by an LE client before the data target ATM address has been resolved. The LE configuration server implements the assignment of individual LE clients to different emulated LANs. There are six types of ATM connections used in LE service, as illustrated in Figure 9.16.

- *Configuration direct VCC* is set up by the LE client to perform the configuration protocol exchanges with the LE configuration server.

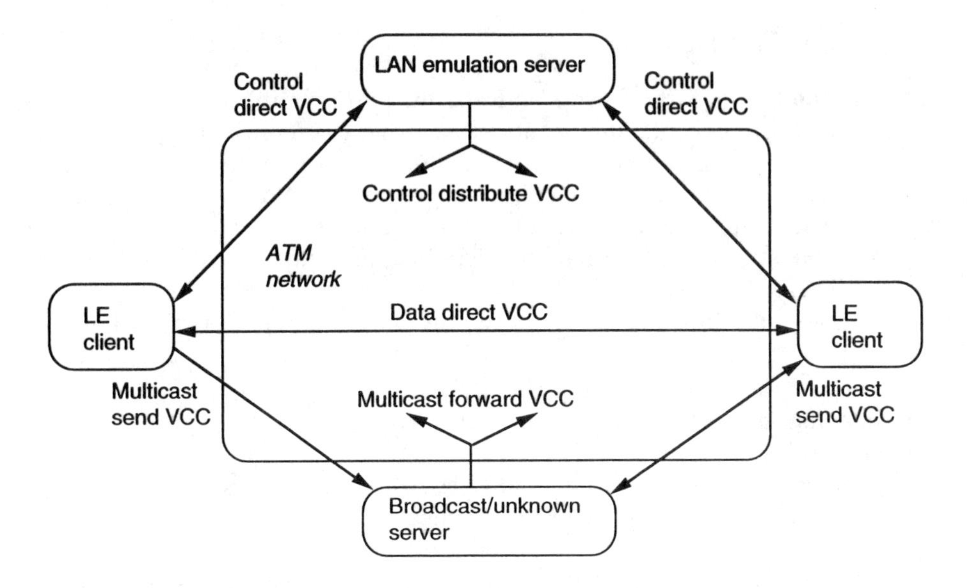

Figure 9.16 Components of LAN emulation service.

- *Control direct VCC* is set up by the LE client as part of the initialization phase and it is maintained by both the LE client and the LES as long as LE client participates in the emulated LAN.
- *Control distribute VCC* is set up by the LES as part of the initialization phase. The LE client is required to accept this VCC. It is maintained by both the LE client and the LES while participating in the emulated LAN.
- *Data direct VCC* is used for an LE client to send user frames to another LE client after the origin LE client learns the address of the destination LE client. If the end station does not have resources to establish a data direct VCC, it uses a multicast send data VCC to request the BUS to forward its frames to the destination LE clients.
- *Multicast send VCC* is established from an LE client to the BUS. It is maintained while the LE client is participating in the emulated LAN.
- *Multicast forward VCC* is a point-to-multipoint VCC with the BUS being the root and all LE clients being the leaves.

An LE client is identified by two addresses: an individual IEEE 48-bit MAC address and one or more 20-byte ATM address of which one is designated as the primary ATM address. An LE client uses its primary ATM address for establishing control direct VCC, multicast send VCC, and as the source address in the control frames. An LES is assigned one or more ATM addresses. This address may be shared with the BUS. Similarly, a BUS may have one or more ATM addresses assigned to it, which may be shared with the LES.

9.3.3.1 LAN Emulation Client

The LE client is a part of each end station participating in an LE service. Each LE client represents one or more users that are identified by their MAC addresses. The main function of the LE client is to provide its users with a MAC connectionless service by using the services of AAL 5 to communicate with other LE clients and remote LE service components. These communications may take place over permanent, switched, or mixed (both permanent and switched) ATM connections.

Various functions performed at the LE client include the following:

- Establishing a connection to the LES;
- Establishing a connection to the BUS;
- Registering with the LES;
- Generating and responding to address resolution requests;
- Transmitting and receiving MAC frames.

Before joining an emulated LAN, the LE client is connected to an LE configuration server for its configuration information, which includes the ATM address of the LES, one or more MAC addresses, emulated media type (IEEE 802.3 or 802.5), and maximum frame size for the media being emulated (i.e., AAL-5 SDU size of 1,516, 4,544, 9,234, or 18,190 bytes). The method used for LE clients to obtain the ATM address of the LE configuration server is presented in Section 9.3.3.2.

Following the configuration, the LE client should have the ATM address of the LES. Using this ATM address, the LE client establishes a control direct VCC with the LES. This connection is mainly used for address resolution. When the LE client has a frame to send to another LE client and does not know its ATM address, it sends a message to the LES requesting the ATM address of the destination station. The LES returns the requested address using the same VCC (i.e., control direct VCC). When the LE client has the ATM address of the destination LE client, it can establish a switched VCC (i.e., data direct VCC) and transmit its frames using this connection. Alternatively, the LE client can send its frame to the BUS at the same time it sends the address resolution request to the LES. The LE client determines the address of the BUS by using the address resolution procedure. After obtaining the ATM address of the bus, the LE client establishes a multicast send VCC and uses this connection to transmit frames to the BUS. Upon receiving a frame from a multicast send VCC, the BUS multicasts the frame to all LE clients in the emulated LAN.

In summary, an LE client first goes through an initialization phase to obtain its configuration information and the addresses of both the LES and BUS. Upon receiving a frame from a higher layer, an LEC first tries to send the frame directly to the destination LE client over a data direct VCC. This would mean that each LEC maintains an address resolution table that provides a mapping between the destination MAC addresses and data direct VCCs. If the destination MAC address is not in this table, it checks whether or not the ATM address corresponding to the MAC address is known. If known, the LE client establishes a data direct VCC to the destination LE client using UNI signaling. If not, the LE client sends an address resolution request to the LES. Optionally, it can, at this time, send the frame to the BUS for multicasting to all LE clients in the emulated LAN. After receiving a response to its address resolution request, the LE client can use signaling to establish a data direct VCC to the destination client.

9.3.3.2 Configuration Server

An LE client needs to obtain the ATM addresses of both the LES and the BUS and establish connections to each server. Obtaining the ATM address of the BUS is relatively straightforward once a connection to the LES is established: send an address resolution request for the broadcast MAC address. The reply from the BUS would include its own ATM address, which is used by the LE client to establish a multicast send VCC to the BUS.

Determining the ATM address of the LES is not as simple. An ATM network does not provide a broadcast service similar to that of legacy LANs. One alternative to make the ATM address of the LES known to LE clients is to configure it manually at each LE client in an emulated LAN. This is not a preferred solution in ATM networks that support a large number of emulated LANs and/or large number of LE clients. The LE configuration service allows emulated LANs to be built dynamically. However, the existence of this server does not provide a solution to the basic problem of determining its ATM address (and the ATM address of the LES).

To remove any configuration burden, an LE client locates the configuration server (LECS) using one of the following methods, attempted in the order they are listed.

1. Get the LECS address via the ILMI: The LE client must issue an ILMI message to obtain the ATM address of the LECS for the UNI it is attached to.
2. Well-known LECS address: If the ILMI attempt fails or the LE client fails to establish a connection with the ATM address obtained through the ILMI, then a well-known ATM address may be used to open a configuration VCC to the configuration server.
3. Configuration direct SVC: This may be established using the broadband low-level code point for ATM LE configuration direct VCC.

4. LECS PVC: If the LE client cannot establish a VCC to the well-known ATM address of the LECS, then the well-known PVC with VPI = 0, VCI = 17 is used for the configuration direct VCC.

9.3.3.3 LE Server

The LES provides an address resolution mechanism for resolving MAC addresses. Various functions performed at the LES include responding to registration frames, forwarding address resolution requests, and managing LE client address registration information.

If the ATM address of the destination LEC is not known at the originating LE client, an address resolution request is sent to the LES over a control direct VCC. The LES forwards this request to all LE clients in the emulated LAN over a control distribute VCC. This address resolution request includes, among other information, the source and destination MAC addresses and the ATM address of the LE client originating the request. All LE clients in the emulated LAN are required to receive and process address resolution requests from the LES. This processing at an LE client involves checking the destination MAC address included in the request and, if it is responsible from that particular MAC address, respond with an address resolution reply that contains its own ATM address. These responses are sent to the LES over control direct VCC. The LES forwards this reply to the LE client that originated the address resolution request over its control direct VCC.

The control distribute VCC is a point-to-multipoint ATM connection with the LES as the root and all LE clients registered in the emulated LAN as the leaves. An LES maintains an address resolution table that maps MAC addresses of LE clients with the corresponding ATM addresses and control direct VCC identifiers. These MAC addresses are primarily the ones that are assigned to ATM end stations participating in the emulated LAN. Some of these end stations may control another set of MAC addresses (i.e., bridges) and not all of these MAC addresses may be registered at the LES. Hence, even if all ATM end stations participating in an emulated LAN are known to the LES, control distribute VCC may still be necessary to be able to reach stations that are attached to some of these ATM end stations.

9.3.3.4 Broadcast/Unknown Server

The BUS mainly broadcasts LAN frames with unknown MAC addresses (at the origin LE client) to all LE clients participating in an emulated LAN. The various functions performed at a BUS include receiving frames with unknown MAC addresses from a multicast send VCC, assembling the frame, and broadcasting over multicast forward VCC.

An LE client obtains the ATM address of the BUS via an address resolution request using its control direct VCC to the LES and establishes a multicast send

VCC to the BUS. The multicast forward VCC is separate from the multicast send VCCs. A multicast forward VCC is a point-to-multipoint ATM connection, with the BUS as the root and the LE clients as the leaves. This point-to-multipoint VCC may be established as a set of point-to-point connections or may be a tree connection, depending on whether or not the network supports multicast.

If an LE client does not know the ATM address that corresponds to the destination MAC address in a LAN frame, it sends an address resolution request to the LES. As an option, it can transmit the frame to the BUS in parallel. A number of frames may be transmitted over a multicast send VCC to the BUS until a response to the address resolution request is received. The main objective in using the BUS to forward these frames is to reduce the total delay in sending LAN frames to an unknown ATM LE client.

The BUS framework is very simple to implement. However, it does not scale well as the size of the emulated LAN increases. In particular, it is a centralized scheme and cells of different frames cannot be interleaved on the multicast forward VCC, since AAL 5 does not allow VC multiplexing. This means that frames are transmitted as cell trains, which necessitates all the cells of a frame to arrive at the BUS before the frame can be transmitted. The main consequence of this is the delay it takes to receive the whole frame. As the number of LE clients and/or traffic in the emulated LAN increases, the delay at the BUS increases, causing the BUS to become a performance bottleneck.

Similarly, as the number of frames broadcast in the network increases, end stations start to receive a large number of frames. Each frame is required to be processed at each end station to decide whether to copy (i.e., the destination MAC address is the same as one of the MAC addresses the station supports) or discard the frame (i.e., no match). Since an end station discards a large portion of the frames it receives, a considerable processing capacity is wasted. It also adds to the cost of the end station, since high-speed filtering and/or large buffers are required to be able to process a high volume of arriving frames.

Another potential problem with this framework is the effective utilization of network resources. In particular, multicast frames are broadcast to all end stations in the emulated LAN over a multicast tree. The consequence of this is that more network resources than needed are used to transmit unicast frames to their destination end stations, wasting bandwidth in the network. An alternative is provided in the LE specification by using point-to-point VCCs from the server to the LE client to forward the frames intelligently (as opposed to using the multicast tree).

9.3.3.5 Data Frame Formats

The LE framework supports IEEE 802.3/Ethernet and IEEE 802.5 (Token Ring) standards. Accordingly, two LE frame formats for each protocol are defined, as illustrated in Figure 9.17.

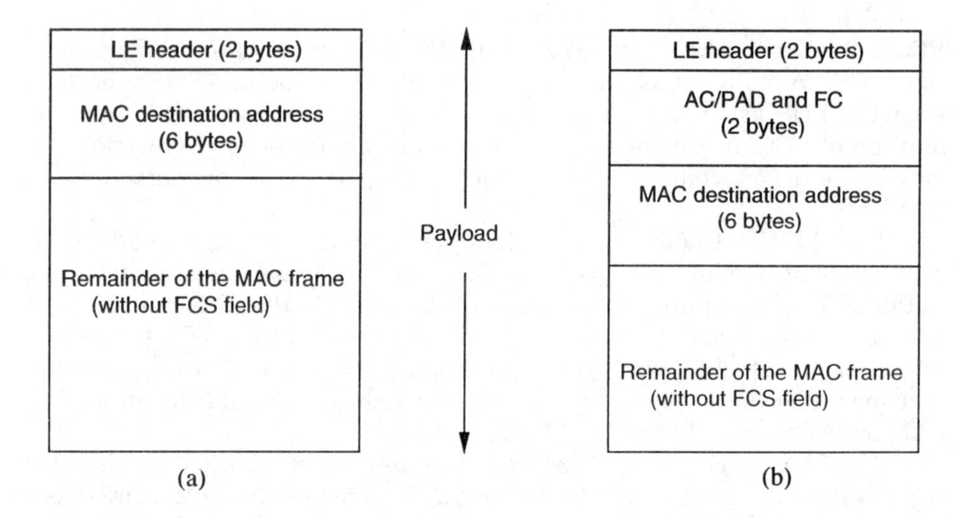

Figure 9.17 LAN emulation frame formats: (a) IEEE 802.3/Ethernet, and (b) IEEE 802.5.

The 2-byte LE header contains the LE client identifier (LECID) used by an LE client to identify and discard frames it has sent to the BUS (which it receives back as the BUS broadcasts all frames to all LE clients in an emulated LAN). LECIDs are assigned uniquely by the LES at the time the LE client first contacts the LES.

The payload contains either an 802.3/Ethernet frame or an IEEE 802.5 frame, excluding the MAC frame check sequence (used for error checking), since error correction capability is already included in AAL 5 through a CRC-32.

9.3.3.6 Summary of ATM Forum LAN Emulation Framework

An emulated LAN consists of a number of LE clients and a single LE service. Communication among LE clients and between LE clients and the LE service is performed over ATM virtual connections (VCCs).

An emulated LAN has two "logical" multicast servers to emulate shared-medium characteristics of legacy LANs. When a station is connected to an emulated LAN, it establishes a circuit (or uses a preestablished circuit) and registers to the LE service its MAC and ATM addresses. The station also registers itself to send and receive broadcast MAC frames and may register for group or functional addresses as well. Once a station establishes a connection to the LES and another one to the BUS, it is ready to process incoming and outgoing frames.

If an LE client knows the ATM address of the destination of its frame, a direct connection may be established between the source and the destination nodes and the frame is transmitted over AAL 5. If the ATM address of the destination is not known, the LE client forwards the frame to the BUS. It also sends an address resolution request to the LES. If the LES knows the ATM address that corresponds to the destination MAC address, it replies to the LE client without any additional processing. If not, then it broadcasts the address resolution request over a control distribute VCC. When an end station responds to the address resolution request to the LES, the LES forwards the response to the LE client requested the address resolution.

If the frame with an unknown ATM address is sent to the BUS, the BUS uses its multicast forward VCC and broadcasts the frame to all LE clients in the emulated LAN or uses its optional intelligent forwarding capability to forward the frame to its destination over a point-to-point VCC.

9.4 CLASSICAL IP AND ARP OVER ATM

IETF RFC 1577 (classical IP and ARP over ATM) describes the initial deployment of ATM within classical IP networks. In this context, the ATM network is viewed as providing a LAN segment replacement for a LAN, local-area backbones between legacy LANs, and dedicated circuits (PVCs) between IP routers. The two areas addressed are the transfer of IP datagrams and ATM address resolution protocol requests and replies over AAL 5. In this context, the ATM network is viewed as a logical IP subnetwork (LIS). The LIS is a closed logical ATM network within which a separate IP protocol administrative entity configures its hosts and routers.

In this scenario, each separate administrative entity configures its end stations (and routers) within a closed LIS. Each LIS operates and communicates independently of other LISs on the same ATM network. Each end station communicates directly to other end stations within the same LIS. Communication to end stations outside of the local LIS is provided via an IP router, which is configured as a member of one or more LISs. Hence, there may be multiple disjoint LISs operating over the same ATM network. The requirements for IP members operating in an ATM LIS are:

- All members have the same IP network/subnet number and address mask.
- All members within an LIS are directly connected to the ATM network.
- All members outside of the LIS are accessed via a router.
- All members of an LIS must have a mechanism for resolving IP addresses to ATM addresses and vice versa when using SVCs.

Implementing IP over ATM directly requires translating an IP address to an ATM address. This is achieved by maintaining an IP-ATM-ARP server, which maintains tables that map IP addresses with corresponding ATM addresses.

This route server is defined in each LIS in order to support switched ATM connections.

The ATM address of the server is configured to each end station in the LIS (alternatively, PVCs are preestablished between each end station and the server). When a new end station is added to an LIS, it sends an ADD HOST message to the server specifying its IP and ATM addresses. This message is sent either over an SVC (since the ATM address of the server is known) or over a PVC (which is preestablished). The server responds with a CONFIRM message, which includes a reserved VCI used by the server to send control messages to the end station. This VCI cannot be used by any station in the network other than the server. Hence, the server has the IP and ATM addresses of all end stations registered in the LIS. Each entry is time-stamped. This time stamp is used to remove aged IP stations from the server. The server may check from time to time to find out if the end station is active. Similarly, end stations may send ADD HOST messages from time to time to refresh their corresponding entries in the server.

When an IP station needs to send a packet to another IP station, it first checks its address cache to see if the ATM address of the destination IP is in the cache table. If it is, then the end station establishes an SVC (if the connection does not already exist) and sends its packet. If the ATM address of the destination IP address is not known, the end station sends an ARP-QUERY message over its ATM connection to the server. Once the server associates the received IP address with the corresponding ATM address, it sends back an ARP-REPLY message over the reserved VCC specifying the destination IP and ATM addresses. Then the end station requests from the network a switched ATM connection using the ATM address of the destination IP host and transfers its data.

In this scenario, we assumed the ATM address of the server is configured to all end stations so that they can establish a VCC to the server to pass their IP and ATM addresses. When these connections are PVC-based, inverse ATM address resolution protocol is used. In this protocol, the end station sends its IP and ATM address over this PVC. The server learns the IP and ATM addresses of the end station and the VCI used by the end station to reach the server. The server then returns its own IP and ATM addresses, as well as the reserved VCC it will use to communicate with the end station. The connection can be terminated by either end.

If the server does not know the corresponding ATM address of the destination IP address, it sends back a negative acknowledgment stating the destination station is not known.

ATM addresses currently do not support multicast addresses. The consequence of this is that there is no mapping available between the two. However, this does not restrict end stations to send or receive broadcast frames.

The IP layer runs on top of the LLC layer. The maximum transmission unit for IP members operating over the ATM network is equal to 9,180 bytes. Together with the 8-byte LLC header, the maximum AAL 5 PDU size is restricted to less than or equal to 9,188 bytes. The ATM layer needs to provide interfaces to different protocols. There are two approaches to multiplexing network interconnect traffic in ATM (cf. RFC 1483). The first method allows multiplexing of multiple protocols over a single VCC. The protocol of the carried frame is identified by prefixing the frame with an LLC header. This encapsulation method is used with PVCs or whenever it is not practical to have separate VCs for each protocol. The identification of VCCs for different protocols must be done at the subscription time (i.e., when they need to be configured). The second method is to use separate VCCs for each protocol supported in the end station. This is mainly used with SVCs. The signaling messages include a low-layer compatibility information element which allows negotiation of the end protocol used between two end stations.

Various addressing models have been proposed to realize IP protocol over ATM: classical IP, wide-area ATM subnet, and peer models. In the classical IP model [6], different LISs are cascaded with network layer protocols. Considering ATM as just another data link layer protocol, VCCs in this model are terminated at each router. That is, every packet transmitted via one or more ATM cells is reassembled at each router it visits. Packet forwarding to the next LIS is carried out by the network layer routing function at the router.

IP protocol entities for end stations and routers periodically exchange IP routing information with each other. Using this information, the source (ES1) recognizes the next router's IP address (RT1) toward the destination (ES2) and resolves its ATM address from its IP address using ARP. ES1 then sets up a connection to RT1 and sends its frame through this connection. Once RT1 receives the frame, it resolves the next hop router's ATM address, sets up a connection (if one does not exist), and forwards the frame. The procedure is repeated until the frame is delivered to ES2.

In this method, intermediate routers have to store arriving IP frames until they resolve the next hop ATM address and set up a connection to that router. This in turn limits the scalability and the efficiency of the method, and requires large buffers as the transmission speeds and the distance between the routers increase.

The wide-area ATM model [27] attempts to address this problem by separating a number of LISs without routers and providing a VCC across them by using the next hop resolution protocol (NHRP) [28].

The NHRP protocol depends on next hop servers (NHS). At least one NHS is located at each LIS and is used to resolve the next hop ATM address in the whole ATM network from any destination network layer address. Each NHS serves a set of end stations that are directly connected to the same LIS. For example, when a source end station attempts to send data to a destination end station in another LIS, it sends an NHRP request packet to the NHS to resolve the ATM address of the destination end station. If NHS in the LIS the connection originated from does not have this information, it forwards a request packet to the destination's NHS, which resolves the end station's ATM address. The address is then returned to the source end station, while the acquired ATM address is cached on intermediate NHSs on the way back from the destination NHS toward the source NHS. The source end station establishes a connection to the destination and forwards its frame.

The address resolution data administered by an NHS is collected by the periodic address registration of each end station directly connected to the LIS. The main problem with this approach is the time involved to resolve the addresses and the delay in setting up a connection.

Another approach considered by IETF is the peer model. In this method, the ATM address includes the hierarchical network prefix and the host address parts, described by each terminal's MAC address. The peer model directly maps multiple network layer protocol addresses algorithmically. End stations can use these ATM addresses to set up ATM connections without going through the ARP. Therefore, it eliminates the ARP delay. However, it requires end stations to support a completely different IP over ATM and ARP from the current standard defined in RFC 1577. Furthermore, this solution requires each ATM switch to handle multiple network layer routing protocols. It also requires network layer addresses to be updated frequently, since end stations move on the same administrative network. This approach was not accepted by the IETF for further consideration.

9.5 CONGESTION CONTROL FRAMEWORK FOR CONNECTIONLESS SERVICE IN ATM

ATM networks are expected to support a variety of services with different QOS requirements. Connectionless service over ATM discussed in this chapter is also referred to as best effort service and ABR service. Their source behavior are often bursty and unpredictable. They are delay-tolerant and do not require explicit service guarantees. Typical examples of connectionless traffic are file transfer and datagram services.

Implicitly, however, they require the CLR to be minimal. These are data applications and require end-to-end data integrity. Higher layer protocols are used to ensure the correct exchange of data packets between end stations. This is achieved by the retransmission of packets that are lost in the network (i.e., dropped in the network due to buffer overflows) and/or received at the destination end station with errors. However, the retransmission of packets reduces the effective resource utilization in the network. As the traffic increases, more and more packets are lost, causing more and more retransmissions by the end stations. More retransmissions increase the traffic even further in the network, eventually causing the network to reach a state in which a high percentage of traffic in the network is retransmissions.

ABR traffic increases resource utilization in the network to increase beyond what can be achieved by the traffic that requires resource reservation (i.e., connection-oriented traffic with stringent delay and/or loss requirements). When the two types of traffic are integrated in the network, it is necessary to design transport services so that ABR traffic does not cause degradation to the service provided to the reserved traffic. One way to achieve this is to have "logically" separate buffers for ABR and reserved traffic and transmit cells from the ABR traffic buffer only when there is no cell waiting in the reserved traffic buffers waiting to be transmitted. This framework guarantees that the service provided to the reserved traffic is not affected by the ABR traffic. Variants of this scheme can also be implemented as long as it is guaranteed with a provisioned probability that ABR traffic will not cause service degradation to the reserved traffic.

The main consequence of this approach is that ABR traffic buffers may start to fill up and start dropping arriving cells generated by connectionless service sources. As discussed above, this would cause retransmission of packets, thereby reducing the effective utilization of available bandwidth in the network for ABR traffic. Accordingly, congestion control/avoidance mechanisms are needed to support ABR traffic in the network efficiently. Some of the desired features of congestion control mechanisms for ABR traffic include simple implementation, use available bandwidth in the network as efficiently as possible (i.e., use it as much as possible while minimizing CLR for this type of traffic), share the available bandwidth in the network fairly among all sources that generate ABR traffic, and use as little processing power and bandwidth as possible in the network for its operation.

In this context, the ATM Forum and other standards bodies have defined two services: unspecified bit rate and available bit rate.

A UBR connection does not have a traffic contract or committed QOS with the network. It is not subject to connection admission control or UPC policing (other than its peak rate). There is no information provided at the UNI to indicate to the source of the connection that a high CLR is observed in the network. A UBR source can transmit its cells at any rate (up to its peak rate), but its cells can be discarded by the network if there is no resource available in the network to carry this traffic.

ABR service is similar to UBR service. The main difference is that the network in this case provides a feedback to the source of the ABR service about its congestion status. In response to the feedback from the network, the ABR sources may reduce or increase their sending rates to achieve high efficiency and low CLR (however, there is no explicit value defined, but negligible CLR is expected).

The ATM Forum is in the process of defining the UNI protocol for the ABR service to address interoperability issues. The current proposal is referred to as Proportional Rate-Control Algorithm (PCRA). It is a rate-based, closed-loop, end-to-end flow control residing at the ATM end stations. A sequence of special cells called RM cells are sent by the source end station (SES) to the destination end station (DES) to collect the congestion status of the forward path (i.e., the forward direction). The frequency of the RM cells is an input parameter. The congestion indicator (CI) bit in the forward RM cells can be marked at the congested intermediate switches along their path. The DES generates a reversed RM cell to the SES carrying the CI bit information. The SES increases or decreases its current rate in response to the reversed RM cells it receives.

Various options may be allowed to improve the performance of this scheme:

1. Virtual source (VS) and virtual destination (VD) can be placed on the path to close the end-to-end loop sooner in order to shorten the response time, thereby isolating a segment of a path.
2. A forward RM cell can carry the actual cell rate (ACR), which is the current rate at which the SAS is operating. If a switch feels it necessary to alter the ACR (significant increase or decrease) it can choose to change the ACR by placing the explicit rate field in the RM cell. The RM cell eventually reaches the DES or VD and is returned to the SES or VS. Once the SES or VS receives the RM cell, the source can update its ACR by the value specified in the explicit field rate.

ABR congestion control framework is discussed in more detail in Chapter 6.

9.6 CONCLUSIONS

This chapter is a review of the requirements and architectures for providing connectionless service in ATM. Allowing existing applications and protocols to run over ATM is an essential step for the universal acceptance of B-ISDNs. ATM is initially being deployed at the local area, and a large number of hubs and ATM LANs are already commercially available.

The ATM Forum's LE service is envisioned to provide a migration path toward universal networking by allowing network users to experiment with ATM networks (with emerging multimedia applications) without requiring any changes to their current application environment.

The ITU-T connectionless service framework has been developed mainly for migrating SMDS services over ATM technology. As public networks start to migrate into B-ISDN, ATM will be used to provide various carrier services, including SMDS. Supporting these services over ATM preserves customer investment on these services.

The IETF IP over ATM initiative treats ATM as a logical IP subnetwork and considers only the application of ATM as a direct replacement for the "wires" and local LAN segments connecting IP end stations and routers operating in the classical LAN-based paradigm. These efforts may be viewed as preliminary steps in providing connectionless service over ATM. ATM Forum LE work is currently defined as a single emulated LAN. All three architectures essentially create an overlay network at different degrees and do not use the unique features of ATM such as QOS (with different degrees as discussed next). Congestion control of connectionless traffic is a crucial piece of work that needs to be completed as soon as possible to support this type of traffic in ATM networks efficiently. Multiprotocol routing in ATM networks is another area that needs to be satisfactorily addressed. Work in this area has started just recently in the ATM Forum and IETF.

The ITU-T's connectionless overlay network is attractive to high-traffic users with multiple destinations (in which case semipermanent VP connections may be used) or low-traffic and occasional users (in which case switched VCCs may be used). CLSs may be used as a virtual router to handle traffic between different protocol LANs. The term *virtual* is used to indicate that, unlike regular routers, there is no dedicated WAN among routers. Instead, the physical network is shared by different applications/connections. CLSs maintain the connectionless overlay network (i.e., they handle connections to other such servers, perform address translation/mapping, and perform routing) and terminate the connectionless protocols, and the relay of information is performed on the basis of information in the CLNAP. This particular protocol is defined over AAL 3/4.

LE defines a layer below the LLC with the same interface (i.e., NDIS or ODI) as an LLC to MAC. The LE layer uses AAL 5. The architecture specifies the LE client, the LES, and the BUS. The LES maintains the address clients in the emulated LAN. The LE client forwards frames (using AAL 5 and the lower layers of the ATM architecture) either to the BUS, which in turn may multicast the frame to all clients, or intelligently forward to its destination. As another option, LE clients may request the ATM address of the destination node, request a connection from the network, and forward the frame over the established connection.

The main advantage of the LE framework is that no changes in the applications are required to run them over ATM networks. This is also the main disadvantage of this framework in that it is not possible to take advantage of the unique (particularly QOS-related) features of ATM. The connectionless overlay method of ITU-T provides the flexibility to use the capabilities of the underlying ATM network. Its main disadvantage is that it is defined over AAL

3/4, which is not expected to be popular in general and almost certainly not in workstations. Requiring IWUs with additional processing complexity that is not necessarily required to cross a WAN (i.e., LAN interconnection) may be viewed as the main disadvantage of this framework.

IETF, at best, views ATM as a subnetwork. Perhaps IP over ATM work is being undertaken mainly because ATM is at a stage where its potential popularity cannot be ignored. Similarly, IP cannot be ignored by the ATM community, since there are several millions of IP hosts worldwide (and growing). The best scenario for ATM would be to support IP applications with no changes while emerging multimedia applications are developed over ATM natively.

However, ATM is not going to be everywhere anytime soon. Taking advantage of this fact, IETF has chosen to extend the IP architecture to support emerging multimedia applications. These applications require resource reservation and connection-oriented network service, while IP is a datagram service. Extending IP to support connection-oriented traffic may no longer be viewed as IP. Although "IP" will be included in the name of what comes out of this task and clearly it will support existing IP applications with no changes, the next generation of IP will not be IP. The outcome of this is expected to delay the deployment of ATM.

References and Bibliography

[1] ITU-T Recommendation F.812, "Broadband Connectionless Bearer Service," 1993.

[2] ITU-T Recommendation I.364, "Support of B-ISDN Connectionless Data Service on B-ISDN," 1993.

[3] RACE project 1035 (Customer Premises Network) Deliverable D17, "Nature, Scope, and Provision of Connectionless Service in the BCPN," December 1991.

[4] LAN Emulation over ATM Specification, ATM Forum (B. Ellington), 1994.

[5] Institute of Electrical and Electronics Engineers, IEEE Project 802: Local Area Network Standards.

[6] Classical IP and ARP over ATM, RFC 1577, Information Sciences Institute (M. Laubach), 1994.

[7] Multiprotocol Encapsulation over ATM Adaptation Layer 5, RFC 1483 (J. Heinanen), 1993.

[8] Abeysundara, B. W., and A. E. Kamal, "High Speed Local Area Networks and Their Performance: a Survey," *Computing Surveys,* Vol. 23, No. 2, 1991, pp. 221–264.

[9] Axner, D. H., "Differing Approaches to Virtual LANs," *Business Comm. Rev.,* Vol. 23, No. 12, 1993, pp. 42–45.

[10] Biagoni, E., E. Cooper, and R. Sansom, "Designing a Practical ATM LAN," *IEEE Network,* Vol. 7, No. 2, 1993, pp. 32–39.

[11] Chao, H. J., D. Ghosal, D. Saha, and S. K. Tripathi, "IP on ATM Local Area Networks," *IEEE Comm. Mag.,* Vol. 32, No. 8, 1994, pp. 52–59.

[12] Cherbonnier, J., J.-Y. Le Boudec, and H. L. Truong, "ATM Direct Connectionless Service," *ICC '93,* 1993, pp. 1859–1863.

[13] Clapp, G. H., "LAN Interconnection Across SMDS," *IEEE Network,* Vol. 5, No. 5, 1991, pp. 25–32.

[14] Clark, D., et al., "An Introduction to Local Area Networks," *Proc. IEEE,* Vol. 66, No. 11, 1978, pp. 1497–1516.

[15] Feldmeier, D., "A Framework of Architectural Concepts for High Speed Communication System," *IEEE JSAC,* Vol. 11, No. 4, 1993, pp. 480–488.
[16] Fischer, W., E. Wallmeier, T. Worster, S. Davis, and A. Hayter, "Data Communications Using ATM: Architectures, Protocols, and Resource Management," *IEEE Comm. Mag.,* Vol. 32, No. 8, 1994, pp. 24–33.
[17] Gerla, M., T.-Y. C. Tai, and G. Galassi, "Internetting LAN's and MAN's to B-ISDN's for Connectionless Traffic Support," *IEEE JSAC,* Vol. 11, No. 8, 1993, pp. 1145–1159.
[18] Le Boudec, J.-Y., A. Meier, R. Oechsle, and H. L. Truong, "Connectionless Data Service in an ATM Based Customer Premises Network," *Computer Networks and ISDN Systems,* 1994, pp. 1409–1424.
[19] Le Boudec, J.-Y., D. A. Pitt, and H. L. Truong, "ATM Connectionless Service Using Computer Networking Methods," *ICC '92,* Genova, Italy, 1992.
[20] Le Boudec, J.-Y., E. Port, and H. L. Truong, "Flight of the Falcon," *IEEE Comm. Mag.,* Vol. 31, No. 2, 1993, pp. 50–66.
[21] Newman, P., "ATM Local Area Networks," *IEEE Comm. Mag.,* 1994, pp. 86–96.
[22] Newman, P., "Traffic Management for ATM Local Area Networks," *IEEE Comm. Mag.,* Vol. 32, No. 8, 1994, pp. 44–50.
[23] Onvural, R. O., B. Ellington, W. Pace, and H. Truong, "Requirements and Architectures for Connectionless Service in ATM Networks," IBM Tech. Rept., 1994.
[24] Sachs, S. R., "Alternative Local Area Network Protocols," *IEEE Comm. Mag.,* Vol. 6, No. 3, 1988, pp. 25–45.
[25] Stallings, W., "Local Networks," *Computing Surveys,* Vol. 16, No. 1, 1984, pp. 3–41.
[26] Truong, H. L., W. W. Ellington, J.-Y. Le Boudec, A. X. Meier, and J. W. Pace, "LAN Emulation on an ATM Network," IBM Tech. Rept., 1994.
[26] Vickers, B. J., and T. Suda, "Connectionless Service for Public ATM Networks," *IEEE Comm. Mag.,* Vol. 32, No. 8, 1994, pp. 34–42.
[27] "IP over ATM: A Framework Document," Internet Draft, January 1994 (R. G. Cole).
[28] "NBMA Next Hop Resolution Protocol," Internet Draft, August 1994 (J. Heinanen).
[29] Perkins, D., "Beyond Classical IP: Integrated and ATM protocol Specification," ATM Forum contribution, September 1994.
[30] Iwata, A., et al., "ATM Connection and Traffic Management Schemes for Multimedia Interworking," *Comm. ACM,* Vol. 38, No. 2, February 1995, pp. 54–71.
[31] Perloff, M., and K. Reiss, "Improvements to TCP Performance in High Speed Networks," *Comm. ACM,* Vol. 38, No. 2, February 1995, pp. 72–90.

Transport Protocols 10

The primary function of a transport protocol is to establish, manage, and terminate end-to-end connections and to provide interfaces to higher layers requesting such functions. The criteria used in implementing a transport protocol mainly depend on the service provided by the network. In virtual-circuit networks, packets are delivered from the sender to the receiver in sequence, usually without any duplication, making it relatively simple to implement the protocol. In the case of datagram networks, the network does not provide much support for ordered delivery, errors, or duplication, necessitating more sophisticated transport protocols.

10.1 TRANSPORT FUNCTIONS

A transport protocol, in general, performs various functions, including:

1. Addressing;
2. Connection establishment and termination;
3. Flow and rate control;
4. Buffering;
5. Multiplexing;
6. Segmentation and reassembly;
7. Handling duplicated packets;
8. Error recovery and control;
9. Priority handling.

10.1.1. Addressing

The transport layer defines a set of transport addresses through which communications take place. The collection of transport addresses forms a networkwide address space. With this framework, users of transport services request a transport connection between local and remote addresses. In order to establish a connection, the local transport layer needs to know the transport address of the service at the remote end (that it is requested to connect to). This address may be known a priori, or may be a well-known address that is rarely changed.

Another possibility is to retrieve the address from a name server. Finally, the required address may be found by using a well-known address to connect to a process and then by exchanging messages with the process to specify what service it is attempting to connect to. In this case, the process chooses an idle address, spawns a new process, and assigns the new address to the spawned process. Then, the process transmits the address to the originator, which in turn establishes a connection with the spawned process.

10.1.2 Connection Establishment and Termination

Communication between two transport entities exists, during which each entity maintains state information regarding its partner. Initially, the two entities create this information at the connection establishment phase and destroy it after the connection is terminated. There are two different techniques used to establish and terminate connections: handshaking and implicit management.

In handshaking, the two end nodes exchange a number of messages to negotiate the resources to be allocated to the connection at call establishment as well as to release reserved resources at connection termination. In the case of implicit connection management, the sender transmits a control packet or a data packet that includes control information to the receiver either to establish or terminate the connection, and timers are used to manage the state of the two end nodes, reducing the number of messages exchanged.

10.1.3 Flow and Rate Control

End-to-end flow control between two transport entities is required to ensure for a particular connection that the resource usage at the receiver for the duration of the connection stays within the negotiated limits set at the connection establishment phase. That is, there are sufficient resources at the receiver to accept and process incoming packets. Furthermore, this type of flow control provides back-pressure speed matching between the sender and the receiver and protects the network to a certain degree from congestion caused by an excessive number of packets.

One of the well-known techniques used in current networks is the sliding window. In this scheme, the receiver specifies the maximum amount of data it can accept, referred to as the *window size*. The window size is determined based on the current availability of resources available at the host (i.e., the number of buffers and processing power) to handle incoming packets. Updates on the window sizes are issued by the receiver and conveyed to the sender as control information. The amount of change can be absolute values or relative increments or decrements. At any given time, the maximum amount of data unacknowledged between the two transport service users is limited by the current window size.

Packets arriving at the host may be dropped at the interface if they arrive faster than they can be processed. The window scheme alone may not be enough to solve this problem.

Furthermore, even if the resources at the receiving end are sufficient *not* to impose any limitation on the rate the sender uses to transmit data during the duration of the connection, the network connecting the two nodes may not have resources to support the sender's traffic flow. In particular, if the sender transmits more packets than what can be handled, then the network will become congested and it will not be able to deliver packets as fast as they arrive. To solve this problem, a rate control mechanism based on the availability of the network resources may be used at the sender. Since the traffic in the network changes dynamically, the network performance delivered for the connection must be monitored and the flow rate should be adjusted accordingly.

10.1.4 Buffering

Buffers at the receiving side are generally used to smooth the incoming traffic flow and to temporarily store packets while the processor is busy handling other tasks. Buffers at the sender side, on the other hand, are used to store packets temporarily until they are acknowledged and are used for flow and rate control. The decision on reserving buffers at the sender, receiver, or both depends on the type of traffic flow that will take place between the two nodes. For low-bandwidth sporadic traffic, it is better not to dedicate any buffers at either host, but rather acquire them dynamically at both ends. For high-bandwidth bursty traffic, it is better to dedicate buffers at the receiving end so that the traffic can flow at the maximum rate allowed by the network. Similarly, for low-bandwidth bursty traffic, it is better to reserve buffers at the sender, and for high-bandwidth smooth traffic, it is better to dedicate buffers at the receiver [48].

Buffer size is also an important design parameter. If all packets are about the same size, a fixed buffer size can be chosen, simplifying the buffer management scheme and reducing the waste of resources. However, if the packet sizes differ significantly for different services, achieving high resource utilization requires the use of variable-size buffers, complicating the buffering scheme.

10.1.5 Multiplexing

There are two forms of multiplexing that take place at the transport layer: upward multiplexing and downward multiplexing. In upward multiplexing, different transport connections are multiplexed onto the same network connection (i.e., virtual circuit). This type of multiplexing is used to achieve better utilization of resources reserved for virtual circuits. For high-bandwidth applications, the amount of bandwidth effectively given to a transport connection is limited by the sliding window scheme. In particular, once the window is full, no new packets can be transmitted until at least one packet is acknowledged by the receiver, and that packet has to stay at the sender buffer for the time it takes for a packet to be received by the receiver plus the time it takes for the acknowledgment packet to be transmitted back and received by the sender. Downward multiplexing provides a solution to this problem by opening multiple connections and distributing traffic among them in a round-robin manner.

10.1.6 Segmentation and Reassembly

The transport layer of the sender receives arbitrary-length packets generated by higher layers, referred to as *transport service data units* (TSDU), and segments them into multiple smaller packets, referred to as *transport protocol data units* (TPDU), to be processed by lower layers. At the receiving side, these TPDUs are reassembled to form original TSDUs. There are various reasons for segmentation. Networks, in general, impose a limit on the maximum packet size transmitted. Also, buffers at both the sender and receiver hosts may be of fixed size (but may differ from host to host) for efficiency in buffer operations. Furthermore, smaller TPDUs may decrease the amount of wasted bandwidth in case of retransmissions, since packet loss probability due to bit errors decreases as the packet size decreases. However, as the packet size decreases, the amount of overhead associated with the segmentation of TSDUs into TPDUs increases, thereby decreasing the effective network throughput and increasing processing times at the hosts.

10.1.7 Handling Duplicated Packets

It is possible for the packets transmitted between two end nodes to be stored in the network for an indefinite period of time, which occurs when the routing algorithms get in a loop due to link failures, inconsistency of routing information at the nodes, or congestion. When a packet inside the network is delayed more than what the sender expects, the packet is retransmitted, causing duplicate packets in the network. Handling duplicate packets at the transport layer of the receiving side is a time-consuming task and can have a significant impact on the correctness of the TSDUs delivered to higher layers.

10.1.8 Error Recovery and Control

Transport services ensure the correct delivery of user data. To achieve this, TPDUs between two transport addresses must be delivered error-free and TSDUs constructed from TPDUs should be the same as the original TSDU. The first process can be achieved by the receiving side checking for bit errors and acknowledging accepted packets to the sender side. In the case of errors or lost packets, the sender will not receive an acknowledgment and will retransmit the packet until it is acknowledged, thereby guaranteeing error-free reception of TPDUs at the receiving side. The second process requires ordering TPDUs, using their sequence numbers to reassemble them to form TSDUs, and handling duplicate packets. Duplicate packets, if discovered, are discarded.

10.1.9 Priority Handling

Different protocol services have different levels of QOS requirements. In order to meet their performance requirements, priorities can be assigned at the transport layer processing in allocating resources such as number of buffers and processing power. This is a particularly important feature in B-ISDN, which integrates various types of applications with different service requirements onto the same network.

10.2 DESIGN ISSUES

B-ISDNs are being built on transmission networks with high data and low bit error rates. This changing environment introduces various limitations to the transport protocols. In particular, higher transmission rates impose various constraints that include:

1. The processing times available to handle packets;
2. The amount of time tolerated to establish and terminate connections;
3. The design of various transport processes such as retransmission techniques, buffer allocation policies, and flow and rate control.

To be able to operate at very high speeds, processing at intermediate network nodes should be kept at a minimum. Hence, as the technology advances rapidly toward meeting the requirements of B-ISDN, the bottleneck of communication networks has been moving from the transmission medium to the communication processors at the end nodes.

It has been widely recognized that there is a definite need for international standards on the transport protocols used in B-ISDNs so that separate networks can be interconnected in the most efficient way. However, the issue of selecting an appropriate transport protocol for B-ISDN has not yet been fully addressed in the standardization committees. In particular, it is not clear whether an existing protocol modified for high speeds through clever tuning or a new protocol specifically designed for high-speed networks will be the transport protocol of choice for B-ISDN. Adopting a modified version of an existing transport protocol, assuming high throughput rates can be achieved, can significantly reduce the amount of time required to standardize and therefore deploy B-ISDNs, whereas the standardization process of a new protocol may take several years to finalize.

Improvements to existing protocols may be achieved by reducing their overheads by, for example:

1. Optimizing the steady-state execution path by removing code required for handling events that occur infrequently, such as error and exception handling;
2. Using larger packets to reduce the time spent on per-packet-specific operations;
3. Packet grouping to reduce the number of acknowledgment and control information;
4. Modifying protocol dynamics in exchanging state information, such as the periodic exchange of messages as opposed to exception handling;
5. Providing operating system support to provide more suitable platforms for protocol processing.

In addition, new protocols designed for high-speed networks might have various improvements over existing protocols that might include:

- Format fields aligned with word boundaries to reduce the memory access times to fields, and perhaps padding packets to integral word boundaries to increase the efficiency of group memory operations, such as block move.
- Distribution of control information to header and trailer to optimize the execution of protocol-specific fields. For example, source and destination addresses may be placed at the header and the checksum field at the trailer. Doing so would make checksum computation parallel with transmission.
- Unique use of control fields to maximize the parallelism of operations and minimize the time spent on determining the use of fields at a specific instant before the control can be passed to corresponding part of the code.
- Fixed-length field definition to reduce the access and processing times of fields.
- Streamlining of steady-state protocol functions to maximize pipeline operations.
- Cutting down per-byte handling time on data movement and buffer management by allowing network interfaces access to main memory, enabling

direct memory access (DMA) operations to take place simultaneously with transmitting buffers, and not using general-purpose memory management scheme.

- Better use of timers.

The performance of a protocol does not only depend on its design; it is also heavily affected by its implementation in the system, both in hardware support and software design. Figure 10.1 classifies different approaches in high-performance protocol implementations [51].

In terms of hardware support, both multiprocessor platforms and very large scale integration (VLSI) implementations are considered. A multiprocessor platform can further be classified into general and special-purpose processors. In the case of general-purpose processors, transputers provide a flexible platform of multiprocessing power well suited to pipelining and parallel implementations of protocol suites. A transputer is a microprocessor with built-in serial communications links used to connect to other transputers and devices.

Special-purpose processors are used to decrease the load on the system processor and to minimize the processing overhead for interrupts, the load in system buses, and the delay of packets in the system. VLSI implementations of parts of protocol suites are being considered in order to achieve very high performance solutions. It might be possible to design protocols with state machines that can be developed completely in VLSI. However, pure VLSI implementations restrict to some extent how much flexibility can be achieved with such protocols in issues related to overall system design. Hardware assist for transport protocols can prove to be most useful for functions such as byte swapping, CRC, timers, and content-addressable memory.

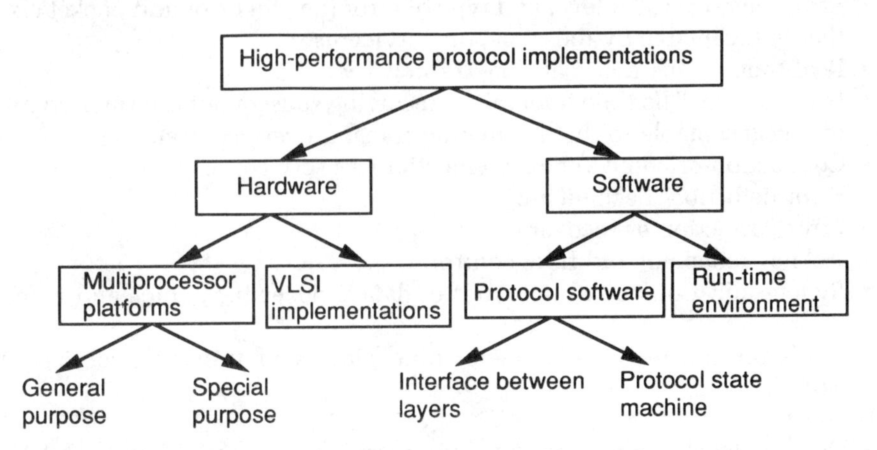

Figure 10.1 Approaches in high-performance protocol implementations.

In terms of software support, optimization in both the protocol software and runtime environment are required to achieve high throughput from the transport layer. It is illustrated in various studies that only 20% of the total time it takes to process a packet is protocol processing, while 80% of the time is spent by the runtime environment: interrupt handling, context switching, and so forth. Hence, there should be a considerable amount of effort concentrated on optimizing the functions provided by the operating system, perhaps using hardware support in such functions.

The OSI protocol architecture defines five different classes of transport services with respect to the reliability of underlying network services, the need for multiplexing, and the QOS parameters related to connection establishment, data transfer, and connection termination such as throughput and delay. Similarly, various service classes are expected to be defined for B-ISDNs. This issue has not yet been finalized by the ITU-T. Various applications in B-ISDNs will have different service requirements. As discussed in Chapter 3, some types of traffic may tolerate some amount of losses but not latency. Therefore, there is no need to do error handling for this type of traffic. Some other types of traffic may be intolerant to errors but can tolerate delays. In this case, the expense of some connection management techniques (i.e., three-way handshaking) may be acceptable. Following the developments in [51] and [52], various performance and functional criteria used in B-ISDN applications can be summarized as follows:

- Throughput: The number of packets (or bits) exchanged between two transport users;
- Delay: The time elapsed between sending a data unit from a transport service user and its reception at the destination transport service user;
- Error tolerance: The level and type of error (i.e., bit error and packet error) that is acceptable by the transport service user;
- Real-time versus non-real time traffic;
- Isochronism: The time interval for receiving subsequent information units that is acceptable to the destination transport service user;
- Connection-oriented versus connectionless service;
- Error detection mechanism;
- Error correction or recovery;
- Acknowledgment and flow control;
- Synchronization of various types of data services (i.e., voice and video).

With these service parameters, nine classes of transport services are defined:

- Class 1: Isochronous, two-way conversation, low throughput, and high tolerance to errors.

- Class 2: Isochronous, two-way conversation, high throughput, and low tolerance to errors.
- Class 3: Isochronous, one-way conversation, and average to low tolerance to errors.
- Class 4: Nonisochronous, real time, high tolerance to errors.
- Class 5: Nonisochronous, real time, average tolerance to errors.
- Class 6: Nonisochronous, real time, no tolerance to errors.
- Class 7: Nonisochronous, non-real time, high tolerance to errors.
- Class 8: Nonisochronous, non-real time, average tolerance to errors.
- Class 9: Nonisochronous, non-real time, no tolerance to errors.

Classes 1, 2, 3, 8, and 9 require connections to be established and terminated for data transfer, while classes 4 to 7 are for connectionless services. Acknowledgments are not used in connectionless services and can be initiated either by the receiver or the sender in connection-oriented services. Checksum is used to determine bit errors either in the control field or information field, or both. Packets received with errors or not received in sequence are required to be retransmitted by the sender in reliable services. Segmentation and reassembly is a transport layer function and its implementation depends on the maximum packet size allowed in the network. Duplicate packet handling is important for class 6 service and selectable by the transport service users in most other cases. Flow control is required to meet the receiver's ability to process packets. Functional requirements of the nine service classes are illustrated in Table 10.1 [51]. If a similar set of service classes is to be standardized by ITU-T and other standardization bodies, each service class can be implemented with a specific set of functions required to service the class, and their state machines can be optimized.

Table 10.1
Protocol Functions Required by the Service Classes

	1	*2*	*3*	*4*	*5*	*6*	*7*	*8*	*9*
Connection management	X	X	X					X	X
Acknowledgment		A			A		A	A	X
Checksum calculation		A	X	A	A	A	A	A	X
Packet retransmission		A			A	X	A	A	X
Duplicate packet handling		A	A	A	A	X	A	A	X
SAR	T	T	T	T	T	T	T	T	T
Flow control	X	X	X					X	X

A = Application selectable.
T = Transport system selectable.
X = Function required.

10.3 FEATURES OF TRANSPORT PROTOCOLS

The most important features developed or proposed in various protocols are defined in this section.

10.3.1 Signaling

Signaling refers to the function of exchanging control information between transport services users in order to manage the connection (i.e., establishment, control, and termination). Two methods used are out-of-band signaling and inband signaling. Packets in inband signaling carrying control information are multiplexed to the same connection used for data transmission, while out-of-band signaling refers to using separate connections for control and data transmission.

Inband signaling, when combined with the implicit connection setup discussed below, can result in reduced latency and is therefore suitable for applications requiring fast response times. The use of inband signaling may, however, increase the processing times of packets, since each packet has to be processed to determine whether it is a control packet or not.

Out-of-band signaling, on the other hand, reduces steady-state processing requirements. Separating signaling protocol from data transfer protocol simplifies the protocol design and implementation, makes it possible for multiple protocols to use a single signaling system (if needed), and allows incorporation of additional services such as security and billing.

Furthermore, the use of out-of-band signaling provides the flexibility of parameter negotiation prior to data transmission. The disadvantage of out-of-band signaling is that data transmission cannot be overlapped with connection management, thereby limiting the applicability of transport protocols that use this type of signaling to establish connections for applications with short holding times.

10.3.2 Handshake

Handshake refers to the method used to manage connections. There are three methods most frequently used in current and proposed protocols:

- In a three-way handshake, sender sends a message to receiver requesting connection establishment. Receiver sends a message back to make sure this is an original request (i.e., to eliminate the possibility of duplicate requests). Sender replies back that this is not a duplicate, establishing the connection. Similar sets of events take place at connection termination.

- A two-way handshake is the same as a three-way handshake except that sender does not need to reply back to receiver to ensure that this is not a duplicate.

In the case of implicit management, a connection is established with the arrival of the first packet to the receiver. Upon receiving a packet, the receiver checks the source address and establishes a new connection. The connection termination takes place using timers. In particular, the receiver tears down the connection after the expiration of a timer started upon reception of a packet. We note that the method used in a protocol to establish a connection can be different from the one used to terminate the connection.

The overhead introduced by connection management depends mainly on the duration of the connection. If the connection stays active for a long period of time, then the overhead introduced is negligible even with three-way handshaking. However, if the connection duration is less than few times the order of the propagation delay, then three-way handshaking may require more time to establish and terminate the connection than the actual time used for data transmission. For such connections, an implicit request is more efficient. On the other hand, an implicit request necessarily uses inband signaling, which, as discussed above, has its disadvantages, while two- or three-way handshaking can be easily implemented by using out-of-band signaling.

10.3.3 Connection Parameters

Connection parameters are the set of values exchanged during the lifetime of the connection (i.e., negotiation at connection establishment phase, updates during data transfer, and selection of the mode of operation). Negotiated and updated parameters include QOS and resource availability parameters: maximum packet size, timeout values, retry counters, buffer sizes, sequence numbers, and so forth. On the other hand, the selection of operation mode is used to choose the type of service requested from the transport layer. For example, mode 1 may correspond to a service with no error and flow control (i.e., quick interaction between terminals connected to a host), mode 2 service may only disable error handling (i.e., packetized voice), and mode 3 may provide both error and flow control (i.e., large file transfers).

10.3.4 Multiplexing

Multiplexing at the transport layer corresponds to placing several transport connections onto one virtual circuit, mainly to increase the virtual-circuit utilization. The multiplexing feature in ATM networks may not be as effective as it is in current networks, since it will increase the processing time at the receiver.

In particular, ATM networks are expected to carry large number of connections simultaneously and the table lookup operations required at the receiver to demultiplex connections can be quite processing-intensive.

10.3.5 Acknowledgment

Acknowledgment packets are used to indicate the acceptance of packets by the receiver. It is not a necessary feature if the data link is fully reliable. If not, then there are two types of methods used:

- Sender-dependent acknowledgment can be initiated by the receiver after receiving a number of packets or can be requested by the sender after transmitting a number of packets.
- Sender-independent acknowledgment can be initiated by the receiver under the control of timers.

Window-based end-to-end flow control and various error handling mechanisms are based on using acknowledgments. Furthermore, acknowledgments are used to update the state information to discard data stored at the sender buffers temporarily for a possible retransmission. Another approach is to exchange messages on a routine and periodic basis, independent of any significant event that may take place. This may in turn simplify protocol processing.

10.3.6 Flow Control Techniques

Flow control techniques at the transport level are used mainly to avoid congestion at the receiver. As discussed in Section 10.1.3, there are two types of methods used at the transport level:

- End-to-end flow control mechanisms, in which the number of packets sent over a period of time is controlled by the receiver using window-type mechanisms. This type of control mechanism takes place between the receiver and the sender and attempts to protect against congestion at the receiver transport entity.
- Rate control, in which the rate packets transmitted by the sender is either negotiated at the connection establishment phase (explicit control) or dynamically using the estimates from the round-trip delays (implicit control), or both. This type of control takes place between the network and the sender and protects against congestion in the network. The transmission rate used is usually the minimum of the rate at which the source can transmit data, the rate at which receiver can process data, and the rate at which network can transport data.

As the transmission speeds increase, the effectiveness of end-to-end flow control mechanisms decreases due to the very large number of bits in transit. In particular, it takes at least twice the propagation delay for the sender to receive an acknowledgment packet. Hence, the window size should be at least twice the propagation delay divided by the packet transmission time. Window sizes smaller than that would restrict the throughput that can be achieved by the transport protocol. However, larger window sizes may increase the possibility of congestion due to a sudden load increase, causing larger round-trip delays, which may in turn result in time-outs and retransmissions.

The window-based scheme works only when the window size is less than or equal to the burst size that can be handled by intermediate network nodes and receivers. As the window size increases, the possibility of packets being dropped at the receiver interface increases if they arrive faster than they can be processed. Controlling the transmission rate is an attempt to solve this problem. Rate control, on the other hand, is not sufficient by itself, since by the time the sender becomes aware of congestion, the complete path may already be congested.

10.3.7 Error Handling

Error handling is necessary for providing reliable transport services. It is not possible to fully eliminate bit errors at the transmission media and cell losses due to buffer overflows. When a damaged frame arrives at the transport layer, it clearly should be discarded. The sender needs to be informed about the discarded (or equivalently accepted) frames to retransmit (or release) buffers. There are three methods used for this purpose. Whenever the receiver detects an error in the received frame, it can send a negative acknowledgment (NACK) packet back to the sender, identifying the frame, after which data are lost. Instead of sending back the point at which an error is detected, a NACK packet that contains the sequence numbers of lost and damaged packets can be sent. This technique, selective reject (SRej), eliminates the problem of retransmitting packets that are delivered correctly to the receiver but are rejected because they are delivered after a lost or errored packet. Another technique used is positive acknowledgment with retransmission (PAR), in which accepted packets (i.e., error-free and in sequence) are acknowledged to the sender. In this case, retransmission of packets is triggered by time-outs at the sender that expire while waiting for acknowledgments sent back on packets received correctly by the receiver.

Upon detection of an error, there are two basic approaches used to handle errors at the sender: go-back-N and selective repeat (SR). In the go-back-N scheme, all subsequent frames received after the detection of a lost or damaged frame are discarded and no acknowledgment is sent back. The sender eventually times out and retransmits all unacknowledged frames. PAR necessarily uses this scheme to recover from errors. Although PAR minimizes the processing times at the receiver, considerable bandwidth can be wasted with the go-back-N scheme, particularly as the transmission rates increase, since a large numbers of packets can be transmitted by the sender before the time-out expires. Accordingly, the go-back-N scheme can be used if errors occur in bursts or a number of packets are grouped and acknowledged together. Another problem with this approach is the determination of good time-out values.

In the selective repeat scheme, all data received correctly after lost or damaged data are accepted by the receiver. The sender eventually retransmits the lost data and the receiver eventually receives the data correctly and sends back the largest sequence number of the data accepted. This scheme requires large buffers and more processing times at the receiver. Hence, the tradeoff between the two approaches is bandwidth and buffer space. One possible approach is to use both methods and then dynamically switch between them depending upon the errors encountered during the duration of the connection [43].

10.3.8 Evaluation of Features

The suitability of the features discussed in Sections 10.3.1 to 10.3.7 for a high-speed environment is summarized here [12].

Out-of-band signaling for control packets is preferable to inband signaling, since it reduces the amount of processing for data packets. Furthermore, out-of-band signaling can support more than one type of data transport on an established connection.

Handshake mechanisms may not be acceptable for connections with short holding times. Implicit schemes, on the other hand, require the connection state to be maintained for a time on the order of three times the maximum end-to-end delay following the last transmitted packet. Considering the large number of connections handled by an end node, this may not be a desirable feature in B-ISDN. In particular, new connections, in this case, may have to be unnecessarily rejected due to lack of resources as the maximum end-to-end delay between two nodes in a network increases. Furthermore, an implicit scheme requires inband signaling as opposed to handshaking that can be implemented out of band. Hence, the overhead associated with handshaking may be justifiable in high-speed networks with large bounds on end-to-end delay.

To establish connections as quickly as possible, it is possible to initiate a connection with a given set of default values of connection parameters. The default values may be based on the information at the sender about the receiver, which necessarily cannot be current, as well as the type and requirements of applications. However, as network dynamics change, it is necessary to update their values during the life of the connection, and the transport protocol will provide an efficient means to support this in B-ISDN.

Multiplexing may be used to use resources at the lower layers more efficiently. However, this advantage is rather nulled by the expense of demultiplexing connections at the other end, particularly as the number of multiplexed connections between the two end nodes increases.

In general, more processing takes place at the receiver than at the sender. Hence, sender-initiated acknowledgment schemes are used more often to reduce the processing burden at the receiver. As an alternative, the receiver can send back acknowledgments on a periodic basis or conditioned on the number of packets received. However, these alternatives require timers or extra processing per connection and may be prohibitive as the number of connections increases.

As discussed previously, the two methods of flow control used are the window-based scheme and rate control. Neither scheme alone is suitable for high-speed networks, requiring both mechanisms to be used simultaneously. Rate control can be based on using timers or interpacket gaps. The latter can be implemented more efficiently and is preferable over timer-based control.

There are various error handling techniques with respective advantages and disadvantages. Go-back-N is a simple technique that requires retransmission of all packets sent after the last accepted packet by the receiver, independent of the fact that some of that data may in fact be received at the receiver correctly. As transmission speeds increase, the amount of data transmitted during a large propagation delay required to receive NACK at the sender can be restrictive to use this scheme. Selective transmission techniques eliminate this problem. However, the disadvantage of this scheme is the amount of processing required at the receiver, particularly reordering packets for insequence delivery. The error recovery scheme of [43], which allows the use of both mechanisms while providing synchronization between the two ends, appears to be the best alternative proposed so far.

10.4 CONVENTIONAL PROTOCOLS

There are two conventional protocols representing the state of the art in standard commercial networks: transmission control protocol (TCP) and ISO transport protocol class 4 (TP4).

10.4.1 TCP

TCP is designed to provide reliable data delivery between two end nodes. It is a connection-oriented protocol, whereby end-to-end connections are established before data transmission. A TCP connection provides full-duplex byte stream connection. Three-way handshaking with inband signaling is used to establish and terminate the connection. TCP uses PAR for acknowledgment and the sender uses timers to retransmit packets that are not acknowledged. The connection parameters are negotiated at the connection establishment phase and can be updated at any time during data transmission. TCP also provides mechanisms to multiplex transport connections on virtual circuits to increase resource utilization. A credit-based adaptive sliding-window mechanism is used for end-to-end flow control. A slow-start algorithm developed in [27] operates as follows. The sender estimates the round-trip delay to the receiver using PAR messages and uses this estimate to determine congestion at the intermediate nodes connecting the two nodes. Upon detecting congestion, the sender dramatically reduces the window size and slowly starts to increase the rate at which it transmits packets back up to its previous levels. This approach reduces the probability of events that cause packet retransmission, thereby increasing the effective throughput of the connection.

10.4.2 ISO/TP4

TP4 is one of the five classes of connection-oriented transport protocols developed by the ISO. TP4 makes no assumptions concerning the reliability of the service delivered by the lower layers and provides reliable transmission over an unreliable network. Unlike TCP, TP4 provides mechanisms for users to negotiate the QOS parameters of their connections, which include throughput, priority, delay, and error rate at the connection establishment phase.

Furthermore, connection parameters that are negotiated at the connection establishment phase can be updated during data transmission. The reliable data service is packet-oriented as opposed to the byte-oriented TCP service. A unique feature of TP4 is the possible splitting of a transport connection into a number of network connections or routes. Three-way handshaking is used to establish connections. However, TP4 assumes the graceful termination is handled by higher layers and accordingly uses a two-way handshake to terminate the connection. All control messages are exchanged using inband signaling. TP4 provides multiplexing of transport connections on network connections. Like TCP, TP4 uses PAR for acknowledgment and the sender uses timers to retransmit packets that it has not received acknowledgment for. The credit-based sliding-window technique is used for end-to-end flow control. Optionally, the receiver uses timers to send unsolicited acknowledgment on accepted packets, thereby decreasing the amount of processing on received packets.

10.5 LIGHTWEIGHT PROTOCOLS

The term *lightweight protocols* is used in [12] to refer to a set of protocols wherein considerable effort is made to shorten the length of instruction paths for data transmission as well as data reception in the case where no errors occur, and to minimize the overhead for communications. The set of protocols briefly reviewed next includes protocols designed specifically for high-speed networks and for various networking environments including distributed systems, LANs with features that can be used in high-speed networks.

10.5.1 Delta-T

The delta-T protocol [49] has been developed at Lawrence Livermore National Laboratory for an environment with a wide range of heterogeneous micro-to-supercomputer systems to support both the request/response transaction style of communication needed for client server interactions and the stream style of communication needed for terminal sessions and data transfer. It provides a reliable, end-to-end, stream-oriented service over a nonreliable datagram service and supports multiplexing at the transport layer. Connections are established using the implicit scheme and inband signaling is used for connection establishment. Perhaps the main contribution of delta-T to lightweight protocol development was to demonstrate the fundamental role of timer mechanisms in safe connection management mechanisms and to develop a pure timer-based connection management scheme. For a given stream, the sender and receiver maintain the connection state and a send timer and receive timer, respectively, when data are being exchanged. A connection between two nodes is established at the sender upon transmission of the first packet and at the receiver upon reception of the packet. Hence, no negotiation of connection parameters can take place at the connection establishment phase. However, the default values of the connection can be updated during data transfer. The receive timer ensures that the state is maintained long enough to detect all duplicate packets, a reliable connection opening, and that the node will not accept packets for a safe interval of time on crash recovery. The send timer ensures that sequence numbers will not be reused until all data or acknowledgment packets with a given sequence number have expired. The send timers and receive timers together ensure that the connection state is maintained long enough to guarantee a graceful connection termination and that acceptable sequence numbers are generated and accepted. To achieve hazard-free communication, the lifetime of each packet must be bounded, and this time, the maximum packet lifetime (MPL), has to be strictly enforced by the network. A time-to-live field in packet headers is decremented during packet routing, retransmission, and acknowledgment to ensure that the lifetimes of packets are bounded. When the value of this field reaches zero, the packet is discarded. Delta-T uses protocol header checksum with optional data checksum for error detection and NACK for error reporting. A sliding-window mechanism is used for end-to-end flow control. Byte-level sequence numbers are used to detect errors and ARQ with go-back-N scheme is used to retransmit missing data at the receiver.

10.5.2 Universal Receiver Protocol (Datakit)

The universal receiver protocol (URP) [17] is designed to efficiently transmit user data independent of the protocol the end user desired. To achieve this, URP uses protocol encapsulation and carries traffic using virtual circuits. In doing so, the overall network performance the user receives from a URP network should be similar to the one that would have been achieved with the original protocol. Low queuing delays require the use of small packets, whereas the effective throughput of the user increases as the packet sizes used in the network increase. URP provides byte stream connections. Therefore, it does not depend on the network choice of packet size for efficient multiplexing. The fundamental unit of transmission at the transport layer is a 9-bit unit. The ninth bit is used to multiplex data and control messages at any time during transmission. However, no multiplexing of the transport connection is allowed, since there is a 1:1 correspondence between transport and network connections. URP networks employ a queue service algorithm that gives priority to short messages without losing the proper sequence of transmission on any virtual circuit. A separate queue is used for each virtual circuit and the queues are served with a modified round-robin discipline. In particular, as each queue is visited, a certain amount of data is transferred from the queue, packetized, and transmitted over the transmission link. The amount of data serviced during each visit depends on the number of active virtual circuits, and it is determined so that each queue receives service during a reasonable period of time that will allow user performance requirements to be met. Since user data are broken into several small pieces, a considerable burden is placed at the receiver. Flow control, error detection, and recovery are optional in URP, which can be selected to match the requirements of applications. URP uses NACK for error reporting, and ARQ with go-back-N for error correction. The connection management is implemented using out-of-band signaling with two-way handshaking. Connection parameters are negotiated at the connection establishment phase and can be updated during the duration of the connection. Furthermore, URP provides various user-selectable options on the operation modes.

10.5.3 Network Block-Transfer Protocol

The network block-transfer protocol (NETBLT) [9] was developed for high-throughput bulk data transfer. NETBLT runs on an unreliable datagram network and was designed to operate efficiently over long-delay links (i.e., satellites). NETBLT uses a protocol header checksum with optional data checksum for error detection. Two-way handshaking with inband signaling is used for connection management. The connection parameters are negotiated at the connection establishment and can be updated during the data transfer. Connections are multiplexed at the transport layer. NETBLT separates the flow and error control functions so that the detection of errors is not tied to flow control. Flow control is achieved by employing both sliding-window and rate control mechanisms. The window size is the amount of buffer reserved at the receiver. The rate control mechanism restricts the amount of data that can be transmitted consecutively, referred to as the *burst size*, and imposes a delay between the retransmission of two bursts, referred to as the *interpacket gap*. Sequence numbers used in NETBLT are packet-based. Since the receiver knows the size of the buffer and the rate at which transmission is taking place, it can determine the amount of time it will take to fill the buffer. Hence, detecting packets that are lost within the networks (and therefore not received) is an easy task. To detect errors, the protocol header checksum with the optional data checksum is applied. A selective-reject scheme is used to report missing packets, and ARQ with selective retransmission for error handling.

10.5.4 Versatile Message Transaction Protocol

The versatile message transaction protocol (VMTP) [8] is designed for distributed systems that rely heavily on request/response interactions. VMTP provides transport communication via message transactions. A message transaction consists of a request message sent by the sender followed by zero or one message sent back by each receiver. VMTP assumes an underlying network service providing datagrams. Furthermore, VMTP does not establish or terminate connections at the transport layer. Instead, it contends that these two functions are redundant in that the users of the transport layer will themselves be maintaining a sequence of communications actions. VMTP uses transaction exchanges to facilitate the communications actions of higher layers and support multiplexing at the transport layer. VMTP at the transport layer can then be considered as providing implicit, two-way handshake control management functions controlled by the higher layers. Once a connection is established, VMTP supports out-of-band message exchange for connection control. Connection parameters are negotiated at the connection establishment phase and can be updated for the duration of the connection. Furthermore, VMTP provides various user-selectable options on the operation modes. Because it is a request/response type of protocol, packets in VMTP are acknowledged implicitly by the response message. Flow control is achieved by controlling the intertransmission times of packets (i.e., delaying the transmission of the burst for the duration of the interpacket gap).

There are three types of message transactions. A group message transaction is used when a client wants to send a transaction message to a group of users. In return, the client may receive multiple responses. A datagram transaction is used when no response is needed from the receiver. Finally, a forward-message session transaction is a transaction in which the transaction send by a client is sent to a server and is then forwarded to another server that responds directly to the client. VMTP uses the protocol header checksum with the optional data checksum for error detection.

10.5.5 Advanced Peer-to-Peer Networking

Advanced peer-to-peer networking (APPN) [3,46] is an IBM networking feature built upon the existing System Network Architecture (SNA) peer-to-peer networks. APPN is not a transport protocol but provides various transport layer functions such as connection establishment and connection termination. Unlike most other lightweight protocols, APPN assumes a fully reliable, virtual-circuit-oriented network service. Accordingly, two-way handshaking is sufficient for fully reliable connection management and the update of connection parameters negotiated at the connection establishment phase is not required during the duration of the connection. Similarly, end-to-end flow control and error handling functions are not required in APPN. The connection establishment and termination take place using out-of-band signaling. Multiplexing of transport connections is not allowed.

10.5.6 Rapid Transport Protocol

The rapid transport protocol (RTP) [32] is a reliable, connection-oriented, lightweight transport protocol that provides its users with the functions of connection-oriented and, optionally, datagram services. It provides stream-oriented services and is designed to be implemented over networks that are not necessarily reliable. Accordingly, RTP performs packet segmentation, reassembly, sequence checking, retransmission, and inorder delivery, and allows multiplexing of transport connections. The high-performance implementation of the protocol is based on various design principles. RTP assumes that the underlying hardware and network are *mostly* reliable and peer communication applications are normally available and ready to communicate successfully. Accordingly, RTP establishes an end-to-end connection with the first packet sent. Based on the assumption that the receiver has enough resources, the sender can send a burst of packets, up to some maximum number, if the first information packet is segmented into smaller packets due to network limitations on the maximum packet sizes. This is a tradeoff that favors fast connection setup and low message latency against the possibility that the receiver does not have enough resources to handle the incoming burst. Upon reception of the first packet, the receiver sends back an acknowledgment that includes a set of connection parameters. Similarly, connections in RTP are released with the last packet transmitted. The sender may or may not require an acknowledgment to this last packet. Hence, RTP provides the user with the flexibility of choosing the degree of acknowledgment service, depending on the requirements of applications.

RTP also provides reliable multicast data services, where a sender can multicast a packet to a number of receivers and receive messages from each one of them. End-to-end flow control is achieved using a credit-based sliding-window mechanism. When no acknowledgment option is selected by the sender, the receiver only reports the gaps that occur due to lost packets in the network for retransmission. If the acknowledgment option is selected, then the receiver sends back acknowledgment packets upon receiving such a request from the sender. The sender can request acknowledgments at any time, for example periodically or upon transmitting one full window. Retransmission can take place selectively or by using the go-back-N scheme for error handling.

10.5.7 Xpress Transfer Protocol

The Xpress transfer protocol (XTP) [43,47] is a reliable, real-time, lightweight protocol that combines the functions of both the transport and network layers into a single layer. XTP benefited from the various concepts developed in different protocols discussed above. The protocol is being designed so that parts of it can be implemented in hardware as a VLSI chipset. The functions of address translation, context creation, flow control, error control, rate control, and host system interfacing can all execute in parallel. XTP supports multiplexing of transport connections, provides a stream-oriented service, and supports both virtual-circuit and datagram services at the transport layer.

The connection parameters are exchanged at the connection establishment phase and can be updated at any time during data transmission. Furthermore, the mode of operation can be changed during the data transmission. XTP also provides mechanisms to multiplex transport connections on virtual circuits to increase resource utilization. A connection is established with the first packet sent, whereas three-way handshaking is used for connection termination. Inband signaling is used for both functions. In addition, two-way handshaking can optionally be used for connection establishment. In addition to using the credit-based sliding-window mechanism for end-to-end control, XTP uses rate control to restrict the size and time spacing of bursts of data from the sender. Error detection is achieved by the protocol header checksum with the optional data checksum for the information field. If the receiver detects an error in the packet received, it informs the sender for retransmission of the packet. To inform the sender about packets received out of sequence, the receiver describes the groups of bytes that were received using two sequence numbers, referred to as *selective acknowledgment.* The first sequence number is the starting byte in the group, whereas the second sequence number is one more than the last byte of the group. This scheme provides enough information for the sender to retransmit packets selectively. However, XTP also defines mechanisms to use a go-back-N retransmission algorithm for error handling.

XTP allows users to define the service types that best suit their applications from a rich set of options provided and supports prioritized communication.

10.6 COMPARISON OF PROTOCOLS

A detailed comparison of APPN, Datakit, Delta-T, NETBLT, TP4, TCP, VMTP, and XTP is given in [12]. XTP and RTP have very similar features and we concentrate only on XTP. Various aspects of the eight protocols defined in sections 10.4 and 10.5 are summarized in tables 10.2 and 10.3 [12].

It has been demonstrated in various studies that a significant portion of the total time spent at the transport layer is consumed by the overhead of computing control parameters and interfacing with the host operating system. Despite their importance, the implementation issues have not received much attention in the literature compared to the design issues.

Advances in technology making it feasible to increase the transmission link speeds to several megabits per second are practically useless if the amount of bandwidth delivered to the users is restricted to only a small fraction of the available bandwidth due to protocol processing at the end nodes. Although in principle all transport protocols provide the user with a similar set of functions, the question arises of how effectively a protocol uses the changing environment of dramatic decreases in bit error rates and tremendous increases in transmission rates effectively to meet the requirements of B-ISDN applications.

TCP is being extended to support high-speed networks [10]. Using a production copy of TCP for UNIX, it was illustrated in [10] that (excluding the connection establishment, termination, and error handling functions) TCP can

Table 10.2
Various Aspects of Lightweight Protocols

Protocol	*Signaling*	*Connection Establishment*	*Connection Termination*	*Connection Parameters*	*Multiplexing*
APPN	Out-of-band	2-way	2-way	S	No
Datakit	Out-of-band	2-way	2-way	S-U-M	No
Delta-T	Inband	Implicit	Timer-based	U	Yes
NETBLT	In Band	2-way	2-way	S-U	Yes
TP4	Inband	3-way	2-way	S-U	Yes
TCP	Inband	3-way	3-way	S-U	Yes
VMTP	Inband	Implicit	Implicit	S-U-M	Yes
XTP	Inband	Implicit	3-way	S-U-M	Yes

S = Setup.
U = Updated.
M = Mode selection.

Table 10.3
Various Aspects of Lightweight Protocols

Protocol	*Ack. Scheme*	*Flow Control*	*Flow Control Technique*	*Error Reporting*	*Error Handling*
APPN	Not required	Not required	Not required	Not required	Abort
Datakit	Explicitly solicited by sender	End-to-end	Adaptive window (optional)	NACK	Go-back-N
Delta-T	Sender-dependent	End to end	Adaptive window	NACK (optional)	Go-back-n
NETBLT	Sender-dependent	End-to-end	Cumulative window or rate control	SRej	SR
TP4	Sender-independent (opt.)	End-to-end	Adaptive window	Not available	PAR
TCP	Sender-dependent	End to end or implicit access control	Adaptive window	Not available	PAR
VMTP	Sender-dependent	Implicit access control	Rate control with interpacket gap	SRej	SR
XTP	Explicitly solicited by sender	Explicit access control	Adaptive window or rate control	SRej	SR and go-back-N

produce up to 800 Mbps of throughput in its steady-state execution path with 4,000-byte frames.

The three-way handshake connection management used in TCP may not be suitable for some B-ISDN applications, particularly when the total data transmission time is less than the time it takes to establish and terminate the connection. Another problem with TCP is the possibility of accepting duplicate packets from a current connection. In particular, the problem occurs if the sequence number space wraps around in less time than the maximum packet length (MPL), which takes place in 30 min at 10 Mbps, 170 sec at 100 Mbps, and 17 sec at 1 Gbps. Yet the IP MPL is on the order of 120 sec, complicating the problem as the transmission speeds increase. It is proposed in [28] to use sender time stamps, transmitted using the TCP echo option introduced in [29] to eliminate the reliability problems due to old duplicate segments on the same connection and allow very high TCP transfer rates.

The largest window size in TCP is 65K, which is not sufficient for high-speed networks. It is proposed in [28] to increase the window size using an implicit scale factor to multiply the window size value stored in the TCP header to determine the actual window size. IETF is working on extending TCP to high-speed networks.

Delta-T is designed to be used in a LAN environment. Like TCP, delta-T has the sequence number wraparound problem. Although implicit connection establishment and termination in delta-T minimizes the amount of time spent on connection management, it also introduces dependency on the MPL bound at connection termination. Furthermore, it is noted that although implicit connection establishment minimizes the delay for the first packet, it does not affect long-term throughput.

The XTP design includes the better parts of all protocols developed previously in addition to some new concepts introduced at various transport functions. The use of the implicit scheme allows fast connection establishment, whereas the three-way handshake connection termination reduces the dependency on the MPL bound. It uses keys to identify connections as opposed to the source and destination ports in TCP, thereby reducing the time it takes to discard connection states. However, the use of the implicit connection establishment scheme together with rate control introduces the problem of overrunning the receiver, since the resource availability at the receiver is not known a priori. Furthermore, XTP segmentation is an expensive operation, since checksum is computed for each packet.

Datakit connection management is based on handshaking that takes place out-of-band. However, the inclusion of call setup information in the transport protocol complicates the implementation of the protocol. Datakit relies on insequence delivery by the network, which is supported in ATM. This eliminates the possibility of duplicate packets. The steady-state execution path in datakit is reduced, since data are not packetized and reliability protocols are simplified.

VMTP is designed to be used in LANs. The transmission rate set initially remains constant for one complete block to be transmitted, although it is possible that the receiver may not be able to process the incoming data at this rate.

10.7 BENCHMARKING TRANSPORT PROTOCOLS

Recently, ISO has initiated a new project entitled *Enhanced Communications Functions and Facilities for OSI Lower Layers* toward the development of new protocols for high-speed networks. The enhancements initially considered are:

- High-data-throughput capability;
- Multicast operation;
- Selectable error control procedures;
- QOS selection and management;
- Out-of-band signaling and synchronization;
- Efficient operation.

The project referred to as *high-speed transport protocol* (HSTP) will perhaps take several years to finalize. Most probably the project will not end up standardizing any of the above protocols. However, such efforts will no doubt benefit from most of the concepts defined and the experience to be gained from the use of various lightweight protocols as well as modifications to existing protocols.

A benchmark is a set of programs that is used as a standard load against which the performance of a computer system can be measured. There are several reasons for performing benchmarks:

- To compare the performance of different systems;
- To make design decisions on hardware and software architectures;
- To determine the most efficient means of using the system;
- To evaluate the effects of equipment upgrade or system tuning.

Although benchmarking has been used widely to evaluate the performance of computer systems and various communication devices such as routers and bridges, there is a lack of benchmarks for high-speed networks. There are various protocols that have been implemented, with more to come. For a new protocol to be standardized for B-ISDN, it is necessary to design a benchmark representative of this new environment. Doing so would provide a standard base to compare the performance characteristics of various protocols and to ensure that the performance required by various existing applications, in addition to the new services expected, are met. Furthermore, different implementations of a given protocol would produce different performance numbers, and the results of benchmarks can be used to evaluate different implementations. A quantitative comparison of various transport protocols and their particular implementations are meaningful, since they all have the same goal: to establish, manage, and disconnect end-to-end connections and to provide interfaces to higher layers requesting such functions.

A benchmark of a transport protocol should include the perspectives of both the transport service users and network operators. Users are most interested in the throughput and the response times provided by the network. Network operators, on the other hand, would like to maximize the utilization of network resources. Similar to other types of benchmarks, protocol benchmark should be representative of the various ways the protocols are used and must be generic enough to compare various protocol implementations. The framework to be developed should be unbiased toward any protocol, should be independent of the network topology, should address the requirements of a representative set of B-ISDN applications, and should include a set of cases to evaluate all transport functions.

The throughput provided to an application is a function of the total time a connection is active; that is, the time interval between the issue of a connection establishment and the time the connection is terminated and the connection state is cleared from all communicating entities. For example, consider the file transfer application. The file transfer time is the time interval between the instant the first byte of the file arrives at the sender transport layer and the time the acknowledgment for the last byte of the file is received by the layer. The total time is affected by the connection management scheme, flow and rate control, error recovery and control, and handling duplicated packets. Various connections may be required to be active to evaluate the multiplexing and the priority-handling capabilities of the protocol. Furthermore, the efficiency of the service provided must be evaluated for short-, medium-, and long-lived applications that may be sensitive to errors or delay, or both. Efficient support of multicast is another desired property in B-ISDN. However, achieving all these in a unique platform is by no means an easy task. Clearly, network congestion is an important factor affecting the performance at the transport layer. The problem is further complicated by the need to incorporate the requirement for these protocols to deal with high bandwidth and large propagation delay. Yet every design would probably look impressive on paper, and there is no easy way to standardize a protocol without meaningful comparisons.

10.8 RELIABILITY IN MULTICAST SERVICES

Multipoint communication allows information transfer within a group of peer entities. Multicast is defined as a subset of multipoint communication, where traffic generated at a node (referred to as the *root*) is distributed to two or more other nodes. This section reviews the protocol functions and mechanisms for reliable multicast data transmission based on the survey given in [53]. Multipoint communication among *N* group members may take place over *N*-multicast connection, with each node in the group being the root of a multicast connection.

Different applications that are best served with multicast communications have different reliability requirements, varying from no reliability to full reliability. During its execution, an application may need some information that is unavailable locally. Furthermore, where the requested information is available may not be known a priori. In this case, the application searching for the particular information may distribute a request message to the other members of the group likely to have the information, hoping that at least one response will be received back from the group. This type of multicast is referred to as *resource location* in distributed systems and often used for transactions in a distributed database or in distributed (i.e., client/server) systems.

When several processes complete their processing of the same application at different sites, a coordinating process may start collecting their results to start a vote. Multicast is used to send that request.

Data replication is a simple way to achieve a crash-tolerant architecture. Active replication consists of executing the same application at different sites. The results are compared at the end of the execution with a vote. Multicast services may be used to update the copy of a particular data at different sites to have a coherent state. This requires the updates to be performed at each site in the same order.

Teleconferencing and video-on-demand are two other applications that are best served by multicast communications. The various functions required to provide these services include location and access of video information; management of distributed and replicated information; and transmission of video, voice, and data from one node to two or more other nodes. Such multimedia services are expected to be the major applications requiring multicast communication in emerging ATM networks.

Depending on the application requirements, different levels of reliability in multicast communication may be provided. For example, database update service requires data integrity among all replicated databases in the network, whereas interactive video service may not at all require (and cannot tolerate the associated delays with) a reliable service. Accordingly, different classes of multicast transmission have been defined to address the different reliability requirements of different applications.

- Connectionless multicast: The sender only multicasts its packets without worrying about their reception by the other group members. The multicast protocol in UNIX 4.2 is an example of connectionless multicast.
- k-reliable multicast: Considering a group of $N + 1$ members (including the sender), a multicast operation is k-reliable if at least k of the N members of the group correctly receive the sender's message (with $0 \leq k \leq N$). 0-reliable service corresponds to the connectionless multicast. For example, 1-reliable service would be used if it is sufficient for at least one group member to receive the message. N-reliable corresponds to a fully reliable service in which all N members of the group are required to receive the message correctly.

In a group of N group members, ordered reliable multicast communication may be used after each group member receives multicast messages from different originating nodes correctly (i.e., multipoint-to-multipoint communication, which can be realized as N-multicast communications). Receivers in this case verify an ordering rule, not necessarily the transmission order, but the order in which data are delivered to the receiving application. There are three different ordering notions reported in the literature: partial ordering, casual ordering, and total ordering.

1. In partial ordering, messages received from each source are delivered to the application in the same order that they are received at the receiver, but not necessarily in the order they were transmitted by each source. There is no ordering rule specified between the messages transmitted by different senders.
2. In casual ordering, messages received from each source are delivered to the application in the same order as the order of transmission by each source, with no ordering rule specified between messages transmitted by different senders.
3. In total ordering, all messages transmitted by any source are delivered in the order they were transmitted among different sources. Total ordering requires a global view of the connection in which messages transmitted from different sources are ordered. For example, to have a consistent view in a replicated database system, the use of total ordering may be necessary (i.e., cash transactions).

As discussed throughout this chapter, messages may be lost or damaged during their transmission. Error control helps to detect and sometimes to correct errors. In ATM networks, there is no retransmission that takes place inside the network; hence, the error control function is performed at the end stations. Two policies used for error correction are positive and negative acknowledgment.

If the positive acknowledgment scheme is used for multicast communication, the network may be forced to carry significant overhead for acknowledgment messages (i.e., each message is acknowledged *N* times, one by each group member). Negative acknowledgment detects gaps in received messages in the receivers and appears to be more efficient for multicast communications. The two schemes can also be mixed so that positive acknowledgments may be sent at the request of the sender, whereas negative acknowledgments may be sent by the receiver whenever the receiver detects a lost or corrupted message.

Flow control in multicast communication is to say the least problematic. When receivers that can access data at different rates participate in a multicast communication, the receiver has no choice but to use flow control parameters based on the values provided by the receiver that can support the lowest incoming traffic rate (i.e., minimum buffers and/or receive rate).

10.9 MULTIMEDIA NETWORKING

As discussed in Chapter 3, multimedia is the blending together of computer technology, audio/video technology, and display technology in an interactive environment. Evolving applications such as desktop conferencing and video-on-demand impose new requirements on the network. ATM provides the basic framework to provide the required bandwidth and to guarantee QOS requirements of multimedia services. Various emerging applications may run natively on top of ATM networks (i.e., MPEG-2 over ATM specification in the ATM Forum). Similarly, there are various activities in different forums to define the specifications of multimedia services over different subnetworks such as priority Token Ring, Ether Star and Token Star, 100-Mbps Ethernet (IEEE 802.3, 100baseT), 100-Mbps VG (IEEE 802.12), and ISDN. These solutions are referred to as *subnetwork-specific solutions* (cf. [54]). Following the classification in [54], subnetwork-independent solutions are referred to as *logical network solutions.* The two approaches are illustrated in Figure 10.2.

While the subnetwork-specific solutions are designed to sit directly on top of the subnetwork, the logical network solutions are an evolution of current data networking structures (OSI layers 2 to 4). IETF has currently enhanced the internet suite of protocols as depicted in Figure 10.3.

At the network layer, the IETF has two parallel activities underway: the experimental internet stream transport protocol, version 2 (ST-II), and the resource reservation protocol (RSVP). Both protocols support multicast, resource reservation, bandwidth management, QOS, and so forth, but they have fundamentally different design points.

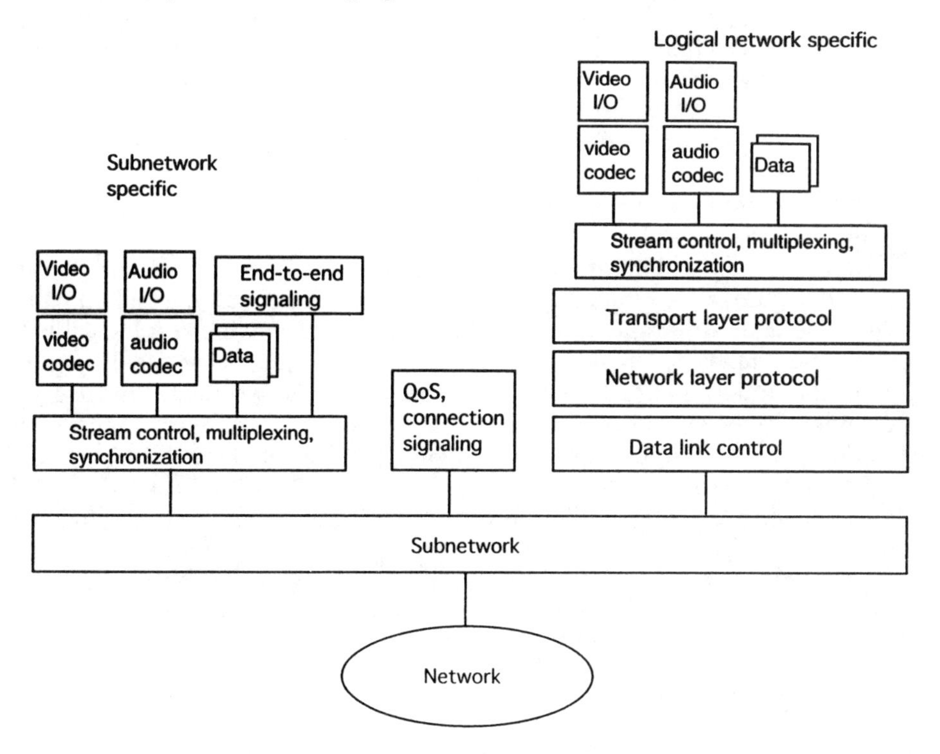

Figure 10.2 Subnetwork-specific and logical solutions.

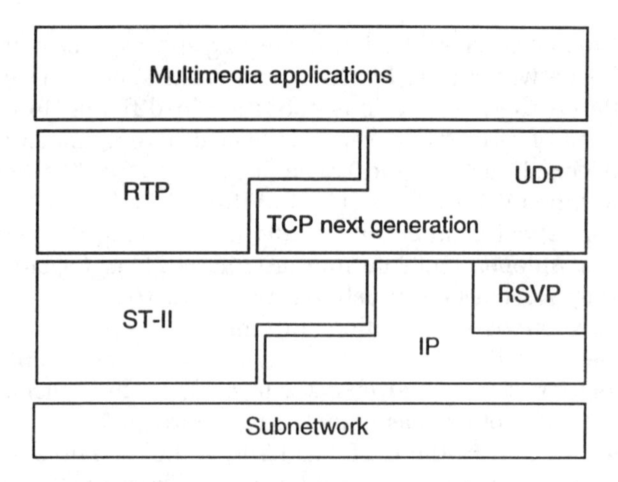

Figure 10.3 IETF multimedia service framework.

ST-II is an experimental protocol defined by RFC 1190. It is a connection-oriented layer 3 network protocol that is defined to coexist with the internet protocol (IP). RSVP builds on connectionless IP and is designed to scale to very large multicast groups.

The real-time transport protocol (RTP) provides framing, multiplexing/demultiplexing, encoding identification, synchronization, error detection, and encryption. It is accompanied by a real-time transport control protocol (RTCP) that performs group management, QOS monitoring, and so forth. The IETF activities in these areas have not yet been completed.

Hence, there will be different solutions available in the near future to run multimedia applications not only over ATM but other networking technologies as well. The main difference between the subnetwork-specific and logical networking approaches is that subnetwork-specific solutions are designed specifically for the underlying technology, whereas logical networking solutions are independent of the specifics of the networking environment. Accordingly, subnetworking-based solutions can take advantage of what the subnetwork can offer. In the case of ATM, this is QOS guarantees, statistical multiplexing, and so forth. The disadvantage of this approach is that only the end stations attached to the subnetwork directly can communicate to each other (without interworking functions). Logical networking approach solves the interoperability problem but has to be a generic solution and it cannot depend on and take advantage of the capabilities provided in the subnetwork.

10.10 SUMMARY

As in low-speed networks, control messages in B-ISDNs to guarantee reliable end-to-end transfer are required to exchange state information between the end nodes and to cause retransmissions as necessary. End-to-end flow control is also required to reduce the possibility of overwhelming the end stations. Various transport functions, being end-to-end, are required to be designed based on the bandwidth-propagation delay product. As this product increases, achieving the maximum possible throughput is restricted by the window size. As the window size increases, the effectiveness of end-to-end scheme decreases. The go-back-N scheme, used effectively in low-speed networks, may not be as efficient to use as the bandwidth delay product increases. By the time a source knows about a packet being lost, the amount of data transmitted since the lost packet can be too large and cause considerable bandwidth to be wasted when retransmitted. Selective retransmission methods solve this problem by transmitting only the lost packets. However, these techniques complicate the receiver design.

Considering the wide range of applications in B-ISDNs, different users may require different capabilities such as quick and not necessarily fully reliable connection or slow but fully reliable connection establishment capabilities. Similar options are required for fully reliable transmission to unreliable transmission, different service priorities, and error reporting and handling. Furthermore, considering the higher speeds, parts of the protocol should be simple enough to be implemented mostly in hardware. Finally, the standardization process needs to be completed as soon as possible.

Although some distance has already been covered, a single transport protocol for B-ISDNs is not expected to be defined any time soon. This is not necessarily a problem for the introduction of ATM networks, since these networks are required to support existing services. Accordingly, the current activities are mostly concentrated on how to support existing protocols such as IP and frame relay in ATM networks.

Another area that has been receiving considerable attention is the design of higher layer protocols for emerging multimedia applications. Logical networking solutions that essentially ignore the capabilities provided in the network require the design of transport (and other) layer protocols to meet application demands. Until ATM becomes widely available in both the local and the wide area, there is not much choice but to use these solutions to enable multimedia applications.

References and Bibliography

[1] Ahmadi, H., and P. Kermani, "Throughput Analysis of a Class of Selective Repeat Protocols in High-Speed Environments," *GLOBECOM '89*, 1989, pp. 930–938.

[2] Bae, J. J., T. Suda, and N. Watanabe, "Evaluation of Protocol Processing Overhead in Error Recovery Schemes for a High Speed Packet Switched Network: Link by Link versus Edge to Edge Schemesn," *IEEE JSAC*, Vol. 9, No. 9, 1991, pp. 1496–1509.

[3] Baratz, A. E., J. P. Gray, P. E. Green, J. M. Jaffe, and D. P. Pozefsky, "SNA Networks of Small Systems," *IEEE JSAC*, Vol. SAC-3, No. 3, May 1985, pp. 416–426.

[4] Bhargava, A., J. F. Kurose, D. Towsley, and G. Vanleemput, "Performance Comparison of

Error Control Schemes in High-Speed Computer Communication Networks," *IEEE JSAC*, Vol. 6, No. 9, December 1988, pp. 1565–1575.
[5] Biersack, E. W., and D. C. Feldmeier, "Transport Protocol Issues for ATM Based Networks," *Proc. 8th European Fibre Optic Comm. and LAN Exposition*, West Germany, 1990.
[6] Bondi, A. B., "On the Bunching of Cell Losses in ATM Based Networks," *GLOBECOM '91*, 1991.
[7] Burda, R., and T. Uhl, "Performance Analysis of Entry-to-Exit Flow Control in a Virtual Channel," *ITC-13*, 1991, pp. 359–363.
[8] Cheriton, D. R., and C. L. Williamson, "VMTP as the Transport Layer for High-Performance Distributed Systems," *IEEE Comm. Mag.*, June 1989, pp. 37–44.
[9] Clark, D., M. L. Lambert, and L. Zhang, *RFC 998*, "NETBLT: a Bulk Data Transfer Protocol," March 1987.
[10] Clark, D. D., V. Jacobson, J. Romkey, and H. Salwen, "An Analysis of TCP Processing Overhead," *IEEE Comm. Mag.*, June 1989, pp. 23–29.
[11] Comer, D. E., "*Internetworking with TCP/IP: Principles, Protocols and Architecture*," Prentice-Hall, 1988.
[12] Doeringer, W. A., D. Dykeman, M. Kaiserwerth, B. W. Meister, H. Rudin, and R. Williamson, "A Survey of Light-Weight Transport Protocols for High-Speed Networks," *IEEE Trans. Comm.*, Vol. 38, No. 11, November 1990, pp. 2025–2039.
[13] Doshi, B. T., P. K. Johri, A. N. Netravali, and K. K. Sabnani, "Retransmission Protocols and Flow Controls in High Speed Packet Networks, *ITC-13*, 1991, pp. 309–314.
[14] Doshi, B., A. Eckberg, V. Saksena, and R. Zoccolillo, "An Overall B-ISDN/ATM Congestion/Flow/Error Control Architecture and Its Predicted Performance," *Proc. ITC-7* Seminar, New Jersey, 1990.
[15] Dykeman, D., and M. Kaiserwerth, "Tutorial: High-Speed Protocols," *3rd IFIP Conf. High-Speed Networking*, 19 March 1991.
[16] Eckberg, A. E., D. T. Luan, and D. M. Lucantoni, "An Approach to Controlling Congestion in ATM Networks," manuscript.
[17] Frazer, A. G., "The Universal Receiver Protocol," in *Protocols for High Speed Networks*, Rudin and Williamson, eds., The Netherlands: Elsevier, 1989.
[18] Fuhrmann, S., and J. Le Boudec, "Burst and Cell Level Models for ATM Buffers," *ITC-13*, 1991, pp. 975–980.
[19] Garner, G. M., "End-to-End Performance Modeling for Virtual Circuit Networks," *ITC-13*, 1991, pp. 371–376.
[20] Gerla, M., and L. Kleinrock, "Flow Control: A Comparative Survey," *IEEE Trans. Comm.*, Vol. COM-28, No. 4, April 1980, pp. 289–310.
[21] Gihr, 0., and P. J. Kuehn, "Comparison of Communication Services with Connection-Oriented and Connectionless Data Transmission," *Proc. Int. Seminar on Computer Networking and Performance Evaluation*, Tokyo, September 1985.
[22] Green, P. E., R. J. Chappuis, J. D. Fisher, P. S. Frosch, and C. E. Wood, "A Perspective on Advanced Peer-to-Peer Networking," *IBM Sys. J.*, Vol. 26, No. 4, 1987, pp. 414–427.
[23] Gunningberg, P., et al., "Application Protocols and Performance Benchmarks," *IEEE Comm. Mag.*, June 1989, pp. 30–36.
[24] Heatley, S., and D. Stockesberry, "Analysis of Transport Measurements Over a Local Area Network," *IEEE Comm. Mag.*, June 1989, pp. 16–22.
[25] Huitema, C., and W. Dabbous, "End-to-End Transmission Control on ATM Networks," manuscript.
[26] Inai, H., T. Nishida, T. Yokohira, and H. Miyahara, "End-to-End Performance Modeling for Layered Communication Protocol," *INFOCOM '90*, 1990, pp. 442–449.
[27] Jacobson, V., "Congestion Avoidance and Control," *SIGCOMM '88*, 1988, pp. 314–329.
[28] Jacobson, V., R. Broden, and L. Zhang, RFC 1185: "TCP Extensions for High Speed Paths," Network Working Group.

[29] V. Jacobson and R. Broden, RFC 1072, "TCP Extensions for Long Delay Paths," Network Working Group.
[30] Jain, R., "A Timeout-Based Congestion Control Scheme for Window Flow-Controlled Networks," *IEEE JSAC*, Vol. SAC-4, No. 7, October 1986, pp. 1162–1167.
[31] Kanakia, H., S. Keshav, and P. P. Mishra, "A Benchmark Suite for Comparing Congestion Control Schemes," manuscript.
[32] Kaplan, M,."Rapid Transportation Protocol," *IBM Research, Tech. Rept.*, 1989.
[33] Kim, Y.-B., A. G. Vacroux, "A Unified End-to-End Control Scheme for Fast Packet Networks at ISO/OSI TSAP," *ICC '91*, 1991, pp. 397–402.
[34] Kitami, T., and I. Tokizawa, "Cell Loss Compensation Schemes Employing Error Correction Coding for Asynchronous Broadband ISDN," *INFOCOM '90*, 1990, pp. 116–123.
[35] La Porta, T. F., and M. Schwartz, "Architectures, Features, and Implementation of High Speed Transport Protocols," *GLOBECOM '91*, 1991.
[36] Lai, W. S., "Protocols for High-Speed Networks," *INFOCOM '90*, 1990, pp. 1268–1269.
[37] Li, S. Q., "Study of Information Loss in Packet Voice Systems," *IEEE Trans. Comm.*, Vol. 37, No. 11, November 1989, pp. 1192–1202.
[38] Mase, K., and S. Shioda, "Real-Time Network Management for ATM Networks," *ITC-13*, 1991, pp. 133–140.
[39] Norros, I., and J. T. Virtamo, "Who Loses Cells in the Case of Burst Scale Congestion," *ITC-13*, 1991, pp. 829–833.
[40] Partridge, C., *RFC 1152*, "Workshop Report; Internet Research Steering Group Workshop on Very High-Speed Networks," April 1990.
[41] Pach, A. R., "A Technique for Calculation of the Optimal Timeout for the Class 4 Transport Protocol in a Packet Switched Network," *ITC-13*, 1991, pp. 291–296.
[42] Sabnani, K., and A. Netravali, "A High Speed Transport Protocol for Datagram/Virtual Circuit Networks," *SIGCOMM '89*, 1989, pp. 146–157.
[43] Sanders, R. M., "The Xpress Transfer Protocol (XTP) A Tutorial," Univ. of Virginia Comp. Sci. Dept., TR-89-10, 1 November 1990.
[44] Shacham, N., "Queueing Analysis of Selective-Repeat ARQ Receiver," *INFOCOM '87*, 1987, pp. 512–520.
[45] Sjodin, P., et al., "Towards Protocol Benchmarks," in *Protocols for High Speed Networks*, H. Rudin and R. Williamson, eds., Amsterdam: Elsevier, 1989, pp. 57–67.
[46] Sultan, R. A., P. Kermani, G. A. Grover, T. P. Barzilai, and A. E. Baratz, "Implementing System/36 Advanced Peer-to-Peer Networking," *IBM Sys. J.*, Vol. 26, No. 4, 1987, pp. 429–451.
[47] Strayer, W. T., B. J. Dempsey, and A. C. Weaver, *Xpress Transfer Protocol*, Addison-Wesley, 1992.
[48] Tanenbaum, A. S., *Computer Networks*, Prentice Hall, 1981.
[49] Watson, R. W., "Delta-T Protocol Specification," Lawrence Livermore Lab., Report UCID-19293, 15 April 1983.
[50] Wright, D. J., and M. To, "Telecommunications Applications of the 1990s and Their Transport Protocol," *IEEE Network Mag.*, March 1990.
[51] Zitterbart, M., and A. N. Tantawy, "Transport Service and Protocols for High Speed Networks," *IBM Research Rept.*, RC 17074 (#75746), 1991.
[52] Zitterbart, M., and B. Stiller, "A Concept for a Flexible High Performance Transport System," *GI Jahrestagung Darmstadt*, Germany, October 1991.
[53] Diot, C,."Reliability in Multicast Services and Protocols: a Survey," *Proc. Int. Conf. Local and Metropolitan Comm. Systems*, Kyoto, Japan, 1994, pp. 295–313.
[54] Lynch, J. J,."Multimedia Networking," *IBM Tech. Rept.*, Research Triangle Park, 1994.
[55] Svobodova, L,."Measured Performance of Transport Service in LANs," *Computer Networks and ISDN Systems*, Vol. 18, 1989, pp. 31–45.

ATM Network Management 11

As the communication networks increase in complexity and in size, it becomes essential to use automated network management tools to assist network administrators in monitoring (i.e., collecting information about various network resources) and control of network elements. Network management is the process of controlling a complex data network so as to maximize its efficiency, productivity, and reliability.

The ATM network management framework is currently defined in three general areas: interface management including UNI (i.e., ILMI), DXI, and LAN emulation; layer management (i.e., OAM framework), and global network management used for total management of ATM networks and services, which is modeled in five subareas.

Interface management deals with the exchange of information at the interface level. It is used primarily for configuration and alarms of ATM interfaces and provides the basic framework for two different devices to interface to one another. Layer management allows testing for continuity and loopback at a segment or an end-to-end virtual-circuit level (both at the VC and the VP levels). It is primarily used for end-to-end circuit management and allows checking a user circuit through the network. Global network management deals with the configuration of an ATM network consisting of one or more switches and the monitoring and control of ATM devices in the network. It provides total management of networks from a top-down perspective.

The OAM and various parts of ILMI are covered in Chapter 2. In this chapter, an overview of the network management framework is followed by a discussion on interface management and the global management of ATM networks.

11.1 NETWORK MANAGEMENT FRAMEWORK

A network management system consists of four parts: management station (or manager), agent, MIB, and network management protocol.

The management station serves as an interface for the administrator with the network management system. It translates administrator's commands into actual monitoring and control of the network elements. Various services provided at the management station include applications for providing data analysis and fault recovery, and a database for holding network management information extracted from the databases of all the managed elements in the network.

Each node in the network (including end stations) that participates in network management contains network management entity (NME) software for performing network management-related tasks. Each NME collects data on resources and stores related statistics locally. NMEs also respond to commands from the network administrator (manager). At least one node in the network is designated as a network management application, which includes an interface to the administrator to manage the network. NMEs are referred to as *agents.* Network management applications (NMAs), under the control of the operator, respond to user commands by passing on the information available in its database. It also issues commands and exchanges information with NMEs in the network.

The third component of the network management framework is the MIB. An MIB is a collection of objects, each representing a particular aspect of a managed agent. For example, a manager performs a monitoring function by retrieving the value of a particular object or changes the settings of a network resource by modifying the value of the corresponding object.

Finally, the communication between the manager and the agents is carried out using a network management protocol. The two standard network management protocols are the IETF's simple network management protocol (SNMP) and ISO's common management information protocol (CMIP). In addition to defining a specific protocol, both SNMP and CMIP define a corresponding database structure specification and a set of data objects (MIBs).

11.2 FUNCTIONAL AREAS OF NETWORK MANAGEMENT

ISO defined the following five functional areas for network management:

1. Configuration management: The facilities that exercise control over, identify, collect data from, and provide data to managed objects for the purpose of assisting in providing for continuous operation of interconnection services;
2. Fault management: The facilities that enable the detection, isolation, and correction of abnormal operation of network resources;
3. Performance management: The facilities needed to evaluate the behavior of managed objects and the effectiveness of communication activities;
4. Security management: Addresses those aspects essential to operating network management system correctly and to protecting managed objects.

5. Accounting management: The facilities that enable charges to be established for the use of managed objects and costs to be identified for the use of those managed objects.

Configuration management is the process of finding and setting up network devices. It is a set of short- and medium-range activities for controlling network resource inventory, tracking vendor performance, maintaining trouble files, evaluating and negotiating service levels, managing and changing security levels, and dealing with costs and charging.

Customers play a major role in configuration management. Their expectations for network service levels and demand for capacity are determined when the network is set up. It is important that these expectations are met in a cost-effective manner. If not, it is necessary to reconfigure the network. Accordingly, configuration management consists of obtaining data from the network and using them to manage all network resources by means of administrative management policies. Applied to an ATM network including end stations, the configuration management requirements include:

- Creating and deleting ATM connections (VPCs and VCCs);
- Getting connection status;
- Determining the number of connections active on an interface;
- Determining maximum number of connections supported on an interface;
- Determining the number of preconfigured connections on an interface;
- Configuring the number of VPI/VCI bits supported;
- Configuring and determining the status of interface address information.

Fault management is the process of locating problems or faults in the network. It is a collection of activities required to maintain the desired network service level. The proper operation of a network is essential in providing network services. However, abnormal conditions such as excessive errors and failure to operate correctly do occur. Such occasional faults need to be determined quickly. During the time the failed components are being repaired and/or replaced to restore the network to its initial state, the network may be reconfigured so that the impact of failed components is minimized. Accordingly, when a fault occurs, it is required, as quickly as possible, to:

- Determine the location of the failed resource and the nature of the fault;
- Isolate and reconfigure the rest of the network from the failure;
- Restore the network to its original state.

Applied to an ATM network including end stations, the fault management requirements include:

- Notification of inability to establish ATM connections;
- Notification of an ATM connection failure;
- Notification of multiple simultaneous connection failures;
- Notification of UNI adjacent peer failure;
- Support of OAM fault management flows.

Performance management is to measure the performance of the network hardware, software, and transmission media. It is a quantitative investigation of the network resources to verify that the service levels are maintained and to support other network management functions such as configuration and fault management. In most cases, it is critical to network applications that the network resources operate within their specified performance limits. Performance management is used to monitor the performance levels of various network resources and to allow adjustments to improve network performance. This function addresses the effectiveness with which a network is delivering services to its users. The performance management requirements are:

- Determine whether or not an ATM connection meets the QOS requirements;
- Determine the number of cells in violation of the traffic contract;
- Support OAM performance management flows;
- Support counter sets appropriate to send and receive operations of the VP and VC levels.

Security management is the process of controlling network access and protecting objects within the network. It involves the protection of the transfer of authorized user information from one location to another. The five main security services for effective security management are:

- Authentication: Verification of the information source (i.e., information came from the expected source).
- Access control: The ability to restrict or control access to a source or network resources. This function attempts to ensure that only authorized users have access to corresponding files, network partitions, and databases.
- Integrity: The verification that the received data is indeed the data that was sent.
- Confidentiality: The property of data such as the receiver's need to know and authorization to release the information.
- Nonrepudiation: The verification that either the receiver or the sender did indeed receive or send the data.

Accordingly, the major functions of a security system include storage protection, data transmission protection, data receipt, and transmission control.

All of these may rely on encryption, which is altering the data in a way that it is meaningless to unauthorized intruders.

Accounting management deals with tracking the usage of network resources by network users and the cost incurred for the service. It is also used to limit the amount of resources allocated to users and informing them of the costs incurred and resources used. Various processes and procedures associated with accounting management include:

- Identification of cost components, including services, personnel, and overhead;
- Establishing charge-back policies, which involves determining whether to charge the user and on what basis;
- Calculation of charge-back from the definition of the charge-back procedures;
- Processing vendor bills, which involves computing the charges and transferring them to the network users.

Applied to an ATM network including end stations, the accounting management requirements may include:

- Record connection QOS;
- Record connection bandwidth;
- Record connection duration;
- Record number of cells successfully transmitted and received;
- Record number of cells received in error;
- Record number of cells received in violation of the traffic contract.

The task of determining communication cost is a complex task. There are several direct as well as indirect factors that make up the cost of a service. Hardware, software, personnel, and overhead used for providing different network services all contribute to the cost. However, the contribution of each factor to a specific service cannot be determined accurately.

11.3 NETWORK MANAGEMENT PROTOCOLS

The manager/agent model presented in Section 11.1 performs network management through the communication between a manager and several agents. The manager controls the overall network management, while the agent adjusts and controls managed objects belonging to it as directed by its manager and reports the results back to the manager. Figure 11.1 illustrates the manager/agent model.

Network management requires a communication protocol to give and take information between a manager and the agents. A management protocol provides a means of communication by providing several functions:

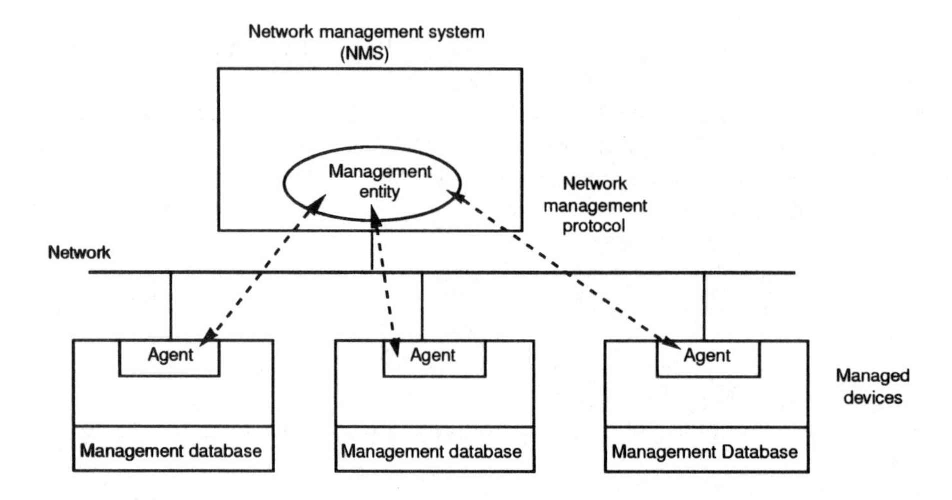

Figure 11.1 Manager/agent model.

- Reading and updating attributes of managed objects;
- Requiring managed objects to perform a specific function;
- Reporting results produced by managed objects;
- Creating and deleting managed objects.

The two standard management protocols SNMP and CMIP have their own formats and syntax to specify the managed objects, and both employ the manager/agent model.

11.3.1 Simple Network Management Protocol

SNMP was developed by IETF and is widely used to manage TCP/IP networks. It is a connectionless protocol. The SNMP agents reside on the managed devices and are designed to operate using minimal system resources, not to slow down the operation of the device significantly from providing network management services. The agent gathers data about the number and type of error messages received, the number of bytes and packets processed by the device, the maximum queue length, and so forth, and stores them in the MIB residing on that device. The processing of the collected data is done by the management application. The management application communicates with the managed devices via SNMP.

SNMP is an application layer protocol. It interfaces at the transport layer with the user datagram protocol (UDP). Accordingly, it can span heterogeneous networks. As an application that utilizes the TCP/IP protocols, SNMP can be used to manage devices on a number of physical topologies by simply allowing TCP/IP to deal with the lower level protocols.

SNMP provides two basic services: fetching and storing variables. It operates so that new variables can be defined in the MIB without requiring any change in the SNMP commands (in contrast to a management protocol, which provides different commands for a variety of different tasks that would have required new commands for newly defined MIB variables). Five SNMP commands are:

1. Get-request: Fetches a value from a specified variable;
2. Get-next request: Can fetch a variable without knowing its exact name; useful for stepping through tables;
3. Get-response: Replies to a fetch command with the requested data;
4. Set-request: Stores a value in a particular variable;
5. Trap: Sends a message in response to a particular event.

SNMP commands and replies contain a community name and data. A community name provides a form of authentication identifying a set of management systems that are allowed to execute SNMP commands, whereas data contain either an SNMP command or response.

11.3.2 Common Management Information Protocol

CMIP is developed by ISO. Unlike SNMP, CMIP is designed to provide generic solutions to overall network management. It is, however, a quite sophisticated protocol and is used mainly in service provider networks. While SNMP is based on a simple command structure, CMIP includes more powerful and difficult-to-develop commands. Accordingly, although CMIP offers more versatility and ease of use, it is not yet as popular as SNMP. In particular, currently there are few commercial applications available that use CMIP.

The ISO framework consists of layer management entities (LME), system management application entities (SMAE), and the CMIP protocol. The LMEs operate at each OSI layer, monitoring the operation of the system at that specific layer. They give focused attention to the layer they are monitoring while they do not give any indication of the performance of the system as a whole. The latter is the function of the SMAE. Each managed device has its own SMAE, which assimilates the information gathered by the LMEs and produces an overall picture of how the device is functioning with respect to its networking functions. The SMAEs of different network devices communicate with each other via CMIP, allowing system administrators to gather and analyze data about the functionality of the network as a whole. This system is sometimes referred to as the common management information service (CMIS), which provides monitoring of the network, control of network devices, and error reporting capabilities.

11.4 MANAGED INFORMATION BASE

The information gathered by the agents is the most important part of the management system, since all other activities are based on it. Thus, much effort has been devoted to defining and standardizing exactly what information is to be gathered and how it is to be stored.

MIB standards define network management variables and their meanings. The variables themselves are based on the structure of management information (SMI). The SMI not only places restrictions on the types of variables allowed to exist in the MIB, but it also specifies the rules for naming them. In particular, the SMI specifies that the MIB variables be defined and referenced using the ISO abstract syntax notation 1 (ASN.1), which consists of a readable notation used for documentation and an encoded representation of the information used by the communication protocols.

The contents of the SNMP MIB are defined using the SMI, while its structure is defined by the ISO as a branch of the global object identifier name-space tree. This tree is jointly administered by ISO and ITU-T. Each node in the tree has a name and a number, both of which are unique at that level of the particular branch of the tree. Through a hierarchical design, the name-space tree allows individual groups authority over certain branches (as illustrated in Figure 11.2).

By following a path from the root to the object, one can determine the globally unique name for that object.

11.5 ATM INTERFACE MANAGEMENT

ATM interface management is used primarily for configuration and obtaining status information of ATM interfaces. Currently defined interfaces by the ATM Forum are UNI, DXI, B-ICI, and LAN emulation UNI (LUNI). The MIBs defined for some of these interfaces are discussed in the next three sections.

11.5.1 User-to-Network Interface

Whether public or private, the two sides of a UNI exchange configuration and status information using a network management protocol. The information is then made available to a network management station by the remotely accessible agents residing in ATM end systems. There are two major organizations defining ATM UNI MIBs: the ATM Forum and IETF. The perspectives that these two organizations have in managing a UNI are different.

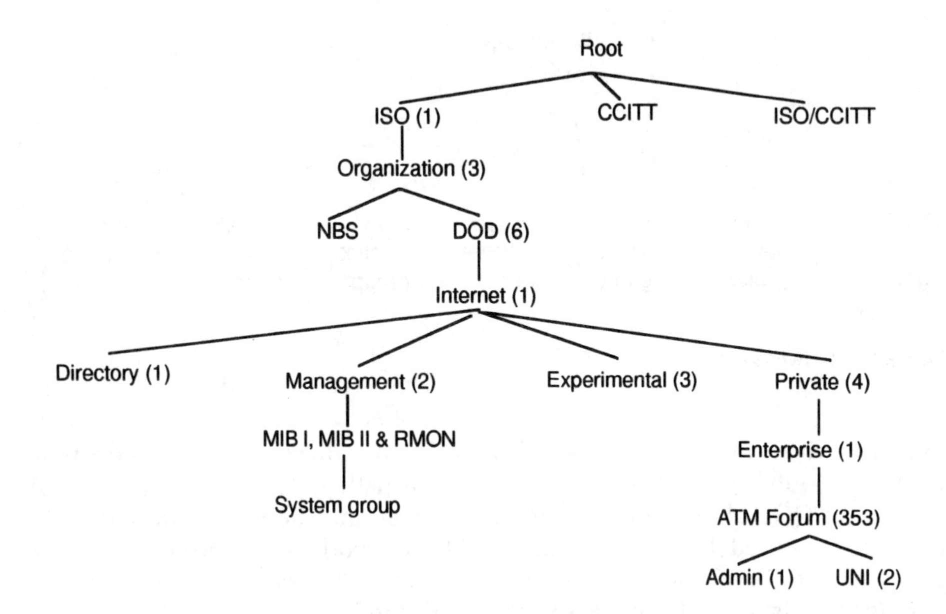

Figure 11.2 Name-space tree.

The ILMI is specified by the ATM Forum and is used for configuration and status information at a specific UNI and does not provide security, fault, or accounting management functions. The ATM Forum framework does not require an agent on either side of the UNI. Instead, it has defined a UME that uses SNMP over AAL 5. The ILMI supports the exchange of management information between UMEs related to both ATM and physical layer parameters. The communication between adjacent UMEs so that one UME can access the MIB information associated with its adjacent UME takes place based on SNMP commands that are transported over AAL 5 and does not use UDP or IP (unlike the original SNMP).

The IETF solution requires IP agents to reside on either side of the interface being managed. It does not make any explicit discussion of the ATM protocol use or define any interaction between the adjacent ATM peers.

The set of managed objects per UNI is referred to as UNI ILMI attributes, which are organized into a standard MIB structure, as illustrated in Figure 11.3.

The ATM statistics group is optional, while all the others are mandatory. The port table contains configuration status information about the physical interface including port index, address, transmission type (e.g., DS-3, OC-3), the type of physical media it is running on (e.g., UTP-3, fiber), its operational status (e.g., up, down, turned off, broken, on-line) and port-specific data that vary with the type of media used. The ATM layer group specifies the maximum number of VPCs and VCCs supported at the UNI. In addition, it includes infor-

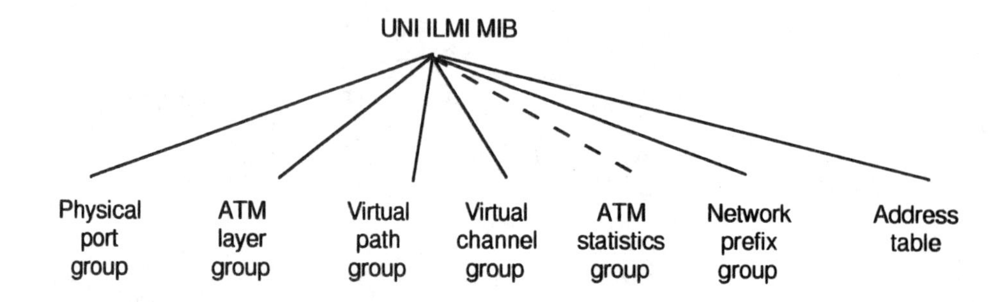

Figure 11.3 UNI ILMI MIB.

mation on the number of configured paths, path indicator bits, and port type (private or public). The VP group contains a path indicator, its operational status, and QOS class supported in each direction. There is a similar set of information stored for virtual circuits. The network prefix group gives the associated prefix of the ATM address and the address table group, which includes the list of addresses associated with the UNI.

Finally, optional ATM layer statistics group attributes include information on the count of the number of assigned (excluding idle) cells received across the UNI, the number of cells dropped due to the outcome of HEC processing, and cells with invalid headers.

11.5.2 Data Exchange Interface

DXI LMI operates in conjunction with the DXI and defines the protocol for exchanging data across the DXI that includes DXI-specific, AAL-specific, and ATM UNI-specific management information. The LMI definition assumes a 1:1 relationship between the DXI and UNI. It is designed to support a management station running SNMP and/or a switch running ILMI.

To recall the DXI specification described in Chapter 6, DXI is the interface between the DTE (i.e., a router) and the DCE (i.e., a data service unit). The SNMP commands (cf. Section 11.3.1) are issued only in the DTE and the DCE responds. DCE can only originate the trap message and responds to commands from the DCE. DXI LMI MIB consists of two objects dealing with the configuration and performance management functions: the DXI configuration group and DFA group. The configuration group includes the configuration information (similar to corresponding ILMI MIB), including the mode of operation (i.e., 1.a, 1.b, 2) and interface identification. The DFA group includes information on the AAL type supported at a particular DFA and the identification of the enterprise authority.

11.5.3 LAN Emulation

LE allows running legacy applications developed on top of legacy networks to run over ATM without any changes to the applications. As discussed in Chapter 9, the LAN emulation framework consists of four components: LE clients, LE servers, BUS, and configuration servers.

The LE network management [20] addresses the configuration, performance, and fault management functions only. The LE configuration management addresses various tasks that include identification of all LE clients currently set up at a managed device; creating and destroying LE clients; forcing LE clients to join and leave emulated LANs; examining and changing various system parameters; and identifying configuration, control, and multicast VCCs.

There are various factors that make it more difficult to observe emulated LANs than to observe legacy LANs [20]. Traffic is spread out over many virtual circuits that are set up and torn down frequently, instead of being concentrated in one physical network segment. The performance of each virtual circuit may be affected by factors outside the control of the emulated LAN hosts (i.e., switch congestions and congestion control mechanisms at the switches). Given these difficulties, the network management station monitors the amount of traffic going to a specific host by enlisting the help of LE clients in collecting this information, collecting and aggregating performance statistics about individual virtual circuits, collecting performance statistics at the ATM port level, and eavesdropping on communication between two LE hosts.

There are various levels of performance management that include performance monitoring of LUNI traffic (particularly LE ARPs), performance management of individual VCs within an emulated LAN, and performance management of ATM network over which an emulated LAN runs.

Finally, LE fault management is concerned with the prevention, detection, and correction of problems in an emulated LAN that are caused by the failure of network elements.

The LE client MIB is organized into groups, each corresponding to a table:

- Interfaces group, which includes emulated LAN, statistics, and server connections groups; ATM addresses group;
- Registered LAN destinations group, which includes MAC address and route descriptor groups;
- LE ARP cache groups that include MAC address and route descriptor translations.

11.6 ATM TOTAL MANAGEMENT

The ATM Forum network management subworking group has been developing an ATM network management reference model defining five management interfaces as illustrated in Figure 11.4.

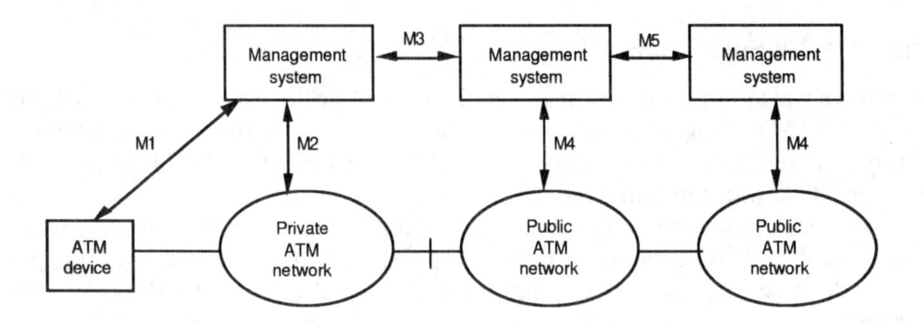

Figure 11.4 The ATM network management reference model.

The five management interfaces are referred to as M1 through M5. The M1 interface supports various functions for the management of the ATM devices. M2 supports functions for the management of private ATM networks, while the M4 interface is specified for managing public ATM networks. The M3 interface allows the two management systems, one in a private network and the other in a public network, to communicate with each other. Similarly, M5 provides communication capabilities between two management systems of public networks.

All five management interfaces will support all five network management functions (i.e., configuration, accounting, fault, security, and performance management). Some of the functional requirements of the configuration management in ATM include the configuration of ATM interfaces (i.e., UNI, B-ICI, DXI) for the maximum ingress and egress bandwidth, the maximum number of active VPCs and VCCs supported, the ATM subscriber address, and the interface identifier. A request to establish a permanent ATM connection (i.e., PVP or PVC) is another configuration management function. This request also includes the connection characteristics such as the VP/VC link termination, traffic descriptor, and QOS requirements. Additional requests will include connection teardown, release of resources, and retrieving link-related configuration information. Other functional requirements may involve the configuration of VPC and VCC termination points, segment end points, configuration, modification, viewing system attributes, control of reports and system functions, and obtaining information about the current configuration and state of a network element/system.

Fault management functions may involve the notification of a failure, logging received failure reports, and isolation of faults. Each failure notification may include a set of parameters such as the identification of the failed components, cause codes, and severity. The interface should also facilitate the reconfiguration of the default alarm severity assignments and overriding new severity assignments.

The performance management function in ATM may involve the performance monitoring of the physical and ATM layers, traffic management, UPC/NPC monitoring, define and reset the threshold values for the collected data, retrieve current counts on various performance metrics of interest such as number of cells with header errors and undefined VPI/VCI, retrieve current counts and reset them, and define and modify the sampling intervals.

11.6.1 ATOM MIB

The ATOM MIB has been designed by IETF and specifies M1, M2, and some of the M3 management interfaces describing how to configure a network, and addresses various AALs and switch-related management. It uses the SNMP version 2 format (cf. RFC 1573), which is not compatible with the first version of the SNMP.

This framework requires an IP agent to reside on either side of the interface being managed and does not make any explicit discussion of ATM protocol use or define any interaction between ATM adjacent peers. The manager obtains management information from the agents using SNMP over UDP/IP.

The current specification of the ATOM MIB defines the information for managing a UNI as well as ATM devices within an ATM network or cross-connect nodes that include:

- ATM interface configuration group;
- DS3 interface group;
- Transmission convergence sublayer group;
- ATM traffic parameter group;
- ATM VP link group;
- ATM VC link group;
- ATM VP cross-connect group;
- ATM VC cross-connect group;
- AAL 5 group.

These objects allow an incoming ATM connection to be mapped to an outgoing connection at a switch.

11.6.2 M3 Interface

M3 is the management interface needed to allow customers to supervise their use of the portion of a public ATM network. Current M3 functions being defined between a private network manager and a public network provider include only the configuration, fault, and performance management functions [18].

With the M3 interface, a customer is provided with a view of a virtual ATM network, showing the customer's logical portion of the public ATM network. Management of the actual internal aspects of a public network (e.g., network routing tables, switching elements, and so forth) is not within the scope of the M3 specification.

The M3 requirements are defined in two classes. Mandatory class I functions identify those requirements by which a public network provider can provide monitoring information on the configuration, fault, and performance management of a specific customer's portion of a public ATM network (i.e., retrieving performance management data for a UNI link, reporting an alarm (or trap) message following the loss of a UNI link from a customer site to the public network).

Class 1 requirements are identified as follows (cf. [18]).

- Class I will be based on SNMP;
- Retrieval of general UNI stack information;
- Retrieval of both the general ATM-level performance information and physical-level performance and status information;
- Retrieval of both the ATM- and physical-level configuration and status information;
- Retrieval of ATM-level VP and VC link configuration and status information;
- Retrieval of both VPC and VCC configuration and status information;
- Retrieval of traffic characterization information; Receive event notifications from the public network provider.

Optional class II M3 functions identify those requirements by which a customer can request the addition, modification, or deletion of virtual connections and subscription information in a public ATM network. The M3 class II requirements are divided into three subgroups to allow for functional implementations that meet the modular needs of customers:

- ATM level subgroup;
- VPC/VCC level subgroup;
- Traffic subgroup.

The first two subgroups contain all the related configuration information, whereas the traffic subgroup contains traffic-related information. At the ATM level, customers are given the ability to modify the ATM-level configuration information. At the VPC/VCC, network elements of interest are the VP and VC links and VPCs and VCCs. The customers are given the ability to modify the configuration and status information of each one of these network elements.

The traffic subgroup provides the customer with the ability to modify ATM traffic descriptors for both VC and VP connections.

M3 MIBs will be based on the SNMP standard. All of the objects currently defined in both class I and class II MIBs supporting, respectively, class 1 and class 2 M3 requirements have been developed by IETF (i.e., ATOM MIB).

11.6.3 M4 Interface

Current M4 interface requirements are defined in the following areas of ATM network management:

- Network configuration management;
- Network fault management;
- Network connection configuration management;
- Network connection fault management;
- Network connection monitoring management;
- Internetwork link management;
- Network planning and performance management.

Network configuration management maintains data representing the physical and logical resources within the network. All network-level alarms are logged and are retrievable through the fault management functions via the M4 interface. These alarms may be generated due to the occurrence of various fault conditions such as changes in the operational state of a connection, connection availability (i.e., intest, failed, dependency). The network connection monitoring management function provides the management systems with the ability to retrieve all performance monitoring data gathered by the individual network elements.

In order to set up, reserve, and release permanent ATM connections and modify the connection parameters, the M4 interface is used to pass commands to the network to carry out resource assignment, connection activation, modification, reconfiguration, and release. With each connection request, various connection-related information provided to the network includes the VPI value of a VP link termination within a specific ATM interface or the VCI value of a VC link termination within a specific VPC, the identity of the supporting ATM interface termination point (in which case the agent selects the VPI value) or the identity of the VPC termination point (in which case the agent selects the VCI value within a VPC), and the connection parameters (PCR, CDVT, SCR, BT, QOS class, and the direction of the connection).

B-ICI-managed entity requirements are used to organize data associated with the B-ICIs terminating on the ATM network elements. It is required that one instance of this managed entity exist for each B-ICI. The attributes of this managed entity are (1) the managed entity identifier, which provides a unique name for the managed entity instance in the ATM NE; (2) the identifier of the underlying physical path termination, which provides a pointer to the identity of the underlying physical path termination for the interface; (3) the maximum number of simultaneously active VPCs supported; (4) the maximum number of simultaneously active VCCs supported; (5) the number of allocated VPI bits; (6) the number of allocated VCI bits; and (7) the carrier network, which identifies the adjacent carrier to which the BICI transmission path is connected.

The B-ISSI-managed entity is used to organize data associated with broadband interswitching system interfaces (BISSI) terminating on the ATM NE. As with the BICI managed entity, one instance of a BISSI managed entity is required to exist for each BISSI terminating on the NE. The attributes of this managed entity are defined the same as for the BICI managed entity, except for the carrier network, since BISSI is the demarcation point between two switches belonging to one service provider.

11.7 CONCLUSIONS

One of the major promises of ATM is the seamless integration from workstation to LANs to private and public WANs. Although the ATM technology starts to appear in all these network elements, one crucial step in the realization of this seamless integration is end-to-end network management capabilities.

ATM network management work has been undertaken by both the ATM Forum and the IETF. These two organizations are not necessarily developing the related specifications (i.e., MIBs) in a fully compliant manner. In addition, there are essentially three network management protocols available today: SNMP, SNMP version II, and CMIP.

SNMP may not be suitable for the management of large networks, since it requires one packet to be sent out to get back one packet of information. It is not well suited to retrieving large volumes of data. Because it uses connectionless service (UDP/IP), it is not assured that critical trap messages reach their destinations. SNMP does not provide sophisticated security mechanisms and, accordingly, is not well suited for network control. In addition, SNMP does not support manager-to-manager communication.

CMIP addresses several of these issues. However, it is a complex protocol that has not yet enjoyed wide deployment. There is also SNMP version II, which includes various improvements over SNMP, particularly in the area of security. SNMP also requires the use of IP agents. Although it is possible to use the SNMP framework without IP (i.e., as in ILMI), it is developed by IETF. How well their specifications can be applied natively on top of ATM is an open issue.

In summary, end-to-end seamless integration of networking based on ATM technology is not expected to ease the development of an end-to-end network management framework, and it appears that network management will remain a complex task for several years to come.

References and Bibliography

[1] Ananthanpillai, R., *Managing Messaging Networks: a Systems Approach*, 1995.
[2] Ball, L. L., *Cost Efficient Network Management*, New York: McGraw-Hill 1992.
[3] Leinwand, A., and K. Fang, *Network Management: a Practical Perspective*, Addison-Wesley, 1993.
[4] Rose, M., and K. McCloghrie, *Concise MIB Definition*, Addison-Wesley, 1993.
[5] Stallings, W., *SNMP, SNMPv2, and CMIP: the Practical Guide to Network Management Standards*, Addison-Wesley, 1993.
[6] Terplan, K., *Communications Networks Management*, Englewood Cliffs, NJ: Prentice Hall, 1992.
[7] RFC 1155, "Structure and Identification of Management Information for TCP/IP Based Internets," 1990 (M. Rose and K. McCloghrie).
[8] RFC 1157, "A Simple Network Management Protocol (SNMP)," Internet Engineering Task Force, 1990 (J. Case, et al.).
[9] RFC 1189, "The Common Management Information Services and Protocols for the Internet (CMOT and CMIP)," 1990 (U. Warrier, et al.).
[10] RFC 1441, "SNMP Version 2," Internet Engineering Task Force, 1992.
[11] ISO 7498-4, OSI Basic Reference Model Part 4: "Management Framework."
[12] ISO 10040, "System Management Overview."
[13] ISO 9595, "Common Management Information Service Definition."
[14] ATM Forum M4 Interface Requirements and Logical MIB: ATM Network Element View, Version 1.0, 1994.
[15] ATM Forum ATM Data Exchange Interface (DXI) Specification, Version 1.1, 1994.
[16] ATM Forum User-to-Network Specification, Version 3.1, 1994.
[17] ATM Forum CMIP Specification for the M4 Interface, 1994.
[18] ATM Forum Customer Network Management (CNM) for ATM Public Network Service (M3 Specification), 1994.
[19] ATM Forum Network Element Management Interface: Functional Requirements and Logical MIB, Version 1.0, 1994.
[20] ATM Forum LAN Emulation Client Management, 1995.

Appendix A: Superposition of Arrival Streams

To obtain its performance metrics when a new connection request arrives, a queue can be analyzed with two arrival streams: one corresponding to that of the new connection and another one to represent the superposition of existing arrival streams on that link. Assuming that the superposed traffic is characterized by a k-state MMBP and the new connection request is characterized with a two-state IBP, the number of equations is now equal to $2k(K + 1)$. For example, with $k = 2$ and $K = 100$, we have 404 equations to solve. It may be possible to further reduce this number simply by superposing all arrivals, including the new connection, into one stream.

For presentation purposes, let us consider N arrival processes, each characterized by a Poisson process with rate $\lambda(i)$, $i = 1, \ldots, N$. Then the probability distribution of the number of customers in the queue can be exactly obtained with a single Poisson arrival stream with rate

$$\lambda = \sum_{i=1}^{N} \lambda(i)$$

Unfortunately, this exact result is not applicable to queuing models that arise in ATM networks. In particular, the superposition of N independent renewal processes is not a Poisson process if one or more of the component processes are not Poisson. Various source models of B-ISDN applications, such as Markov modulated Poisson process (MMPP) and MMBP, are not even renewal processes and cannot be modeled with Poisson processes, since the squared coefficients of variations of the interarrival times of cells is greater than 1 and possesses correlation. Clearly, one can still assume that the superposed process is Poisson for modeling purposes and analyze the queue accordingly, particularly as the number of individual components increases. However, the discrepancy between the performance metrics of the original queue with non-Poisson arrival streams and the queue with a Poisson arrival stream increases to unacceptable levels as the queue utilization increases.

Various techniques, mostly approximate, have been proposed to obtain the superposition of arrival streams of B-ISDN applications. In most cases, the superposed traffic is approximately represented as a two-state MMPP or MMBP.

A.1 MARKOV MODULATED POISSON PROCESS

The MMPP has been extensively used to model various B-ISDN sources, such as voice and video, as well as characterizing the superposed traffic. It has the property of capturing both the time-varying arrival rates and correlation between the interarrival times. In addition to characterizing the desired properties of B-ISDN applications, these models are analytically tractable and produce fairly accurate results.

An MMPP is a doubly stochastic Poisson process. The arrivals occur in a Poisson manner with a rate that varies according to a k-state Markov chain, which is independent of the arrival process. Accordingly, an MMPP is characterized by the transition rate matrix of its underlying Markov chain and arrival rates. Let i be the state of the Markov chain, $i \in \{1, \ldots, k\}$; σ_{ij} the transition rate of going from state i to state j, $i \neq j$; and λ_i the arrival rate when the Markov chain is in state i, $\lambda_i > 0$. Define

$$\sigma_i = -\sum_{j=1; i \neq j}^{k} \sigma_{ik}$$

In matrix form, we have

$$Q = \begin{bmatrix} -\sigma_1 & \sigma_{12} & \cdots & \sigma_{1k} \\ \sigma_{21} & -\sigma_2 & \cdots & \sigma_{2k} \\ \cdot & \cdot & \cdot & \cdot \\ \sigma_{k1} & \sigma_{k2} & \cdots & -\sigma_k \end{bmatrix} \quad \Lambda = \begin{bmatrix} \lambda_1 & 0 & \cdots & 0 \\ 0 & \lambda_2 & \cdots & 0 \\ \cdot & \cdot & \cdot & \cdot \\ 0 & 0 & \cdots & \lambda_k \end{bmatrix}$$

Assuming that Q does not depend on time t, the steady-state probability vector π of Q is the solution of the following system of equations:

$$\pi Q = 0; \qquad \sum_{i=1}^{k} \pi_i = 1 \tag{A.1}$$

Let X_m denote the time between the $(m-1)$st and mth arrivals. The distribution of X_m depends on the states of the Markov process at both arrival instants, say states i and j. Between the two arrival instants, a geometrically distributed number of transitions occurs from state i to state j via several steps (i.e., one or more other states are visited) during which there is no arrival. This period is finally followed by a transition from state i to state j in which an arrival occurs when the process is in state j. Let:

- J_m = state of the Markov process at the time the mth arrival occurs.
- $F_{ij}(x) = \Pr\{J_m = j, X_m \le x | J_m - 1 = i\}$, $m \ge 2$, the probability that the time between the mth and the $(m-1)$st arrivals is less than or equal to x, and the process is in state j when the mth arrival occurs given that the $(m-1)$st arrival occurred when the process was in state i. When the time origin does not correspond to an arrival epoch, the probabilities $\Pr\{J_1 = j, X_1 \le x | J_0 = i\}$ need to be defined (i.e., $m = 1$).
- $F(x) = \{F_{ij}(x)\}$; $i, j = 1, \ldots, k$.

Then the sequence $\{(J_m, X_m), m \ge 0\}$ is a Markov renewal process with transition probability matrix $F(x)$, where

$$F(x) = \int_0^x \exp[(Q - \Lambda)u]du \, \Lambda \tag{A.2}$$

with

$$F(\propto) = (\Lambda - Q)^{-1} \Lambda \tag{A.3}$$

To obtain the conditional moments of the time between the $(m-1)$st and the mth arrivals, we first note that the Laplace transform of $F(\propto)$, $f^*(s)$, is given by

$$f^*(s) = E\{\exp(-sX)\} = (sI + \Lambda - Q)^{-1} \Lambda \tag{A.4}$$

The rth moment of X_m, μ_m^r, can then be obtained from the rth derivative of $f^*(s)$, which yields

$$\mu_m^r = r!\{(\Lambda - Q)^{-1} \Lambda\}^{m-1}(\Lambda - Q)^{-(r+1)}\Lambda, \; i \ge 1, \; k \ge 1 \tag{A.5}$$

Similarly, the joint Laplace transform matrix $f^*(s_1, s_2, \ldots, s_m)$ of $X_1, X_2, \ldots, X_m$, $m > 1$ is given as follows:

$$f^*(s_1, s_2, \ldots, s_m) = E\{\exp - \sum_{l=1}^{m} s_l X_l) = \prod_{j=1}^{m} \{(s_j I + \Lambda - Q)^{-1} \Lambda\} \tag{A.6}$$

Then $\mu_{1;m+1} = E\{X_1 X_{m+1}\}$ is obtained from $f^*(s_1, s_2, \ldots, s_m)$:

$$\mu_{1;m+1} = (\Lambda - Q)^{-2} \Lambda \{(\Lambda - Q)^{-1} \Lambda\}^{-(m+1)} (\Lambda - Q)^{-2} \Lambda \tag{A.7}$$

Then, the r-step correlation matrix $E(X_1 - E\{X_1\})(X_{r+1} - E\{X_{r+1}\})$, $r \ge 1$, is given by

$$(\Lambda - Q)^{-2}\Lambda\{(\Lambda - Q)^{-1}\Lambda\}^{m-1}\{I - (\Lambda - Q)^{-1}\Lambda\}(\Lambda - Q)^{-2}\Lambda \qquad \text{(A.8)}$$

In order to fully characterize the MMPP process, the initial state of the process P needs to be defined. Two possible definitions of practical importance are:

- MMPP starts at an arbitrary arrival epoch, that is, $P = \pi\Lambda/\pi\lambda$, where $\lambda = (\lambda_1, \ldots, \lambda_1)^T$, referred to as *environment stationary* MMPP.
- P is the stationary vector of $F(\infty)$, referred to as *interval stationary* MMPP.

Given the initial probability vector P of the MMPP, its m-step transition probability matrix—that is, its probability of being in state j at the time the mth arrival occurs, $j = 1, \ldots, k$—is given by

$$P\{(\Lambda - Q)^{-1}\,\Lambda\}^m \qquad \text{(A.9)}$$

Moments of the interarrival time distribution and the correlation matrix are similarly determined by appropriately unconditioning with P.

Let us now consider a two-state MMPP with parameters Q and Λ, where

$$Q = \begin{bmatrix} -\sigma_1 & \sigma_1 \\ \sigma_2 & -\sigma_2 \end{bmatrix}, \qquad \Lambda = \begin{bmatrix} \lambda_1 & 0 \\ 0 & \lambda_2 \end{bmatrix}$$

Furthermore, let:

T_i = time until the next arrival given that an arrival occurred when the process is in state i;
t_i = time until the next event given that an arrival occurred when the process is in state i.

Consider the process immediately after an arrival occurs at state i. Due to the memoryless property of the exponential distribution, the remaining time at state i is exponentially distributed with rate σ_i. The next event is either an arrival or a transition from state i to the other state, with the distribution of the time until the next event being the minimum of the two random variables. Since the two events are exponentially distributed respectively with rates σ_i and λ_i, the time until the next event is also exponentially distributed with rate $(\sigma_i + \lambda_i)$. Furthermore, the next event is an arrival with probability (hereafter referred to as w.p.) $\lambda_i/(\sigma_i + \lambda_i)$ or a state transition w.p. $\sigma_i/(\sigma_i + \lambda_i)$. Then

$$T_1 = \begin{cases} t_1 & \text{w.p. } \lambda_1/(\sigma_1 + \lambda_1) \\ t_1 + T_2 & \text{w.p. } \sigma_1/(\sigma_1 + \lambda_1) \end{cases} \qquad T_2 = \begin{cases} t_2 & \text{w.p. } \lambda_2/(\sigma_2 + \lambda_2) \\ t_2 + T_1 & \text{w.p. } \sigma_2/(\sigma_2 + \lambda_2) \end{cases} \qquad \text{(A.10)}$$

The Laplace transform of T_1, $A_1(s)$, and T_2, $A_2(s)$ is obtained as follows: From (A.10), we have

$$A_1(s) = \frac{\sigma_1 + \lambda_1}{\sigma_1 + \lambda_1 + s}\frac{\lambda_1}{\sigma_1 + \lambda_1} + \frac{\sigma_1 + \lambda_1}{\sigma_1 + \lambda_1 + s}\frac{\sigma_1}{\sigma_1 + \lambda_1}A_2(s)$$

$$A_2(s) = \frac{\sigma_2 + \lambda_2}{\sigma_2 + \lambda_2 + s}\frac{\lambda_2}{\sigma_2 + \lambda_2} + \frac{\sigma_2 + \lambda_2}{\sigma_2 + \lambda_2 + s}\frac{\sigma_2}{\sigma_2 + \lambda_2}A_1(s)$$

After some manipulation,

$$A_1(s) = \frac{\sigma_1\lambda_2 + \lambda_1(\sigma_2 + \lambda_2 + s)}{(\sigma_1 + \lambda_1 + s)(\sigma_2 + \lambda_2 + s) - \sigma_1\sigma_2} \tag{A.11}$$

$$A_2(s) = \frac{\sigma_2\lambda_1 + \lambda_2(\sigma_1\lambda_1 + s)}{(\sigma_1 + \lambda_1 + s)(\sigma_2 + \lambda_2 + s) - \sigma_1\sigma_2} \tag{A.12}$$

Given $A_1(s)$ and $A_2(s)$, and $(\pi_1, \pi_2) = \{\sigma_2/(\sigma_1 + \sigma_2), \sigma_1/(\sigma_1 + \sigma_2)\}$, the Laplace transform of the probability distribution function of the unconditional interarrival time distribution $A(s)$ is given by

$$A(s) = \pi_1 A_1(s) + \pi_2 A_2(s) \tag{A.13}$$

$$= \frac{1 + \dfrac{(\sigma_2\lambda_1^2 + \sigma_1\lambda_2^2)s}{(\sigma_2\lambda_1 + \sigma_1\lambda_2)(\sigma_2\lambda_1 + \sigma_1\lambda_2 + \lambda_1\lambda_2)}}{1 + \dfrac{(\sigma_1 + \sigma_2 + \lambda_1 + \lambda_2)s + s^2}{\sigma_2\lambda_1 + \sigma_1\lambda_2 + \lambda_1\lambda_2}}$$

The mean $E(T)$ and the squared coefficient of variation $c^2(T)$ of the interarrival time of a two-state MMPP are obtained from the first two derivatives of $A(s)$ (i.e., the first two moments) and are

$$E(T) = \frac{\sigma_1 + \sigma_2}{\sigma_2\lambda_1 + \sigma_1\lambda_2} \tag{A.14}$$

$$c^2(T) = 1 + \frac{2\sigma_1\sigma_2(\lambda_1 - \lambda_2)^2}{(\sigma_2\lambda_1 + \sigma_1\lambda_2 + \lambda_1\lambda_2)(\sigma_1 + \sigma_2)^2} \tag{A.15}$$

Furthermore, the one-step correlation of the interval times of cells $C(1)$ is

$$C(1) = \frac{E\{(T_r - E\{T_r\})(T_{r+1} - E\{T_{r+1}\})\}}{\mathrm{Var}(T)} \tag{A.16}$$

$$= \frac{\lambda_1\lambda_2(\lambda_1 - \lambda_2)\sigma_1\sigma_2}{c^2(T)\{\sigma_1 + \sigma_2\}^2(\lambda_1\lambda_2 + \lambda_1\sigma_2 + \lambda_2\sigma_1)^2}$$

We note that an IPP is a special case of MMPP with $k = 2$ and $\lambda_2 = 0$. Furthermore, if $\lambda_1 = \lambda_2$, then MMPP is reduced to a Poisson process with rate λ_1.

A.1.1 Superposition of MMPPs

The superposition of MMPPs is also an MMPP. Consider N MMPPs, each with parameters Q_i and Λ_i. Then the transition rate matrix Q and the arrival rate matrix of Λ of the superposed process are

$$Q = Q_1 \oplus Q_2 \oplus \ldots \oplus Q_N \quad \text{and} \quad \Lambda = \Lambda_1 \oplus \Lambda_2 \oplus \ldots \oplus \Lambda_N$$

where $\oplus$ denotes the Kronecker sum defined below. We note that both Q and Λ are $k \times k$ matrices, where

$$k = \prod_{i=1}^{N} k_i$$

if Q_i and Λ_i are $k_i \times k_i$ matrices, $i = 1, \ldots, N$. The Kronecker sum of two matrices Q_1 and Q_2 is defined as follows:

$$Q_1 \oplus Q_2 = (Q_1 \otimes I_{Q_2}) + (I_{Q_1} \otimes Q_2)$$

where I_{Q_i}, $i = 1, 2$, is an identity matrix of the same order as matrix Q_i and $\otimes$ denotes the Kronecker product, which is defined for two matrices $C = \{c_{ij}\}$ and $D = \{d_{ij}\}$ as

$$C \oplus D = \begin{bmatrix} c_{11}D & c_{12}D & \ldots & c_{1m}D \\ \ldots & \ldots & \ldots & \ldots \\ c_{n1}D & c_{n2}D & \ldots & c_{nm}D \end{bmatrix}$$

Equivalently, the state of the superposition of N MMPPs is $(i_1, i_2, \ldots, i_N)$, where i_j denotes the state of the jth component MMPP. The total arrival rate at state $(i_1, i_2, \ldots, i_N)$ is the sum of the arrival rates of component processes, which depends on state i_j; that is,

$$\lambda(i_1, i_2, \ldots, i_N) = \sum_{j=1}^{N} \lambda_{i_j}$$

The transition rate from state $(i_1, i_2, \ldots, i_j = k, \ldots, i_N)$ to state $(i_1, i_2, \ldots, i_j = m, \ldots, i_N)$ is given by the rate of going from state $i_j = k$ to state $i_j = m$ in the jth component MMPP.

As the number of component processes increases, the number of states of the superposed process increases exponentially. To reduce the complexity of solving queues with a large number of arrival streams, the superposed process may be approximated by a simpler process that captures important characteristics of the original process as closely as possible. The simplest model that has the potential to approximate an MMPP accurately with a large number of phases is the two-phase MMPP defined by four parameters (λ_1, λ_2) and (σ_1, σ_2). Then the problem is reduced to choosing the parameters of the two-state MMPP using the four metrics of the superposed process. For example, let us assume that the first four moments of the superposed process are known. Then, using the equations of the first four moments of the two-state MMPP and the given values, we have four equations and four unknowns, which can be solved to obtain the required unknowns. However, in general, there is no guarantee that there is a two-state MMPP that matches the first four moments of the superposed process exactly. Even if there is an exact match, we have a nonlinear system of equations to work with. Hence, most often the matching has to be done approximately.

Let us now consider the counting process of cells in which the interarrival time is distributed according to a two-state MMPP, and let $M(t)$ and $J(t)$ respectively denote the number of arriving cells during $(0, t]$, $t > 0$, and the state of the MMPP at time t, $J(t) = 1, 2$. Then the process $\{M(t), J(t)\}$ has the Markovian property with its Laplace transform function $\phi(z, s)$ given as follows:

$$\phi(z, s) = [sI - Q - (z - 1)\Lambda]^{-1}e$$

The mean m, its variance v, and the third moment μ_3 of the arrival process are given as follows:

$$m = \frac{\lambda_1\sigma_2 + \lambda_2\sigma_1}{\sigma_2 + \sigma_1}$$

$$v = \frac{(\lambda_1 - \lambda_2)^2\sigma_1\sigma_2}{(\sigma_2 + \sigma_1)^2}$$

$$\mu_3 = \frac{\lambda_1^3\sigma_2 + \lambda_2^3\sigma_1}{\sigma_1 + \sigma_2}$$

Furthermore, in order to capture the correlation in the arrival stream, let $r(t)$ denote the covariance function, which characterizes the dependence between the arrival instants at two different instants in time. The following equality holds for a large class of doubly stochastic Poisson processes.

$$\frac{\operatorname{var}\{m(T)\}}{E\{m(T)\}} = 1 + \frac{2\int_0^\infty r(t)dt}{m} \quad \text{as } T \text{ goes to infinity}$$

Furthermore, let τ denote a time constant defined by

$$\tau = (1/\nu) \int_0^\infty r(t)dt$$

Then the exponential covariance approximation $r_{app}(t) = \nu \exp\{-t/\tau\}$, which matches $r(0)$, and the covariance integral is a good approximation to $r(t)$ over a wide range of conditions.

For a two-state MMPP, τ is given by

$$\tau = \frac{1}{\sigma_1 + \sigma_2}$$

Then, given m, ν, μ_3, and τ, the parameters of the two-state MMPP are obtained by

$$\sigma_1 = \frac{1}{\tau(1 + \eta)} \qquad \sigma_2 = \frac{\eta}{\tau(1 + \eta)}$$
$$\lambda_1 = m + \sqrt{v/\eta} \quad \lambda_2 = m - \sqrt{v/\eta}$$

where

$$\eta = 1 + \frac{\eta}{2}(\delta - \sqrt{4 + \delta^2}) \quad \text{and} \quad \delta = \frac{m_3 - 3m\nu - m^3}{\nu^{1.5}}$$

With this framework, it is not necessary to construct the superposed process and obtain its parameters m, ν, μ_3, and τ in order to obtain the parameters of the two-state MMPP. Instead, observing that $M(t)$ is the counting process and that the arrival streams are independent with parameters m_i, ν_i, μ_{i3} and τ_i, $i = 1, \ldots, N$, with

$$V = \sum_{i=1}^{n} \nu_i$$

we have

$$m = \sum_{i=1}^{N} m_i \quad \nu = \sum_{i=1}^{N} \nu_i \quad \mu_3 \sum_{i=1}^{N} \mu_{i3} \quad \text{and} \quad \tau = \sum_{i=1}^{N} \frac{\nu_i}{V} \tau_i$$

which simplifies the procedure.

A.2 MARKOV MODULATED BERNOULLI PROCESS

Time in MMBPs is discretized into fixed-length slots. The probability that a slot contains a cell is a Bernoulli process with a parameter that varies according to an r-state Markov process, which is independent of the arrival process. At the end of each slot, the Markov process moves from state i to state j w.p. p_{ij} or stays at state i w.p. p_{ii} such that

$$\sum_{j=1}^{r} p_{ij} = 1$$

for all $i = 1, \ldots, r$. At state i, a slot contains a cell w.p. α_i and no cell w.p. $1 - \alpha_j$. The arrival probabilities of cells and the underlying Markov process are assumed to be independent of each other. An MMBP is characterized by the transition probability matrix P and the diagonal matrix Λ of arrival probabilities:

$$P = \begin{bmatrix} p_{11} & \cdots & p_{1r} \\ \cdots & \cdots & \cdots \\ p_{r1} & \cdots & p_{rr} \end{bmatrix} \qquad \lambda = \begin{bmatrix} \alpha_1 & 0 & 0 \\ \cdots & \cdots & \cdots \\ 0 & 0 & \alpha_r \end{bmatrix}$$

To determine the generating function $T(z)$ of the interarrival time of cells in an MMBP, let π_j denote the steady-state probability that MMBP is in state j and $\pi = (\pi_1, \ldots, \pi_r)$ and b_j denote the conditional probability that the MMBP is in state j, given that a cell arrival occurs and $b = (b_1, \ldots, b_r)$. Then π is the solution of the system of equations $\pi P = \pi$ and $b = \pi\Lambda / \pi\underline{\lambda}$ with $\underline{\lambda} = (\alpha_1, \ldots, \alpha_r)^T$. $T(z)$ can be obtained in a similar way to that of MMPP, which is derived next for $r = 2$. In general, we have for $r \geq 2$,

$$T(z) = bz\{I - zP(I - \Lambda)\}^{-1}P\underline{\lambda}$$

The nth derivative $T^{(n)}(z)$ of $T(z)$ is given by

$$T^{(n)}(z) = n!b[\{I - zP(I - \Lambda)\}^{-1}P(I - \Lambda)^{n-1}\{I - zP(I - \Lambda)\}^{-1}$$
$$[I + zP(I - \Lambda)\{I - zP(I - \Lambda)\}^{-1}]P\underline{\lambda}$$

Then the nth moment of the cell interarrival time $E\{T^n\}$ is equal to $T^{(n)}(z)$ evaluated at $z = 1$; that is,

$$E\{T^n\} = T^{(n)}(1) = n!b\{I - P(I - \Lambda)\}^{-(n+1)}\{P(I - \Lambda)\}^{n-1}P\underline{\lambda}$$

Furthermore, the probability that the interarrival time is equal to m slots $P(T = m)$ is

$$P(T = m) = b\{P(I - \Lambda)\}^{n-1}P\underline{\lambda}$$

Finally, let t_n denote the time between the $(n-1)$st and the nth arrival. Then we have

$$E\{t_n t_{n+k}\} = b\{I - P(I - \Lambda)\}^{-2} P \Lambda T^{k-1} \{I - P(I - \Lambda)\}^{-2} P \underline{\lambda}$$

from which the correlation coefficient of the interarrival time of an MMBP for lag k ψ_j can be calculated as follows:

$$\psi_k = \frac{E\{t_n t_{n+k}\} - E^2\{t_n\}}{\mathrm{Var}\{t_n\}}$$

Let us now consider a two-state MMBP with the probability matrix P and the arrival probabilities at each state Λ:

$$P = \begin{bmatrix} p & 1-p \\ 1-q & q \end{bmatrix} \qquad \Lambda = \begin{bmatrix} \alpha & 0 \\ 0 & \beta \end{bmatrix}$$

Given that the Markov process is in state 1 (2), it will remain in the same state in the next slot w.p. p (q) or change state w.p. $1 - p$ $(1 - q)$. The steady-state probability that the MMBP is in state i, π_i, is the solution of the system of equations:

$$[\pi_1, \pi_2]P = [\pi_1, \pi_2] \text{ and } \pi_1 + \pi_2 = 1 \qquad \text{(A.17a)}$$

That is,

$$\pi_1 = (1-q)/(2-p-q) \text{ and } \pi_2 = (1-p)/(2-p-q) \qquad \text{(A.17b)}$$

The probability that a slot contains a cell (i.e. source utilization) is given as follows:

$$\rho = \pi_1 \alpha + \pi_1 \beta = \frac{(1-q)\alpha + (1-p)\beta}{2-p-q}$$

Next we obtain the z-transform of the interarrival times of cells. Let T_i be the time interval to the next arrival, given that the Markov process is in state i, and T be the interarrival time of a cell. Consider the time an arrival occurs when the Markov process is in state 1. In the next slot:

- MMBP may remain in state 1 and an arrival may occur, which happens w.p. $p\alpha$.
- MMBP may move to state 2 and an arrival may occur, which happens w.p. $(1 - p)\beta$.

- MMBP may remain in state 1 and no arrival occurs, which happens w.p. $p(1 - \alpha)$.
- MMBP may move to state 2 and no arrival occurs, which happens w.p. $(1 - p)(1 - \beta)$.

Hence, we have

$$T_1 = \begin{cases} 1 & \text{w.p.} \quad p\alpha + (1-p)\beta \\ 1 + T_1 & \text{w.p.} \quad p(1-\alpha) \\ 1 + T_2 & \text{w.p.} \quad (1-p)(1-\beta) \end{cases} \quad \text{and} \quad T_2 = \begin{cases} 1 & \text{w.p.} \quad q\beta + (1-q)\alpha \\ 1 + T_2 & \text{w.p.} \quad q(1-\beta) \\ 1 + T_1 & \text{w.p.} \quad (1-q)(1-\alpha) \end{cases} \tag{A.18}$$

After some manipulation, the z-transforms of T_1 and T_2 are

$$T_1(z) = \frac{(1-\beta)\alpha(1-p-q)z^2 + [p\alpha + (1-p)\beta]z}{(1-\beta)(1-\alpha)(p+q-1)z^2 - [q(1-\beta) + p(1-\alpha)]z + 1} \tag{A.19}$$

$$T_2(z) = \frac{(1-\alpha)\beta(1-p-q)z^2 + [q\beta + (1-q)\alpha]z}{(1-\beta)(1-\alpha)(p+q-1)z^2 - [q(1-\beta) + p(1-\alpha)]z + 1} \tag{A.20}$$

Now let a_i denote the probability that MMBP is in state i and an arrival occurs. Using the independence of the arrival and Markov process, we have

$$a_1 = \pi_1\alpha; \qquad a_2 = \pi_2\beta \tag{A.21}$$

Then, given that an arrival occurs, the probability that MMBP is in state i, b_i, is equal to

$$b_1 = a_1/(a_1 + a_2); \; b_2 = a_2/(a_1 + a_2) \tag{A.22}$$

By appropriately unconditioning the conditional interarrival times T_i, $i = 1, 2$, the z-transform of the cell interarrival time is given as follows:

$$T(z) = \frac{c_2z^2 + c_1z}{d_2z^2 + d_1z + d_0} \tag{A.23}$$

where

$$c_2 = (1-p-q)[(1-q)(1-\beta)\alpha^2 + (1-p)(1-\alpha)\beta^2]$$
$$c_1 = (1-q)\alpha[p\alpha + (1-p)\beta] + (1-p)\beta[q\beta + \alpha(1-q)]$$
$$d_2 = (1-\alpha)(1-\beta)(p+q-1)[(1-q)\alpha + (1-p)\beta]$$
$$d_1 = -[(1-q)\alpha + (1-p)\beta][q(1-\beta) + p(1-\alpha)]$$
$$d_0 = (1-q)\alpha + (1-p)\beta$$

In general, the kth moment of the cell interarrival time with $B = (b_1, b_2)$ is equal to

$$E(T^k) = k!B(I - F)^{-k}F^{k-1}e \tag{A.24}$$

where $e = (1, 1)^T$ and

$$F = \begin{bmatrix} p(1-\alpha) & (1-p)(1-\beta) \\ (1-q)(1-\alpha) & q(1-\beta) \end{bmatrix}$$

The mean $E(T)$ and squared coefficient of variation c^2 of the interarrival times of cells is then obtained from the respective derivatives of $T(z)$:

$$E(T) = \frac{2-p-q}{(1-q)\alpha + (1-p)\beta} \tag{A.25}$$

$$c^2 = \frac{2[(1-q)\alpha + (1-p)\beta]}{(1-q)\alpha + (1-p)\beta + \alpha\beta(p+q-1)} - \frac{(1-q)\alpha + (1-p)\beta}{2-p-q} + \frac{2[(1-p)\alpha + (1-q)\beta][(1-q)\alpha + (1-p)\beta](p+q-1)}{(2-p-q)^2[(1-q)\alpha + (1-p)\beta + \alpha\beta(p+q-1)]} - 1 \tag{A.26}$$

The autocorrelation of the interarrival time of an MMBP can be obtained in a similar way to that of MMPP. Let t_{ij} denote the time interval between a particular slot when the arrival process is in state i and ending at a slot when the next arrival occurs and the arrival process is in state j. Then,

$$t_{11} = \begin{cases} 1 & \text{w.p.} \quad \alpha p \\ 1 + t_{11} & \text{w.p.} \quad (1-\alpha)p \\ 1 + t_{21} & \text{w.p.} \quad (1-\beta)(1-p) \end{cases} \qquad t_{21} = \begin{cases} 1 & \text{w.p.} \quad \alpha(1-q) \\ 1 + t_{11} & \text{w.p.} \quad (1-\alpha)(1-q) \\ 1 + t_{21} & \text{w.p.} \quad (1-\beta)q \end{cases} \tag{A.27}$$

$$t_{12} = \begin{cases} 1 & \text{w.p.} \quad \beta(1-p) \\ 1 + t_{12} & \text{w.p.} \quad (1-\alpha)p \\ 1 + t_{22} & \text{w.p.} \quad (1-\beta)(1-q) \end{cases} \qquad t_{22} = \begin{cases} 1 & \text{w.p.} \quad \beta q \\ 1 + t_{12} & \text{w.p.} \quad (1-\alpha)(1-q) \\ 1 + t_{22} & \text{w.p.} \quad (1-\beta)(1-q) \end{cases} \tag{A.28}$$

Let S_n denote the state of the arrival process when the nth arrival occurs, T_n the interarrival time between the $(n-1)$st and the nth arrivals, and T_{nj} the interarrival time between the $(n-1)$st and the nth arrivals, while the nth arrival occurs in state j. Defining

$$A_{ij} = E\{z^{T_{nj}} | S_{n-1} = i\}$$

and using the definitions of t_{ij} and T_{nj}, we have

$$A_{ij} = E\{z^{t_{ij}}\},\ i, j = 1, 2$$

Then

$$A_{11}(z) = \alpha pz + (1-\alpha)pzA_{11}(z) + (1-\beta)(1-p)zA_{21}(z) \quad (A.29)$$
$$A_{21}(z) = \alpha(1-q)z + (1-\alpha)(1-q)zA_{11}(z) + (1-\beta)qzA_{21}(z) \quad (A.30)$$
$$A_{12}(z) = \beta(1-p)z + (1-\alpha)pzA_{12}(z) + (1-\beta)(1-p)zA_{22}(z) \quad (A.31)$$
$$A_{22}(z) = \beta qz + (1-\alpha)(1-q)zA_{12}(z) + (1-\beta)qzA_{22}(z) \quad (A.32)$$

Let us further define

$$B_j(z) = E\{z^{T_n}|S_{n-1} = j\} \text{ and } C_i(z_1, z_2) = E\{z_1^{T_{n-1}} z_2^{T_n}|S_{n-2} = \mathrm{I}\}$$
$$A(z) = \begin{bmatrix} A_{11}(z) & A_{12}(z) \\ A_{21}(z) & A_{22}(z) \end{bmatrix},\ B(z) = \begin{bmatrix} B_1(z) \\ B_2(z) \end{bmatrix}, \text{ and } C(z_1, z_2) = \begin{bmatrix} C_1(z_1, z_2) \\ C_2(z_1, z_2) \end{bmatrix}$$

Then

$$C_i(z_1, z_2) = \sum_{j=1}^{2} A_{ij}(z_1)B_j(z_2)$$

Seeing that $C(z_1, z_2) = A(z_1)\, B(z_2)$ and rewriting (A.29) to (A.32) in matrix form, we have

$$\begin{bmatrix} 1-(1-\alpha)pz & -(1-\beta)(1-p)z \\ -(1-\alpha)(1-q)z & 1-(1-\beta)qz \end{bmatrix} \begin{bmatrix} A_{11}(z) & A_{12}(z) \\ A_{21}(z) & A_{22}(z) \end{bmatrix} = \begin{bmatrix} \alpha pz & \beta(1-p)z \\ \alpha(1-q)z & \beta qz \end{bmatrix}$$

or equivalently,

$$[I - zP(I - \Lambda)]A(z) = P\Lambda z \quad (A.33)$$

The term $P(I - \Lambda)$ represents a transition without an arrival, whereas $P\Lambda$ represents a transition with an arrival.

From (A.33), $A(z)$ can be expressed as

$$A(z) = [I - zP(I - \Lambda)]^{-1}P\Lambda z \quad (A.34)$$

Seeing that $B_i(z) = A_{i1}(z) + A_{i2}(z)$ and defining $e = (1, 1)^T$,

$$B(z) = [I - zP(I - \Lambda)]^{-1}P\Lambda ez \quad (A.35)$$

From the definition of $C_i(z_1, z_2)$, we have

$$E\{z_1^{T_{n-1}} z_2^{T_n}\} = [P(S_{n-2} = 1)P(S_{n-2} = 2)] \begin{bmatrix} C_1(z_1, z_2) \\ C_2(z_1, z_2) \end{bmatrix} \tag{A.36}$$

$$= \left[\frac{\alpha(1-q)}{\alpha(1-q) + \beta(1-p)} \; \frac{\beta(1-p)}{\alpha(1-q) + \beta(1-p)} \right] A(z_1)B(z_2) \tag{A.37}$$

Then

$$E\{T_{n-1}T_n\} =$$

$$\left[\frac{\alpha(1-q)}{\alpha(1-q) + \beta(1-p)} \; \frac{\beta(1-p)}{\alpha(1-q) + \beta(1-p)} \right] \frac{\partial A(z_1)}{\partial z_1} \frac{\partial B(z_2)}{\partial z_2} \bigg|_{z_1=1;\ z_2=1} \tag{A.38}$$

$$= \left[\frac{\alpha(1-q)}{\alpha(1-q) + \beta(1-p)} \; \frac{\beta(1-p)}{\alpha(1-q) + \beta(1-p)} \right]$$

$$[I - P(I - \Lambda)]^{-2} P\Lambda [I - P(I - \Lambda)]^{-2} P\Lambda e$$

The autocorrelation function of the interarrival time of cells with lag 1 is then given by

$$\psi_1 = \frac{\text{Covariance}\{T_{n-1}T_n\}}{\text{Variance}\{T_n\}} = \frac{E\{T_{n-1}T_n\} - E\{T_{n-1}\}E\{T_n\}}{\text{Variance}\{T_n\}} \tag{A.39}$$

$$= \frac{\alpha\beta(\alpha - \beta)^2(1-p)(1-q)(p+q-1)^2}{c^2(2-p-q)^2[\alpha(1-q) + \beta(1-p) + \alpha\beta(p+q-1)]^2}$$

We note that when $\alpha = \beta$, MMBP has, in essence, only one state and becomes a Bernoulli process. If either α or β is equal to zero, then MMBP degenerates to an IBP. In both of these special cases, the autocorrelation of the interarrival times of cells is equal to zero, which can easily be validated from the autocorrelation function with lag 1.

A.2.1 Superposition of MMBPs

The superposition of MMBPs is a *switched batch Bernoulli process* (SBBP). Time in an SBBP is divided into slots of equal length. The arrivals during a slot occur as a batch process with the batch size distribution varying according to a k-state Markov chain.

For simplicity of presentation, the superposed process is discussed only for the superposition of N MMBPs, each with two states and parameters P_i and Λ_i. Let i_j denote the state of the jth component MMBP process. Then the superposed process has 2^N states in which each state $(i_1, i_2, \ldots, i_N)$ denotes the states of the N MMBP processes.

In particular, at the end of a slot, each component process in state 1 moves to state 2 w.p. $(1 - p_i)$ or stays at state 1 w.p. p_i. The respective probabilities for component processes that are in state 2 are equal to $(1 - q_i)$ and q_i. Furthermore, define $p(i_j \rightarrow i_j^*)$ *as the probability that the* jth component process is in state i_j^*, given that it was in state i_j during the previous slot, and $p(i \rightarrow i')$ as the probability that the SBBP is in state i', given that it was in state i during the previous slot. Then

$$p(i_j \rightarrow i_j') = \prod_{j=1}^{N} p(i_j \rightarrow i_j^*) \tag{A.40}$$

The probabilities $p(i \rightarrow i')$ define the elements of the transition probability matrix of the superposed process. In order to characterize the SBBP completely, let B_i be the random variable denoting the batch size distribution when the superposed process is in state $i = (i_1, i_2, \ldots, i_N)$. Let $\gamma(i_j)$ denote the probability that a cell arrival occurs at the component process j; $\gamma(i_j) = \alpha_j$ if the process is in state 1 or $\gamma(i_j) = \beta_j$ if it is in state 2. At the two extremes, none of the component processes may generate a cell, that is, $B_i = 0$, which occurs w.p.

$$\prod_{j=1}^{N} \{1 - \gamma(i_j)\}$$

or each component process may generate a cell, that is, $B_i = N$, w.p.

$$\prod_{j=1}^{N} \gamma(i_j)$$

When m cells generated, m out of N sources generate cells, while no arrivals occur from the remaining $N - m$ sources. There are $N!/\{m!\ (N - m)!\}$ different combinations of m out of n sources generating cells for this to happen. Let S_l denote the set of m-tuples of indexes and $S = \{1, \ldots, N\}$. Furthermore, let $q \in S_l$ be the indexes of l-component processes that generate cells, whereas those that do not generate a cell are given in $N - q$. The probability distribution of B_i can now be given as follows:

$$\Pr\{B_i = l\} = \sum_{r \in S_l} \prod_{i_j \in q} \gamma(i_j) \prod_{i_j \in (S-q)} \{1 - \gamma(i_j)\} \qquad l = 0, \ldots, N \tag{A.41}$$

Then the superposed process is a 2^N-state SBBP with transition probability matrix $P = \{p(i \rightarrow i')\}$ and batch size distribution $\Pr\{B_i = l\}$ for each state i and $l = 0, \ldots, n$.

Two-State MMBP to Approximate an Unknown Function with Known Moments

Let us now consider the problem of characterizing an unknown distribution with a set of known parameters by a two-state MMBP. As discussed above, a two-state MMBP is characterized by the parameters p, q, α, and β. Hence, four parameters of the unknown distribution need to be known to estimate these four parameters. If we have less than four, then it is necessary to set the values of some of the parameters arbitrarily. In this case, there are infinitely many choices. For example, if we have two known parameters, then α and β can both be set to 1 and the two other parameters of MMBP, p and q, are calculated from the respective equations of the two known parameters.

The following method uses the mean number of cells per slot, ρ, the squared coefficient of variation of the interarrival time, c^2, and the autocorrelation coefficients $\psi_1(T)$ and $\psi_1(C)$, respectively denoting the lag 1 autocorrelation for the interarrival time and the number of arrivals. The parameters ρ, c^2, and $\psi_1(T)$ for MMBP are defined in (A.25), (A.26), and (A.39). $\psi_i(C)$ for lag i is given by

$$\psi_1(C) = \frac{(1-p)(1-q)(\alpha-\beta)^2(p+q-1)^i}{\{(1-q)\alpha+(1-p)\beta\}[(1-q)(1-\alpha)+(1-p)(1-\beta)]} \tag{A.42}$$

Fitting an MMBP for an Unknown Distribution

Obtain α as the root of the quadratic equation $A\alpha^2 + B\alpha + C = 0$, where

$$\begin{aligned}
A &= 4c^2(1-\rho)\psi_1(C)\psi_1(T) + (c^2+\rho-1)\{(c^2+\rho-1) - 2(1-\rho)\psi_1(C)\} \\
B &= 4c^2(1-\rho^2)\psi_1(C)\psi_1(T) + (c^2+\rho-1)^2\rho + [(c^2-1)^2-\rho^2](1-\rho)\psi_1(C) \\
C &= 4c^2\rho(1-\rho)\psi_1(C)\psi_1(T)
\end{aligned} \tag{A.43}$$

Given α, ρ, c^2, and $\psi_1(T)$,

$$\beta = \frac{\rho(c^2+\rho-1)[(1-\rho)\psi_1(C)-\alpha+\rho] + 2(\alpha-\rho)\rho(1-\rho)\psi_1(C)}{(1-\rho)\psi_1(C)[(c^2+\rho+1)\alpha-2\rho] - (\alpha-\rho)(c^2+\rho-1)} \tag{A.44}$$

$$p = 1 - \frac{(\alpha-\rho)(\rho-\beta) - \rho(1-\rho)\psi_1(C)}{(\alpha-\beta)(\rho-\beta)} \tag{A.45}$$

$$q = 1 - \frac{(\alpha-\rho)(\rho-\beta) - \rho(1-\rho)\psi_1(C)}{(\alpha-\beta)(\alpha-\rho)} \tag{A.46}$$

We note that α is the root of a quadratic equation, which in general has two roots. However, one root has the same value as β. It is possible that the above set of equations produces an unfeasible value; for example, $p < 0$. This is mainly because the original distribution is not an MMBP and the estimation by the two-state MMBP may not produce a consistent set of equations with the parameters of the unknown distribution. Alternatively, the parameters of the MMBP can be approximated as follows: Let $\psi_1^{est}(T)$ denote the autocorrelation coefficient of the cell interarrival time for lag i, estimated using the fitted MMBP and

$$\epsilon(k) = \sum_{i=1}^{k} |y_1(T) - y_1^{est}(T)|$$

where $|a - b|$ denotes the absolute value of the difference.

1. Set $\alpha = 0$, $\delta = 1E - 6$, $\epsilon(k) = 1E + 30$.
2. Set $\alpha = \alpha + \delta$. If $\alpha > 1$, then stop. The parameters of MMBP are estimated.
3. Calculate p, q, and β using (A.44), (A.45), and (A.46) with the current value of α.
4. Compute $\psi_1^{est}(T)$ from (A.42) and $\epsilon(1)$. If the new value of $\epsilon(1)$ is smaller than the previous value, then reset $\epsilon(1)$ to the new value. Go to step 2.

Tests carried out in [Ch 4, 111] illustrate that the algorithm produces fairly accurate results.

A.3 NUMERICAL ANALYSIS

Let us now consider a single queuing system with a single server and a finite buffer B. The service time is constant due to fixed cell size. The buffer receives cells according to a process modeled either as an MMPP or an SBBP, giving rise to an MMPP/D/1/B or SBBP/D/1/B queue, respectively. Note that MMBP is a special case of SBBP and that the following discussion also applies to MMBP/D/1/B queues.

In an SBBP/D/1/B queue, let (i, n) denote the state of the queue, where i is the phase of the SBBP and n is the number of cells in the queue immediately after a departure. We have $i = 1, \ldots, k$ and $n = 0, \ldots, B - 1$. The queue is analyzed under the following assumptions:

- Time is slotted with service time being equal to one slot. A cell can start receiving service only at the beginning of the slot.
- Arrivals occur within a slot, whereas departures occur at slot boundaries.
- The maximum number of cells that can arrive during a slot is bounded by m.

Furthermore, let $p(i \rightarrow j)$ denote the transition probability from state i of SBBP to state j and $a(i, l)$ denote the probability that l cells arrive during a slot when the SBBP is in state i at the beginning of the slot. Then the transition probability matrix of the queue is

$$Q_d = \begin{bmatrix} A(0) & A(1) & .. & A(m) & 0 & .. & 0 & 0 & 0 \\ A(0) & A(1) & .. & A(m) & 0 & .. & 0 & 0 & 0 \\ 0 & A(0) & .. & A(m-1) & A(m) & .. & 0 & 0 & 0 \\ .. & .. & .. & .. & .. & .. & .. & .. & .. \\ 0 & 0 & .. & 0 & 0 & .. & A(0) & A(1) & D(2) \\ 0 & 0 & .. & 0 & 0 & .. & 0 & A(0) & D(1) \end{bmatrix} \quad \text{(A.47)}$$

The matrices A_r and D_r are $k \times k$ matrices whose elements are defined as follows:

$$A(r)[i, j] = a(i, j)p(i \to j)$$

$$D(r)[i, j] = \left(\sum_{s=r}^{m} a(i, s)\right)p(i \to j)$$

Let $\pi = (\pi_0, \pi_1 \ldots, \pi_{B-1})$ denote the steady-state queue length distribution of an SBBP/D/1/B queue. $\pi_j = (\pi_{j1}, \ldots, \pi_{jk})$ is a $1 \times k$ vector whose elements π_{jp} denote the steady-state probability of having j cells in the queue, and the SBBP is in phase p upon departure of a cell. Then π is the solution of the linear system of equations $\pi Q_d = \pi$ and $\pi e = 1$, where $e = (1, 1, \ldots, 1)^T$.

The transition rate matrix of an MMPP/D/1/B queue has a similar form to that of an SBBP/D/1/B queue. The main difference between the two is that it is not possible for more than one cell to arrive in MMPP simultaneously, whereas one or more cells can arrive during a slot in the latter. The transition rate matrix of an MMPP/D/1/B queue is

$$Q_c = \begin{bmatrix} A(0) & A(1) & 0 & 0 & 0 & .. & 0 & 0 & 0 \\ A(0) & A(1) & 0 & 0 & 0 & .. & 0 & 0 & 0 \\ 0 & A(0) & A(1) & A(2) & 0 & .. & 0 & 0 & 0 \\ .. & .. & .. & .. & .. & .. & .. & .. & .. \\ 0 & 0 & .. & 0 & 0 & .. & A(0) & A(1) & A(2) \\ 0 & 0 & .. & 0 & 0 & .. & 0 & A(0) & A(1) \end{bmatrix} \quad \text{(A.48)}$$

Let $\pi = (\pi_0, \pi_1, \ldots, \pi_{B-1})$ denote the steady-state queue length distribution of an MMPP/D/1/B queue. As before, $\pi_j = (\pi_{j1}, \ldots, \pi_{jk})$ is a $1 \times k$ vector whose elements π_{jp} denote the steady-state probability of having j cells in the queue, and the MMPP is in phase p upon departure of a cell. Then π is the solution of the linear system of equations $\pi Q_c = 0$ and $\pi e = 1$, where $e = (1, 1, \ldots, 1)^T$.

Various efficient algorithms that explore the special structure of the system of equations have been developed to obtain the distribution π. In particular, both matrices Q_d and Q_c are block Hessenberg matrices, which have no blocks below the one parallel to the diagonal. Furthermore, a mapping can be provided from the continuous-time Markov chains to their discrete-time counterparts so that the steady-state queue length distribution obtained from either one is the same. In particular, the following mapping is applicable to any transition rate matrix. Let $A(j, j')[i, i']$ denote the $[i, i']$th element of the matrix $A(j, j')$. Define diagonal matrices $E(j)[i, i]$ as

$$E(j)[i, i] = \sum_{m;m'} A(j, m)[i, m']$$

Then the steady-state queue length distribution of the continuous-time Markov chain is the same as that of the discrete-time Markov chain with transition probabilities:

$$T(j, j')[i, i'] = \{A(j, j')E^{-1}(j)\}[i, i']$$

In general, a block upper Hessenberg matrix has the following form. Though generally it is not always the case, for the two systems considered here, each $A(j, j')$ is a square matrix.

$$H = \left[\begin{array}{ccccccccc} A(0,0) & A(0,1) & .. & .. & .. & .. & .. & .. & A(0,s) \\ A(1,0) & A(1,1) & .. & .. & .. & .. & .. & .. & A(1,s) \\ 0 & A(2,1) & A(2,2) & .. & .. & .. & .. & .. & A(2,s) \\ .. & .. & .. & .. & .. & .. & .. & .. & .. \\ .. & 0 & 0 & 0 & 0 & .. & A(s-1,s-2) & A(s-1,s-1) & A(s-1,s) \\ .. & 0 & .. & 0 & 0 & .. & 0 & A(s,s-1) & A(s,s) \end{array}\right]$$

A.3.1 Block Forward Substitution

Consider a continuous-time Markov chain with a block upper Hessenberg form. Given the matrix, the equation $\pi H = 0$ can be written as follows:

$$\sum_{i=0}^{j+1} \pi_i A(i, j) = 0, j = 0, \ldots, s - 1 \tag{A.49}$$

$$\sum_{i=0}^{s} \pi_i A(i, j) = 0, j = s \tag{A.50}$$

Let us for a moment assume that π_0 is known. Then we have

$$\pi_0 A(0, 0) + \pi_1 A(1, 0) = 0 \Rightarrow \pi_1 = -\pi_0 A(0, 0)A^{-1}(1, 0)$$

$$\pi_0 A(0, 1) + \pi_1 A(1, 1) + \pi_2 A(2, 1) = 0 \Rightarrow \pi_2 = -\pi_0 A(0, 1)A^{-1}(2, 1) - \pi_1 A(1, 1)A^{-1}(2, 1) \tag{A.51}$$

$$\Rightarrow \pi_2 = -\pi_0 A(0, 1)A^{-1}(2, 1) + \pi_0 A(0, 0)A^{-1}(1, 0)A(1, 1)A^{-1}(2, 1) \tag{A.52}$$

Continuing in this manner, subvectors π_i can all be determined in terms of π_0. The procedure, referred to as a *block forward substitution,* at step i performs the following operation:

$$\pi_{i+1} = -\sum_{j=0}^{i} \pi_j A(j, k)\{A^{-1}(i + 1, i) \tag{A.53}$$

This operation requires the subdiagonal blocks to be nonsingular, which is the case with the ATM models considered here.

π_0 is assumed to be known to recursively calculate the other π_j's, which is not known a priori. Note that the last column is not used in the recursive procedure. At the end of the recursion, we have all π_j's expressed in terms of π_0. Then the last equation can be used to determine the value of π_0. The last step is to compute the values of π_j from π_0.

In the case of the MMPP/D/1/B model, each recursion equation consists of three matrices $A(0)$, $A(1)$, and $A(2)$. The last equation, $j = B - 1$, has two matrices $A(1)$ and $A(2)$, from which π_0 is obtained.

A survey of various numerical techniques developed to solve Hessenberg matrices in general and models of ATM networks in particular can be found in [Ch 4, 140].

A.3.2 Matrix Geometric Solution

Consider the transition probability matrix of a discrete-time Markov chain, which is a finite square upper-block Hessenberg matrix H with $K.B$ rows (and columns) and π denoting its steady-state queue length distribution. Then

$$\pi_{j-1} = \pi_j R_j \quad \text{for } j = 1, \ldots, B-1 \text{ where } R_j \text{ is given by } R_j = A(j, j-1)(I - C_j)^{-1}$$
$$\text{with } C_j = A(j-1, j-1) + R_{j-1}\{A(j-2, j-1) + R_{j-2}(A(j-3, j-1) + \ldots + R_1(A(0, j-1)) \ldots)\}$$

with I denoting the identity matrix and

$$\pi_{B-1} = \pi_{B-1} D, \text{ where } D = A(B-1, B-1) + R_{B-1}A(B-2, B-1) + R_{B-2}\{A(B-3, B-1) + \ldots + R_1(A(0, B-1)) \ldots)\}$$

The inversion $(I - C_j)^{-1}$ can be performed either directly, which is subject to possibly large roundoff errors, or as follows, which has the advantage of guaranteeing accuracy. Let

$$\begin{matrix} U_0 = C \\ U_m = U_0^2;\ j \geq 1 \end{matrix} \quad \text{and} \quad \begin{matrix} V_0 = I \\ V_m = V_{m-1}U_{m-1} + V_{m-1};\ j \geq 1 \end{matrix} \tag{A.54}$$

Then $V_m = (I - C)^{-1}$ as m goes to infinity.

To obtain the performance metrics of interest, the following approach does not require the calculation of the steady-state queue length distribution explicitly. In particular, let π_j, $j = 0, \ldots, B-1$, be the solution of the system of linear equation $\pi H = \pi$. Then the normalization constant G is given by $G = \pi_{B-1}S_{B-1}$, where the sequence of column vectors S_j satisfies $S_0 = e$ and $S_j = R_jS_{j-1} + e$ with $e = (1, 1, \ldots, 1)^{-1}$. Similarly, suppose the algorithm is requested to yield the steady-state expected value L of some function q of the Markov chain. Then $L = \pi_{B-1}L_{B-1}/G$, where the sequence of column vectors L_j corresponding to the jth component of π satisfies $L_0 = q(0)$ and $L_j = R_jL_{j-1} + q(j)$. As already presented above, with L corresponding to the normalizing constant, the ith element of $q(j)$, $q(j)[i]$, is equal to 1 for all i and j. This immediately follows from the fact that $\pi e = 1$. Similarly, the cell loss probability is defined as follows, where $a(i, s)$ is the probability of s cells arriving in a slot when the SBBP is in phase I.

$$\text{cell loss probability} = \frac{\text{average number of cells lost per second}}{\text{utilization of offered load}}$$
$$= \sum_{j;i} \pi_j \sum_{s \geq 1} a(i, s)\{j + s - 1 - (B-1)\}$$
$$/\text{utilization of offered load}$$

Hence, $q(j)[i]$ for the cell loss probability is given as

$$q(j)[i] = \sum_{s \geq 1} a(i, s)\{j + s - 1 - (B-1)\}/\text{utilization of offered load}$$

As another example, the mean queue length MQL calculated at departure instants is

$$\text{MQL} = \sum_{j=1}^{B-1} j\pi_j$$

In this case, $q(j)[i] = j$. The steps of the algorithm can now be summarized as follows (Ch 4, [88]):

{initialization}
  $R_j = 0$ for $j \leq 0$
  $S_0 = e$
  {specific performance metric of interest}
  $L_0 = q(0)$
  {main loop}
  do $j = 1$ to $B - 1$
    begin
    $R_j = A(j, j-1)(I - C_j)^{-1}$
    $S_j = R_j S_{j-1} + e$
    {specific performance metric}
    $L_j = R_j L_{j-1} + q(j)$
    end
  {final values}
  $D = A(B-1, B-1) + R_{B-1}\{A(B-2, B-1) + R_{B-2}(A(B-3, B-1) + \ldots + R_1(A(0, B-1)) \ldots)\}$
  Solve $\pi_{B-1} = \pi_{B-1} D$ for π_{B-1}.
  {normalization constant}
  $G = \pi S_{B-1}$
  specific performance metric}
  $L = \pi_{B-1} L_{B-1} / G$

Appendix B: Derivation of Equivalent Capacity

A two-state model with active and silent states is used to characterize each source. Furthermore, it is assumed that the duration at each state is exponentially distributed and independent of each other. Let:

R = peak bit rate of the connection;
b = average duration of the active period;
ρ = source utilization (i.e., probability that the source is in an active state);
μ = transition rate out of active state (i.e., $\mu = 1/b$);
λ = transition rate out of the silent state (i.e., $\lambda = \rho/\{b(1-\rho)\}$);
c = link speed;
X = buffer size.

Furthermore, let $P_i(t, x)$ denote the probability that the source is in state i and the buffer content is x at time t; $i = 1$ if the source is active and $i = 0$ if it is idle. Then the evaluation of the single queue under consideration is described by the following two equations:

$$P_0(t + dt, x) = (1 - \lambda dt)P_0(t, x + c\,dt) + \mu dt P_1(t, x + (c - R)dt) \qquad \text{(B.1)}$$
$$P_1(t + dt, x) = \lambda dt P_0(t, x + c\,dt) + (1 - \mu\,dt)P_1(t, x + (c - R)dt)$$

After some manipulation, (B.1) yields the following two partial differential equations:

$$\frac{\partial P_0(t, x)}{\partial t} - c\frac{\partial P_0(t, x)}{\partial x} = -\lambda P_0(t, x) + \mu P_1(t, x) \qquad \text{(B.2)}$$
$$\frac{\partial P_1(t, x)}{\partial t} - (c - R)\frac{\partial P_1(t, x)}{\partial x} = \lambda P_0(t, x) - \mu P_1(t, x)$$

The stationary behavior of the system is then characterized by the vector $F(x) = [F_0(x), F_1(x)]^T$, where

$$F_i(x) = \lim_{t \to \infty} P_i(t, x), \; i = 0, 1$$

which is the solution of the following system of equations:

$$\begin{bmatrix} -c & 0 \\ 0 & -(c-R) \end{bmatrix} F'(x) = \begin{bmatrix} -\lambda & \mu \\ \lambda & -\mu \end{bmatrix} F(x) \tag{B.3}$$

where $F'(x)$ is the derivative of $F(x)$ with respect to x. The solution of this system, $F(x)$, is then given by

$$F(x) = \alpha_0 \begin{bmatrix} \mu \\ \lambda \end{bmatrix} + \alpha_1 \begin{bmatrix} (R-c)/c \\ 1 \end{bmatrix} \exp\{-x(c-\rho R)/[b(1-\rho)(R-c)c]\} \tag{B.4}$$

where the constants α_0 and α_1 are to be obtained from the boundary conditions. Assuming $R > c$, the buffer cannot be empty. Hence, $F_1(0) = 0$. Similarly, seeing that the buffer cannot be full when the server is idle, we have $F_0(X^-) = 1 - \rho$. Using these two boundary conditions and (B.4), we have

$$\alpha_0 = b(1-\rho)^2/\Delta \quad \text{and} \quad \alpha_1 = -\rho(1-\rho)c)/\Delta \tag{B.5}$$

where

$$\Delta = (1-\rho)c - \rho(R-c)\exp\{-X(c-\rho R)/[b(1-\rho)(R-c)c]\}$$

Then the queue length distribution $\Pr\{Q < x\} = F_0(x) + F_1(x)$ is given as follows:

$$\Pr\{Q \le x\} = \begin{cases} c(1-\rho)/\Delta - \{\rho(1-\rho)R/\Delta\}\exp\{-x(c-\rho R)/[b(1-\rho)(R-c)c]\} & x < X \\ 1 & \text{otherwise} \end{cases} \tag{B.6}$$

The steady-state overflow probability p is equal to the probability that the source is active and the buffer is full, which can be obtained from the identity $\pi_1 = p + F_1(X^-)$, where $\pi_1 = \rho$ is the probability that the source is active. Hence, we have

$$p = \frac{\rho(c-\rho R)\exp\{-X(c-\rho R)/[b(1-\rho)(R-c)c]\}}{(1-\rho)c - \rho(R-c)\exp\{-X(c-\rho R)/[b(1-\rho)(R-c)c]\}} \tag{B.7}$$

Using this framework, the amount of bandwidth required by a connection is obtained in isolation as the answer to the following question. If a connection with parameters (R, m, b) is input to a link with buffer capacity X, what should be the transmission rate for this link to achieve a desired buffer overflow probability ϵ? To find the conditional probability that the buffer is overflowing, it is assumed that $\epsilon = \rho p$. Then, with $\delta = (c-\rho R)/[b(1-\rho)(R-c)c]$, (B.7) becomes

$$(c - \rho R)\exp\{-\delta X\} = \epsilon\{(1 - \rho)c - \rho(R - c)\exp\{-\delta X\} \tag{B.8}$$

Using the fact that the overflow probability is required to be less than or equal to ϵ, (B.8), and the identity $\epsilon = \rho p$, we have

$$e^{-\delta X} \leq \frac{\epsilon(1 - \rho)c}{(c - \rho R) + \epsilon\rho(R - c)} \tag{B.9}$$

where c is referred to as the equivalent capacity of the connection. Since (B.9) includes both exponential and rational functions, an explicit solution for c is not possible to derive. Furthermore, a numerical solution of the equation is time-consuming, considering the real-time requirements to perform the processing. Instead, an upper bound on c is obtained by substituting $(1 - \rho)c/\{c - \rho R\} + \epsilon\rho(R - c)\} = 1$ in (B.9), which is reduced to the following quadratic equation of c:

$$\alpha b(1 - \rho)c^2 + \{X - \alpha b(1 - \rho)R\}c - X\rho R = 0$$

where $\alpha = \ln(1/\epsilon)$. Then the equivalent capacity is

$$c = R\frac{y - X + \sqrt{(y - X)^2 + 4X\rho y}}{2y} \tag{B.10}$$

with $y = \alpha b(1 - \rho)R$. With this framework, the total bandwidth of n multiplexed connections is equal to the sum of the equivalent capacities of individual connections c_i; that is,

$$C = \sum_{i=1}^{n} c_i$$

However, C may significantly overestimate the required bandwidth for the aggregate traffic, since the interaction between individual connections is not taken into consideration. To capture the effect of multiplexing, the Gaussian approximation (see Appendix C) is used together with the equivalent capacities. In particular, the total bandwidth required for the aggregate traffic of n connections C is given by

$$C = \min\{m + \alpha'\sigma, \sum_{i=1}^{n} c_i\} \tag{B.11}$$

Note that the mean m_i and the variance σ_i^2 of the connection bit rate with the above flow model are respectively equal to $R_i b_i$ and $m_i(R_i - m_i)$, and that

$$m = \sum_{i=1}^{n} m_i, \ \sigma^2 = \sum_{i=1}^{n} \sigma_i^2$$

Appendix C: Gaussian Approximation

In this method, each connection is characterized by its average bit rate m_i and standard deviation σ_i. With n multiplexed connections, the problem is determining the total bandwidth required by n connections, c_0, so that the probability that the instantaneous aggregate bit rate exceeding c_0 is less than a given value ϵ.

Let A be a random variable denoting the aggregate bit rate of n multiplexed connections. Then the problem is to determine the value of c_0 such that $\Pr\{A > c_0\} < \epsilon$. Assuming that the aggregate bit rate distribution is Gaussian, we have

$$\Pr\{A > c_0\} = \Pr\{(A - m)/\sigma > (c_0 - m)/\sigma\} \approx \Pr\{A_{01} > (c_0 - m)/\sigma)\} = \Pr\{A_{m\sigma} > c_0\} \quad \text{(C.1)}$$

where

$$m = \sum_{i=1}^{n} m_i, \; \sigma^2 = \sum_{i=1}^{n} \sigma_i^2$$

A_{01} and $A_{m\sigma}$ are Gaussian random variables with mean and standard deviation (0, 1) and (m, σ), respectively. We then have

$$\Pr\{A > c_0\} \approx \Pr\{A > m + \alpha\sigma\} \approx \epsilon \quad \text{(C.2)}$$

Hence, $c_0 \approx m + \alpha\sigma$. The parameter α is the inverse of the Gaussian distribution. Various formulas have been developed to obtain its value approximately. An accurate approximation is

$$\alpha = \sqrt{2\ln(1/e) - \ln 2\pi} \quad \text{(C.3)}$$

For most practical cases of interest, c_0 is an upper bound on the actual bandwidth required for the aggregate traffic to have a cell loss probability of less than or equal to ϵ. This is mainly due to the fact that the buffer size is not considered in the calculation of c_0 and that in reality the instantaneous aggregate bit rate is allowed to exceed c_0 for some period of time until the buffer becomes full, thereby absorbing some of the inaccuracy introduced with this method.

However, the aggregate bit rate often exhibits a longer tail than that of Gaussian distribution when the Gaussian assumption for the aggregate bit rate does not hold. Then $\Pr\{A > c_0\}$ is larger than ϵ when c_0 is calculated from (C.2). To solve this problem, the approach is extended as follows. Instead of assuming that they are Gaussian, the distributions of bit rates of each connection are embedded into a Gaussian envelope. In particular, with m_i and σ_i being the parameters of connection i, the parameters of the Gaussian envelope (m_i^*, σ_i^*) are

$$m_i^* = m_i + a\sigma_i, \; \sigma_i^* = b\sigma_i \tag{C.4}$$

Then the amount of bandwidth reserved for n multiplexed connections for some constants a and b is given by

$$c_0 = \sum_{i=1}^{n} m_i + a\sigma_i + \alpha b \sqrt{\sum_{i=1}^{n} \sigma_i^2} \tag{C.5}$$

However, the values of a and b are not known a priori. A method to estimate their values so that c_0 is indeed an upper bound is given in Ch 4, [80]. However, this method requires that the activity of each source be monitored, and it is based on a set of measurements taken for each connection, limiting its practicality as a call admission procedure.

Appendix D: Fast Buffer Reservation

In this scheme, a bursty source is characterized by a two-state Markovian model. When a source becomes active, a prespecified number of buffer slots in the link buffer is reserved for the duration of its active period. At the end of the active period, all reserved slots are released. This process is repeated throughout the duration of the connection. The technique uses marked cells to specify the transitions between active and silent periods. Let:

B = number of buffer slots available at the link buffer;
B_i = number of buffer slots to be reserved for connection i;
b_i = number of slots currently in use;
s_i = state of connection i (active or idle).

The operation of the fast buffer reservation scheme is given in Section 4.3.3.2. To decide if a new connection can be multiplexed with the existing connections on a link, the probability of requiring more buffer slots than are available is calculated and referred to as the *excess demand probability*. If this probability is greater than a predefined value, the new connection is rejected; otherwise it is accepted. To calculate the excess demand probability, let:

λ_i = peak bit rate of connection i;
μ_i = average bit rate of connection i;
x_i = random variable representing the number of buffer slots needed by connection i.

x_i is either equal to 0 (in a silent state) or B_i (in an active state). We have

$$p_i = \Pr\{x_i = B_i\} = \mu_i/\lambda_i \text{ and } \Pr\{x_i = 0\} = 1 - \mu_i/\lambda_i$$

Using the peak-to-link-rate ratio, the number of buffer slots required by an active source is assumed to be equal to

$$B_i = \lceil L\lambda_i/R \rceil \quad \text{(D.1)}$$

where $\lceil z \rceil$ denotes the smallest integer greater than or equal to z, L is the total number of buffer slots at the link buffer, and R is the link rate. Consider now a link carrying n connections with buffer demands $x_1, \ldots, x_n$. The total buffer demand X is the sum of n random variables; that is,

$$X = \sum_{i=1}^{n} x_i$$

Assuming that x_i's are mutually independent, the probability generation function of X, $f_x(z)$, is

$$f_X(z) = \prod_{i=1}^{n} \{(1 - p_i) + p_i z^{B_i}\} = C_0 + C_1 z + C_2 z^2 + \ldots + C_k z^k \qquad \text{(D.2)}$$

where

$$k = \sum_{i=1}^{n} B_i$$

Let $\Pr\{X = j\} = C_j$ denote the probability that the total buffer demand is equal to j. Then the excess demand probability is equal to

$$1 - \sum_{i=1}^{L} C_i$$

and the problem is reduced to obtaining the values of C_j's.

Consider a new connection request with a buffer demand x_{n+1} to be multiplexed on a link with n existing connections. With connection $n + 1$, (D.2) becomes

$$f_X^*(z) = \prod_{i=1}^{n+1} \{(1 - p_i) + p_i z^{B_i}\} = \{(1 - p_{n+1}) + p_{n+1} z^{B_{n+1}}\} \prod_{i=1}^{n} \{(1 - p_i) + p_i z^{B_i}\} \qquad \text{(D.3)}$$

That is, the new coefficients C_j' are calculated from the old coefficients (i.e., obtained without the new connection):

$$C_j' = (1 - p_{n+1}) C_j + p_{n+1} C_{j-B_{n+1}}, \qquad \text{(D.4)}$$

for all j. Equivalently, when the buffers of connection i are released, the new coefficients are calculated from the current ones as follows:

$$C_j' = C_j (1 - p_i) + C_{j-B_i} / p_i \qquad \text{(D.5)}$$

for all j. In (D.4) and (D.5), for the simplicity of notation we have $C_j = 0$ for $j < 0$.

Although the excess demand probability can be calculated recursively using the above set of equations in real time, certain anomalies are observed to occur in this framework. Consider two connections, both of which require all the buffer slots when active, and that $p_1 = 0.9$, $p_2 = 0.01$. The excess buffer demand probability with the two connections is equal to 0.009. However, when the second connection becomes active, there is no buffer space available w.p. 0.9, meaning that only 10% of its bursts will succeed. If the excess demand probability is greater than 0.009, then the second connection will be accepted, which is clearly unacceptable.

This problem is addressed as follows. Instead of using the excess demand probability, use the probability that the number of buffer slots requested is not available at the time when source i transmits its burst. This probability is bounded above by the probability $\Pr\{X - x_i > L - B_i\}$, referred to as the contention probability of connection i. To obtain the contention probability for connection i,

$$
\begin{aligned}
&\Pr\{X = L\} = (1 - p_i)\Pr\{X - x_i = L\} + p_i\Pr\{X - x_i = L - B_i\} \\
&\Rightarrow \Pr\{X > L\} = (1 - p_i)\Pr\{X - x_i > L\} + p_i\Pr\{X - x_i > L - B_i\} \\
&\Rightarrow \Pr\{X - x_i > L - B_i\} = (1/p_i)\Pr\{X > L\} - \{(1 - p_i)/p_i\}\Pr\{X - x_i > L\} \\
&\Rightarrow \Pr\{X - x_i > L - B_i\} \leq (1/p_i)\Pr\{X > L\}
\end{aligned}
\tag{D.6}
$$

Hence, as long as p_i is not too small, the excess demand probability is not too much larger than the contention probability. However, the two differ significantly as p_i decreases. When p_i is small, tight bounds on the contention probability can be obtained with some additional computation [143]:

$$
\begin{aligned}
\Pr\{X - x_i > j\} = {}& \{1/(1 - p_i)\}\sum_{h=0}^{k-1}\{-p_i/(1 - p_i)\}^h\Pr\{X > j - hB_i\} \\
& + \{-p_i/(1 - p_i)\}^k\Pr\{X > j - hB_i\}
\end{aligned}
\tag{D.7}
$$

where k is chosen such that $\{p_i/(1 - p_i)\}^k$ is sufficiently small.

Appendix E: Flow Approximation to Cell Loss Rate

Consider a VBR source alternating between active and silent periods characterized only by its peak p_i and average m_i bit rates. No assumptions on the distributions of the two periods are required, except that the probability of source i being active is equal to m_i/p_i and idle w.p. $1 - (m_i/p_i)$.

Consider a link with N independent sources, a transmission rate of C cells/sec and a buffer size of M. Let r_i and R be the random variables denoting respectively the cell arrival distribution of connection i and the aggregate traffic from N connections; that is,

$$R = \sum_{i=1}^{N} r_i$$

Given this parameterization, a continuous cell stream with rate R arrives at the queue and departs at rate C in the flow model of the system under consideration. With q denoting the number of buffer slots available, the rate at which cells are lost L is given as follows:

$$L = \sum_{X>C} (X - C)\Pr(R = X)\{1 - \Pr(q > 0)|R = X)\} \qquad \text{(E.1)}$$

That is, if $X > C$ and the buffer is full, then the rate at which cells are lost is equal to $(X > C)$.

The main building block of this method is that as the lengths of the active periods increase while their peak and average bit rates stay the same, the probability that there is a buffer slot available, given that the aggregate cell arrival rate is X, $\Pr\{q > 0|R = X\}$, approaches zero for $X > C$. Then (E.1) is reduced to (E.2), which defines a supremum of L with respect to the burst length.

$$OF = \sum_{X>C} (X - C)\Pr(R = X) \qquad \text{(E.2)}$$

Using (E.2), an upper bound on the cell loss probability PV is equal to

$$PV = OF / \sum_{i=1}^{N} m_i \tag{E.3}$$

Let Ω_k be the set of indexes of connections $\{i_1, \ldots, i_r\}$, such that

$$\sum_{j=1}^{r} P_j > C$$

and let $\Omega = \{\Omega_k\}$ be the set of such vectors. With r_i equal to either p_i or 0, (E.2) can be rewritten as

$$OF = \sum_{\Omega_k \in \Omega} \left\{ \left(\sum_{j \in \Omega_k} p_j \right) \prod_{j \in \Omega_k} m_j / p_j \prod_{j \in \Omega'_k} (1 - m_j / p_j) \right\} \tag{E.4}$$

where $\Omega'_k = \{1, \ldots, N\} - \Omega_k$, and, for the simplicity of notation, the product is equal to 1 when Ω'_k is an empty set.

Equation (E.4), based on only the peak and average cell generation rates of connections, can be used for call admission decisions. However, based on the information available, obtaining the exact value of PV may not be feasible in real time, particularly as N increases.

The following approximation to (E.4) meets the real-time requirements. Let

$$T = \sum_{i=1}^{N} p_i$$

be the sum of the peak rates of connections. Then

$$OF = -\int_{C}^{T} (X - C) \mathrm{d} \Pr\{R > X\} = \int_{C}^{T} \Pr\{R > X\} \tag{E.5}$$

Applying Chernoff bound for some positive s, we have

$$\Pr\{R > X\} < E\{e^{sR}\} / e^{sX} \tag{E.6}$$

Substituting (E.6) into (E.5) and letting T go to infinity,

$$OF < E\{e^{sR}\} \int_{C}^{T} e^{-sX} \mathrm{d}x < E\{e^{sR}\} / (s e^{sC}) \tag{E.7}$$

Then $f(s) = E\{e^{sR}\} / (s e^{sC})$ is a parametric upper bound of OF.

$$f(s) = E\{e^{sR}\}/(se^{sC}) = E\{e^{s(R-C)}\}/s \tag{E.8}$$

$f(s)$ is a convex function and has the minimum value at $s = s^*$, which is the root of the following equation with $T > C$.

$$\partial \ln E\{e^{sR}\}/\partial s - 1/s - C = 0 \tag{E.9}$$

However, with r_n equal to 1 w.p. m_n/p_n and 0 w.p. $1 - m_n/p_n$, and using the definition of R, $E\{e^{sR}\}$ is

$$E\{e^{sR}\} = \prod_{n=1}^{N} E\{e^{sr_n}\} = \prod_{n=1}^{N} (m_i/p_i)\{e^{sp_n} - 1\} + 1 \tag{E.10}$$

With (E.9) and (E.10), s^* can now be obtained as the root of (E.11):

$$\sum_{i=1}^{N} \frac{m_i e^{sp_i}}{(m_i/p_i)\{e^{sp_i} - 1\} + 1} - \frac{1}{s} - C = 0 \tag{E.11}$$

With s^* being obtained from (E.11), an upper bound on PV is

$$PV = \frac{\prod_{n=1}^{N} (m_i/p_i)\{e^{-s^* p_n} - 1\} + 1}{s^* e^{s^* C} \sum_{i=1}^{N} m_i} \tag{E.12}$$

PV can now be used to make call admission decisions similar to the other techniques discussed above. There are two drawbacks of this approach. First, it does not use the buffer size in decision making, thereby restricting the amount of statistical gains that can be achieved. Secondly, the procedure does not distinguish between the cell loss requirements of individual connections. Cell loss probability for each connection PV_n can be obtained by modifying (E.2) to obtain OF_n as follows.

$$OF_n = \sum_{X>C} (X - C)\Pr(R = X)p_n/X, \quad n = 1, \ldots, N \tag{E.13}$$

Then the call admission procedure can compare the calculated value of PV_n for a given current load on a link with the QOS requirement of the connection and make the admit/reject decision.

Appendix F: Nonparametric Approach

The nonparametric approach is based on the peak and average cell rates of connections and does not require any knowledge on the distribution of the arrival process.

Consider a link with a transmission capacity of C cells/sec with a slot length (i.e., time to transmit one cell) equal to $1/C$. Furthermore, let R_i' and a_i' denote the peak and average cell rate of connection i. Given an observation period of r slots, the new parameters R_i and a_i are defined based on the peak and average rates during the observation period as $R_i = \lceil rR_i'/C \rceil$ and $a = \lceil ra_i'/C \rceil$.

If the link (or the VP) with n connections multiplexed onto it is provisioned for a CLR of P_{CLR}, then

$$P_{\mathrm{CLR}} \leq U(n, r) = \frac{\sum_{k=0}^{\infty} \max(0, k-r)\theta_1 * \theta_2 * \ldots * \theta_n(k)}{\sum_{k=0}^{\infty} k\theta_1 * \theta_2 * \ldots * \theta_n(k)}$$

where $\theta_1 * \theta_2 * \ldots * \theta_n(k)$ denotes the n-fold convolution of $\theta_i(j)$'s, with $\theta_i(j)$ denoting the distribution of the number of cells arriving (i.e., maximum or none) during r slots; that is

$$\theta_i(j) = \begin{cases} a_i/R_i & \text{if } j = R_i \\ 1 - a_i/R_i & \text{if } j = 0 \\ 0 & \text{otherwise} \end{cases}$$

This framework allows a simple call admission procedure to be used. A new connection arriving on a link with n multiplexed connections, calculate $U(n+1, r)$. If $U(n+1, r)$ is greater than the desired loss ratio, then the new connection is accepted; otherwise it is rejected.

In order to implement this nonparametric approach in real time, a recursive scheme to calculate $U(n, r)$ is developed. Define a state vector $S = [S(0), S(1), \ldots]$, where

$$S(m) = \sum_{k=m}^{\infty} \sum_{i=k}^{\infty} \theta_1 * \theta_2 * \ldots * \theta_n(i), \quad m \geq 0$$

Furthermore, let

$$A = \sum_{i=1}^{n} a_i$$

be the average number of cells arriving from all n multiplexed connections. Then

$$U(n, r) = S(r + 1)/A$$

where $S(r + 1)$ with a new connection $n + 1$ is calculated from $S(j)$, $i = 0, \ldots, r$, and from the parameters of the new connection as

$$\{1 - (a_{n+1}/R_{n+1})\}S(r + 1) = \begin{cases} (a_{n+1}/R_{n+1})S(r + 1 - R_{n+1}) & r + 1 \geq R_{n+1} \\ (a_{n+1}/R_{n+1})[(R_{n+1} - r + 1) + S(0)] & r + 1 < R_{n+1} \end{cases}$$

Then $U(n + 1, r)$, used in the call admission decision, is given by $S(r + 1)/(A + a_{n+1})$.

Appendix G: Heavy Traffic Approximation

Heavy traffic approximation is based on the asymptotic behavior of the tail of the queue length distribution in an infinite capacity queue, with constant service times and a Markovian cell arrival process governed by a probability matrix $P(z)$.

The tail behavior of the steady-state queue length distribution of such a queuing system is characterized by the smallest root, z^*, outside the unit circle of the determinant. Then, for sufficiently large i,

$$P\{\text{queue length} > i\} = \alpha(1/z^*)^i$$

where α is an unknown constant that is difficult to obtain.

$P\{\text{queue length} > i\} = p$ can be used to make call admission decisions. In particular, the arrival process is general enough to model the superposition of N sources into a queue. If the probability p with the new connection included in the superposed process is less than or equal to the desired loss ratio, then the new connection is accepted; otherwise it is rejected. Then the problem is reduced to estimating the values of z^* and α.

Let us first consider an on/off source with parameters R denoting its peak rate, m its average rate, and b the average burst length. The utilization of this source is given by $\rho = m/R$. The constant α is set to the total utilization at the queue with N multiplexed connections; that is

$$\alpha = \sum_{i=1}^{N} m_i/R_i$$

With the assumption that the on and off periods are exponentially distributed, z^* is approximated as

$$z^* = 1 + \frac{1 - \alpha}{\sum_{i=1}^{N} m_i(1 - \rho_i)^2 b_i}$$

Furthermore, the approximation is extended to sources with arbitrary on and off period distributions. In this case, we have

$$z^* = 1 + \frac{2(1 - \alpha)}{\sum_{i=1}^{N} m_i (1 - \rho_i)^2 b_i [c_i^2(\text{on}) + c_i^2(\text{off})]}$$

where $c_i^2(\text{on})$ and $c_i^2(\text{off})$ are the squared coefficients of variation of the on and off periods, respectively.

Appendix H: Leaky Bucket Analysis

Consider a buffered leaky bucket with a finite buffer C and token pool size K. Tokens are generated in fixed intervals of N slots, where a slot is defined by the peak rate of the source (i.e., 53×8/peak rate). Given the parameters of the leaky bucket, C, K, and N, the performance metrics of interest are:

- Cell loss probability and delay characteristics at the buffer;
- Accuracy of the leaky bucket in policing the user traffic:
  - The longest burst of cells that can pass unimpeded through the leaky bucket;
  - The interdeparture time characteristics of cells.

The cell loss probability as well as the average number of slots between two consecutive cell departures from the leaky bucket depend only on the sum of the sizes of the token pool and the cell queue, not on their individual sizes. However, for a given total value of $C + K$, the waiting time at the queue increases as C increases, and vice versa.

Let us now consider the leaky bucket scheme with a two-phase MMBP arrival process. The state of the system is the triple (c, k, p), conditioned on just before the arrival of a token, where:

c = number of cells waiting in the queue, $0 \leq c \leq C$;
k = number of tokens waiting in the token pool, $0 \leq k \leq K$;
p = phase of the arrival process, $p = 1, 2$.

Since $k = 0$ if $c > 0$ and $c = 0$ if $k > 0$, the state of the leaky bucket (c, k) can be represented by a single-state variable x by the mapping $x = K - k + c$, $0 \leq x \leq C + K$. In this case, $x = 0$ corresponds to the case where $k = K$ and $c = 0$, $x = C + K$ corresponds to $c = C$ and $k = 0$, and so on. Then the state of the system is (x, p).

The system can be solved numerically using the set of linear equations $\pi Q = 0$, $\pi e = 1$, where Q is the rate matrix. The rate matrix is determined from the number of arrivals that occurs in N slots and the probability that the arrival process moves from phase j to k. Let $a_{jk}(i, l)$ denote the probability that i cell arrivals occur and the arrival process moves from phase j to k during a period of l slot. To determine $a_{jk}(i, N)$, let e_{ij} denote the probability that the arrival process moves from phase i to phase j during a single slot and that no arrival occurs during the slot, and let d_i denote the probability that an arrival occurs during a slot given that the arrival process starts in phase i. Then the matrices $E = \{e_{ij}\}$ and $D = \{d_i\}$ of a two-state MMBP with parameters (p, q, α, β) are

$$E = \begin{bmatrix} p(1-\alpha) & (1-p)(1-\beta) \\ (1-q)(1-\alpha) & q(1-\beta) \end{bmatrix} \quad D = \begin{bmatrix} p\alpha + (1-p)\beta \\ q\beta + (1-q)\alpha \end{bmatrix} \tag{H.1}$$

We have $a_{jk}(0, 1) = e_{jk}$, $j, k = 1, 2$; whereas $B = \{a_{jk}(1, 1)\}$ is given as:

$$B = \begin{bmatrix} p\alpha & (1-p)\beta \\ (1-q)\alpha & q\beta \end{bmatrix} \tag{H.2}$$

If we define $A_i = \{a_{jk}(i, N)\}$, then the matrix A_i is obtained from the ith element of the first row of the matrix R^N, where

$$R = \begin{bmatrix} E & B & 0 & 0 & \dots & 0 & 0 \\ 0 & E & B & 0 & \dots & 0 & 0 \\ \cdot & \cdot & \cdot & \cdot & \cdot & \cdot & \cdot \\ 0 & 0 & \cdot & \cdot & \cdot & E & B \\ 0 & 0 & 0 & 0 & 0 & 0 & 1 \end{bmatrix} \tag{H.3}$$

Then the rate matrix Q is

$$\begin{bmatrix} A_0 & A_1 & A_2 & \dots & \dots & A^*_{C+K} \\ A_0 & A_1 & A_2 & \dots & \dots & A^*_{C+K} \\ 0 & A_0 & A_1 & \dots & \dots & A^*_{C+K-1} \\ \cdot & \cdot & \cdot & \cdot & \cdot & \cdot \\ 0 & 0 & 0 & \cdot & A_0 & A^*_1 \end{bmatrix} \tag{H.4}$$

with $A^*_i = \{a^*_{jk}(i, N)\}$, $1 < i < C + K$, where $a^*_{jk}(i, N)$ is the probability that i or more cells arrive and the arrival process moves from phase j to k during N slots. These probabilities are

$$a^*_{jk}(i, N) = \begin{cases} 1 - \sum_{l=0}^{i-1} a_{jk}(l, N) - \sum_{\substack{l=0 \\ k \neq k'}}^{C+K} a_{jk}'(l, N) & 0 < i \le N;\ 1 \le j;\ k \le 2 \\ 0 & i > N \end{cases} \tag{H.5}$$

where $a_{jk}(i, 1) = 0$, $l > N$, $j, k = 1, 2$.

If we assume that the steady-state probability vector

$$\pi = \{\pi_0^1, \pi_0^2, \pi_1^1 \pi_1^2, \ldots, \pi_{C+K}^1, \pi_{C+K}^2\}$$

is obtained numerically as the solution of the linear system of equations $\pi Q = \pi$, the performance metrics of interest of the buffered leaky bucket can now be obtained as follows.

Let $P_{\text{loss}}^{\text{cell}}$ and $P_{\text{loss}}^{\text{token}}$ respectively denote the probability that a cell upon arrival finds the buffer full and lost and the probability that a token upon arrival finds the token pool full and lost. A token is dropped whenever it arrives to a full token pool. Then

$$P_{\text{loss}}^{\text{token}} = \pi_0^1 + \pi_0^2 \tag{H.6}$$

Every token that is not dropped upon arrival eventually leaves the token pool. Similarly, every cell that is not lost upon arrival eventually leaves the buffer. Furthermore, the expected number of arrivals during N slots (i.e., interarrival times of tokens) is equal to $N\rho$, where ρ is the expected number of cells to arrive in a slot. Recall that ρ for an MMBP is equal to

$$\rho = \frac{(1 - q)\alpha + (1 - p)\beta}{2 - p - q}$$

We have

$$(1 - P_{\text{loss}}^{\text{cell}})N\rho = 1 - P_{\text{loss}}^{\text{token}} \tag{H.7}$$

Equivalently,

$$P_{\text{loss}}^{\text{cell}} = 1 - \frac{(1 - P_{\text{loss}}^{\text{token}})}{N\rho} \tag{H.8}$$

The rate at which cells depart from the leaky bucket is equal to

$$\lambda = \rho(1 - P_{\text{loss}}^{\text{cell}})$$

Equivalently, the mean number of slots between the interdeparture times of cells is equal to

$$(\rho(1 - P_{\text{loss}}^{\text{cell}}))^{-1}$$

By definition, the mean number of cells L waiting in the buffer is

$$L = \sum_{i=K+1}^{C+K} (i - K)(P_i^1 + P_i^2) \tag{H.9}$$

Appendix I: ATM Standards

ATM standardization is necessary and provides the basic architectural and operational framework for building ATM networks. The ATM standards have been developed by ITU-T with contributions from various national standardization organizations such as ANSI and ETSI. The ATM Forum, on the other hand, is a consortium of more than 700 companies established to speed up the development and deployment of ATM products through interoperability specifications. This appendix describes these organizations and provides a summary of various ATM standards and specifications produced by them.

I.1 INTERNATIONAL TELECOMMUNICATIONS UNION

ITU is a United Nations agency with activities including the regulation, standardization, and development of international telecommunications. ITU is organized into three sectors that reflect its main activities:

- ITU Telecommunication Standardization Sector (ITU-T);
- ITU Radiocommunication Sector (ITU-R);
- ITU Telecommunication Development Sector (ITU-D).

ITU-R is responsible for international radio regulation and for recommendations on technical and operational matters in radiocommunication. ITU-D's aim is to contribute to the growth and development of telecommunications throughout the world, with particular emphasis on the requirements of developing countries. ITU-T (formerly CCITT) is the body that studies technical, operating, and tariff questions and adopts recommendations on these areas with a view to standardizing telecommunications on a worldwide basis. Where telecommunications technical standards are concerned, ITU-T is clearly the standards body. ATM development was started by this body.

Various recommendations that have been developed by ITU-T include:

- General network planning, network operation, and network architecture;
- Terminals (e.g., data modems, multiplexers), systems (e.g., switching, signaling, and transmission), networks (e.g., telephone, data, ISDN, B-ISDN);

- Services and applications (e.g., multimedia, network, and service interworking);
- Operations and maintenance (e.g., telecommunication management network);
- Tariff principles and accounting rates.

In order to achieve its objectives, ITU-T works through world telecommunication standardization conferences, advisory groups, standardization bureaus, and standardization study groups.

Standardization work is done in study groups. Currently, there are 15 study groups studying different aspects of telecommunications, as illustrated in Table I.1.

Each study group is divided into a number of working parties (WP), each of which has a specific area of work. Working parties are further divided into expert teams, ad hoc groups, and so on. For example, there are four working parties within study group XIII:

- WP-1: Networking capabilities and internetworking;
- WP-2: B-ISDN (ATM) aspects;
- WP-3: Transport networks and layer 1 access (physical layer);
- WP-4: Performance aspects.

Furthermore, WP-2 is subdivided into six subworking parties (SWP) as follows:

Table I.1
Study Groups of the Standardization Sector

Study Group	
1	Service definition
2	Network operation
3	Tariff and accounting principles
4	Network maintenance
5	Protection against electromagnetic environment effects
6	Outside plant
7	Data networks and open system communications
8	Terminals for telematic services
9	Television and sound transmission
10	Languages for telecommunications applications
11	Switching and signaling
12	End-to-end transmission performance of networks and terminals
13	General network aspects
14	Modems and transmission techniques for data, telegraph, and telematic services
15	Transmission systems and equipment

- SWP-1: ATM layer;
- SWP-2: AAL;
- SWP-3: Operations and maintenance;
- SWP-4: Traffic management;
- SWP-5: Connectionless data service;
- SWG-6: Integrated video services (IVS).

Various ITU-T recommendations that have been produced by ITU-T are summarized in Tables I.2 and I.3.

I.2 ATM Forum

The ATM Forum was formed in October 1991. Current membership exceeds 700 organizations worldwide—computer vendors, LAN and WAN vendors, switch vendors, local and long-distance carriers, government and research agencies, and potential ATM users. Established originally in the United States, the ATM Forum now includes committees in Europe and the Pacific rim.

The main mission of the ATM Forum is to speed up the development and deployment of ATM products through interoperability specifications. Accordingly, the ATM Forum is not a standards organization. Instead, it produces implementation agreements based on international standards where standards are available. In other words, early deployment of ATM products requires that specifications be available much earlier than the target days of standards bodies, and the goal of the ATM Forum is to fill specification gaps, select options, and set parameters produced by international standards.

The ATM Forum cannot take its specifications to any standards organizations as contributions. Instead, Forum specifications have been contributed to various standards bodies by the member companies, who are also members of national standards bodies. However, there is always the possibility that Forum specifications will be incompatible with international standards and become a "de facto" standard.

The ATM Forum is a nonprofit mutual benefit corporation. It is managed by its board of directors elected by the member companies. Currently, there are five committees, as listed in Table I.4.

All interoperability specifications are produced by the subworking groups of the Technical Committee. Currently, there are nine subworking groups: signaling, broadband intercarrier interface, physical layer, traffic management, private NNI, LAN emulation, service aspects and applications (SAA), network management, and testing.

The B-ICI group defines a carrier-to-carrier interface to provide a basic framework for facilitating end-to-end national and international carrier service. This requires the specification of various physical layer interfaces and the

Table I.2
ITU-T Recommendations

Recommendation	*Scope*
I.113	*Vocabulary of terms for broadband aspects of ISDN* Provides the definitions of terms considered essential in understanding the principles of B-ISDN
I.121	*Broadband aspects of ISDN* Presents the basic principles of B-ISDN
I.151	*B-ISDN ATM functional characteristics* Presents a high-level review of the ATM layer functions
I.211	*B-ISDN service aspects* Classifies broadband services into a number of service classes and presents examples of services in each class
I.311	*B-ISDN general network aspects* Defines and presents a high-level review of ATM transport network hierarchy and the specification of each layer including signaling
I.321	*B-ISDN protocol reference model and its applications* Defines the B-ISDN reference architecture
I.327	*B-ISDN functional architecture* Presents a review of B-ISDN functions, where they are located, and where they are distributed in the B-ISDN
I.356	*QOS configuration and principles*
I.361	*B-ISDN ATM layer specification* Addresses the ATM cell structure, header coding, and the ATM protocol structure
I.362	*B-ISDN AAL functional description* Presents the AAL functional organization and the basic principles of AALs
I.363	*B-ISDN AAL specification* Provides the details of various types of AALs defined to support different B-ISDN service classes
I.364	*Support of broadband connectionless data service on B-ISDN* Describes the support of a connectionless data service based on AAL type 3/4 on B-ISDN
I.371	*Traffic control and congestion control in B-ISDN* Presents the objectives and mechanisms of traffic control and congestion control and defines traffic descriptors, traffic parameters, and traffic contract
I.374	*Network capabilities to support for multimedia*
I.413	B-ISDN UNI Presents the reference configuration for the B-ISDN UNI and various examples of physical realizations
I.432	*B-ISDN UNI—physical layer specification* Defines the physical media and the transmission system and discusses the implementation of related OAM functions at the UNI
I.555	*Interworking*
I.610	*B-ISDN operation and maintenance principles and functions* Identifies the minimum set of functions required to operate and maintain physical layer and ATM layer aspects of B-ISDN UNI as well as the individual VP and VP connections routed through a B-ISDN

Table I.3
ITU-T Recommendations on B-ISDN Signaling

Recommendation	*Scope*
Q.2931	*B-ISDN signaling* Defines the signaling procedures and functions used at UNI
Q.2100	*Q.SAAL.0 B-ISDN signaling AAL overview description* Describes the different components that make up the AAL functions necessary to support B-ISDN signaling
Q.2110	*Q.SAAL.1 SSCOP specification* Specifies the peer-to-peer protocol for the transfer of information and control between any pair of SSCOP entities
Q.2130	*Q.SAAL.2 B-ISDN signaling AAL-SSCF for the support of signaling at the UNI* Defines a mapping between the SSCOP of the signaling AAL and the Q.2931 layer at the UNI
Q.2140	*Signaling at the NNI* Defines a mapping between the SSCOP of the signaling AAL and the Q.2931 layer between network nodes and between networks
Q.2120	*Metasignaling* Defines the metasignaling protocol used to establish and maintain signaling connections at the UNI

Table I.4
The ATM Forum Organization

ATM Forum Committees	*Functions*
Technical Committee	Produces interoperability specifications
Market, Awareness, and Education Committee North American, European, Asia Pacific	Promotes the ATM technology within both the industry and the end user community: • Designs and promotes end user interaction • Raises public awareness of ATM technology (and the ATM Forum) • Publicizes the efforts of the ATM Forum through news releases
Enterprise Round Table	A user group that provides feedback and input to the Technical Committee in its efforts to develop Forum specifications in response to well-understood and analyzed multi-industry requirements

protocols and procedures to support the transport and multiplexing of multiple services for intercarrier delivery.

The main focus of the physical layer group is the development of specifications for ATM transmission on different types of transmission mediums, including fiber, unshielded twisted pair, shielded twisted pair, coax, and copper.

The P-NNI group is defining the private switching-system-to-switching-system interface in which a switching system may consist of a single switch (switch-to-switch interface) or maybe a subnetwork (network-to-network interface). The scope of the work includes P-NNI signaling and P-NNI routing frameworks.

The network management group is focused on the specification of managed objects in ATM networks and information flows between management systems based on existing standards whenever they are available.

SAA is chartered to define specifications that enable new and existing applications such as audio-visual services and circuit emulation to use AAL services.

The traffic management group works on traffic aspects of ATM networking that include specifications of application traffic parameters, conformance of user traffic, development of QOS guidelines, and definitions of service classes.

Based on the specifications produced by other groups, the testing group works on designing interoperability, conformance, and performance test suites.

The LAN emulation group is working towards defining a LAN emulation architecture to emulate connectionless service required to support existing LAN applications without any changes over connection-oriented ATM networks.

Finally, the signaling group works on additional features, procedures, and functions needed to address the signaling requirements of existing and emerging high-bandwidth applications.

I.3 American National Standards Institute

The Exchange Carriers Standards Association (ECSA) is the ANSI-accredited body for developing standards and technical reports related to interfaces for U.S. telecommunications networks. Its working group is called Committee T1—Telecommunications. T1 develops positions on related subjects under consideration in ITU-T. Specifically, T1 focuses on those functions and characteristics associated with interconnection and interoperability of telecommunications networks at interfaces with end users, carriers, and information and enhanced service providers. To carry out its work, ECSA has established six technical subcommittees, as listed in Table I.6.

I.4 EUROPEAN TELECOMMUNICATION STANDARDS INSTITUTE

ETSI produces standards for the European telecommunications market. The activities of ETSI are aimed at building upon ITU-T standards. The 11 ETSI technical committees are listed in Table I.8.

Table I.5
Some of the Specifications Produced by ATM Forum

Specification	*Scope*
UNI 3.1	*ATM Forum UNI specification* Includes specifications of physical layer interfaces, the ATM layer, ILMI, and UNI signaling
B-ICI	*B-ISDN intercarrier interface* Defines a carrier-to-carrier interface which specifies various physical layer interfaces and supports multiplexing of different services which include SMDS, frame relay, circuit emulation, and cell relay
DXI	*Data exchange interface* Defines data link control and physical layers as well as ILMI specifications which allow existing routers to interwork with ATM networks without requiring special hardware

Table I.6
Committee T1

Technical Subcommittee	*Responsibilities*
T1A1	Performance and signal processing
T1E1	Network interfaces and environmental considerations
T1M1	Internetwork operations, administration, maintenance, and provisioning
T1P1	Systems engineering, standards planning, and program management
T1S1	Service architecture and signaling
T1X1	Digital hierarchy and synchronization

Table I.7
Approved ANSI Standards

Standard	*Scope*
T1.624.1993	*B-ISDN UNI: Rates and formats specification* Defines the mapping of ATM cells into DS-3 and SONET payloads
T1.627.1993	*B-ISDN ATM functionality and specification* Defines the ATM layer (following I.361) including extended interpretations of traffic management and further explanations of the protocol model
T1.629.1993	*B-ISDN AAL 3/4 common part functionality and specification* Defines the AAL 3/4 functionality (following I.363)
T1.630.1993	*B-ISDN: Adaptation layer for CBR services functionality and specification* Defines AAL 1 (following I.363), including specifics of emulating North American DS-1 circuit function, interface, and management
T1.633	*Frame relay bearer service interworking* Follows I.555
T1.634	*Frame Relay service specific convergence sublayer* Follows I.365
T1.635	*B-ISDN AAL type 5* Defines AAL 5 (following I.363)

Table I.8
ETSI Technical Committees

Network aspects
Business telecommunications
Signaling protocols and switching
Transmission and multiplexing
Terminal equipment
Equipment engineering
Radio equipment and systems
Special mobile group
Paging systems
Satellite earth stations
Advanced testing methods

Reference

[1] IEEE Comm. Mag., Special Issue on Standards and Global Impact, Vol. 32, No. 1, January 1994.

Glossary

B-ISUP AE	B-ISUP application entity
CSMA/CD	Carrier Sense Multiple Access/Collision Detection
AAL	ATM adaptation layer
ABR	available bit rate; actual cell rate
ACR	available cell rate
ADPCM	adaptive differential PCM
AE	application element
AEI	application entity identifier
AFI	authority and format identifier
ANSI	American National Standards Institute
API	application programming interface
APPN	advanced peer-to-peer networking
ARMA	autoregressive moving average process
ASE	application service element
ATBCL	average time between cell losses
ATF	access termination function
ATM	asynchronous transfer mode
ATM-PDU	ATM physical data unit
ATOM	ATM output buffer modular
B-ICI	broadband intercarrier interface
B-ISDN	broadband integrated services digital network
B-ISUP	B-ISDN user part
B-NT	broadband network termination
B-TA	broadband terminal adapter
B-TE	broadband terminal equipment
BASize	buffer allocation size
BCC	bearer connection control
BCN	broadcast channel number
BER	bit error ratio
BHCA	busy hour call attempt
BIB	backward indicator bit
BISSI	broadband interswitching system interface

BOM	beginning of message
BSN	backward sequence number
BT	burst tolerance
Btag	beginning tag
BUS	broadcast/unknown server
CAC	connection admission control
CBO CBR	constant bit rate
CCS	common channel signaling
CD	compact disc
CD-ROM	compact disc–read-only memory
CDV	cell delay variation
CDVT	CDV tolerance
CER	cell error ratio
CES	circuit emulation service
CI	congestion indicator; connection identifier
CIB	CRC indication bit
CIF	common intermediate format
CIR	cell insertion ratio
CLNAP	connectionless network access protocol
CLNIP	connectionless network interface protocol
CLP	cell loss priority
CLR	cell loss ratio
CLS	connectionless server
CLSF	connectionless service function
CMIP	common management information protocol
CMIS	common management information service
CN	congestion notification
COM	continuation of a message
CPCS	common part convergence sublayer
CPCS-CI	CPCS congestion indication
CPCS-LP	CPCS loss priority
CPCS-UU	CPCS user-to-user indication
CPI	common part indicator
CRC	cyclic redundancy check
CRM	cell rate margin
CRS	cell relay service
CS	convergence sublayer
CSI	convergence sublayer indicator
CTD	cell transfer delay
DCC	data country code
DCE	data communication equipment
DCR	Dynamically Controlled Routing

DCT	discrete cosine transform
DES	destination end station
DFA	DXI frame address
DFD	displaced frame difference
DNHR	Dynamic Nonhierarchial Routing
DPC	destination point code
DPCM	differential pulse code modulation
DSP	domain-specific part
DSU	data service unit
DTE	data terminal equipment
DTL	designated transit list
DUP	data user part
DXI	data exchange interface
EBCN	explicit backward congestion notification
ECSA	Exchange Carriers Standards Association
EFCN	explicit forward congestion notification
EOM	end of a message
ES	end system
Etag	ending tag
ETSI	European Telecommunications Standards Institute
EWMA	exponentially weighted moving average
FCS	frame check sequence
FEC	forward error correction
FIB	forward indicator bit
FIFA	first in, first allocated
FIFO	first in, first out
FRS	frame relay service
GCRA	generic cell rate algorithm
GFC	generic flow control
HDTV	high-definition TV
HEC	header error control
HEL	Header extension length
HOL	head of line
HSTP	high-speed transport protocol
I/O	input/output
IAM	initial address message
IAR	IAM reject message
IBP	interrupted Bernoulli process
ICD	international code designator
ICIP	intercarrier service protocol
IDI	initial domain identifier
IDP	initial domain part

IE	information element
IEEE	Institute of Electrical and Electronics Engineers
IETF	Internet Engineering Task Force
ILMI	interim local management interface
IP	
IP	intelligent peripheral
IP	internet protocol
IPP	interrupted Poisson process
ISDN	integrated services digital network
ISO	International Standards Organization
ISP	intermediate service part
ISUP	ISDN user part
ITU	International Telecommunications Union
ITU-D	ITU Telecommunication Development Sector
ITU-R	ITU Radiocommunication Sector
ITU-T	ITU Telecommunication Standardization Sector (formerly CCITT)
IWU	interworking units
JPEG	Joint Photographic Experts Group
JW	jumping window
LAN	local-area network
LCN	logical channel number
LCT	last cell compliance time
LECID	LE client identifier
LECS	LE configuration server
LI	length indicator
LIS	logical IP subnetwork
LLC	logical link layer
LME	layer management entity
LMI	local management interface
LSU	link state update
LUNI	LAN emulation UNI
MAC	medium access control
MAN	metropolitan-area network
MaxBUN	maximum bandwidth unused
MBS	maximum burst size
MC	motion compensation
MCR	minimum cell rate
MCDV(*i*)	maximum CDV for traffic class *i*
MCLR(*i*)	maximum CLR for traffic class *i*
MCTD(*i*)	maximum CTD for traffic class *i*
MIB	management information base

MID	multiplexing identification
MinBUN	minimum bandwidth unused
MMPP	Markov modulated Poisson process
MPEG	Motion Picture Experts Group
MPL	maximum packet length; maximum packet lifetime
MSU	message signal unit
MTP	message transfer part
MTU	maximum transmission unit
MX	moving window
NACK	negative acknowledgment
NDIS	network driver interface specification
NETBLT	network block-transfer protocol
NFS	network file system
NHRP	next hop resolution protocol
NHS	next hop server
NI	network indicator
NME	network management entity
NNI	network-to-network interface
NPC	network parameter control
NRT-VBR	non-real-time VBR
NSAP	network service access point
NTF	network termination functions
OAM	operations, administration, and maintenance
OC	optical carrier
ODI	open data link interface
OMAP	operations maintenance and administration part
OPC	origin point code
OSI	Open Systems Interconnection
P-NNI	private NNI
PAR	positive acknowledgment with retransmission
PARIS	packetized automated routing integrated system
PBX	private branch exchange
PC	personal computer
PCI	protocol control information
PCM	pulse code modulation
PCR	peak cell rate
PCRA	Proportional Rate-Control Algorithm
PDH	plesiochronous digital hierarchy
PDU	protocol data unit
PES	packetized elementary system
PI	protocol identifier
PLCP	physical layer convergence protocol

PM	performance monitoring
PMD	physical media dependent
PRM	Protocol Reference Model
PS	program stream
PT	payload type
PTI	payload type indicator
PVC	permanent virtual connection
QCIF	quarter-CIF
RACE	Research on Advanced Communication in Europe
Res	reserved field
RM	resource management
RSVP	resource reservation protocol
RT-VBR	real-time VBR
RTCP	real-time transport control protocol
RTP	rapid transport protocol; real-time transport protocol
RTS	residual time stamp
SAA	service aspects and applications
SAAL	signaling AAL
SACF	single association control function
SAO	single association object
SAP	service access point
SAR	segmentation and reassembly sublayer
SBBP	switched batch Bernoulli process
SC	switching center
SCCP	signaling connection control part
SCP	service control point
SCPS	synchronous composite packet switching
SCR	sustainable cell rate
SDH	synchronous digital hierarchy
SDU	service data unit
SES	source end station
SI	service indicator
SID	signaling identifier
SIF	signaling information field
SIO	service information field
SIP	SMDS interface protocol
SLC	signaling link code
SLS	signaling link selection
SMAE	system management application entities
SMDS	switched multimegabit data service
SMI	structure of management information
SMS	switched management system

SN	sequence number
SNA	system network architecture
SNI	subscriber network interface
SNMP	simple network management protocol
SONET	synchronous optical network
SP	signaling point
SRej	selective reject
SRTS	synchronous residual time stamp
SS7	signaling system 7
SSCF	service-specific coordination function
SSCOP	service-specific connection-oriented protocol
SSCS	service-specific convergence sublayer
SSM	single-segment message
SSP	service switching point
STP	shielded twisted pair
STP	signaling transfer point
STS-1	Synchronous Transport Signal level 1
SVC	switched virtual connection
SWP	subworking party
TC	transmission convergence
TCAP	transaction capabilities application part
TCP	transmission control protocol
TJW	triggered jumping window
TP4	transport protocol class 4
TPDU	transport protocol data unit
TS	transport stream
TSDU	transport service data unit
TUP	telephone user part
UBR	unspecified bit rate
UDP	user datagram protocol
UME	UNI management entity
UNI	user-to-network interface
UPC	usage parameter control
URP	universal receiver protocol
UTP-3	unshielded twisted pair—category 3
VBR	variable bit rate
VC	virtual channel
VCC	virtual channel connection
VCI	virtual channel identifier
VCR	video cassette recorder
VD	virtual destination
VF	variance factor

VFN	vendor feature node
VLSI	very large scale integration
VMTP	versatile message transaction protocol
VP	virtual path
VP-AIS	VP alarm indication signal
VP-FERF	VP far-end receive failure
VPC	virtual path connection
VPI	virtual path identifier
VS	virtual source
WAN	wide-area network
WP	working party
w.p.	with probability
XTP	Xpress transfer protocol

About the Author

Dr. Raif O. Onvural received his Ph.D. at the joint Computer Studies and Operations Research program at North Carolina State University in 1987. He is currently a manager at IBM, Research Triangle Park, in the High Speed Networking group. His department is involved with different aspects of emerging networking technologies which include ATM, multimedia, and wireless. He is also IBM's coordinator to the ATM Forum and has participated in the Forum meetings since February 1993.

Dr. Onvural is the cochair of the Second International Workshop on Queueing Networks with Finite Capacity, the First International Conference on LAN Interconnection, TRICOMM '93 on ATM Networks, and the Sixth International Conference on the IFIP Performance Analysis of Communication Networks. He is the coeditor of the special issue of *Annals of OR,* on methodologies used in high-speed networks, queuing networks with finite capacity, published by North Hollands, and of *ATM Networks,* published by Plenum. He has also been a guest editor for the special issue of *Performance Evaluation,* on queuing networks with finite capacity. He has given tutorials on various aspects of ATM in a number of IFIP and IEEE conferences.

His current research interests include the design of algorithms for congestion control, service integration, path selection, multicast services in ATM networks, and performance modeling and architecture of high-speed networks in general and ATM networks in particular. He is a member of IFIP WG 6.3, ACM, Sigmetrics, IEEE, and ORSA.

Index

The Artech House Telecommunications Library

Vinton G. Cerf, Series Editor

Advanced Technology for Road Transport: IVHS and ATT, Ian Catling, editor

Advances in Computer Communications and Networking, Wesley W. Chu, editor

Advances in Computer Systems Security, Rein Turn, editor

Advances in Telecommunications Networks, William S. Lee and Derrick C. Brown

Analysis and Synthesis of Logic Systems, Daniel Mange

An Introduction to International Telecommunications Law, Charles H. Kennedy and M. Veronica Pastor

An Introduction to U.S. Telecommunications Law, Charles H. Kennedy

Asynchronous Transfer Mode Networks: Performance Issues, Raif O. Onvural

ATM and Information Superhighways, Cesare´ Carre, Marc Boisseau, and Jean-Marie Munier

ATM Switching Systems, Thomas M. Chen and Stephen S. Liu

A Bibliography of Telecommunications and Socio-Economic Development, Heather E. Hudson

Broadband: Business Services, Technologies, and Strategic Impact, David Wright

Broadband Network Analysis and Design, Daniel Minoli

Broadband Telecommunications Technology, Byeong Lee, Minho Kang, and Jonghee Lee

Cellular Radio: Analog and Digital Systems, Asha Mehrotra

Cellular Radio Systems, D. M. Balston and R. C. V. Macario, editors

Client/Server Computing: Architecture, Applications, and Distributed Systems Management, Bruce Elbert and Bobby Martyna

Codes for Error Control and Synchronization, Djimitri Wiggert

Communications Directory, Manus Egan, editor

The Complete Guide to Buying a Telephone System, Paul Daubitz

Computer Telephone Integration, Rob Walters

The Corporate Cabling Guide, Mark W. McElroy

Corporate Networks: The Strategic Use of Telecommunications, Thomas Valovic

Current Advances in LANs, MANs, and ISDN, B. G. Kim, editor

Digital Cellular Radio, George Calhoun

Digital Hardware Testing: Transistor-Level Fault Modeling and Testing, Rochit Rajsuman, editor

Digital Signal Processing, Murat Kunt

Digital Switching Control Architectures, Giuseppe Fantauzzi

Distributed Multimedia Through Broadband Communications Services, Daniel Minoli and Robert Keinath

Disaster Recovery Planning for Telecommunications, Leo A. Wrobel

Document Imaging Systems: Technology and Applications, Nathan J. Muller

EDI Security, Control, and Audit, Albert J. Marcella and Sally Chen

Electronic Mail, Jacob Palme

Enterprise Networking: Fractional T1 to SONET, Frame Relay to BISDN, Daniel Minoli

Expert Systems Applications in Integrated Network Management, E. C. Ericson, L. T. Ericson, and D. Minoli, editors

FAX: Digital Facsimile Technology and Applications, Second Edition, Dennis Bodson, Kenneth McConnell, and Richard Schaphorst

FDDI and FDDI-II: Architecture, Protocols, and Performance, Bernhard Albert and Anura P. Jayasumana

Fiber Network Service Survivability, Tsong-Ho Wu

Fiber Optics and CATV Business Strategy, Robert K. Yates *et al.*

A Guide to Fractional T1, J. E. Trulove

A Guide to the TCP/IP Protocol Suite, Floyd Wilder

Implementing EDI, Mike Hendry

Implementing X.400 and X.500: The PP and QUIPU Systems, Steve Kille

Inbound Call Centers: Design, Implementation, and Management, Robert A. Gable

Information Superhighways: The Economics of Advanced Public Communication Networks, Bruce Egan

Integrated Broadband Networks, Amit Bhargava

Intelcom '94: The Outlook for Mediterranean Communications, Stephen McClelland, editor

International Telecommunications Management, Bruce R. Elbert

International Telecommunication Standards Organizations, Andrew Macpherson

Internetworking LANs: Operation, Design, and Management, Robert Davidson and Nathan Muller

Introduction to Document Image Processing Techniques, Ronald G. Matteson

Introduction to Error-Correcting Codes, Michael Purser

Introduction to Satellite Communication, Bruce R. Elbert

Introduction to T1/T3 Networking, Regis J. (Bud) Bates

Introduction to Telecommunication Electronics, Second Edition, A. Michael Noll

Introduction to Telephones and Telephone Systems, Second Edition, A. Michael Noll

Introduction to X.400, Cemil Betanov

Land-Mobile Radio System Engineering, Garry C. Hess

LAN/WAN Optimization Techniques, Harrell Van Norman

LANs to WANs: Network Management in the 1990s, Nathan J. Muller and Robert P. Davidson

Long Distance Services: A Buyer's Guide, Daniel D. Briere

Measurement of Optical Fibers and Devices, G. Cancellieri and U. Ravaioli

Meteor Burst Communication, Jacob Z. Schanker

Minimum Risk Strategy for Acquiring Communications Equipment and Services, Nathan J. Muller

Mobile Communications in the U.S. and Europe: Regulation, Technology, and Markets, Michael Paetsch

Mobile Information Systems, John Walker

Narrowband Land-Mobile Radio Networks, Jean-Paul Linnartz

Networking Strategies for Information Technology, Bruce Elbert

Numerical Analysis of Linear Networks and Systems, Hermann Kremer *et al.*

Optimization of Digital Transmission Systems, K. Trondle and Gunter Soder

Packet Switching Evolution from Narrowband to Broadband ISDN, M. Smouts

Packet Video: Modeling and Signal Processing, Naohisa Ohta

Personal Communication Systems and Technologies, John Gardiner and Barry West, editors

The PP and QUIPU Implementation of X.400 and X.500, Stephen Kille

Practical Computer Network Security, Mike Hendry

Principles of Secure Communication Systems, Second Edition, Don J. Torrieri

Principles of Signaling for Cell Relay and Frame Relay, Daniel Minoli and George Dobrowski

Principles of Signals and Systems: Deterministic Signals, B. Picinbono

Private Telecommunication Networks, Bruce Elbert

Radio-Relay Systems, Anton A. Huurdeman

Radiodetermination Satellite Services and Standards, Martin Rothblatt

Residential Fiber Optic Networks: An Engineering and Economic Analysis, David Reed

Secure Data Networking, Michael Purser

Service Management in Computing and Telecommunications, Richard Hallows

Setting Global Telecommunication Standards: The Stakes, The Players, and The Process, Gerd Wallenstein

Smart Cards, José Manuel Otón and José Luis Zoreda

Super-High-Definition Images: Beyond HDTV, Naohisa Ohta, Sadayasu Ono, and Tomonori Aoyama

Television Technology: Fundamentals and Future Prospects, A. Michael Noll

Telecommunications Technology Handbook, Daniel Minoli

Telecommuting, Osman Eldib and Daniel Minoli

Telemetry Systems Design, Frank Carden

Telephone Company and Cable Television Competition, Stuart N. Brotman

Teletraffic Technologies in ATM Networks, Hiroshi Saito

Terrestrial Digital Microwave Communications, Ferdo Ivanek, editor

Toll-Free Services: A Complete Guide to Design, Implementation, and Management, Robert A. Gable

Transmission Networking: SONET and the SDH, Mike Sexton and Andy Reid

Transmission Performance of Evolving Telecommunications Networks, John Gruber and Godfrey Williams

Troposcatter Radio Links, G. Roda

Understanding Emerging Network Services, Pricing, and Regulation, Leo A. Wrobel and Eddie M. Pope

UNIX Internetworking, Uday O. Pabrai

Virtual Networks: A Buyer's Guide, Daniel D. Briere

Voice Processing, Second Edition, Walt Tetschner

Voice Teletraffic System Engineering, James R. Boucher

Wireless Access and the Local Telephone Network, George Calhoun

Wireless Data Networking, Nathan J. Muller

Wireless LAN Systems, A. Santamaría and F. J. López-Hernández

Wireless: The Revolution in Personal Telecommunications, Ira Brodsky

Writing Disaster Recovery Plans for Telecommunications Networks and LANs, Leo A. Wrobel

X Window System User's Guide, Uday O. Pabrai

For further information on these and other Artech House titles, contact:

Artech House
685 Canton Street
Norwood, MA 02062
617-769-9750
Fax: 617-769-6334
Telex: 951-659
e-mail: artech@artech-house.com

Artech House
Portland House, Stag Place
London SW1E 5XA England
+44 (0) 171-973-8077
Fax: +44 (0) 171-630-0166
Telex: 951-659
e-mail: artech-uk@artech-house.com